Lecture Notes in Artificial Intel

Edited by R. Goebel, J. Siekmann, and W. Wahlster

Subseries of Lecture Notes in Computer Science

João Gama   Vítor Santos Costa
Alípio Mário Jorge
Pavel B. Brazdil (Eds.)

Discovery Science

12th International Conference, DS 2009
Porto, Portugal, October 3-5, 2009
Proceedings

 Springer

Volume Editors

João Gama
LIAAD-INESC Porto L.A./Faculty of Economics, University of Porto, Portugal
E-mail: jgama@fep.up.pt

Vítor Santos Costa
CRACS-INESC Porto L.A./Faculty of Science, University of Porto, Portugal
E-mail: vsc@dcc.up.pt

Alípio Mário Jorge
LIAAD-INESC Porto L.A./Faculty of Science, University of Porto, Portugal
E-mail: amjorge@fep.up.pt

Pavel B. Brazdil
LIAAD-INESC Porto L.A./Faculty of Economics, University of Porto, Portugal
E-mail: pbrazdil@liaad.up.pt

Library of Congress Control Number: 2009935064

CR Subject Classification (1998): I.2, I.2.4, I.2.6, K.3.1, H.5, H.2.8, J.1, J.2

LNCS Sublibrary: SL 7 – Artificial Intelligence

ISSN   0302-9743
ISBN-10   3-642-04746-7 Springer Berlin Heidelberg New York
ISBN-13   978-3-642-04746-6 Springer Berlin Heidelberg New York

springer.com

© Springer-Verlag Berlin Heidelberg 2009
Printed in Germany

Typesetting: Camera-ready by author, data conversion by Scientific Publishing Services, Chennai, India
Printed on acid-free paper   SPIN: 12761876   06/3180   5 4 3 2 1 0

Preface

We are pleased to present the proceedings of the 12th International Conference on Discovery Science (DS 2009), held in Porto, Portugal, October 3–5, 2009. DS 2009 was collocated with ALT 2009, the 20th International Conference on Algorithmic Learning Theory, continuing the successful DS conference series. DS 2009 provided an open forum for intensive discussions and the exchange of new ideas among researchers working in the area of discovery science. The scope of the conference included the development and analysis of methods for automatic scientific knowledge discovery, machine learning, intelligent data analysis, and theory of learning, as well as their applications. We were honored to have a very strong program. Acceptance for the conference proceedings was very competitive. There were 92 papers submitted, with the authors coming from roughly 20 different countries. All papers were reviewed by three senior researchers followed by an extensive discussion. The program committee decided to accept 23 long papers (an acceptance rate of 25%) and 12 regular papers. The overall acceptance rate was 38%. The contributed papers cover a wide range of topics, from discovery in general to data mining in particular.

In addition to the technical papers, we were delighted to have five prestigious invited speakers and two tutorials. Fernando Pereira, University of Pennsylvania, USA, presented new fundamental questions that should be investigated in natural language processing in web mining. Hector Geffner, from Pompeu Fabra University, Spain, discussed learning methods for solving complete planning domains. Jiawei Han, University of Illinois at Urbana-Champain, USA, presented the challenges in learning from massive links in heterogeneous information networks. Sanjoy Dasgupta, University of California, USA, discussed the cost-benefit tradeoff in active learning. Yishay Mansour, Tel Aviv University, Israel, discussed issues in generalization across domains, a very significant challenge for many machine learning applications. The tutorial presented by Concha Bielza and Pedro Larrañaga, Universidad Politécnica de Madrid, Spain, focused on issues in a topic that will influence science and technology for the next decade: computational intelligence for neuroscience. The tutorial presented by Howard Hamilton and Fabrice Guillet concentrated on the never-ending problem of interestingness measures for data mining systems and systems designed to simulate or automate scientific processes that commonly face the problem of determining when a result is interesting.

We wish to express our gratitude to all authors of submitted papers, the program committee for their effort in reviewing, discussing, and commenting on the submitted papers, the members of the Discovery Science Steering Committee and especially its Chair, Einoshin Suzuki, the program chairs of ALT 2009, Sandra Zilles and Gábor Lugosi, for their efficient collaboration, and Frank Holzwarth from Springer for his efficient support with the Springer Conference Management System.

We gratefully acknowledge the work of the Publicity Chair, Pedro Pereira Rodrigues, and all those who collaborated in the local organization: Carlos Ferreira, Raquel Sebastião, Marcos Domingos, Rita Ribeiro, Orlando Oashi, and Nuno Escudeiro. Vítor Morais was invaluable in helping us navigate the Conference Management System.

We acknowledge the collaboration of the Department of Computer Science, Faculty of Science, University of Porto, where the event took place; the Faculty of Economics, University of Porto; and INESC-Porto LA. We would like to express a word of gratitude for the financial support given by the University of Porto, the Fundação para a Ciência e Tecnologia, the Artificial Intelligence and Decision Support Laboratory (LIAAD), the Center for Research in Advanced Computer Systems (CRACS), SAS, and the Portuguese Artificial Intelligence Society (APPIA). We also extend our gratitude to Yahoo! Research Barcelona, sponsor of the Carl Smith Award for the best student paper.

October 2009   João Gama
                                                  Vítor Santos Costa
                                                     Alípio Jorge
                                                     Pavel Brazdil

Organization

General Chair

Pavel Brazdil — University of Porto, Portugal

Steering Committee Chair

Einoshin Suzuki — Kyushu University, Japan

Conference Chair

João Gama — University of Porto, Portugal

Program Chair

João Gama	University of Porto, Portugal
Vítor Santos Costa	University of Porto, Portugal
Alípio Jorge	University of Porto, Portugal

Publicity Chair

Pedro Pereira Rodrigues — University of Porto, Portugal

Local Organization

Carlos Ferreira	Polytechnic Institute of Porto, Portugal
Marcos Domingues	University of Porto, Portugal
Nuno Escudeiro	Polytechnic Institute of Porto, Portugal
Orlando Ohashi	University of Porto, Portugal
Raquel Sebastião	University of Porto, Portugal
Rita Ribeiro	University of Porto, Portugal

Program Committee

Achim Hoffmann	University of New South Wales, Australia
Albert Bifet	Universitat Politècnica de Catalunya, Spain
Alfredo Cuzzocrea	University of Calabria, Italy
Alneu de Andrade Lopes	University of São Paulo, Brazil
Amílcar Cardoso	University of Coimbra, Portugal
André Carvalho	University of São Paulo, Brazil

Masayuki Takeda	Kyushu University, Japan
Michael Berthold	University of Konstanz, Germany
Michael May	Fraunhofer IAIS Bonn, Germany
Mohand-Said Hacid	University Claude Bernard Lyon 1, France
Nada Lavrac	Jožef Stefan Institute, Slovenia
Nuno Fonseca	University of Porto, Portugal
Paulo Azevedo	University of Minho, Portugal
Pedro Larrañaga	Polytechnic University of Madrid, Spain
Ricard Gavaldà	Universitat Politècnica de Catalunya, Spain
Ross King	University of Wales, UK
Rui Camacho	University of Porto, Portugal
Sarabjot Anand	University of Warwick, UK
Sašo Džeroski	Jožef Stefan Institute, Slovenia
Shinichi Morishita	University of Tokyo, Japan
Siegfried Nijssen	Katholieke Universiteit Leuven, Belgium
Simon Colton	Imperial College London, UK
Solange Rezende	University of São Paulo, Brazil
Sriraam Natarajan	University of Wisconsin Madison, USA
Stan Matwin	University of Ottawa, Canada
Stefan Kramer	Technische Universität München, Germany
Takashi Washio	Osaka University, Japan
Tamás Horváth	University of Bonn and Fraunhofer IAIS, Germany
Tapio Elomaa	Tampere University of Technology, Finland
Tetsuhiro Miyahara	Hiroshima City University, Japan
Tu Bao Ho	Advanced Institute of Science and Technology, Japan
Vincent Corruble	Université Pierre et Marie Curie, France
Yücel Saygin	Sabanci University, Turkey

External Reviewers

Andreas Hapfelmeier
Aneta Ivanovska
Annalisa Appice
António Oliveira
Bruno Feres de Souza
Carlos Ferreira
Catarina Silva
Cédric Herpson
Constanze Schmitt
Debora Medeiros
Dragi Kocev
Frederik Janssen
Hiroshi Sakamoto
Ilija Šubasic
Ivica Slavkov

Jana Schmidt
Jan-Nikolas Sulzmann
Jörg Wicker
Jussi Kujala
Lisa di Jori
Mathieu Roche
Marianne L. Mueller
Michelangelo Ceci
Michele Berlingerio
Michele Coscia
Murilo Naldi
Nuno Cardoso
Nuno Castro
Panče Panov
Pascal Poncelet

Pedro Ferreira
Pedro Pereira Rodrigues
Prapaporn Rattanatamrong
Sang-Hyeun Park

Sandra Bringay
Tetsuji Kuboyama
Timo Aho

Sponsoring Institutions

Table of Contents

Inference and Learning in Planning
(Extended Abstract)

Hector Geffner

ICREA & Universitat Pompeu Fabra
C/Roc Boronat 138, E-08018 Barcelona, Spain
hector.geffner@upf.edu
http://www.tecn.upf.es/~hgeffner

Abstract. Planning is concerned with the development of solvers for a wide range of models where actions must be selected for achieving goals. In these models, actions may be deterministic or not, and full or partial sensing may be available. In the last few years, significant progress has been made, resulting in algorithms that can produce plans effectively in a variety of settings. These developments have to do with the formulation and use of general inference techniques and transformations. In this invited talk, I'll review the inference techniques used for solving individual planning instances from scratch, and discuss the use of learning methods and transformations for obtaining more general solutions.

1 Introduction

The problem of creating agents that can decide what to do on their own has been at the center of AI research since its beginnings. One of the first AI programs to tackle this problem, back in the 50's, was the General Problem Solver (GPS) that selects actions for reducing a difference between the current state and a desired target state [1]. Ever since then, this problem has been tackled in a number of ways in many areas of AI, and in particular in the area of Planning.

The problem of selecting actions for achieving goals, however, even in its most basic version – deterministic actions and complete information – is computationally intractable [2]. Under these assumptions, the problem of finding a plan becomes the well-known problem of finding a path in a directed graph whose nodes, that represent the possible states of the system, are exponential in the number of problem variables.

Until the middle 90's in fact, no planner or program of any sort could synthesize plans for large problems in an effective manner from a description of the actions and goals. In recent years, however, the situation has changed: in the presence of deterministic actions and full knowledge about the initial situation, classical planning algorithms can find plans quickly even in large problems with hundred of variables and actions [3,4]. This is the result of new ideas, like the automatic derivation of heuristic functions [5,6], and a established empirical methodology featuring benchmarks, comparisons, and competitions. Moreover,

J. Gama et al. (Eds.): DS 2009, LNAI 5808, pp. 1–12, 2009.

many of these planners are *action selection mechanisms* that can commit to the next action to do in real-time without having to construct a full plan first [7].

These developments, however, while crucial, do not suffice for producing autonomous agents that can decide by themselves what to do in environments where the two assumptions above (deterministic actions, complete information) do not apply. The more general problem of selecting actions in uncertain, dynamic and/or partially known environments arises in a number of contexts (a rover in Mars, a character in a video-game, a robot in a health-care facility, a softbot in the web, etc.), and has been tackled through a number of different methodologies:

1. *programming-based:* where the *desired behavior is encoded explicitly* by a human programmer in a suitable high-level language,
2. *learning-based:* where the *desired behavior is learned automatically* from trial-and-error experience or information provided by a teacher, or
3. *model-based:* where *the desidered behavior is inferred automatically* from a suitable description of the actions, sensors, and goals.

None of these approaches, however, or a combination of them, has resulted yet in a solid methodology for building agents that can display a robust and flexible behavior in real time in partially known environments. Programming agents by hand puts all the burden in the programmer that cannot anticipate all possible contingencies, leading to systems that are brittle. Learning methods such as reinforcement learning [8], are restricted in scope and do not deal with the problem of incomplete state information. Finally, traditional model-based methods, when applied to models that are more realistic than the ones underlying classical planning, have difficulties scaling up.

Planning in Artificial Intelligence represents the model-based approach to autonomous behavior: a planner is a solver that accepts a model of the actions, sensors, and goals, and produces a controller that determines the actions to do given the observations gathered (Fig. 1). Planners come in a great variety, depending on the types of models they target. Classical planners address deterministic state models with full information about the initial situation [9]; conformant planners address state models with non-deterministic actions and incomplete information about the initial state [10,11], POMDP planners address stochastic state model with partial observability [12], and so on.

In all cases, the models of the environment considered in planning are intractable in the worst case, meaning that brute force methods do not scale up. Domain-independent planning approaches aimed at solving these planning models effectively must thus *recognize* and *exploit* the structure of the individual

Fig. 1. Model-based approach to intelligent behavior: the next action to do is determined by a controller derived from a model of the actions, sensors, and goals

problems that are given. The key to exploiting this structure is *inference*, as in other AI models such as Constraint Satisfaction Problems and Bayesian Networks [13,14]. In the paper, we will go over the inference techniques that have been found computationally useful in planning research and identify areas where they could benefit from learning techniques as well. In this sense, planners solve problems from scratch by combining search and inference, and do not get any better as more instances from a given domain are solved. Learning should thus help planners to automatically extract domain knowledge that could be used to solve other domain instances more effectively, and in principle, without any search at all.

The paper is organized as follows. We consider the model, language, and inference techniques developed for classical planning, conformant planning, and planning with sensing, in that order. We focus on inference techniques of two types: heuristic functions and transformations. We then consider the use and role of inductive learning methods in planning, in particular, when plan strategies for a whole domain, and not for a single domain instance, are required.

2   Classical Planning

Classical planning is concerned with the selection of actions in environments that are *deterministic* and whose initial state is *fully known*. The model underlying classical planning can be described as a state space containing

- a finite and discrete set of states S,
- a *known initial state* $s_0 \in S$,
- a set $S_G \subseteq S$ of goal states,
- actions $A(s) \subseteq A$ applicable in each $s \in S$,
- a *deterministic transition function* $s' = f(a, s)$ for $a \in A(s)$, and
- *uniform action costs* $c(a, s)$ equal to 1.

A solution or *plan* in this model is a sequence of actions $a_0, \ldots, a_n$ that generates a state sequence $s_0, s_1, \ldots, s_{n+1}$ such that a_i is applicable in the state s_i and results in the state $s_{i+1} = f(a_i, s_i)$, the last of which is a goal state.

The cost of a plan is the sum of the action costs, which in this setting, corresponds to plan length. A plan is optimal it is has minimum cost, and the cost of a problem is the cost of an optimal plan.

Domain-independent classical planners accept a compact description of the above models, and automatically produce a plan (an optimal plan if the planner is optimal). This problem is intractable in the worst case, yet currently large classical problems can be solved using heuristic functions derived from the problem encodings.

A simple but still common language for encoding classical planning problems is Strips [9]. A problem in Strips is a tuple $P = \langle F, O, I, G \rangle$ where

- F stands for set of all *atoms* (boolean vars),
- O stands for set of all *operators* (actions),

- $I \subseteq F$ stands for the *initial situation*, and
- $G \subseteq F$ stands for the *goal situation*.

The actions $o \in O$ are represented by three sets of atoms from F called the Add, Delete, and Precondition lists, denoted as $Add(o)$, $Del(o)$, $Pre(o)$. The first, describes the atoms that the action o makes true, the second, the atoms that o makes false, and the third, the atoms that must be true for the action o to be applicable.

A Strips problem $P = \langle F, O, I, G \rangle$ encodes the state model $\mathcal{S}(P)$ where

- the states $s \in S$ are *collections of atoms* from F,
- the initial state s_0 is I,
- the goal states s are those for which $G \subseteq s$,
- the actions a in $A(s)$ are the ones in O such that $Prec(a) \subseteq s$, and
- the next state is $s' = f(a, s) = (s \setminus Del(a)) \cup Add(a)$.

All areas in Planning, and in particular Classical Planning, have become quite empirical in recent years, with competitions held every two years [15], and hundreds of benchmark problems available in PDDL, a standard syntax for planning that extends Strips [16].

The classical planners that scale up best can solve large problems with hundreds of fluents and actions [17,18]. These planners do not compute optimal solutions and cast the planning problem P as an *heuristic search problem* over the state space $\mathcal{S}(P)$ that defines a directed graph whose nodes are the states, whose initial node is the initial state, and whose target nodes are the states where the goals are true [19]. This graph is never made explicit as it contains a number of states that is exponential in the number of fluents of P, but can be searched quite efficiently with current heuristics.

Heuristic functions $h(s)$ provide an estimate of the cost to reach the goal from any state s, and are derived automatically from a relaxation (simplification) of the problem P [20]. The relaxation most commonly used in planning, called the delete-relaxation and denoted as P^+, is obtained by removing the delete lists from the actions in P. While finding the *optimal* solution to the relaxation P^+ is still NP-hard, finding just one *solution* is easy and can be done in low polynomial time.

The *additive heuristic*, for example, estimates the cost $h(p; s)$ of achieving the atoms p from s through the equations [19]:

$$h(p; s) = \begin{cases} 0 & \text{if } p \in s \\ h(a_p; s) & \text{otherwise} \end{cases}$$

where a_p is a *best support* for p in s defined as

$$a_p = \operatorname{argmin}_{a \in O(p)} h(a; s)$$

$O(p)$ is the set of actions that add p in P, and $h(a; s)$ is

$$h(a; s) = cost(a) + \sum_{q \in Pre(a)} h(q; s) .$$

The cost of achieving the goal G from s is then defined as

$$h_{add}(s) = \sum_{p \in G} h(p; s) .$$

The heuristic h_{add} is not admissible (it's not a lower bound) but is informative and its computation involves the solution of a shortest-path problem in *atom space* as opposed to *state space*. A plan $\pi^+(s)$ for the relaxation P^+ can be obtained from the heuristic $h_{add}(s)$ by simply collecting the *best supports* recursively backwards from the goal [21]. This is actually the technique used in the state-of-the-art planner LAMA [18], winner of the 2008 International Planning Competition [15], that defines the heuristic $h(s)$ as the cost of this 'relaxed plan', and uses it in problems where action costs are not uniform. The search algorithm in LAMA is (greedy) best first search with the evaluation function $f(s) = h(s)$ and two open lists rather one, for giving precedence to the actions applicable in the state s that are most relevant to the goal according to $\pi^+(s)$; the so-called helpful actions [7].

3   Incomplete Information

The good news about classical planning is that it works: large problems can be solved quite fast, and the sheer size of a problem is not an obstacle to its solution. The bad news is that the assumptions underlying classical planning are too restrictive. We address now the problems that arise from the presence of *uncertainty* in the initial situation. The resulting problems are called *conformant* as they have the same form as classical plans, namely plain action sequences, but they must work for each of the initial states that are possible.

An example that illustrates the difficulties that arise from the presence of incomplete information in the initial situation is shown in Fig. 2. It displays a robot that must move from an uncertain initial location I, shown in gray, to the target cell G that must be reached with certainty. The robot can move one cell at a time, without leaving the grid: moves that would leave the agent out of the grid have no effects. The problem is very much like a classical planning

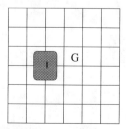

Fig. 2. A problem involving incomplete information: a robot must move from an uncertain initial location I shown in gray, to the target cell G with certainy. For this, it must locate itself into a corner and then head to G.

problem except for the uncertain initial situation I. The solutions to the problem, however, are quite different. Indeed, the best *conformant plan* for the problem must move the robot to a corner first, and then head with certainty to the target G. For example, for being certain that the robot is at the left lower corner of the grid, the robot can move *left* three times, and *down* three times. Notice that this is the opposite of reasoning by cases; indeed, the best action to do from each of the possible initial locations is not to move left or right, but up or right. Yet such moves would not help the robot reach the goal with certainty.

The model for the conformant planning problem is the model for classical planning but with the initial state s_0 replaced by a non-empty *set* S_0 of possible initial states. The Strips syntax for the problem $P = \langle F, O, I, G \rangle$ is also extended to let I stand for a *set of clauses* and not just a set of atoms, and O to include actions with effects L, positive or negative, that are *conditional* on a set of literals $L_1, \ldots, L_n$, written as $L_1, \ldots, L_n \to L$, where each L_i and L are positive or negative literals.

Conformant planning problems are no longer path-finding problems over a directed graph whose nodes are the *states* of the problem, but rather path-finding problems over a directed graph whose nodes are *sets of states*, also called *belief states* [22]. Belief states express the states of the world that are deemed possible to the agent. Thus, while in classical planning, the size of the (state) space to search is exponential in the number of variables in the problem; in conformant planning, the size of the (belief) space to search is exponential in the number of states. Indeed, conformant planning is harder than classical planning, as even the verification of conformant plans is NP-hard [23].

Conformant planners such as Contingent-FF, MBP, and POND [24,25,26], address the search in belief space using suitable belief representations such as OBDDs, that do not necessarily blow up with the number of states deemed possible, and heuristics that can guide the search for the target beliefs. Another approach that has been pursued recently, that turned out to be the most competitive in the 2006 Int. Planning Competition, is to automatically transform the conformant problems P into classical problems $K(P)$ that are solved by off-the-shelf classical planners.

The translation $K(P) = K_{T,M}(P)$ of a conformant problem P involves two parameters: a set of *tags* T and a set of *merges* M [27]. A tag t is a set (conjunction) of literals in P whose status in the initial situation I is not known, and a merge $m \in M$ is a collection of tags $t_1, \ldots, t_n$ that stands for the DNF formula $t_1 \vee \cdots \vee t_n$. Tags are assumed to represent consistent assumptions about I, i.e. $I \not\models \neg t$, and merges represent disjunctions of assumptions that follow from I; i.e. $I \models t_1 \vee \cdots \vee t_n$.

The fluents in $K_{T,M}(P)$, for the conformant problem $P = \langle F, O, I, G \rangle$ are of the form KL/t for each $L \in F$ and $t \in T$, meaning that "it is known that if t is true in the initial situation, L is true". In addition, $K_{T,M}(P)$ includes extra actions, called *merge actions*, that allow the derivation of a literal KL (i.e. KL/t with the "empty tag", expressing that L is known unconditionally) when KL/t' has been obtained for each tag t' in a merge $m \in M$ for L.

Formally, for a conformant problem $P = \langle F, O, I, G \rangle$, the translation defines the *classical problem* $K_{T,M}(P) = \langle F', O', I', G' \rangle$ where

$$F' = \{KL/t, K\neg L/t \mid L \in F\}$$
$$I' = \{KL/t \mid \text{if } I \models t \supset L\}$$
$$G' = \{KL \mid L \in G\}$$
$$O' = \{a : KC/t \to KL/t,\ a : \neg K\neg C/t \to \neg K\neg L/t$$
$$\mid a : C \to L \text{ in } P\} \cup \{\bigwedge_{t \in m} KL/t \to KL \mid m \in M_L\}$$

with t ranging over T and with the preconditions of the actions a in $K_{T,M}(P)$ including the literal KL if the preconditions of a in P include the literal L.

When $C = L_1, \ldots, L_n$, the expressions KC/t and $\neg K\neg C/t$ are abbreviations for $KL_1/t, \ldots, KL_n/t$ and $\neg K\neg L_1/t, \ldots, \neg K\neg L_n/t$ respectively. A rule $a : C \to L$ in P gets mapped into "support rules" $a : KC/t \to KL/t$ and "cancellation rules" $a : \neg K\neg C/t \to \neg K\neg L/t$; the former "adds" KL/t when the condition C is known in t, the latter undercut the persistence of $K\neg L/t$ except when (a literal in) C is known to be false in t.

The translation $K_{T,M}(P)$ is *sound*, meaning that the classical plans that solve $K_{T,M}(P)$ yield valid conformant plans for P that can be obtained by just dropping the merge actions. On the other hand, the *complexity* and *completeness* of the translation depend on the choice of tags T and merges M. The $K_i(P)$ translation, where i is a non-negative integer, is a special case of the $K_{T,M}(P)$ translation where the tags t are restricted to contain at most i literals. $K_i(P)$ is exponential in i and complete for problems with *conformant width* less than or equal to i. The planner T_0 feeds the $K_1(P)$ translation into the classical FF planner [7] and was the winning entry in the Conformant Track of the 2006 IPC [28].

4   Sensing and Finite-State Controllers

Most often problems that involve *uncertainty* in the initial state of the environment or in the action effects, also involve some type of *feedback* or *sensors* that provide partial state information. As an illustration of a problem of this type, consider the simple grid shown on the left of Fig. 3, where an agent starting in some cell between A and B, mut move to B, and then to A. In this problem, while the exact initial location of the agent is not known, it is assumed that the marks A and B are *observable*.

The solutions to problems involving observations can be expressed in many forms: as contingent plans [24], as policies mapping beliefs into actions [12], and as *finite-state controllers*. A finite-state controller that solves the problem above is shown on the right of Fig. 3. An arrow $q_i \to q_j$ between one controller state q_i and another (or the same) controller state q_i labeled with a pair O/a means to do action a and switch to state q_j, when o is observed in the state q_i. Starting in the controller state q_0, the controller shown tells the agent to move right until

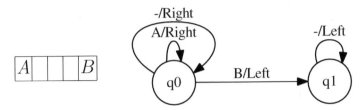

Fig. 3. *Left:* A problem where an agent, initially between A and B, must move to B and then back to A. *Right:* A finite-state controller that solves the problem.

observing B, and then to move left until observing A or B (the observation '-' means no observation).

Finite-state controllers such as the one displayed above have two features that make them more appealing than contingent plans and POMDP policies: they are often very *compact*, and they often quite *general* too. Indeed, the problem above can be changed in a number of ways and the controller shown would still work. For example, the *size of the grid* can be changed from 1×5 to $1 \times n$, the agent can be placed *initially* anywhere in the grid (except at B), and the actions can be made *non-deterministic* by the addition of 'noise'. This generality is well beyond the power of contingent plans or exact POMDP policies that are tied to a particular state space. For these reasons, finite-state controllers are widely used in practice, from controlling non-playing characters in video-games [29] to mobile robots [30,31]. Memoryless controllers or policies [32] are widely used as well, and they are nothing but finite-state controllers with a single state. The additional states provide finite-state controllers with memory that allows different actions to be taken given the same observation.

The benefits of finite-state controllers, however, come at a price: unlike contingent trees and POMDP policies, they are usually not derived automatically from a model but are written by hand; a task that is not trivial even in the simplest cases. There have been attempts for deriving finite-state controllers for POMDPs with a given number of states [33,34,35], but the problem can be solved approximately only, with no correctness guarantees.

Recently, we have extended the translation-based approach to conformant planning presented above [27], to derive finite-state controllers [36]. For this, the *control problem* P is defined in terms of a *conformant problem* with no preconditions, extended with a set O of *observable fluents*. The solution to the problem P is defined in terms of *finite state controllers* C_N with a given number N of *controller states*. This rules out sequential plans as possible solutions, as they would involve a number of controller states equal to the number of time steps in the plan.

The controller C_N is a set of tuples $t = \langle i, o, a, k \rangle$ that tell the agent to do a and switch to state q_k when the observation is o and the controller state is q_i. The key result is that a finite-state controller C_N that solves P can be obtained from the *classical plans* of a *classical problem* P_N obtained by a suitable translation from P, O, and N. The key idea in the translation is to replace each action a in P by an action $a(t)$, for each $t = \langle i, o, a, k \rangle$, so that the effects $C \to C'$ of a in

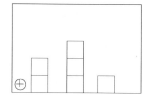
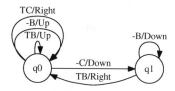

Fig. 4. *Left:* Problem where visual-marker (circle on the lower left) must be moved on top of a green block. The observations are whether the cell currently marked contains a green block (G), a non-green block (B), or neither (C); and whether this cell is at the level of the table (T) or not (–). *Right:* Finite-state controller that solves the problem for any number and arrangement of blocks.

P become effects $q_i, o, C \rightarrow \neg q_i, q_k, C'$ of $a(t)$ in P_N. That is, the effects of the action a are made conditional on the observation o and state q_i in the actions $a(t)$ where $t = \langle i, o, a, k \rangle$.

Fig. 4 shows a more challenging problem solved in this way, resulting in a very compact and general controller. In the problem, shown on the left, a visual-marker (a circle on the lower left) must be moved on top of a green block . The observations are whether the cell currently marked contains a green block (G), a non-green block (B), or neither (C); and whether this cell is at the level of the table (T) or not ('-'). The visual marker can be moved one cell at a time in the four directions. This is a problem à la Chapman or Ballard, that have advocated the use of deictic representations of this sort [37,38]. The finite-state controller that results for this problem is shown on the right. Interestingly, it is a very compact and general controller: it involves two states only and can be used to solve the same problem for any number and arrangement of blocks. See [36] for details.

5   Learning and Generalized Policies

We illustrated above that it is possible to obtain from a concrete problem P, a finite-state controller that not only solves P but many variations too, including changes in the initial situation and action effects, and changes in the number of objects and size of the state space. This generalization does not follow from an *inductive* approach over many problem instances, but from a *deductive* approach over a single instance upon which the solution is guaranteed to be correct. The generalization is achieved from a *change in the representation of the solution:* while solutions to P that take the form of contingent plans or POMDP policies would not generalize to problems that involve a different state space, solutions that are expressed as compact finite-state controllers, often do. In principle, these techniques can be used to derive finite-state controllers for solving any instance of a given domain such as Blocks. Such general strategies exist; indeed, one such strategy for Blocks is to put all blocks on the table, and then build the desired towers in order, from the bottom up. Of course, this strategy is not optimal, and indeed no compact optimal strategy for Blocks exists. Yet, the approach presented above does not handle problems of this type. For this, first, observations must be defined on fluents that are not primitive in the problem, and which thus, must be conveniently discovered, like the fluent *above*, the transitive closure of the *on* predicate, that comes very handy in Blocks. Second, the resulting pool of observable fluents becomes then too large, so that the resulting translation P_N into a classical planning problem cannot even be constructed; the translation is indeed exponential in the number of observables. Interestingly, *inductive approaches* have been shown to be able to generate general strategies for domains like Blocks [39], and moreover, some of these inductive approaches do not require any background knowledge and work just with the definition of the planning domain and a small set of solved instances [40,41]. A interesting challenge for the future is the combination of *inductive* and *deductive* approaches for the derivation of general policies able to solve any instance of a given planning domain without search. From the discussion above, it seems that inductive methods are good for selecting informative features, while deductive methods are good for assembling these features into correct general policies. In this sense, planning appears to be an ideally rich application domain where learning adequate representations appears to be possible and critical for achieving both efficiency and generality.

Acknowledgments. This is joint work with a number of students and colleagues, in particular Blai Bonet and Hector Palacios. I thank Joao Gamma and the organizers of DS-09 for the invitation. The author is partially supported by grant TIN2006-15387-C03-03 from MEC, Spain.

References

1. Newell, A., Simon, H.: GPS: a program that simulates human thought. In: Feigenbaum, E., Feldman, J. (eds.) Computers and Thought, pp. 279–293. McGraw-Hill, New York (1963)
2. Bylander, T.: The computational complexity of STRIPS planning. Artificial Intelligence 69, 165–204 (1994)
3. Blum, A., Furst, M.: Fast planning through planning graph analysis. In: Proceedings of IJCAI 1995, pp. 1636–1642. Morgan Kaufmann, San Francisco (1995)
4. Kautz, H., Selman, B.: Pushing the envelope: Planning, propositional logic, and stochastic search. In: Proc. AAAI, pp. 1194–1201 (1996)
5. McDermott, D.: Using regression-match graphs to control search in planning. Artificial Intelligence 109(1-2), 111–159 (1999)
6. Bonet, B., Loerincs, G., Geffner, H.: A robust and fast action selection mechanism for planning. In: Proceedings of AAAI 1997, pp. 714–719. MIT Press, Cambridge (1997)
7. Hoffmann, J., Nebel, B.: The FF planning system: Fast plan generation through heuristic search. Journal of Artificial Intelligence Research 14, 253–302 (2001)
8. Sutton, R., Barto, A.: Introduction to Reinforcement Learning. MIT Press, Cambridge (1998)
9. Fikes, R., Nilsson, N.: STRIPS: A new approach to the application of theorem proving to problem solving. Artificial Intelligence 1, 27–120 (1971)

10. Goldman, R.P., Boddy, M.S.: Expressive planning and explicit knowledge. In: Proc. AIPS 1996 (1996)
11. Smith, D., Weld, D.: Conformant graphplan. In: Proceedings AAAI 1998, pp. 889–896. AAAI Press, Menlo Park (1998)
12. Cassandra, A., Kaelbling, L., Littman, M.L.: Acting optimally in partially observable stochastic domains. In: Proc. AAAI, pp. 1023–1028 (1994)
13. Dechter, R.: Constraint Processing. Morgan Kaufmann, San Francisco (2003)
14. Pearl, J.: Probabilistic Reasoning in Intelligent Systems. Morgan Kaufmann, San Francisco (1988)
15. International Planning Competitions (2009),
    http://icaps-conference.org/index.php/main/competitions
16. McDermott, D., Ghallab, M., Howe, A., Knoblock, C., Ram, A., Veloso, M., Weld, D., Wilkins, D.: PDDL – The Planning Domain Definition Language. Technical Report CVC TR-98-003/DCS TR-1165, Yale Center for Computational Vision and Control, New Haven, CT (1998)
17. Helmert, M.: The Fast Downward planning system. Journal of Artificial Intelligence Research 26, 191–246 (2006)
18. Richter, S., Helmert, M., Westphal, M.: Landmarks revisited. In: Proc. AAAI, pp. 975–982 (2008)
19. Bonet, B., Geffner, H.: Planning as heuristic search. Artificial Intelligence 129(1–2), 5–33 (2001)
20. Pearl, J.: Heuristics. Addison-Wesley, Reading (1983)
21. Keyder, E., Geffner, H.: Heuristics for planning with action costs revisited. In: Proc. ECAI 2008 (2008)
22. Bonet, B., Geffner, H.: Planning with incomplete information as heuristic search in belief space. In: Proc. of AIPS 2000, pp. 52–61. AAAI Press, Menlo Park (2000)
23. Haslum, P., Jonsson, P.: Some results on the complexity of planning with incomplete information. In: Biundo, S., Fox, M. (eds.) ECP 1999. LNCS (LNAI), vol. 1809, pp. 308–318. Springer, Heidelberg (2000)
24. Hoffmann, J., Brafman, R.: Contingent planning via heuristic forward search with implicit belief states. In: Proc. ICAPS, pp. 71–80 (2005)
25. Bertoli, P., Cimatti, A., Roveri, M., Traverso, P.: Planning in nondeterministic domains under partial observability via symbolic model checking. In: Proc. IJCAI 2001 (2001)
26. Bryce, D., Kambhampati, S., Smith, D.E.: Planning graph heuristics for belief space search. Journal of AI Research 26, 35–99 (2006)
27. Palacios, H., Geffner, H.: From conformant into classical planning: Efficient translations that may be complete too. In: Proc. 17th Int. Conf. on Planning and Scheduling, ICAPS 2007 (2007)
28. Bonet, B., Givan, B.: Results of the conformant track of the 5th int. planning competition (2006),
    http://www.ldc.usb.ve/~bonet/ipc5/docs/results-conformant.pdf
29. Buckland, M.: Programming Game AI by Example. Wordware Publishing, Inc. (2004)
30. Murphy, R.R.: An Introduction to AI Robotics. MIT Press, Cambridge (2000)
31. Mataric, M.J.: The Robotics Primer. MIT Press, Cambridge (2007)
32. Littman, M.L.: Memoryless policies: Theoretical limitations and practical results. In: Cliff, D. (ed.) From Animals to Animats 3. MIT Press, Cambridge (1994)
33. Meuleau, N., Peshkin, L., Kim, K., Kaelbling, L.P.: Learning finite-state controllers for partially observable environments. In: Proc. UAI, pp. 427–436 (1999)

34. Poupart, P., Boutilier, C.: Bounded finite state controllers. In: Proc. NIPS, pp. 823–830 (2003)
35. Amato, C., Bernstein, D., Zilberstein, S.: Optimizing memory-bounded controllers for decentralized pomdps. In: Proc. UAI (2007)
36. Bonet, B., Palacios, H., Geffner, H.: Automatic derivation of memoryless policies and finite-state controllers using classical planners. In: Proc. ICAPS 2009 (2009)
37. Chapman, D.: Penguins can make cake. AI Magazine 10(4), 45–50 (1989)
38. Ballard, D., Hayhoe, M., Pook, P., Rao, R.: Deictic codes for the embodiment of cognition. Behavioral and Brain Sciences 20 (1997)
39. Khardon, R.: Learning action strategies for planning domains. Artificial Intelligence 113, 125–148 (1999)
40. Martin, M., Geffner, H.: Learning generalized policies from planning examples using concept languages. Appl. Intelligence 20(1), 9–19 (2004)
41. Yoon, S., Fern, A., Givan, R.: Inductive policy selection for first-order MDPs. In: Proc. UAI 2002 (2002)

Mining Heterogeneous Information Networks by Exploring the Power of Links

Jiawei Han

Department of Computer Science
University of Illinois at Urbana-Champaign
hanj@cs.uiuc.edu

Abstract. Knowledge is power but for interrelated data, knowledge is often hidden in massive links in heterogeneous information networks. We explore the power of links at mining heterogeneous information networks with several interesting tasks, including link-based object distinction, veracity analysis, multidimensional online analytical processing of heterogeneous information networks, and rank-based clustering. Some recent results of our research that explore the crucial information hidden in links will be introduced, including (1) Distinct for object distinction analysis, (2) TruthFinder for veracity analysis, (3) Infonet-OLAP for online analytical processing of information networks, and (4) RankClus for integrated ranking-based clustering. We also discuss some of our on-going studies in this direction.

1 Introduction

Social, natural, and information systems usually consist of a large number of interacting, multi-typed components. Examples of such systems include communication and computer systems, the World-Wide Web, biological networks, transportation systems, epidemic networks, criminal rings, and hidden terrorist networks. All the above systems share an important common feature: they are *networked systems*, *i.e.*, individual agents or components interact with a specific set of components, forming large, interconnected, and heterogeneous (*i.e.*, multi-typed) networks. Without loss of generality, we call such interconnected, multi-typed networks or systems as **heterogeneous information networks**. Clearly, heterogeneous information networks are ubiquitous and form a critical component of modern information infrastructure.

Despite their prevalence in our world, we have only recently recognized the importance of studying information networks as a whole. Hidden in these networks are the answers to important questions. For example, is there a collaborated plot behind a network intrusion, and how can we identify its source in communication networks? How can a company derive a complete view of its products at the retail level from interlinked social communities? These questions are highly relevant to a new class of analytical applications that query and mine massive information networks for pattern and knowledge discovery, data and information integration, veracity analysis and deep understanding of the principles of information networks.

J. Gama et al. (Eds.): DS 2009, LNAI 5808, pp. 13–30, 2009.
© Springer-Verlag Berlin Heidelberg 2009

Searching for information and knowledge inside networks, particularly large networks with thousands of nodes is a complex and time-consuming task. Unfortunately, the lack of a general analytical and access platform makes sensible navigation and human comprehension virtually impossible in large-scale networks. Fortunately, information networks contains massive nodes and links associated with various kinds of information. Knowledge about such networks is often hidden in massive links in heterogeneous information networks but can be uncovered by the development of sophisticated knowledge discovery mechanisms.

In this paper, we outline some of our recent studies that explore the power of links at mining heterogeneous information networks, including link-based object distinction, veracity analysis, multidimensional online analytical processing of heterogeneous information networks, and rank-based clustering. Such studies show that powerful data mining mechanisms can be used for analysis and exploration of large-scale information networks and systematic development of such network mining methods is an important task in future research.

The remaining of the paper is organized as follows. Section 2 introduces object distinction analysis, Section 3 on veracity analysis, Section 4 on OLAP information networks, and Section 5 on integrated ranking-based clustering. We summarize our study in Section 6.

2   Distinguishing Objects with Identical Names by Information Network Analysis

People retrieve information from different databases on the Web, such as DBLP, Amazon shopping, and AllMusic. One disturbing problem is that different objects may share identical names. For example, there are 72 songs and 3 albums named "Forgotten" in allmusic.com; and there are over 200 papers in DBLP written by at least 14 different Wei Wang's. Users are often unable to distinguish them, because the same object may appear in very different contexts, and there is often limited and noisy information associated with each appearance.

The task of *distinguishing objects with identical names* is called *object distinction* analysis. Given a database and a set of references in it referring to multiple objects with identical names, the task is to split the references into clusters, so that each cluster corresponds to one real object. This task is the opposite of a popular problem called *reference reconciliation* (or *record linkage, duplicate detection*), which aims at merging records with different contents referring to the same object, such as two citations referring to the same paper. There have been many approaches developed for record linkage analysis [2], which usually use some efficient techniques [4] to find candidates of duplicate records (*e.g.*, pairs of objects with similar names), and then check duplication for each pair of candidates. Different approaches are used to reconcile each candidate pair, such as probabilistic models of attribute values and textual similarities [2].

Compared with record linkage, objection distinction is a very different problem. First, because the references have identical names, textual similarity is useless. Second, each reference is usually associated with limited information, and thus it is difficult to make good judgement based on it. Third and most importantly, because different references to the same object appear in different contexts, they seldom share common or similar attribute values. Most record linkage approaches are based on the assumption that duplicate records should have equal or similar values, and thus cannot be used on this problem.

Although the references are associated with limited and possibly inconsistent information, the linkages among references and other objects still provide crucial information for grouping references. For example, in a publication database, different references to authors are connected in numerous ways through authors, conferences and citations. References to the same author are often linked in certain ways, such as through their coauthors, coauthors of coauthors, and citations. These linkages provide important information, and a comprehensive analysis on them may likely disclose the identities of objects.

We developed a methodology called Distinct [11] that can distinguish object identities by fusing different types of linkages with differentiating weights, and using a combination of distinct similarity measures to assess the value of each linkage. Because the linkage information is usually sparse and intertwined, Distinct combines two approaches for measuring similarities between records in a relational database: (i) *set resemblance between the neighbor tuples* of two records (the *neighbor tuples* of a record are the tuples linked with it); and (ii) *random walk probability* between two records in the graph of relational data. These two approaches are complementary: one uses the neighborhood information, and the other uses connection strength of linkages. Moreover, since there are many types of linkages among references, each following a join path in the database schema, and different types of linkages have very different semantic meanings and different levels of importance, Distinct uses support vector machines (SVM) to learn a model for weighing different types of linkages. When grouping references, the references to the same object can be merged and considered as a whole. Distinct uses agglomerative hierarchical clustering, which repeatedly merges the most similar pairs of clusters. It combines *average-link* (average similarity between all objects in two clusters) and *collective similarity* (considering each cluster as a single object) to measure the similarity between two clusters, which is less vulnerable to noise.

Distinct uses supervised learning to determine the pertinence of each join path and assign a weight to it. In order to do this, a training set is needed that contains equivalent references as positive examples and distinct references as negative ones. Instead of manually creating a training set which requires much labor and expert knowledge, Distinct constructs the training set automatically, based on the observation that the majority of entities have distinct names in most applications. Take the problem of distinguishing persons as an example. A person's name consists of the first and last names. If a name contains a rather rare first name and a rather rare last name, this name is very likely to be unique. We can find many such names in a database and use them to construct training sets. A pair of references to an object with a unique name can be used as a positive example, and a pair of references to two different objects can be used as a negative example.

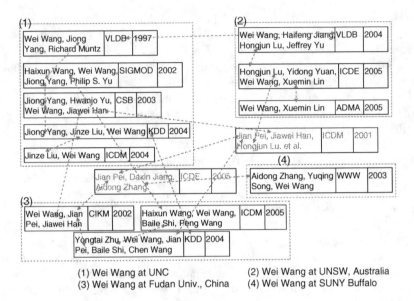

(1) Wei Wang at UNC   (2) Wei Wang at UNSW, Australia
(3) Wei Wang at Fudan Univ., China   (4) Wei Wang at SUNY Buffalo

Fig. 1. Papers by four different Wei Wang's

Example 1: Distinguishing people or objects with identical names.
There are more than 200 papers in DBLP written by at least 14 different Wei
Wang's, each having at least two papers. A mini example is shown in Fig. 1,
which contains some papers by four different Wei Wang's and the linkages among
them. Users are often unable to distinguish them, because the same person or
object may appear in very different contexts, and there is often limited and noisy
information associated with each appearance.

We report our empirical study on testing the effectiveness of the proposed
approach. Distinct is tested on the DBLP database. First, authors with no more
than 2 papers are removed, and there are 127,124 authors left. There are about
616K papers and 1.29M references to authors in *Publish* relation (authorship).
In DBLP we focus on distinguishing references to authors with identical names.

We first build a training set using the method illustrated above, which contains
1000 positive and 1000 negative examples. Then SVM with linear kernel is applied.
We measure the performance of Distinct by precision, recall, and f-measure. Suppose the standard set of clusters is C^*, and the set of clusters by Distinct is C. Let
TP (true positive) be the number of pairs of references that are in the same cluster
in both C^* and C. Let FP (false positive) be the number of pairs of references in
the same cluster in C but not in C^*, and FN (false negative) be the number of
pairs of references in the same cluster in C^* but not in C.

$$precision = \frac{TP}{TP + FP}, \quad recall = \frac{TP}{TP + FN}.$$

f-measure is the harmonic mean of precision and recall.

Table 1. Names corresponding to multiple authors

Name	#author	#ref	Name	#author	#ref
Hui Fang	3	9	Bing Liu	6	89
Ajay Gupta	4	16	Jim Smith	3	19
Joseph Hellerstein	2	151	Lei Wang	13	55
Rakesh Kumar	2	36	Wei Wang	14	141
Michael Wagner	5	29	Bin Yu	5	44

Table 2. Accuracy for distinguishing references

Name	precision	recall	f-measure
Hui Fang	1.0	1.0	1.0
Ajay Gupta	1.0	1.0	1.0
Joseph Hellerstein	1.0	0.810	0.895
Rakesh Kumar	1.0	1.0	1.0
Michael Wagner	1.0	0.395	0.566
Bing Liu	1.0	0.825	0.904
Jim Smith	0.888	0.926	0.906
Lei Wang	0.920	0.932	0.926
Wei Wang	0.855	0.814	0.834
Bin Yu	1.0	0.658	0.794
average	0.966	0.836	0.883

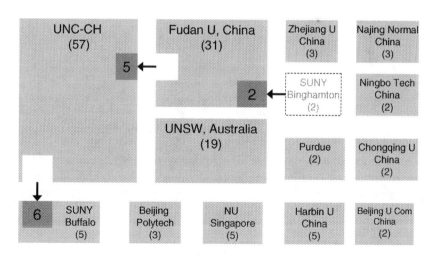

Fig. 2. Groups of references of "Wei Wang"

We test Distinct on real names in DBLP that correspond to multiple authors. 10 such names are shown in Table 1, together with the number of authors and number of references. For each name, we manually divide the references into groups according to the authors' identities, which are determined by the authors' home pages or affiliations shown on the papers.

We use Distinct to distinguish references to each name, with min-sim set to 0.0005. Table 2 shows the performance of Distinct for each name. In general, Distinct successfully group references with high accuracy. There is no false positive in 7 out of 10 cases, and the average recall is 83.6%. In some cases references to one author are divided into multiple groups. For example, 18 references to "Michael Wagner" in Australia are divided into two groups, which leads to low recall.

We visualize the results about "Wei Wang" in Fig. 2. References corresponding to each author are shown in a gray box, together with his/her current affiliation and number of references. The arrows and small blocks indicate the mistakes made by Distinct. It can be seen that in general Distinct does a very good job in distinguishing references, although it makes some mistakes because of the linkages between references to different authors.

3   Truth Discovery with Multiple Conflicting Information Providers

Information networks nowadays are fed with tremendous amounts of data from numerous information sources. These sources may provide *conflicting* information about the *same* entity, and pieces of information on the web could be already outdated when being read. This problem will only go worse since more information will be available on the web, and such conflicting information could become norm instead of exception. Therefore, it is necessary to provide trustable analysis of the truthfulness of information from multiple information providers and automatically identify the correct information. Such truth validation analysis is called veracity analysis.

Example 2. Veracity analysis on the authors of books provided by online bookstores. People retrieve all kinds of information from the web everyday. When shopping online, people find product specifications and sales information from various web sites like Amazon.com or ShopZilla.com. When looking for interesting DVDs, they get information and read movie reviews on web sites such as NetFlix.com or IMDB.com. Almost for any kind of products, there exist hundred of sale agents and information providers, if not more. *Is the information provided on the Web always trustable?* Unfortunately, the answer is negative. There is no guarantee for the correctness of information on the web. Even worse, different web information providers often present conflicting information. For example, we have found that there are multiple versions for the sets of authors of the same book, titled "Rapid Contextual Design" (ISBN: 0123540518), provided by different online bookstores, as shown in Table 3. From the image of the book cover, one can see that *A1 Books* provides the most accurate information. On the other hand, the information from *Powell's books* is incomplete, and that from *Lakeside books* is incorrect.

To analyze such a problem, the data sets can be viewed as a heterogeneous information network, consisting of three types of objects: (i) information providers (*e.g.*, online bookstores), (ii) objects (*e.g.*, books), and (iii) stated facts about

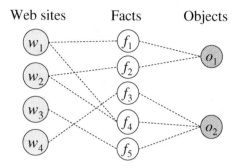

Fig. 3. A network snapshot of information providers, objects, and stated facts

Table 3. Conflicting information about book authors

Web site	Authors
A1 Books	Karen Holtzblatt, Jessamyn Burns Wendell, Shelley Wood
Powell's books	Holtzblatt, Karen
Cornwall books	Holtzblatt-Karen, Wendell-Jessamyn Burns, Wood
Mellon's books	Wendell, Jessamyn
Lakeside books	WENDELL, JESSAMYNHOLTZBLATT, KARENWOOD, SHELLEY
Blackwell online	Wendell, Jessamyn, Holtzblatt, Karen, Wood, Shelley
Barnes & Noble	Karen Holtzblatt, Jessamyn Wendell, Shelley Wood

the object (*i.e.*, claimed set of authors). One such mini-example is shown in Figure 3, which contains five stated facts about two objects provided by four web sites. Each web site provides at most one fact for an object.

There have been many studies on ranking web pages according to authority (or popularity) based on hyperlinks, such as Authority-Hub analysis [7], and PageRank [8]. However, top-ranked web sites may not be the most accurate ones. For example, according to our experiments the bookstores ranked on the very top ones by Google (which are *Barnes & Noble* and *Powell's books*) contain more errors on book author information than some small bookstores (*e.g.*, *A1 Books*) that provide more accurate information.

The problem of discovery of trustable information based on those provided multiple information providers is called the *veracity analysis* problem. It can be stated as follows: *Given a large amount of conflicting information about many objects, which is provided by multiple web sites (or other types of information providers), veracity analysis is to discover the true fact about each object.* Here the word *"fact"* is used to represent something that is claimed as a fact by some web site, and such a fact can be either true or false. Notice that here we only investigate the facts that are either the properties of objects (*e.g.*, weights of laptop computers), or the relationships between two objects (*e.g.*, authors of books).

Our solution is a TruthFinder framework [12], that finds confidently the true facts and trustworthy web sites. The method examines the relationships between information providers and the information they provided, with the following two major heuristics: (1) *an assertion that several information providers agree on is usually more trustable than that only one provider suggests*; and (2) *an information provider is* trustworthy *if it provides many pieces of true information, and a piece of information is likely to be true if it is provided by many trustworthy web sites*. The method links three types of information: (i) the information providers, (ii) stated facts on different entities, and (iii) the corresponding entities, into a heterogeneous information network, and performs an in-depth information network analysis. It starts with no bias on a particular piece of information, but uses the above heuristics to derive initial weights on the trustworthiness of the stated facts and information providers. Then it consolidates the trustworthiness by an iterative enhancement process with weight-propagation and consolidation across this information network. The process is similar to the page ranking process proposed in the PageRank and HITS algorithms [1,7] but the weight to be iteratively revised is the trustworthiness probability rather than authority score. We tested TruthFinder on the book author information provided by book sellers on the Web. The method successfully finds facts about who are the true set of authors and who are the trustable information providers.

Table 4 shows the accuracy of TruthFinder in comparison with that of Voting and Barnes & Noble on determining the authors for 100 randomly selected books, where *Voting* is a simple voting among multiple information providers, *i.e.*, considering the fact provided by a majority of web sites as the true fact. The result shows that TruthFinder achieves high accuracy at finding trustable information.

The TruthFinder methodology, though interesting, has two disadvantages: (1) it takes only one version of truth, and does not recognize there could be *multiple versions of truth*: the judgement of an event or an opinion could be rather different from people to people, *e.g.*, the view on a candidate in an election could be rather different but could be clustered into two to three views; and (2) it does not consider *truth may have timeliness*: *e.g.*, a player could win first but then lose, and depending on when the news was delivered, you may get rather different results.

Our on-going research is to overcome these two limitations and perform veracity analysis in information networks. First, we assume that there are *multiple*

Table 4. Performance comparison on a set of books among three methods: Voting, TruthFinder, and Barnes & Noble

Type of error	Voting	TruthFinder	Barnes & Noble
correct	71	85	64
miss author(s)	12	2	4
incomplete names	18	5	6
wrong first/middle names	1	1	3
has redundant names	0	2	23
add incorrect names	1	5	5
no information	0	0	2

versions of truth, each associated with one cluster of information providers. Second, we adopt the model of *timeliness of truth*, *i.e.*, truth may change with time in a dynamic, interconnected world. Moreover, we take higher priority on the most recent claim as the *up-to-date truth*. However, there are still information providers that deliver false information. Based on these assumptions, we are building a multiversion truth model using an integrated link analysis, information aggregation, and clustering, and consolidate the trustworthiness by an iterative enhancement process with weight-propagation across this information network.

4   On-Line Analytical Processing (OLAP) of Heterogeneous Information Networks

In relational database and data warehouse systems, On-Line Analytical Processing (OLAP) has become a powerful component in multidimensional data analysis. By constructing data cubes [5] over the underlying data and providing easy navigation, OLAP gives users the capability of interactive, multi-dimensional and multi-level analysis over a vast amount of data, with a wide variety of views. Certainly, for more complicated information network data, such capability is greatly needed. *"Can we OLAP information networks?"* In our recent study [3], we address this problem and aim for developing effective and scalable methods for on-line multidimensional analysis of heterogeneous information networks.

There are four major research challenges in Information Network OLAP (*i.e.*, Infonet-OLAP): (1) multi-dimensionality: each node/link in a network contains valuable, multi-typed, and multidimensional information, such as multi-level concepts/abstraction, textual contents, spatiotemporal information, and other properties; (2) scalability: information networks are often very large with millions of nodes and edges; (3) flexibility: for the same set of data, different users/ applications may like to view and analyze the network dramatically differently, which may lead to the efficient formation and exploration of very different information networks; and (4) quality: information networks may contain noisy, inconsistent, and inter-dependent data.

The concept of Infonet-OLAP can be briefly introduced using *bibliographic networks* extracted from the DBLP website (http://dblp.uni-trier.de). DBLP is an online bibliographic database for computer science conference proceedings and journals, indexing more than one million publications and more than 10,000 proceedings and journals. Each entry at DBLP contains (at least) the following pieces of information, $P : (\langle A_1, \ldots, A_k \rangle, T, V, Y)$, indicating that paper P is coauthored by k researchers, $A_1, \ldots, A_k$, with title T, and published at venue V in year Y. This entry consists of multidimensional information: *Authors, Title, Venue,* and *Time,* which can be viewed at multiple levels of abstraction and in a multi-dimensional space. For example, an author could be a junior author vs. a senior one; a prolific one vs. a nonproductive one; and a venue can be viewed similarly, such as a database venue vs. an AI one, a long-history one vs. a new one, and a highly reputed one vs. a low quality one. One also could apply a concept hierarchy to grouping papers according to their contents. Entries in such a database form a gigantic information network. Moreover, by linking with ACM Digital Library, Citeseer, and Google Scholar, citation information can be integrated as well.

This bibliographic network contains huge amounts of rich, heterogeneous, multidimensional, and temporal information. Users may like to view and analyze the network from different angles, which may lead to the *"formation" of different network views*, e.g., coauthor network, conference network, citation network, and author-theme network. Moreover, some may want to examine a network including only the *selected* research themes (e.g., data mining); whereas others may want to analyze hot topics in highly regarded (*i.e.*, highly *ranked*) conferences. Furthermore, some may like to *roll-up* authors to find how junior and senior researchers collaborate; whereas others would like to see the *evolution* of a theme. Clearly, different applications may require the extraction and analysis of multiple, different information networks involving time-variant, multidimensional, heterogeneous entities. A single, homogeneous, static network view cannot satisfy such flexible needs. The focus of Infonet-OLAP is to provide a general OLAP platform, rather than developing yet another specific network mining algorithm or theory.

Let us examine dimensions at first. Actually, there are two types of dimensions in Infonet-OLAP. The first one, called *informational dimension* (or *Info-Dim*, for short), utilizes informational attributes attached at the whole snapshot level. Suppose the following concept hierarchies are associated with *venue* and *time*:

- *venue*: *conference* → *area* → *all*,
- *time*: *year* → *decade* → *all*;

The role of these two dimensions is to organize snapshots into groups based on different perspectives, e.g., *(db-conf, 2004)* and *(sigmod, all-years)*, where each of these groups corresponds to a "cell" in the OLAP terminology. They control what snapshots are to be looked at, without touching the inside of any single snapshot.

Figure 4 shows such an example where the roll-up is first performed on the dimension *venue* to *db-conf* in individual year (*i.e.*., merging the graphs of $\langle SIGMOD, 2004 \rangle$, ..., $\langle VLDB, 2004 \rangle$ to $\langle db\text{-}conf, 2004 \rangle$) and then on the dimension *time* to $\langle db\text{-}conf, all\text{-}years \rangle$.

Second, for the subset of snapshots within each cell, one can summarize them by computing a measure as we did in traditional OLAP. In the Infonet-OLAP context, this gives rise to an *aggregated graph*. For example, a summary network displaying total collaboration frequencies can be achieved by overlaying all snapshots together and summing up the respective edge weights, so that each link now indicates two persons' collaboration activities in the DB conferences of 2004 or during the whole history of SIGMOD.

The second type of dimension, called *topological dimension* (or *Topo-Dim* for short), is provided to operate on nodes and edges within individual

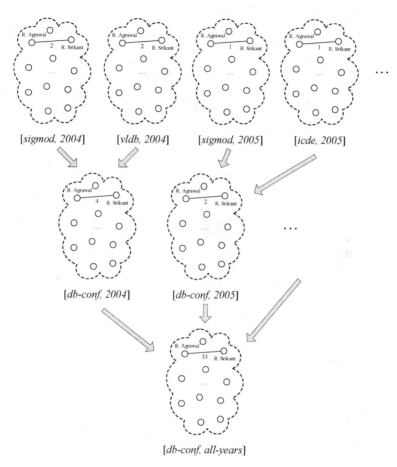

[sigmod, 2004]   [vldb, 2004]   [sigmod, 2005]   [icde, 2005]

[db-conf, 2004]   [db-conf, 2005]

[db-conf, all-years]

Fig. 4. An I-OLAP Scenario on the DBLP network

networks. Take the DBLP database for instance, suppose the following concept hierarchy

- *authorID: individual → department → institution → all*

is associated with the node attribute *authorID*, then it can be used to group authors from the same institution into a "generalized" node, and a new network thus formed will depict interactions among these groups as a whole, which summarizes the original network and hides specific details.

Figure 5 shows such an example on DBLP where the roll-up is performed on the dimension *authorID* to the level *institution*, which merge all persons in the same institution as one node and constructing a new summary graph at the institution level. In the "generalized network", an edge between Stanford and University of Wisconsin will aggregate all collaboration frequencies incurred between Stanford authors and Wisconsin authors. Notice that a roll-up from the individual level to the institution level is achieved by consolidating multiple nodes into one, which shrinks the original network.

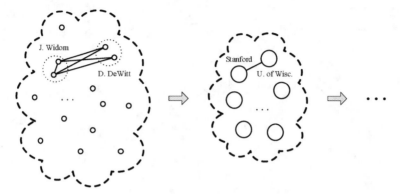

Fig. 5. A T-OLAP Scenario on the DBLP network

The OLAP semantics accomplished through Info-Dims and Topo-Dims are rather different. The first is called *informational OLAP* (abbr. *I-OLAP*), and the second *topological OLAP* (abbr. *T-OLAP*). For roll-up in I-OLAP, the characterizing feature is that, snapshots are just different observations of the same underlying network, and thus when they are all grouped into one cell in the cube, it is like *overlaying* multiple pieces of information, *without changing* the objects whose interactions are being looked at.

For roll-up in T-OLAP, we are no longer grouping snapshots, and the reorganization switches to happen inside individual networks. Here, *merging* is performed internally which "zooms out" the user's focus to a "generalized" set of objects, and a new information network formed by such *shrinking* might greatly alter the original network's topological structure.

As to measures, in traditional OLAP, a measure is calculated by aggregating all the data tuples whose dimensions are of the same values (based on concept hierarchies, such values could range from the finest un-generalized ones to "all/*", which form a multi-level cuboid lattice); casting this to our scenario here:

First, in infonetOLAP, the aggregation of graphs should also take the form of a graph, *i.e.*, an *aggregated graph*. In this sense, graph can be viewed as a special existence, which plays a dual role: as a data source and as an aggregated measure. Of course, other measures that are not graphs, such as node count, average degree, diameter, *etc.*, can also be calculated; however, we do not explicitly include such non-graph measures in our model, but instead treat them as derived from corresponding graph measures.

Second, due to the different semantics of I-OLAP and T-OLAP, aggregating data with identical Info-Dim values groups information among the snapshots, whereas aggregating data with identical Topo-Dim values groups topological elements inside individual networks. As a result, we will give a separate measure definition for each case in below.

A general framework of Infonet-OLAP is presented in [3], however, a systematic study on flexible and efficient implementation of different OALP operations on information networks is still an interesting issue for future research.

5 RankClus: Integrated Ranking-Based Clustering

Besides applying some typical knowledge discovery functions, some process may involve induction on the entire or a substantial portion of the information networks. For example, in order to partition an interconnected, heterogeneous information network into a set of clusters and rank the nodes in each cluster, we have recently develop a RankClus framework [9], which integrates clustering and ranking together to effectively cluster information networks into multiple groups and rank nodes in each group based on certain nice properties (such as authority). Interestingly, such clustering-ranking can be performed based on the links only, even without using the citation information nor the keyword/text information contained in the conferences and/or publication titles. We outline the method in more detail below.

In DBLP, by examining authors, research papers, and conferences, one can group conferences in the same fields together to form conference clusters, group authors based on their publication venues into author clusters and in the meantime rank authors and conferences based on their corresponding authorities. Such an integrated clustering and ranking framework would be an ideal feed for Infonet-OLAP.

Ranking and clustering can provide overall views on information network data, and each has been a hot topic by itself. However, in a large information network, ranking objects globally without considering the clusters they belong to often leads to dumb results, e.g., ranking authors/papers in database/data_mining (DB/DM) and hardware/computer_architcture (HW/CA) conferences together, as shown in Figure 6(a), may not make much sense. Similarly, presenting a huge cluster with numerous entities without distinction is dull as well. Therefore, we propose an integrated approach, called RankClus, to perform ranking and clustering together. Figure 6(b) shows that RankClus can generate meaningful clusters and rankings, even at the expert level, without referring to any citation or content information in DBLP. These clustering and ranking results could be potential input to Infonet-OLAP for effective data mining. RankClus also shows the importance of developing a general Infonet-OLAP framework that allows users to explore interactively with underlying networks for subtle knowledge discovery. An isolated clustering or ranking algorithm is often not enough for such tasks.

According to RankClus, we view the DBLP network as a *bi-type information network* with *conferences* as one type and *authors* as the other. Given two types of object sets X and Y, where $X = \{x_1, x_2, \ldots, x_m\}$, and $Y = \{y_1, y_2, \ldots, y_n\}$, graph $G = \langle V, E \rangle$ is called a bi-type information network on types X and Y, if $V(G) = X \cup Y$ and $E(G) = \{\langle o_i, o_j \rangle\}$, where $o_i, o_j \in X \cup Y$.

Let $W_{(m+n) \times (m+n)} = \{w_{o_i o_j}\}$ be the adjacency matrix of links, where $w_{o_i o_j}$ equals to the weight of link $\langle o_i, o_j \rangle$, which is the number of observations of the link, we thus use $G = \langle \{X \cup Y\}, W \rangle$ to denote this bi-type information network. In the following, we use X and Y denoting both the object set and their type name. For convenience, we decompose the link matrix into four blocks,

$$W = \begin{pmatrix} W_{XX} & W_{XY} \\ W_{YX} & W_{YY} \end{pmatrix}, \text{ each denoting a sub-network of objects between types of}$$

the subscripts.

Rank	Conf.	Rank	Authors
1	DAC	1	Alberto L. Sangiovanni-Vincentelli
2	ICCAD	2	Robert K. Brayton
3	DATE	3	Massoud Pedram
4	ISLPED	4	Miodrag Potkonjak
5	VTS	5	Andrew B. Kahng
6	CODES	6	Kwang-Ting Cheng
7	ISCA	7	Lawrence T. Pileggi
8	VLDB	8	David Blaauw
9	SIGMOD	9	Jason Cong
10	ICDE	10	D. F. Wong

(a) Top-10 ranked conf's/authors in the mixed conf. set

Rank	Conf.	Rank	Authors
1	VLDB	1	H. V. Jagadish
2	SIGMOD	2	Surajit Chaudhuri
3	ICDE	3	Divesh Srivastava
4	PODS	4	Michael Stonebraker
5	KDD	5	Hector Garcia-Molina
6	CIKM	6	Jeffrey F. Naughton
7	ICDM	7	David J. DeWitt
8	PAKDD	8	Jiawei Han
9	ICDT	9	Rakesh Agrawal
10	PKDD	10	Raghu Ramakrishnan

(b) Top-10 ranked conf's/authors in DB/DM set

Fig. 6. RankClus Performs High-Quality Clustering and Ranking Together

Given a bi-type network $G = \langle \{X \cup Y\}, W \rangle$, if a function $f : G \to (r_X, r_Y)$ gives rank score for each object in type X or type Y, where $\sum_{x \in X} r_X(x) = 1$ and $\sum_{y \in Y} r_Y(y) = 1$. We call f a *ranking function* on network G. Similarly, we can define *conditional rank* and *within-cluster rank*, as follows. Given target type X, and a cluster $X' \subseteq X$, sub-network $G' = \langle \{X' \cup Y\}, W' \rangle$ is defined as a vertex induced graph of G by a vertex subset $X' \cup Y$. *Conditional rank* over Y, denoted as $r_{Y|X'}$, and *within-cluster rank* over X', denoted as $r_{X'|X'}$, are defined by the ranking function f on the sub-network G': $(r_{X'|X'}, r_{Y|X'}) = f(G')$. Conditional rank over X, denoted as $r_{X|X'}$, is defined as the propagation score of $r_{Y|X'}$ over network G:

$$r_{X|X'}(x) = \frac{\sum_{j=1}^{n} W_{XY}(x, j) r_{Y|X'}(j)}{\sum_{i=1}^{m} \sum_{j=1}^{n} W_{XY}(i, j) r_{Y|X'}(j)}. \tag{1}$$

Given a bi-type network $G = \langle \{X \cup Y\}, W \rangle$, the target type X, and K, a specified number of clusters, our goal is to generate K clusters $\{X_k\}, k = 1, 2, \ldots, K$ on X, as well as the within-cluster rank for type X and conditional rank for type Y to each cluster, i.e., $r_{X|X_k}$ and $r_{Y|X_k}, k = 1, 2, \ldots, K$.

In order to generate effective clusters based on authoritative ranking, we need to specify a few rules based on prior knowledge. For example, for the DBLP network, we have the following three empirical rules that will affect our clustering results.

Rule 1: Highly ranked authors publish many papers in highly ranked conferences.

According to this rule, each author's score is determined by the number of papers and their publication forums, i.e., $r_Y(j) = \sum_{i=1}^{m} W_{YX}(j,i)r_X(i)$. This implies when author j publishes more papers, there are more nonzero and high weighted $W_{YX}(j,i)$, and when the author publishes papers in a higher ranked conference i, which means a higher $r_X(i)$, the score of author j will be higher. At the end of each step, $r_Y(j)$ is normalized by $r_Y(j) \leftarrow \frac{r_Y(j)}{\sum_{j'=1}^{n} r_Y(j')}$.

Rule 2: Highly ranked conferences attract many papers from many highly ranked authors.

According to this rule, the score of each conference is determined by the quantity and quality of papers in the conference, which is measured by their authors' ranking scores, i.e., $r_X(i) = \sum_{j=1}^{n} W_{XY}(i,j)r_Y(j)$. This implies when there are more papers appearing in conference i, there are more non-zero and high weighted $W_{XY}(i,j)$; and if the papers are published by higher ranked author j, the rank score for j, which is $r_Y(j)$, is higher, and thus the higher score the conference i will get. The score vector is then normalized by $r_X(i) \leftarrow \frac{r_X(i)}{\sum_{i'=1}^{m} r_X(i')}$. Notice that the normalization will not change the ranking position of an object, but it gives a relative importance score to each object. When considering the co-author information, the scoring function can be further refined by the third heuristic:

Rule 3: Highly ranked authors usually co-author with many authors or many highly ranked authors.

Based on this rule, we can revise the above two equations as

$$r_Y(i) = \alpha \sum_{j=1}^{m} W_{YX}(i,j)r_X(j) + (1-\alpha) \sum_{j=1}^{n} W_{YY}(i,j)r_Y(j). \qquad (2)$$

where $\alpha \in [0,1]$ is a weighting coefficient, which could be learned by a training set.

Notice such rules should be worked out by experts based on their research experience in a specific field. For example, in PubMed domain, people may emphasize more on journal publications than conference ones. Also, subtlety exists in certain rules. For example, for Rule 1, a conference/journal will not be reputed if it attracts papers *only* from a tiny group of "prolific" authors because this tiny group may set up its own venue and publish many papers *only* there. They could be "prolific" but few other prolific authors would like to join. Thus neither the venue nor the authors can be "reputed" according to the rule.

Traditional graph clustering methods usually tackle one network only during a clustering process. In contrast, with Rule 3, RankClus is able to combine two information networks, i.e., author-conference bipartite network and co-author network, together for better clustering and ranking.

Putting these rules together, RankClus first randomly generates an initial clustering for target objects. It then performs the following three steps repeatedly until the clustering does not change significantly: (1) rank each cluster, by calculating conditional rank for types Y and X and within-cluster rank for type

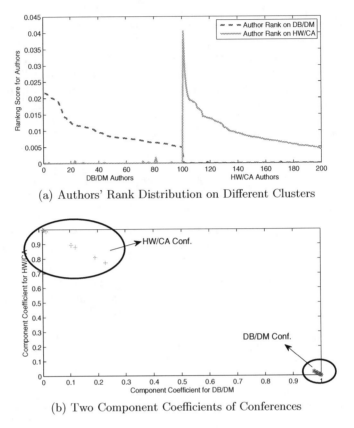

(a) Authors' Rank Distribution on Different Clusters

(b) Two Component Coefficients of Conferences

Fig. 7. Effectiveness of RankClus at Clustering and Ranking of the DBLP Data

X; (2) estimate coefficients in the mixture model component; and (3) adjust membership in clusters.

Our implementation and testing of the RankClus method have demonstrated the high promise of this approach. For processing efficiency, it is an order of magnitude faster than the well-cited SimRank algorithm [6] because SimRank has to calculate the pairwise similarity between every two objects of the same type, whereas RankClus uses conditional ranking as the measure of clusters, and only need to calculate the distances between each object and the cluster center. For effectiveness, Figure 7(a) shows that conditional rank can be used as cluster features: DB/DM authors rank high with respect to DB/DM conferences, but rank extremely low with respect to HW/CA conferences. Figure 7(b) is the scatter plot for each conference's two component coefficients. The figure shows that component coefficients can be effectively used as object attributes: the two kinds of conferences are separated clearly under the new attributes.

Our on-going work is to extend the RankClus framework from the following three perspectives: (1) The patterns described above are mined without using citation or title information. By adding title, citation, and/or abstract information, one can uncover the network structure at rather deep levels of research topic hierarchies. These hierarchies will provide additional dimensions to Infonet-OLAP and enhance discovery-driven OLAP. (2) Infonet-OLAP also introduces new challenges to RankClus. There could be multiple information networks available. For instance, we could add author-article network and citation network into the above example for better ranking. In our current implementation, RankClus can only integrate two networks together. Our recent study extends this framework into NetClus star-network schema [10], and subsequent studies will accommodate multiple networks, multiple hierarchies and constraints in the RankClus framework. (3) To facilitate easy exploration (drill-down and roll-up) of clustering results in information networks with different granularity, it is preferred to have clustering consistent across different levels of abstraction. The principles of such clustering design will be examined for a collection of multi-resolution information networks.

6   Conclusions

In this paper, we have outlined our new research progress on mining knowledge from heterogeneous information networks. We show that heterogeneous information networks are ubiquitous and with broad applications. Moreover, knowledge is often hidden in massive links in heterogeneous information networks, and thus it is necessary to perform a systematic study on how ro explore the power of links at mining heterogeneous information networks. We presented in several interesting link-mining tasks, including link-based object distinction, veracity analysis, multidimensional online analytical processing of heterogeneous information networks, and rank-based clustering. We show some interesting results of our recent research that explore the crucial information hidden in links will be introduced, including four new methods: (1) Distinct for object distinction analysis, (2) TruthFinder for veracity analysis, (3) Infonet-OLAP for online analytical processing of information networks, and (4) RankClus for integrated ranking-based clustering. We also show that mining heterogeneous information networks is an exciting research frontier and there is much space to be explored in future research.

Acknowledgement

The work was supported in part by the U.S. National Science Foundation grants IIS-08-42769 and BDI-05-15813, NASA grant NNX08AC35A, and the Air Force Office of Scientific Research MURI award FA9550-08-1-0265. Any opinions, findings, and conclusions expressed here are those of the authors and do not necessarily reflect the views of the funding agencies. This paper is a summarization of several pieces of work in our research. The author would like to express his sincere thanks to all the collaborators, especially Chen Chen, Yizhou Sun, Xifeng Yan, Xiaoxin Yin, Philip Yu, and Feida Zhu.

References

1. Brin, S., Page, L.: The anatomy of a large-scale hypertextual web search engine. In: Proc. 7th Int. World Wide Web Conf (WWW 1998), Brisbane, Australia, April 1998, pp. 107–117 (1998)
2. Chaudhuri, S., Ganjam, K., Ganti, V., Motwani, R.: Robust and efficient fuzzy match for online data cleaning. In: Proc. 2003 ACM-SIGMOD Int. Conf. Management of Data (SIGMOD 2003), San Diego, CA (June 2003)
3. Chen, C., Yan, X., Zhu, F., Han, J., Yu, P.S.: Graph OLAP: Towards online analytical processing on graphs. In: Proc. 2008 Int. Conf. on Data Mining (ICDM 2008), Pisa, Italy (December 2008)
4. Gravano, L., Ipeirotis, P., Jagadish, H., Koudas, N., Muthukrishnan, S., Srivastava, D.: Approximate string joins in a database (almost) for free. In: Proc. 2001 Int. Conf. Very Large Data Bases (VLDB 2001), Rome, Italy, September 2001, pp. 491–500 (2001)
5. Gray, J., Chaudhuri, S., Bosworth, A., Layman, A., Reichart, D., Venkatrao, M., Pellow, F., Pirahesh, H.: Data cube: A relational aggregation operator generalizing group-by, cross-tab and sub-totals. Data Mining and Knowledge Discovery 1, 29–54 (1997)
6. Jeh, G., Widom, J.: SimRank: a measure of structural-context similarity. In: Proc. 2002 ACM SIGKDD Int. Conf. on Knowledge Discovery and Data Mining (KDD 2002), Edmonton, Canada, July 2002, pp. 538–543 (2002)
7. Kleinberg, J.M.: Authoritative sources in a hyperlinked environment. J. ACM 46, 604–632 (1999)
8. Page, L., Brin, S., Motwani, R., Winograd, T.: The pagerank citation ranking: Bringing order to the web. Technical Report, Computer Science Dept., Stanford University (1998)
9. Sun, Y., Han, J., Zhao, P., Yin, Z., Cheng, H., Wu, T.: RankClus: Integrating clustering with ranking for heterogeneous information network analysis. In: Proc. 2009 Int. Conf. on Extending Data Base Technology (EDBT 2009), Saint-Petersburg, Russia (March 2009)
10. Sun, Y., Yu, Y., Han, J.: Ranking-based clustering of heterogeneous information networks with star network schema. In: Proc. 2009 ACM SIGKDD Int. Conf. on Knowledge Discovery and Data Mining (KDD 2009), Paris, France (June 2009)
11. Yin, X., Han, J., Yu, P.S.: Object distinction: Distinguishing objects with identical names by link analysis. In: Proc. 2007 Int. Conf. Data Engineering (ICDE 2007), Istanbul, Turkey (April 2007)
12. Yin, X., Han, J., Yu, P.S.: Truth discovery with multiple conflicting information providers on the web. IEEE Trans. Knowledge and Data Eng. 20, 796–808 (2008)

Learning on the Web

Fernando C.N. Pereira

University of Pennsylvania, USA

It is commonplace to say that the Web has changed everything. Machine learning researchers often say that their projects and results respond to that change with better methods for finding and organizing Web information. However, not much of the theory, or even the current practice, of machine learning take the Web seriously. We continue to devote much effort to refining supervised learning, but the Web reality is that labeled data is hard to obtain, while unlabeled data is inexhaustible. We cling to the iid assumption, while all the Web data generation processes drift rapidly and involve many hidden correlations. Many of our theory and algorithms assume data representations of fixed dimension, while in fact the dimensionality of data, for example the number of distinct words in text, grows with data size. While there has been much work recently on learning with sparse representations, the actual patterns of sparsity on the Web are not paid much attention. Those patterns might be very relevant to the communication costs of distributed learning algorithms, which are necessary at Web scale, but little work has been done on this.

Nevertheless, practical machine learning is thriving on the Web. Statistical machine translation has developed non-parametric algorithms that learn how to translate by mining the ever-growing volume of source documents and their translations that are created on the Web. Unsupervised learning methods infer useful latent semantic structure from the statistics of term co-occurrences in Web documents. Image search achieves improved ranking by learning from user responses to search results. In all those cases, Web scale demanded distributed algorithms.

I will review some of those practical successes to try to convince you that they are not just engineering feats, but also rich sources of new fundamental questions that we should be investigating.

J. Gama et al. (Eds.): DS 2009, LNAI 5808, p. 31, 2009.

Learning and Domain Adaptation

Yishay Mansour

Blavatnik School of Computer Science,
Tel Aviv University
Tel Aviv, Israel
mansour@tau.ac.il

Abstract. Domain adaptation is a fundamental learning problem where one wishes to use labeled data from one or several source domains to learn a hypothesis performing well on a different, yet related, domain for which no labeled data is available. This generalization across domains is a very significant challenge for many machine learning applications and arises in a variety of natural settings, including NLP tasks (document classification, sentiment analysis, etc.), speech recognition (speakers and noise or environment adaptation) and face recognition (different lighting conditions, different population composition).

The learning theory community has only recently started to analyze domain adaptation problems. In the talk, I will overview some recent theoretical models and results regarding domain adaptation.

This talk is based on joint works with Mehryar Mohri and Afshin Rostamizadeh.

1 Introduction

It is almost standard in machine learning to assume that the training and test instances are drawn from the same distribution. This assumption is explicit in the standard PAC model [19] and other theoretical models of learning, and it is a natural assumption since when the training and test distributions substantially differ there can be no hope for generalization. However, in practice, there are several crucial scenarios where the two distributions are similar but not identical, and therefore effective learning is potentially possible. This is the motivation for *domain adaptation*.

The problem of domain adaptation arises in a variety of applications in natural language processing [6,3,9,4,5], speech processing [11,7,16,18,8,17], computer vision [15], and many other areas. Quite often, little or no labeled data is available from the *target domain*, but labeled data from a *source domain* somewhat similar to the target as well as large amounts of unlabeled data from the target domain are at one's disposal. The domain adaptation problem then consists of leveraging the source labeled and target unlabeled data to derive a hypothesis performing well on the target domain.

The first theoretical analysis of the domain adaptation problem was presented by [1], who gave VC-dimension-based generalization bounds for adaptation in

J. Gama et al. (Eds.): DS 2009, LNAI 5808, pp. 32–34, 2009.

classification tasks. Perhaps, the most significant contribution of that work was the definition and application of a distance between distributions, the d_A distance, that is particularly relevant to the problem of domain adaptation and which can be estimated from finite samples for a finite VC dimension, as previously shown by [10]. This work was later extended by [2] who also gave a bound on the error rate of a hypothesis derived from a weighted combination of the source data sets for the specific case of empirical risk minimization. More refined generalization bounds which apply to more general tasks, including regression and general loss functions appear in [12]. From an algorithmic perspective, it is natural to re-weight the empirical distribution to better reflect the target distribution; efficient algorithms for this re-weighting task were given in [12].

A more complex variant of this problem arises in sentiment analysis and other text classification tasks where the learner receives information from *several* domain sources that he can combine to make predictions about a target domain. As an example, often appraisal information about a relatively small number of domains such as *movies*, *books*, *restaurants*, or *music* may be available, but little or none is accessible for more difficult domains such as *travel*. This is known as the *multiple source adaptation problem*. Instances of this problem can be found in a variety of other natural language and image processing tasks.

The problem of adaptation with multiple sources was introduced and analyzed [13,14]. The problem is formalized as follows. For each source domain $i \in [1, k]$, the learner receives the distribution of the input points Q_i, as well as a hypothesis h_i with loss at most ϵ on that source. The task consists of combining the k hypotheses h_i, $i \in [1, k]$, to derive a hypothesis h with a loss as small as possible with respect to the target distribution P. Unfortunately, a simple convex combination of the k source hypotheses h_i can perform very poorly; for example, there are cases where *any* such convex combination would incur a classification error of a half, even when each source hypothesis h_i makes no error on its domain Q_i (see [13]). In contrast, *distribution weighted combinations* of the source hypotheses, which are combinations of source hypotheses weighted by the source distributions, perform very well. In [13] it was shown that, remarkably, for any fixed target function, there exists a distribution weighted combination of the source hypotheses whose loss is at most ϵ with respect to *any* mixture P of the k source distributions Q_i. For the case that the target distribution P is arbitrary, generalization bounds, based on *Rényi divergence* between the sources and the target distributions, were derived in [14].

References

1. Ben-David, S., Blitzer, J., Crammer, K., Pereira, F.: Analysis of representations for domain adaptation. In: Proceedings of NIPS 2006 (2006)
2. Blitzer, J., Crammer, K., Kulesza, A., Pereira, F., Wortman, J.: Learning bounds for domain adaptation. In: Proceedings of NIPS 2007 (2007)
3. Blitzer, J., Dredze, M., Pereira, F.: Biographies, Bollywood, Boom-boxes and Blenders: Domain Adaptation for Sentiment Classification. In: ACL 2007 (2007)

4. Chelba, C., Acero, A.: Adaptation of maximum entropy capitalizer: Little data can help a lot. Computer Speech & Language 20(4), 382–399 (2006)
5. Daumé III, H., Marcu, D.: Domain adaptation for statistical classifiers. Journal of Artificial Intelligence Research 26, 101–126 (2006)
6. Dredze, M., Blitzer, J., Talukdar, P.P., Ganchev, K., Graca, J., Pereira, F.: Frustratingly Hard Domain Adaptation for Parsing. In: CoNLL 2007 (2007)
7. Gauvain, J.-L., Chin-Hui: Maximum a posteriori estimation for multivariate gaussian mixture observations of markov chains. IEEE Transactions on Speech and Audio Processing 2(2), 291–298 (1994)
8. Jelinek, F.: Statistical Methods for Speech Recognition. MIT Press, Cambridge (1998)
9. Jiang, J., Zhai, C.: Instance Weighting for Domain Adaptation in NLP. In: Proceedings of ACL 2007 (2007)
10. Kifer, D., Ben-David, S., Gehrke, J.: Detecting change in data streams. In: Proceedings of the 30th International Conference on Very Large Data Bases (2004)
11. Legetter, C.J., Woodland, P.C.: Maximum likelihood linear regression for speaker adaptation of continuous density hidden markov models. In: Computer Speech and Language, pp. 171–185 (1995)
12. Mansour, Y., Mohri, M., Rostamizadeh, A.: Domain adaptation: Learning bounds and algorithms. In: COLT (2009)
13. Mansour, Y., Mohri, M., Rostamizadeh, A.: Domain adaptation with multiple sources. In: Proceedings of NIPS 2008 (2008)
14. Mansour, Y., Mohri, M., Rostamizadeh, A.: Multiple source adaptation and the Rényi divergence. In: Uncertainty in Artificial Inteligence, UAI (2009)
15. Martínez, A.M.: Recognizing imprecisely localized, partially occluded, and expression variant faces from a single sample per class. IEEE Trans. Pattern Anal. Mach. Intell. 24(6), 748–763 (2002)
16. Pietra, S.D., Pietra, V.D., Mercer, R.L., Roukos, S.: Adaptive language modeling using minimum discriminant estimation. In: HLT 1991: Proceedings of the workshop on Speech and Natural Language, pp. 103–106 (1992)
17. Roark, B., Bacchiani, M.: Supervised and unsupervised PCFG adaptation to novel domains. In: Proceedings of HLT-NAACL (2003)
18. Rosenfeld, R.: A Maximum Entropy Approach to Adaptive Statistical Language Modeling. Computer Speech and Language 10, 187–228 (1996)
19. Valiant, L.G.: A theory of the learnable. Communication of the ACM 27(11), 1134–1142 (1984)

The Two Faces of Active Learning

Sanjoy Dasgupta

University of California, San Diego

The active learning model is motivated by scenarios in which it is easy to amass vast quantities of unlabeled data (images and videos off the web, speech signals from microphone recordings, and so on) but costly to obtain their labels. Like supervised learning, the goal is ultimately to learn a classifier. But like unsupervised learning, the data come unlabeled. More precisely, the labels are hidden, and each of them can be revealed only at a cost. The idea is to query the labels of just a few points that are especially informative about the decision boundary, and thereby to obtain an accurate classifier at significantly lower cost than regular supervised learning.

There are two distinct narratives for explaining when active learning is helpful. The first has to do with efficient search through the hypothesis space: perhaps one can always explicitly select query points whose labels will significantly shrink the set of plausible classifiers (those roughly consistent with the labels seen so far)? The second argument for active learning has to do with exploiting cluster structure in data. Suppose, for instance, that the unlabeled points form five nice clusters; with luck, these clusters will be pure and only five labels will be necessary!

Both these scenarios are hopelessly optimistic. But I will show that they each motivate realistic models that can effectively be exploited by active learning algorithms. These algorithms have provable label complexity bounds that are in some cases exponentially lower than for supervised learning. I will also present experiments with these algorithms, to illustrate their behavior and get a sense of the gulf that still exists between the theory and practice of active learning.

This is joint work with Alina Beygelzimer, Daniel Hsu, John Langford, and Claire Monteleoni.

J. Gama et al. (Eds.): DS 2009, LNAI 5808, p. 35, 2009.

An Iterative Learning Algorithm for Within-Network Regression in the Transductive Setting

Annalisa Appice, Michelangelo Ceci, and Donato Malerba

Dipartimento di Informatica, Università degli Studi di Bari
via Orabona, 4 - 70126 Bari - Italy
{appice,ceci,malerba}@di.uniba.it

Abstract. Within-network regression addresses the task of regression in partially labeled networked data where labels are sparse and continuous. Data for inference consist of entities associated with nodes for which labels are known and interlinked with nodes for which labels must be estimated. The premise of this work is that many networked datasets are characterized by a form of autocorrelation where values of the response variable in a node depend on values of the predictor variables of interlinked nodes. This autocorrelation is a violation of the independence assumption of observation. To overcome to this problem, the lagged predictor variables are added to the regression model. We investigate a computational solution for this problem in the transductive setting, which asks for predicting the response values only for unlabeled nodes of the network. The neighborhood relation is computed on the basis of the node links. We propose a regression inference procedure that is based on a co-training approach according to separate model trees are learned from both attribute values of labeled nodes and attribute values aggregated in the neighborhood of labeled nodes, respectively. Each model tree is used to label the unlabeled nodes for the other during an iterative learning process. The set of labeled data is changed by including labels which are estimated as confident. The confidence estimate is based on the influence of the predicted labels on known labels of interlinked nodes. Experiments with sparsely labeled networked data show that the proposed method improves traditional model tree induction.

1 Introduction

A data network (also called networked data) consists of entities, generally of the same type such as web-pages or telephone accounts, which are associated with the nodes of the network and which are interlinked with other nodes via various explicit relations (or edges) such as hyperlinks between web-pages or people calling each other. Over the past few years data networks such as sensor networks, communication networks, financial transaction networks and social networks have become ubiquitous in everyday life. This ubiquity of data networks motivates the recent focus of research in data mining to extend traditional inference techniques in order to learn in data networks.

J. Gama et al. (Eds.): DS 2009, LNAI 5808, pp. 36–50, 2009.

Several issues challenge the task of learning in network data, the most important being the consideration of various forms of *autocorrelation* which may affect data networks. Different definitions of autocorrelation are in use depending on the field of study which is being considered and not all of them are equivalent. Here autocorrelation is defined as the property that a value observed at a node depends on the values observed at neighboring nodes in the network. Autocorrelation has been justified in several ways, such as Tobler's first law of geography [20] and the homophily's principle [14], according to the specific application domain.

The major difficulty due to the autocorrelation is that the independence assumptions, which typically underlies machine learning methods, are no longer valid. For example, the violation of the instance independence has been identified as the main responsible of poor performance of traditional machine learning methods [16]. To remedy the negative effects of the violation of independence assumptions, autocorrelation has to be explicitly accommodated in the learned models.

In predictive models, where response variables depend on both predictor variables and an error term, autocorrelation can be expressed in three different ways, by correlating: 1) the error terms of neighboring nodes; 2) the response variables of neighboring nodes; 3) the response variable with the predictor variables of neighboring nodes. In spatial data analysis, these three types of predictive models are respectively known as spatial error models, spatial lag models and spatial cross-regressive models [18]. As observed in [2], the first two types of models are *global* in scope, in the sense that an error term or a dependent variable at a location (node of a network) has a spillover effect on all other locations, while cross-regressive models are *local* in scope, since the effects are confined to the neighbors of each observation. In spatial data analysis, cross-regressive models make more sense from a theoretical point of view [1] and present the additional advantage of being easier to use. In this paper we face the problem of learning predictive (regression) models in data networks and we deal with the autocorrelation issue by considering cross-regressive models.

The consideration of partially labeled data networks, where labeled entities are possibly interlinked to unlabeled entities and vice-versa, adds a further degree of complexity, since it is difficult *to separate data into training and test sets*. Indeed, labeled data would serve as training data and subsequently as background knowledge necessary for labeling entities in the test set. This consideration motivates the investigation of the learning problem in a setting different from the classical inductive one, where the prediction model is built by considering only a finite set of labeled data (training set) and it is then used to make prediction on any possible instance. In this work, we consider the *transductive* setting [21], where both labeled and unlabeled data are used to build the model and predictions are confined to unlabeled data available when learning starts. More precisely, the idea behind *transductive inference* (or *transduction*) is to analyze both the labeled data L and the unlabeled data U to build a model which predicts (exlusively) data in U as accurately as possible. Therefore, in the transductive setting, difficulties due to the separation of training and test set are overcome.

The data mining task considered in this work is *transductive within-network regression*, which is a variant of the classification task recently investigated in [11] for categorical labels. Given a fully described network (nodes and edges) for which continuous labels are provided for only some of the nodes, the goal is to determine labels of the rest of the nodes in the network. We propose a learning algorithm, named ITL (Iterative Transductive Learner), which capitalizes on the strengths of both model tree induction and transductive learning to effectively solve the given problem when labels of data networks are originally sparse and possibly scarce. The specific contributions of this work are highlighted as follows:

1. The combination of iterative transductive learning with the co-training paradigm in order to both generate cross-regressive models and bootstrap from a small set of labeled training data via a large set of unlabeled data.
2. Prediction of continuous labels is based on model trees [3], which do not impose any *a priori* global structure (e.g., linear) of the regression surface. Model trees are build on two views of data, as required by the co-training paradigm. Each model tree labels the unlabeled data for the other during the learning process.
3. The use of co-training paradigm allows us to learn two different model trees: a model tree that identifies the correlation between the label of an entity and the attribute values of the same entity and a model tree that identifies the correlation between the label of an entity and the attribute values in the neighborhood of the entity. Each model tree labels the unlabeled data for the other during the learning process.
4. We present some procedures to estimate the confidence of predicted label(s) through consulting the influence of the labeling of unlabeled entities in a model tree based re-prediction of the labeled entities which are interlinked to the unlabeled one(s).
5. We demonstrate that our approach is robust to both sparse labeling and low label consistency, performing well consistently across a range of data network where traditional model tree induction fails.

The rest of the paper is organized as follows. In Section 2, we review related work. We present the formal definition of the task in Section 3 and our proposed method in Section 4. Section 5 describes the experimental methodology and results. Finally, Section 6 concludes the work.

2   Related Work

Regression inference in data network is still a challenging issue in machine learning and data mining. Due to the recent efforts of various researchers, numerous algorithms have been designed for modeling a partially labeled network and providing estimates of unknown labels associated with nodes. Anyway, at the best of our knowledge, these algorithms address the prediction problem only in the classification case, that is, when labels are categorical.

Currently, the main research in this area is in the thrust of network learning and graph mining. Network learning assumes that data for inference are already in the form of a network and exploits the structure of the network to allow the collective inference. Collective inference allows to infer various interrelated values simultaneously. It is used in network learning since it permits to estimate neighboring labels which influence one another [12,9,19]. Since exact inference is known to be an NP-hard problem and there is no guarantee that data network satisfy the conditions that make exact inference tractable for collective learning, most of the research in collective learning has been devoted to the development of approximate inference algorithms.

Some of the popular approximate inference algorithms are the iterative inference, the Gibbs sampling, the loopy belief propagation and the mean-field relaxation labeling. An outline of strengths and weakness of these algorithms is reported in [19]. In general, one of the major advantages of collective learning lies in its powerful ability to learn various kinds of dependency structures (positive vs. negative autocorrelation, different degrees of correlation and so on) [10]. However, as pointed out in [15], when the labeled data is very sparse, the performance of collective classification might be largely degraded due to the lack of sufficient neighbors. This is overcome by incorporating informative "ghost edges" into the networks to deal with sparsity issues [13,15].

Interestingly learning problems similar to the tasks addressed in network learning have been recently addressed outside the areas of network learning and graph mining. This second area of work has not been cast as a network learning problem, but rather in the area of semi-supervised learning in a transductive setting [21] where a corpus of data without links is given. The basic idea is to connect data into a weighted network by adding edges (in various ways) based on the similarity between entities and to estimate a function on the graph which guarantees the consistency with the label information and the smoothness over the whole graph [23]. The constraint on smoothness implicitly assumes positive autocorrelation in the graph, that is, nearby nodes tend to share the same class labels (i.e., homophily).

A prominent achievement in semi-supervised learning is represented by the co-training paradigm [4] where independent views, i.e., distinct sets of attributes, of labeled and unlabeled data are available for deriving separate learners. Predictions of each learner of unlabeled data are then used to augment the training set of the other within an iterative learning process. Co-training is already used to design regression algorithms in semi-supervised learning. Brefeld et al. [5] use co-training to formulate a semi-supervised least square regression algorithm, where co-training is casted as a regularized risk minimization problem in Hilbert spaces. Several data views are obtained for learning from different instance descriptions, views, and/or different kernel functions. Zhou and Li [22] apply co-training to learn k-NN regression by adopting a single attribute set but considering distinct distance measures for the two hypotheses.

3   Problem Definition and Notations

A network is a set of entities connected by edges. Each entity is called node of the network. A number (which is usually taken to be positive) called weight is associated with each edge. In a general formulation, a network can be represented as a (weighted) graph that is a set of nodes and a ternary relation which represent both the edges between nodes and the weight associated to each edge. Formally,

Definition 1 (Data Network). *A data network N is a pair (V, E), where:*

1. *V is a set of nodes, and*
2. *E is a set of weighted edges between nodes, that is,*

$$E = \{\langle u, v, w \rangle | u, v \in V, w \in \mathbb{R}^+\}.$$

In this work, each node of the network is associated with a data observation $(\mathbf{x}, y) \in \mathbf{X} \times Y$. $\mathbf{X}$ is a feature space spanned by m predictor variables X_i with $i = 1, \ldots, m$ while Y is the possibly unknown response variable (or label) with a range in $\mathbb{R}$. Additionally, labels are typically sparse in the network, that is, nodes for which labels are known may be interlinked with nodes for which labels must be estimated. In several real cases, labels are also scarce since the manual annotation of large data sets can be very costly. In this data context, the problem of regression consists in predicting the labels of unlabeled nodes as accurate as possible. The regression problem is formulated in network learning as follows.

Given:

1. the labeled node set $L \subset \mathbf{X} \times Y$;
2. the projection of the unlabeled (working) node set $U = V - L$ on $\mathbf{X}$;
3. the ternary relation $E \subset V \times V \times \mathbb{R}^+$;
4. the neighborhood function $\eta_E : V \longmapsto 2^{V \times \mathbb{R}^+}$ such that:

$$\eta_E(u) = \{(v_1, w_1), \ldots, (v_k, w_k)\} \text{ with } (u, v_i, w_i) \in E, i = 1 \ldots k$$

Find an estimate $\hat{Y}$ for the unknown value of response variable Y for each node $u \in U$ such that $\hat{y}_u$ is as accurate as possible.

An algorithmic solution to this problem is reported in the next Section. The learner receives full information (including labels) on the nodes of L and partial information (without labels) on the nodes of U as well as weighted edges in E and is asked to predict the labels of the nodes of U. The algorithm is formulated in the original distributional-free transductive setting [21] and requires that both L and U are sampled from the node set V without replacement. This means that, unlike the standard inductive setting, the nodes in the labeled (and unlabeled) set are supposed to be mutually dependent based on the existence of a link (transitively) connecting them. Vapnik introduced an alternative transductive setting which is distributional, since both T and W are assumed to be drawn independently and identically from some unknown distribution. As shown in [21](Theorem 8.1), error bounds for learning algorithms in the distribution-free setting apply to the more popular distributional transductive setting. This justifies our focus on the distributional-free setting.

Algorithm 1. Top-level description of the Iterative Transductive Learner in Co-training style.

1: **ITL**(L, U, E)
2: **Input**
3: the labeled node set $L \subset \mathbf{X} \times Y$;
4: the projection of the unlabeled node set $U = V - L$ on $\mathbf{X}$;
5: the ternary relation $E \subset V \times V \times \mathbb{R}^+$;
6: **Output**
7: an estimate of unknown labels of nodes in U;
8: **begin**
9: $\overline{L} \leftarrow$ laggedPredictorVariables(L, V, E); $\overline{U} \leftarrow$ laggedPredictorVariables(U, V, E);
10: $L_0 \leftarrow L$; $U_0 \leftarrow U$; $L_1 \leftarrow \overline{L}$; $U_1 \leftarrow \overline{U}$;
11: $i \leftarrow 1$;
12: **repeat**
13:     $change \leftarrow$ false;
14:     $t_0 \leftarrow$learn(L_0); $t_1 \leftarrow$learn(L_1);
15:     $P_0 \leftarrow$predictConfidentLabels(t_0, U_0, L_0, E); $P_1 \leftarrow$predictConfidentLabels(t_1, U_1, L_1, E);
16:     **if** $P_0 \neq \oslash$ or $P_1 \neq \oslash$ **then**
17:         $change \leftarrow$ true;
18:         **for** $e \in P_0$ **do**
19:             $L_1 \leftarrow L_1 \cup \{\langle \text{instance}(e, U_1), \hat{y_e}\rangle\}$; $U_1 \leftarrow U_1 - \{\text{instance}(e, U_1)\}$;
20:         **end for**
21:         **for** $\overline{e} \in P_1$ **do**
22:             $L_0 \leftarrow L_0 \cup \{\langle \text{instance}(\overline{e}, U_0), \hat{y_{\overline{e}}}\rangle\}$; $U_0 \leftarrow U_0 - \{\text{instance}(\overline{e}, U_0)\}$;
23:         **end for**
24:     **end if**
25: **until** not$(++i \geq$ MAX_ITERS AND $change)$;
26: $U \leftarrow$label$(t_0, L, U, t_1, \overline{L}, \overline{U})$;
27: **end**

---

4   The Algorithm

The Iterative Transductive Learner (ITL) addresses the problem of predicting the unknown continuous labels of nodes which are distributed in a sparsely labeled network. ITL iteratively induces two distinct model trees in a co-training style. Labels predicted from one model tree which are estimated as confident extend the labeled set to be used by the other model tree learner for the next iteration of the learning process. The model trees which are induced in the last iteration of ITL are used to predict the unknown network labels. Details on the learning in co-training style, the evaluation of the confidence of predicted labels and the labeling of unlabeled nodes are reported in the next subsections.

4.1   Iterative Learning in Co-training Style

The top level description of ITL is reported in Algorithm 1. Let $N(V, E)$ be the sparsely labeled network whose unknown labels have to be predicted, ITL takes as input: the attribute-value observations (with labels) associated with the labeled node set $L \subset V$, the attribute-value observations (without labels) associated with the unlabeled node set $U \subset V$ ($U = V - L$), and the relation E, and predicts the unknown labels for the nodes of U. ITL is iterative and keeps with the main idea of co-training by inducing at each iteration, two distinct regression models from different views of the attribute-value data associated with the node set. The former is a regression model which includes the predictor variables measured at the currently labeled nodes of the network, the latter is a cross-regressive model which includes the *lagged* predictor variables measured in the neighborhood of the nodes.

In Algorithm 1, the function *laggedPredictorVariables()* is in charge of constructing $\overline{L}$ ($\overline{U}$) that is the lagged view of the data associated with the node set L (U). This lagged view of a dataset is obtained by projecting data over the lagged predictor variables in place of the original predictor variables. Formally,

Definition 2 (Lagged variable). *Let $N = (V, E)$ be a network and X be a variable that is measured at the nodes of V. For each node $u \in V$, the lagged variable $\overline{X}$ is assigned with the aggregate of the values which are measured for X at nodes falling in the neighborhood $\eta_E(u)$.*

In particular, by considering the case that X_i is continuous, then,

$$\overline{x_{i_u}} = \frac{\sum\limits_{(v,w)\in\eta_E(u)} (x_{i_v} \times w)}{\sum\limits_{(v,w)\in\eta_E(u)} w}, i = 1 \ldots m. \tag{1}$$

Further extension of ITL would allow to deal with discrete predictor variables.

By initially assigning $L_0 = L, U_0 = U, L_1 = \overline{L}$ and $U_1 = \overline{U}$, ITL learns the regression model t_0 from L_0 and the regression model t_1 from L_1 (see the function *learn()* in Algorithm 1). The basic learner employed to induce both t_0 and t_1 is the model tree learner presented in [3]. The choice of a model tree learner is motivated by the capability of model trees of do not imposing any a-priory defined global form of regression surface, but assuming a functional form at local level. t_0 and t_1 are then used to predict the unknown labels ($\hat{y}$) of the nodes falling in U_0 and U_1, respectively.

Labels which are confidently predicted (see function *predictConfidentLabels()* in Algorithm 1) are assigned to the corresponding nodes in U_1 (U_0). New labeled nodes are then moved from U_1 to L_1 (U_0 to L_0). The function *instance()* is in charge of passing from the original data view to the lagged data view of a node, and vice-versa. In particular, if u belongs to L (U), *instance*($u, \overline{L}$) (*instance*($u, \overline{U}$)) returns $\overline{u}$ in $\overline{L}$ ($\overline{U}$). Similarly, if $\overline{u}$ belongs to $\overline{L}$ ($\overline{U}$), instance($\overline{u}, L$) (instance($\overline{u}, U$)) returns the corresponding u in L (U).

The learning process stops when the maximum number of learning iterations, MAX_ITERS, is reached, or there is no unlabeled node which is confidently moved from the unlabeled set to the labeled set. Model trees which are learned in the last iteration of the learning process are used to definitely label working observations (see function *label()* in Algorithm 1). Details of *predictConfidentLabels()* and *label()* are provided in the next subsections.

Algorithm 2. Predict and estimate confidence of labels according to Single Label Confidence Estimate.

1: **determineConfidentlyLabeledNodes**$(U_i, L_i, E, t_i) \Longrightarrow P_i$
2: **Input**
3: the node set $U_i \subset \mathbf{X} \times Y$ and the node set $L_i \subset \mathbf{X} \times Y$;
4: the ternary relation $E \subset U_i \cup L_i \times U_i \cup L_i \times \mathbb{R}^+$;
5: the model tree t_i induced from L_i;
6: **Output**
7: $P_i \subseteq U_i$ such that P_i includes only the nodes of U_i whose predicted labels are estimated as confident;
8: **begin**
9: $P_i \leftarrow \oslash$;
10: **for** u in U_i **do**
11:     $\hat{y_u} \leftarrow \text{response}(t_i, u)$;
12:     $t'_i \leftarrow \text{learn}(L_i \cup \{\langle u, \hat{y_u}\rangle\})$;
13:     $pos \leftarrow 0$; $neg \leftarrow 0$;
14:     **for** v in $\eta_E(u)|_{L_i}$ **do**
15:         **if** $(y_v\text{-response}(t'_i, v))^2 - (y_v\text{-response}(t_i, v))^2 \geq 0$ **then**
16:             $pos \leftarrow pos + 1$;
17:         **else**
18:             $neg \leftarrow neg + 1$;
19:         **end if**
20:     **end for**
21:     **if** $pos \geq neg$ **then**
22:         $P_i \leftarrow P_i \cup \{\langle u, \hat{y_u}\rangle\}$;
23:     **end if**
24: **end for**
25: **end**

4.2   Evaluating the Confidence of Predicted Labels

A model tree is used to predict the unknown labels in the network. The confidence of each estimated label is evaluated in order to identify the most confident labels. Intuitively, confident labels are with the following property. The error performed by a model tree in re-predicting the labeled node set should decrease the most if the most confidently labeled nodes are utilized in the learning process. According to this property, we have designed two alternative mechanisms, named Single Label Confidence Evaluation (SLCE) and Multi Label Confidence Evaluation (MLCE), which provide an estimate of the confidence of the labels which are predicted in ITL.

Single Label Confidence Evaluation

The SLCE estimates the confidence of predicted labels one by one (see Algorithm 2). The confidence is estimated by a model tree that is learned from a training set consisting of the nodes which are currently labeled in the network and the unlabeled node whose predicted label has to be estimated. The confidence of this label corresponds to the confidence of this model tree in re-predicting the labeled nodes which are interlinked (as neighbors) to the unlabeled one. Formally, let:

1. t_i (with $i = 0, 1$) be the model tree induced from the labeled node set L_i,
2. $u \in U_i$ be a node falling in the unlabeled set U_i (with $i = 0, 1$),
3. $\hat{y}_u$ the label predicted from t_i for u,
4. $\eta_E(u)|_{L_i} = \{v \in L_i | (u, v, w) \in E\}$ be the set of labeled nodes v which are neighbors of u in L_i according to E, and
5. t'_i (with $i = 0, 1$) be the model tree induced from $L_i \cup \{\langle u, \hat{y}_u \rangle\}$,

the confidence of $\hat{y}_u$ is evaluated according to the influence of $\hat{y}_u$ on the known labels of nodes falling in $\eta_E(u)|_{L_i}$. In particular, for each $v \in \eta_E(u)|_{L_i}$, ϵ_v is the result of subtracting the squared error performed by t'_i in determining the label of v from the squared error performed by t_i in determining the label of v,

$$\epsilon_v = (y_v - \text{response}(t'_i, v))^2 - (y_v - \text{response}(t_i, v))^2 \qquad (2)$$

with y_v be the response that originally labels v in L_i at the current iteration of ITL. The function $response(t_i, v)$ returns the label predicted for v by t_i, while $response(t'_i, v)$ returns the label predicted for v by t'_i, respectively. By defining:

$$Pos = |\{v \in \eta_E(u)|_{L_i} | \epsilon_v \geq 0\}| \quad Neg = |\{v \in \eta_E(u)|_{L_i} | \epsilon_v < 0\}|, \qquad (3)$$

with $| \cdot |$ the cardinality of a set $\cdot$, the label $\hat{y}_v$ is estimated as confident if $Pos \geq Neg$, un-confident otherwise.

Multi Label Confidence Evaluation

The MLCE firstly groups unlabeled nodes of the network in possibly overlapping clusters and then estimates the confidence of the entire *cluster* of predicted labels, cluster by cluster (see Algorithm 3). For each labeled node $v \in L_i$, a cluster, $\eta_E(v)|_{U_i}$ is constructed by including the unlabeled neighbors of v which are determined in U_i according to E. By using t_i to assign a label to the nodes falling $\eta_E(v)|_{U_i}$, the labeled node set $\widehat{\eta_E(v)}|_{U_i}$ is constructed from $\eta_E(v)|_{U_i}$ (see function *assignLabel()* in Algorithm 3), as follows:

$$\widehat{\eta_E(v)}|_{U_i} = \{\langle u, \text{response}(t_i, u)\rangle | u \in \eta_E(v)|_{U_i}\} \qquad (4)$$

where $response(t_i, u)$ is the label predicted for u by t_i. Let t'_i be the model tree induced from $L_i \cup \widehat{\eta_E(v)}|_{U_i}$, predicted labels of $\widehat{\eta_E(v)}|_{U_i}$ are estimated as confident iff:

$$(y_v - \text{response}(t'_i, v))^2 - (y_v - \text{response}(t_i, v))^2 \geq 0, \qquad (5)$$

un-confident, otherwise.

Algorithm 3. Predict and estimate confidence of labels according to Multi Label Confidence Estimate.

1: **determineConfidentlyLabeledNodes**$(U_i, L_i, E, t_i) \Longrightarrow P_i$
2: **Input**
3: the node set $U_i \subset \mathbf{X} \times Y$ and the node set $L_i \subset \mathbf{X} \times Y$;
4: the ternary relation $E \subset U_i \cup L_i \times U_i \cup L_i \times \mathbb{R}^+$;
5: the model tree t_i induced from L_i;
6: **Output**
7: $P_i \subseteq U_i$ such that P_i includes only the nodes of U_i whose predicted labels are estimated as confident;
8: **begin**
9: $P_i \leftarrow \oslash$;
10: **for** v in L_i **do**
11:     $\widehat{\eta_E(v)|_{U_i}} \leftarrow$assignLabel$(T_i, \eta_E(v)|_{U_i})$;
12:     $t_i' \leftarrow$learn$(L_i \cup \widehat{\eta_E(v)|_{U_i}})$;
13:     **if** (response(t_i', v)-y$(v))^2$-(response(t_i, v)-y$(v))^2 \geq 0$ **then**
14:         $P_i \leftarrow P_i \cup \widehat{\eta_E(v)|_{U_i}}$;
15:     **end if**
16: **end for**
17: **end**

4.3   Predicting the Unlabeled Nodes in the Network

Model trees t_0 and t_1 which are learned in the last iteration of ITL are used to predict the final labels $\hat{Y}$ to be associated with originally unlabeled nodes of U. Let $\bar{u} \in \overline{U}$ be the lagged data view of the unlabeled node $u \in U$, then:

$$\hat{y}_u = \frac{\omega_0 \times \text{response}(t_0, u) + \omega_1 \times \text{response}(t_1, \bar{u})}{\omega_0 + \omega_1} \text{ with } u \in U. \tag{6}$$

where ω_0 and ω_1 are computed on the basis of the mean square error (mse) of each model tree on the original labeled set (L and $\overline{L}$, respectively). Details are provided in Algorithm 4.

5   Experiments

We demonstrate that ITL is robust to both sparse labeling and low label consistency and it improves traditional model tree induction across a range of several geographical data networks.

Dataset Description

GASD (USA Geographical Analysis Spatial Dataset) [17] contains 3,107 observations on USA county votes cast in 1980 presidential election. Specifically, it contains the total number of votes cast in the 1980 presidential election per county (response attribute), the population in each county of 18 years of age

---

Algorithm 4. Assigning a final label to the unlabeled nodes of the network.

1: **label**$(t_0, L, U, t_1, \overline{L}, \overline{U})$
2: **Input**
3: the model tree t_0 induced on the feature space $\mathbf{X}$;
4: the set $L \subseteq \mathbf{X} \times Y$;
5: the unlabeled set $U \subseteq \mathbf{X}$;
6: the model tree induced on the feature space $\overline{\mathbf{X}}$;
7: the labeled set $\overline{L} \subseteq \overline{\mathbf{X}} \times Y$;
8: the unlabeled set $\overline{U} \subseteq \overline{\mathbf{X}}$;
9: **Output**
10: $U' = \{\langle u, \hat{y_u} \rangle | u \in U, \hat{y_u} \text{ is the final label predicted for } u\}$;
11: **begin**
12: $U \leftarrow \oslash$;
13: $m_0 \leftarrow \text{mse}(t_0, L)$; $m_1 \leftarrow \text{mse}(t_1, \overline{L})$;
14: **if** $m_0 > m_1$ **then**
15:     $\omega_0 \leftarrow 1$; $\omega_1 \leftarrow m0/m1$;
16: **else**
17:     $\omega_0 \leftarrow m1/m0$; $\omega_1 \leftarrow 1$;
18: **end if**
19: **for** $u \in U$ **do**
20:     $\overline{u} \leftarrow \text{instance}(u, \overline{U})$; $\hat{y_u} \leftarrow \frac{\text{response}(t_0, u) \times \omega_0 + \text{response}(t_1, \overline{u}) \times \omega_1}{\omega_0 + \omega_1}$;
21:     $U' \leftarrow U \cup \{\langle u, \hat{y_u} \rangle\}$;
22: **end for**
23: **end**

---

or older, the population in each county with a 12th grade or higher education, the number of owner-occupied housing units, the aggregate income, the XCoord and YCoord spatial coordinates of the county. **Forest Fires** is public available for research at UCI Machine Learning Repository[1]. The details are described in [7]. It collects 512 forest fire observations from the Montesinho natural park in the northeast region of Portugal. The data, collected from January 2000 to December 2003, include the burned area of the forest in ha[2] (response variable), the Fine Fuel Moisture Code (FFMC), the Duff Moisture Code (DMC), the Drought Code (DC), the Initial Spread Index (ISI), the temperature in Celsius degrees, the relative humidity, the wind speed in km/h, the outside rain in mm/m^2, the XCoord and YCoord spatial coordinates within the Montesinho park map. **NWE** (North-West England) contains census data collected in the European project SPIN![3]. Data are census data concerning North West England area that is decomposed into censual sections or wards for a total of 1011 wards. Census data provided by 1998 Census is available at ward level. We consider percentage of mortality (response variable) and measures of deprivation level in the ward according to index scores such as, Jarman Underprivileged Area Score, Townsend score, Carstairs score and the Department of the Environments Index,

---

[1] http://archive.ics.uci.edu/ml/
[2] 1ha/100 = 100 m^2.
[3] http://www.ais. fraunhofer.de/KD/SPIN/project.html

the XCoord and YCoord spatial coordinates of the ward centroid. By removing observations including null values, only 979 observations are used in this experiments. Finally, **Sigmea-Real** [8] collects 817 measurements of the rate of herbicide resistance of two lines of plants (response variables), that is, the transgenic male-fertile (MF) and the non-transgenic male-sterile (MS) line of oilseed rape. Predictor variables of this study are the cardinal direction and distance from the center of the donor field, the visual angle between the sampling plot and the donor field, and the shortest distance between the plot and the nearest edge of the donor field, the XCoord and YCoord spatial coordinates of the plant.

Experimental Setting

Each geo-referenced dataset D is mapped into a data network $N = (V, E)$ that includes a node $u \in V$ for each observation $(x_1, \ldots, x_n, y, xCoord, yCoord) \in D$ and associates u with $(x_1, \ldots, x_n, y)$. Let u and v be two distinct nodes in V, there is an edge from u to v labeled with w in N (i.e., $(u, v, w) \in E$) iff v is one of the k nearest neighbors of u. The Euclidean distance is computed to determine neighbors. Notice that the neighboring relation defined above is not necessarily symmetric, v may be a k nearest neighbor of u, but not necessarily vice-versa. Additionally, u is a neighbor of u. In this paper, several data networks are constructed from the same dataset by varying $k = 5, 10, 15$. They are denoted as N_5, N_{10} and N_{15}. In each data network, the weight w is defined according to a continuous function of Euclidean distance [6] as follows:

$$ w = e^{-\frac{dist(u,v)^2}{b_u^2}} \text{ with } b_u = \max_{v \in k-nearestNeighbors(u,V)} dist(u, v), \qquad (7) $$

If u and v are associated with observations taken at the same geographical site, the weighting of observations collected at this site would be unity. The weighting of other observations will decrease according to a Gaussian curve as the Euclidean distance between u and v increases.

For each data network, experiments are performed in order to: 1) validate the actual advantage of the iterative transductive learner over the basic model tree learner in labeling the unlabeled nodes of a sparsely and scarcely labeled network (ITL vs t_0 and ITL vs t_1), 2) evaluate the advantages of a co-training implementation in the transductive learning (ITL vs ITL*), and 3) compare performance of SLCE and MLCE in estimating the confidence of labels (SLCE vs MLCE). t_0 (t_1) denotes the model tree which is induced from the original set of labeled data by considering the predictor variables (lagged predictor variables) only, ITL* is the iterative transductive learner without co-training, that is, no lagged view of data is considered in the learning process, ITL is the iterative transductive learner with co-training.

The empirical comparison is based on the mean square error (MSE) that is estimated according to a K-fold cross validation. K is set to 10 in experiments performed with GASD dataset, and $K = 5$ in experiments performed with Forest Fires, NWE and Sigmea Real. For each trial, algorithms to be compared are trained on a single fold and tested on the hold-out K - 1 folds, which form the working set. The comparative statistics is computed by averaging the MSE error over the K-folds (Avg.MSE). It is noteworthy that, unlike the standard cross-validation approach, here only one fold is used for the training set. In this way we can simulate datasets with a small percentage of labeled cases (the training set) and a large percentage of unlabeled data (the working set), which is the usual situation for a transductive learning. ITL is run with $MAX\_ITERS = 5$.

Results

The Avg. MSE performed by both the transductive learner and the inductive learner is reported in Table 1. Results suggest several conclusions. First, they confirm that ITL performs generally better than the basic model tree learners (ITL improves both t_0 and t_1 in accuracy) by profitably employing a kind of iterative learning to bootstrap from a small set of labeled training data via a large set of unlabeled data. The exception is represented by Sigmea Real (MF) that is the only dataset where the baseline inductive learner t_0 always outperforms ITL. Our justification is that the worse performance of ITL may depend on the fact that this dataset exhibits about 65% of observations which are labeled as zero which leads to a degradation of both predictive capability of the learner that operates with the aggregate data view in the co-training and capability of identifying confident labels. This is confirmed by the fact accuracy of

Table 1. Avg.MSE: Inductive learners (t0 and t1) vs. iterative transductive learner without co-training (ITL*) and with co-training (ITL)

Avg.MSE		N_5		N_{10}		N_{15}	
		SLCE	MLCE	SLCE	MLCE	SLCE	MLCE
GASD	t0	0.15174	0.15174	0.15174	0.15174	0.15174	0.15174
	t1	0.15879	0.15879	0.17453	0.17453	0.17419	0.17419
	ITL*	0.13582	0.13239	0.13643	0.13387	0.13606	0.13468
	ITL	**0.13006**	**0.12965**	**0.13156**	**0.13162**	**0.13387**	**0.13128**
Forest Fires	t0	81.16599	81.16599	81.16599	81.16599	81.16599	81.16599
	t1	64.68706	64.68706	64.71256	64.71257	64.80331	64.80332
	ITL*	81.45897	82.33336	81.04362	74.48984	81.30787	80.89551
	ITL	**64.44121**	**63.88140**	**64.73077**	**64.28176**	**63.92594**	**64.41118**
NWE	t0	0.00255	0.00255	0.00255	0.00255	0.00255	0.00255
	t1	0.00250	0.00250	0.00252	0.00252	0.00256	0.00256
	ITL*	0.00253	0.00252	0.00254	0.00258	0.00252	0.00253
	ITL	**0.00245**	**0.00244**	**0.00248**	**0.00248**	**0.00247**	**0.00248**
SigmeaReal (MF)	t0	2.35395	2.35395	2.35395	2.35395	2.35395	2.35395
	t1	2.57045	2.57045	2.51061	2.51061	2.51991	2.51991
	ITL*	2.36944	2.36336	2.36579	2.35532	2.37024	2.35966
	ITL	**2.44036**	**2.43278**	**2.40851**	**2.42544**	**2.46547**	**2.43397**
SigmeaReal(MS)	t0	5.87855	5.87855	5.87855	5.87855	5.87855	5.87855
	t1	5.80157	5.80157	6.12569	6.12569	6.08601	6.08601
	ITL*	5.87364	5.87389	5.87374	5.87781	5.87346	5.87340
	ITL	**5.75021**	**5.61658**	**5.85322**	**5.85275**	**5.74338**	**5.87403**

cross-regressive model tree t_1 is significantly worse than the accuracy of the classical model tree t_0. Second, the co-training improves the accuracy of the iterative transductive learner (ITL vs ITL*) by combining cross-regressive models with traditional regression models. Finally, the comparison between MLCE and SLCE suggests that the accuracy of ITL is improved by the use of MLCE when the data network includes nodes with a low number of neighbors ($k = 5$), while the accuracy of ITL is generally improved by the use of SLCE when the data network includes nodes with higher number of neighbors ($k = 15$).

6   Conclusions

In this paper we investigate the task of regression in labeled networked data where labels are sparse and continuous. We assume that data present a form of autocorrelation where the value of the response variable in a node depends on the values of the predictor variables of interlinked nodes. For this reason, we consider the contribution of lagged predictor variables in the regression model. We investigate a computational solution in the transductive setting, which asks for predicting the response values only for unlabelled nodes of the network. The neighborhood relation used in the transductive setting is computed on the basis of the node links. The solution is based on co-training, since separate model trees are learned from attribute values of labeled nodes and attribute values aggregated in the neighborhood of labeled nodes, respectively. Two distinct procedures have been proposed to evaluate confidence of the predicted labels. Experiments with several sparsely labeled networked data are performed. Results show that the proposed method improves accuracy of traditional model tree induction.

Acknowledgment

This work is fully supported by the Project "Scoperta di conoscenza in domini relazionali" funded by the University of Bari. The authors thank Saso Dzeroski for kindly providing Sigmea Areal dataset.

References

1. Abreu, M., de Groot, H., Florax, R.: Space and growth: A survey of empirical evidence and methods. Region and Development, 12–43 (2005)
2. Anselin, L.: Spatial externalities, spatial multipliers and spatial econometrics. International Regional Science Review (26), 153–166 (2003)
3. Appice, A., Dzeroski, S.: Stepwise induction of multi-target model trees. In: Kok, J.N., Koronacki, J., Lopez de Mantaras, R., Matwin, S., Mladenič, D., Skowron, A. (eds.) ECML 2007. LNCS (LNAI), vol. 4701, pp. 502–509. Springer, Heidelberg (2007)
4. Blum, A., Mitchell, T.M.: Combining labeled and unlabeled data with co-training. In: COLT, pp. 92–100 (1998)

5. Brefeld, U., Gärtner, T., Scheffer, T., Wrobel, S.: Efficient co-regularised least squares regression. In: Cohen, W.W., Moore, A. (eds.) 23th International Conference on Machine Learning, ICML 2006. ACM International Conference Proceeding Series, vol. 148, pp. 137–144. ACM, New York (2006)
6. Charlton, M., Fotheringham, S., Brunsdon, C.: Geographically weighted regression. In: ESRC National Centre for Research Methods NCRM Methods Review Papers NCRM/006 (2005)
7. Cortez, P., Morais, A.: A data mining approach to predict forest fires using meteorological data, pp. 512–523. APPIA (2007)
8. Demšar, D., Debeljak, M., Lavigne, C., Džeroski, S.: Modelling pollen dispersal of genetically modified oilseed rape within the field. In: Abstracts of the 90th ESA Annual Meeting, The Ecological Society of America, p. 152 (2005)
9. Gallagher, B., Tong, H., Eliassi-Rad, T., Faloutsos, C.: Using ghost edges for classification in sparsely labeled networks. In: Proceeding of the 14th ACM SIGKDD International Conference on Knowledge Discovery and Data Mining, KDD 2008, pp. 256–264. ACM, New York (2008)
10. David, J., Jennifer, N., Brian, G.: Why collective inference improves relational classification. In: Proceedings of the 10th ACM SIGKDD International Conference on Knowledge Discovery and Data Mining, KDD 2004, pp. 593–598. ACM, New York (2004)
11. Macskassy, S.A., Provost, F.: A Brief Survey of Machine Learning Methods for Classification in Networked Data and an Application to Suspicion Scoring. In: Airoldi, E.M., Blei, D.M., Fienberg, S.E., Goldenberg, A., Xing, E.P., Zheng, A.X. (eds.) ICML 2006. LNCS, vol. 4503, pp. 172–175. Springer, Heidelberg (2007)
12. Macskassy, S., Provost, F.: Classification in networked data: a toolkit and a univariate case study. Machine Learning 8, 935–983 (2007)
13. Macskassy, S.A.: Improving learning in networked data by combining explicit and mined links. In: Proceedings of the 22nd Conference on Artificial Intelligence, AAAI 2007, pp. 590–595. AAAI Press, Menlo Park (2007)
14. McPherson, M., Smith-Lovin, L., Cook, J.: Birds of a feather: Homophily in social networks. Annual Review of Sociology 27, 415–444 (2001)
15. Jennifer, N., David, J.: Relational dependency networks. Journal of Machine Learning Research 8, 653–692 (2007)
16. Neville, J., Simsek, O., Jensen, D.: Autocorrelation and relational learning: Challenges and opportunities. In: Proceedings of the Workshop on Statistical Relational Learning (2004)
17. Pace, P., Barry, R.: Quick computation of regression with a spatially autoregressive dependent variable. Geographical Analysis 29(3), 232–247 (1997)
18. Rey, S.J., Montouri, B.D.: U.s. regional income convergence: a spatial econometric perspective. Regional Studies (33), 145–156 (1999)
19. Sen, P., Namata, G., Bilgic, M., Getoor, L., Gallagher, B., Eliassi-Rad, T.: Collective classification in network data. AI Magazine 29(3), 93–106 (2008)
20. Tobler, W.: Cellular geography. In: Gale, S., Olsson, G. (eds.) Philosophy in Geography (1979)
21. Vapnik, V.: Statistical Learning Theory. Wiley, New York (1998)
22. Zhou, Z.-H., Li, M.: Semisupervised regression with cotraining-style algorithms. IEEE Transaction in Knowledge Data Engineering 19(11), 1479–1493 (2007)
23. Zhu, X., Ghahramani, Z., Lafferty, J.D.: Semi-supervised learning using gaussian fields and harmonic functions. In: Fawcett, T., Mishra, N. (eds.) Proceedings of the 20th International Conference on Machine Learning, ICML 2003, pp. 912–919. AAAI Press, Menlo Park (2003)

Detecting New Kinds of Patient Safety Incidents

James Bentham[1] and David J. Hand[1,2]

[1] Department of Mathematics, Imperial College, UK
[2] Institute for Mathematical Sciences, Imperial College, UK

Abstract. We present a novel approach to discovering small groups of anomalously similar pieces of free text.

The UK's National Reporting and Learning System (NRLS) contains free text and categorical variables describing several million patient safety incidents that have occurred in the National Health Service. The groups of interest represent previously unknown incident types. The task is particularly challenging because the free text descriptions are of random lengths, from very short to quite extensive, and include arbitrary abbreviations and misspellings, as well as technical medical terms. Incidents of the same type may also be described in various different ways.

The aim of the analysis is to produce a global, numerical model of the text, such that the relative positions of the incidents in the model space reflect their meanings. A high dimensional vector space model of the text passages is produced; TF-IDF term weighting is applied, reflecting the differing importance of particular words to a description's meaning. The dimensionality of the model space is reduced, using principal component and linear discriminant analysis. The supervised analysis uses categorical variables from the NRLS, and allows incidents of similar meaning to be positioned close to one another in the model space. Anomaly detection tools are then used to find small groups of descriptions that are more similar than one would expect. The results are evaluated by having the groups assessed qualitatively by domain experts to see whether they are of substantive interest.

1   Introduction

The UK's National Patient Safety Agency, or NPSA, has collected data that describe more than three million patient safety incidents, where a patient safety incident is defined as 'an unintended or unexpected incident that could have or did lead to harm to patients receiving National Health Service (NHS) care'.

The incidents are extremely varied in nature, ranging from the common and mundane to the very rare and unusual. Common incident types include events such as patient falls, errors made when giving medication and injuries to women during childbirth. Some of the less common incident types will already be well known to the medical profession: for example, some patients have an allergy to latex, and therefore latex gloves cannot be worn during operations on these patients. Other types of incidents are less well known, and have only come to light because of qualitative analysis of the NRLS. This analysis has allowed the

J. Gama et al. (Eds.): DS 2009, LNAI 5808, pp. 51–65, 2009.

NPSA to issue alerts, such as [6] for example, which contains guidance on vinca alkaloids.

The size of the dataset means that it is not possible to examine each entry manually to try to find these previously unknown incident types. The problem therefore lends itself to the use of data mining techniques.

In this paper we present a method that discovers these groups of interest semi-automatically. We model the data numerically and use an anomaly detection algorithm to find unexpected local clusters of similar objects. These clusters represent the groups of interest.

2   Data

The data are stored in the National Reporting and Learning System (NRLS). Each National Health Service (NHS) Trust in England and Wales is required to have its own incident reporting system; the NRLS is an amalgamation of the Trusts' data. The Trusts have different incident reporting systems, and therefore the mapping between the Trusts' variables and the variables in the NRLS is quite complex: this leads to some NRLS variables being only sparsely populated.

The NRLS has 73 variables, most of which are categorical: there are various incident type variables, details of the patients' ages, the severity of the incident, and so on.

The data are generally entered by medical professionals, most commonly doctors and nurses, although midwives, ambulance drivers and other health professionals also enter data. It is possible for patients and other members of the public to enter data, but this is much less common. The data are often written on paper forms which are transferred to the computer systems later by clerical staff, which can reduce the quality of the NRLS data.

One of the variables is quite different from the others, as it allows the person entering the data to write a free text description of the incident. This is potentially a very rich source of information and is the focus of the work we present here. There are, however, particular difficulties associated with the analysis of text, both in general, and associated with this dataset in particular.

Whilst the free text is potentially very informative, problems arise because of the freedom that the staff have when entering the data, because of the range of knowledge and experience of the staff concerned, and the time they have available to write the entries. The free text data are of extremely variable quality, as can be seen in Table 1. The lengths of the incident descriptions vary from a single word to several long paragraphs; this includes descriptions that are simply an entry such as 'xxxx'. Other descriptions are one or two words long and are not particular informative: 'patient fell' is a common entry.

Spelling mistakes are very common, both those due to misconceptions of the correct spelling, e.g. 'recieve' instead of 'receive', but also because of typographical errors. The staff entering the data use many abbreviations, e.g., 'pt' for 'patient' or '?' for 'possible'. The NHS is one the world's largest bureaucracies, and has developed its own terms and conventions: the incident descriptions

Table 1. Free Text Examples

Patient waited 48 hours or more, for surgery. 5 days wait. #NOF.
- Enucleation of odontogenic keralocyst in maxilla right. - Carnoy's solution placed in cyst cavity on ribbon gauze and irrigated. - swollen upper and lower lip right + ? small chemical burn lower lip. Seen by Professor.
At the start of a cardiothoracic case before anaesthetising a patient with Dr [Staff Name] for a large cardiac case I rang blood bank to order four units of blood for our patient . I spoke to a lady on the phone whose name I did not take . I explained that we were about to start the case in theatre and she told me that she would ring theatre 7 when the blood was ready and we would send somebody to get it from the fridge in blood bank . When the patient went onto cardiopulmonary bypass she needed blood fairly quickly when blood bank was rang then nobody there was aware of the request and the patient had to be given type specific blood as no cross matched had taken place as requested .
Overcapacity - total number of babies on the unit 35 , capacity is 30 . Delivery suite aware that NNU is closed . .
Overcapacity - total number of babies on the unit 31 , capacity is 30 . .
Pt given 2mg Lorazepam via wrong route - oral route instead of IM . .
Pt found on floor .
Fall.
doHover(this);" onMouseOut="doUnHover(this);" 0065956 : b Number b/span, normal time allocated to block the list (1 - 1 1 / 2 hours) was exceeded , there by emergency list was delayed . Other specialties were updated though unhappy , i.e . orthopaedics postponing a case .
Result not reported . - Test=GQIB . [Person name] entered result in APEX on 26 / 04 / 05 as this was POSITIVE raised to level Q to go to the MEDQ . This sample was not picked up on weekly WFE as it is written in AI test . Also WFE on Sp . Rec . do not include lefel Q. Also report would not print . Referred to [Person name 2] .

include many acronyms and initialisms as well as conventions like writing 'RIP' for 'death' or 'dead'.

The quality of the grammar is also variable, and the use of cases is unpredictable: some entries are all in capitals, whilst others are all in lower case.

3 Model

The difficulty of the problem will be apparent from the description of the data. This section describes our modelling approach.

One way to try to find small groups of potential interest is to use anomaly detection algorithms. These algorithms find groups of points that have density that is higher than would be expected given the background density. However, to do this we need to place the incident descriptions into an appropriate model space. In the following sections we develop a global, numerical model of the text. The relative positions of the incident descriptions in the model space reflect as closely as possible the meanings of the descriptions rather than simply which words appear in them.

3.1 Vector Space Model

The basic model that we produce for the text is the well known vector space model. Each incident description is represented by a vector of length equal to the size of the overall vocabulary used in all of the incident descriptions. We have been provided with samples of 25,000 incidents by the NPSA; our samples have vocabularies of between 19,000 and 28,000 'words'. We remove all symbols other than the letters 'a' to 'z', converting any upper case letters to lower case. Each token is therefore a sequence of one or more letters. Variations such as 'Patient', 'patient' and 'PATIENT' are represented by a single token.

The basic datasets are therefore matrices of size 25,000 x c.20,000, or alternatively, the incidents are represented as points in a c.20,000 dimensional space. Each entry in the matrix is the number of times that a particular word appears in an incident description.

The vector space model is quick and straightforward to calculate, but there are several disadvantages. One is that any information contained in the word ordering is lost. Natural language involves interactions between words: there is a difference in meaning between 'patient fall' and 'patient did not fall' that would not be captured well by the vector space model (particularly given the term weighting described in Section 3.2). The vector space model does not take into account the similarities in meaning between different words: 'fall' and 'fell' have the same root, but would be treated as separate variables in the vector space model. Nevertheless, we use the vector space model as the basis for our analysis, noting that more elaborate versions could lead to superior performance.

3.2   Term Weighting

In a piece of text such as 'patient fell on the floor', it is clear that certain words carry more of the meaning of the sentence than others. These words should therefore be weighted more highly than less interesting words such as 'the'. We adopt the commonly used TF-IDF term weighting scheme, as described in [5]. The terms are weighted according to the frequency of their appearance in a particular description, and their rarity over the whole sample. If the original vector space matrix is $\mathbf{X}$, with i descriptions and j variables, the TF weighted matrix $\mathbf{X_{TF}}$ will have entries,

$$x_{\mathbf{TF},i,j} = \frac{x_{i,j}}{x_{i,\bullet}}$$

i.e., the TF matrix is the original matrix with each entry divided by the sum of its row. This means that each document carries the same weight, regardless of its length.

The IDF, or inverse document frequency, reflects the rarity of a word's appearance. The IDF is a vector $\mathbf{y_{IDF}}$, of length j, with

$$y_{\mathbf{IDF},j} = \frac{1}{\sum_i I_{x_{i,j} \neq 0}}$$

where I is an indicator variable which takes a value of unity if a particular word appears at least once in a description. The $\mathbf{y_{IDF}}$ vector is normalised. In the literature [4], the normalisation function used is generally the logarithm, but in our work we have found that the square root produces superior results (see Section 5.1).

Each column of the $\mathbf{X_{TF}}$ matrix is multiplied by the square root of its corresponding $\mathbf{y_{IDF}}$ value, to form the final TF-IDF weighted vector space matrix.

3.3   High Dimensionality and Dimensionality Reduction

We now have a 25,000 x c.20,000 termweighted data matrix. One approach would be to analyse this dataset immediately to find anomalously dense local clusters of incident descriptions. However, high dimensional spaces behave rather differently from lower dimensional spaces. This issue is described in depth in [2], but in short, as dimensionality increases, the relative contrast,

$$C = \frac{d_{orig,max} - d_{orig,min}}{d_{orig,min}} \to 0$$

where $d_{orig,max}$ is the maximum distance between any point and the origin, and $d_{orig,min}$ is the minimum distance between any point and the origin. This effect means that in high dimensional spaces it is difficult to discriminate between nearest and furthest neighbours, and it is not possible to calculate a meaningful measure of the local density of points. It is therefore not possible to find groups of potential interest using anomaly detection algorithms without carrying out some sort of dimensionality reduction.

The dimensionality reduction must reduce the model space sufficiently that problems caused by high dimensionality are avoided but not so much that there is not enough information in the positions of the data points to reflect the meaning of the incident descriptions. In general, the dimensionality reduction aims to rotate the original model space in such a way that all the information is retained, but that it is described by far fewer dimensions than the original model space. The other, uninformative, dimensions can be discarded.

Unsupervised Methods. Unsupervised methods such as principal component analysis rotate the data based on the inherent properties of the data. In the case of principal components, the model space is rotated such that the dimensions explain the variation in the data in descending order, subject to the constraint that the variables must be orthogonal. The dimensions that explain relatively little of the variance can then be removed.

The disadvantage of using an unsupervised method for dimensionality reduction is that although combinations of highly weighted words that appear together frequently will be found, there is no information by which the model space can be rotated so that incident descriptions that mean the same thing but that use different words will be in adjacent parts of the model space.

Even so, principal component analysis has a role to play in the dimensionality reduction. It has been noted in [4] that dimensionality reduction of vector space models using principal component analysis actually improves the performance of some information retrieval systems. It might be expected that removing information would impair performance, but it appears that using principal component analysis can improve the model by removing noise, possibly by finding combinations of words that represent commonly used, meaningful constructions in English, whilst removing combinations of words that appear together by chance and that might produce spurious results in an information retrieval system. We therefore reduce the dimensionality using principal component analysis before using a supervised method.

Table 2. Categorical Variables

Variable	Examples	Number of Categories
Location, Level 1	General/Acute Hospital Mental Health Unit	12
Location, Level 2	Dental Surgery Outpatient Department	26
Incident Category Level 1	Medical Device Patient Accident	16
Incident Category Level 2	Diagnosis - Wrong Slips, Trips, Falls	87
Specialty, Level 1	Surgical Specialty Mental Health	16
Specialty, Level 2	Gastroenterology Haematology	83

Supervised Methods. Supervised methods such as linear discriminant analysis use external information related to the data; the rotation is carried out to optimise some relationship between the data and this information. For example, for linear discriminant analysis each data point is assigned a class; the model space is then rotated such that a measure of the separation of classes is maximised, again preserving orthogonality between the dimensions. The dimensions that do not separate the classes well can be discarded.

In order to carry out supervised dimensionality reduction, it is necessary for the incident descriptions to be classified into categories; these categories must reflect the meaning of the descriptions. We tried this but found that it is very arduous to do manually, so it is fortunate that each of the incident descriptions has been classified by the person entering the data. Six incident type variables have been used to calculate linear discriminants; they are described in Table 2, with examples of the entries for each variable.

3.4   Final Model

We start with a 25,000 x c.20,000 termweighted matrix. In order to make the calculation of principal components feasible within a reasonable time-frame, and to fit in with constraints on computing power, the dataset is reduced to 5,000 dimensions. Slightly counter-intuitively we select the 5,000 variables with the *lowest* values of IDF, i.e., the 5,000 most common words. Although rarer words are proportionately more interesting than more common words, most of the meaning of the descriptions is carried by the 5,000 most common words. There is also the potential advantage that many of the least common 'words' are spelling mistakes, and these will be removed by this variable selection. Of course, in the case where rare words discriminate between interesting groups and the remainder of the data, these groups will not be identified by our analysis.

Principal components are calculated. The first 2,000 principal components are retained, and this dataset is used to calculate linear discriminants, using the six different categorical variables described in Table 2. The first 15 linear discriminants are selected as the final model for analysis using anomaly detection tools, unless there are fewer than 16 classes for the categorical variable that is being used, in which case all of the linear discriminants are used. The final datasets are therefore generally 25,000 x 15 matrices.

4   Anomaly Detection

We use an anomaly detection tool called $PEAKER$, which is described in depth in [1] and [7].

We calculate a relatively simple measure of the density at each point $\mathbf{x}$,

$$\hat{f}(\mathbf{x}) = \hat{f}(\mathbf{x}; K) = \left[\frac{1}{K} \sum_{i \in \{N\}} d(\mathbf{x}, \mathbf{x_i}) \right]^{-1}$$

where

$$\{N\} = \{\mathbf{x} : \mathbf{x} \in N(\mathbf{x}; K)\}$$

i.e., N is the set of K nearest neighbours to $\mathbf{x}$. The smoothing of the density estimate can be varied by increasing or decreasing the value of K. Any measure of distance can be used to calculate $d(\mathbf{x}, \mathbf{x_i})$: the Euclidean or Manhattan distance metrics are two well-known options (the choice of distance metric has an effect on the relative contrast described in Section 3.3)

The $PEAKER$ algorithm is based on the concept of peaks, defined as points with a higher density estimate than a number of their nearest neighbours. A point is defined as a peak with $M(\mathbf{x_i}) = m$ iff

$$\hat{f}(\mathbf{x_i}) > \hat{f}(\mathbf{x_j}), \forall \mathbf{x_j} \in N_m(\mathbf{x_i})$$

and

$$\hat{f}(\mathbf{x_i}) \leq \hat{f}(x^{(m+1)})$$

where N_m is the set of m nearest neighbours to $\mathbf{x_i}$, and $\mathbf{x}^{(\mathbf{k})}$ is the kth nearest neighbour to $\mathbf{x_i}$. The value $(m + 1)$ therefore describes the size of the group. The group of potential interest which will be returned to us by the algorithm comprises the peak, $\mathbf{x_i}$ and its m nearest neighbours, $N_m(\mathbf{x_i})$.

Minimum and maximum values of M are set. Values of $m_{min} = 5$ and $m_{max} = 99$ were set following discussions with the NPSA about appropriate group sizes.

For samples of 25,000 incidents, around 150 to 250 peaks are generally identified, although for one sample 850 were found.

5   Assessing Results

Given the nature of the problem, the results must be assessed qualitatively; this presents some challenges. The results are only useful if they allow the NPSA to create advice and instructions to send to Trusts. However, the NPSA has only limited resources to assess the results; furthermore, they will be discouraged if the results turn out to be predominantly or uniformly incoherent or uninteresting, and there is a risk that they would abandon a type of analysis that is experimental and unfamiliar to them. We therefore need to ensure that the results that they are presented with are limited in quantity and of a certain quality.

We have more time available than the NPSA to assess the quality of the results and filter out uninteresting groups. However, we have much less knowledge of the subject matter, and can only use our general knowledge and common sense. This will limit the quality of our filtering process. In addition, due to the number of different parameters that can be varied in modelling the data, we cannot possibly assess the results for all of the combinations manually; we therefore need to find a quantitative measure that we can calculate and assess quickly, which acts as a proxy for the quality of the peaks.

These issues are likely to be encountered in most data mining projects, where the results can only be assessed subjectively by the third party who owns the data, but who may have only a limited amount of time to assess the results, and may not in fact know *a priori* what information they are expecting or hoping for from the data.

Our process for assessing the data has three stages. Firstly, given that there are many different parameter combinations that can be used in the model building process, the volume of results produced is very large. A quantitative proxy measure of model quality is used to find the combination of parameters that produces a model of the text that most closely reflects the meaning of the incident descriptions.

The number of groups produced by *PEAKER* for these parameter combinations is still too great for them to be sent to the NPSA. The results are assessed qualitatively, and those groups that are clearly incoherent or uninteresting are discarded. This filtering method is the second stage in the process.

Finally, groups that appear to be of potential interest can be passed to the NPSA for a final assessment. The NPSA's medical experts examine the groups: if they find any of the groups to be interesting, the NPSA will be able to draft advice and instructions to NHS Trusts.

5.1   Proxy Measure

The proxy measure is based on the following assumption. If *PEAKER* can be used to identify known groups of incidents from within a larger random sample of incidents, this provides evidence that the model is of a higher quality than one where the known groups cannot be found, i.e., that incident descriptions that mean the same thing but that use different vocabulary should be in close proximity in the model space.

Specifically, what was done was to take a random sample of 3,000 incidents from the low dimensional representation of the 25,000 incident dataset. The positions in the model space were then calculated for seven known groups. These groups are described in Table 3.

The groups were provided by the NPSA, who had found them during manual analysis of the NRLS. The final datasets were therefore low dimensional representations of 3,419 incidents. *PEAKER* was used to find groups of

Table 3. Known Groups of Incidents

Type Of Incident	Code	Size
Anaesthetics	An	100
Chest drains	Ch	7
Latex	Lt	7
Methotrexate	Mt	5
Obstetrics	Ob	100
Self harm	Sl	100
Sexual safety	Sx	100

potential interest in these datasets. We defined a known group as being found if the following conditions were met:

- A group was identified that contained at least six incidents
- At least 50% of the incidents in the identified group came from the known group
- There were more incidents in the identified group from that known group than any other known group

These ad-hoc conditions appear to produce satisfactory results. We used the proxy method to compare the large volumes of results using different parameter settings. These include, for example, models calculated using different types of supervised dimensionality reduction, or anomaly detection using different values of the K parameter in $PEAKER$. Table 4 shows the results using two different normalising functions to calculate the IDF term weighting. The upper half of the table shows whether the known groups can be found (denoted by a tick), using the square root as the normalising function (see Section 3.2). The lower half of the table shows the results for the logarithm. It can be seen, for example, that the known groups involving latex, obstetrics, self harm and sexual safety were found using the square root for normalisation and variable 6 for dimensionality reduction. These results may be compared with those for normalisation using the logarithm (and variable 6 for dimensionality reduction), where no known groups were found.

5.2   Manual Assessment

Once these optimal parameter settings had been found, analysis was carried out on six datasets that are known to be of particular interest to the NPSA. These are:-

- Medical Devices (Med)
- Surgical Speciality (Sur)
- Treatment Procedure (Tre)
- Diagnostic Services (Dia)
- Accident and Emergency (AE)
- Incidents Involving Death or Severe Harm (DS)

The analysis proceeded as previously, with a low dimensional numerical representation of the text calculated and $PEAKER$ used to find groups of potential interest. We then placed the groups into four broad categories, to reduce to a reasonable number the groups that are passed to the NPSA, and to create a numerical measure of the quality of the results. These are:

- A : coherent, and using varying vocabulary
- B : coherent, potentially interesting but using similar vocabulary
- C : coherent, but already known and using similar vocabulary
- D : incoherent

Table 4. Comparison Of Normalisation Functions Using Proxy Measure

	Variable	Groups						
		An	Ch	Lt	Mt	Ob	Sl	Sx
√	1	✗	✗	✗	✗	✓	✗	✓
	2	✗	✗	✓	✗	✗	✗	✓
	3	✗	✗	✓	✗	✓	✓	✓
	4	✗	✗	✗	✗	✗	✗	✓
	5	✗	✗	✓	✗	✓	✗	✓
	6	✗	✗	✓	✗	✓	✓	✓
Log	1	✗	✗	✗	✗	✗	✗	✗
	2	✗	✗	✗	✗	✗	✗	✓
	3	✗	✗	✗	✗	✗	✓	✓
	4	✗	✗	✗	✗	✓	✓	✓
	5	✗	✗	✗	✗	✗	✗	✗
	6	✗	✗	✗	✗	✗	✗	✗

Table 5. Numbers Of Incidents Filtered Into Each Category By Manual Assessment

Category	Variable	Datasets					
		Med	Sur	Tre	Dia	AE	DS
A	1	8	0	28	58	1	17
	2	52	24	14	35	7	36
	3	X	97	X	71	106	64
	4	48	109	70	139	152	89
	5	66	X	42	X	X	57
	6	70	39	83	81	X	65
B	1	0	0	0	0	0	0
	2	0	0	0	3	0	0
	3	X	0	X	6	0	0
	4	0	0	4	2	0	0
	5	0	X	0	X	X	0
	6	0	0	0	0	X	0
C	1	8	65	58	102	20	12
	2	14	85	76	77	31	18
	3	X	180	X	130	111	53
	4	13	162	83	181	75	66
	5	19	X	90	X	X	32
	6	7	113	83	111	X	29
D	1	134	204	141	125	171	121
	2	25	72	54	94	85	56
	3	X	13	X	15	12	28
	4	802	48	45	21	42	54
	5	36	X	49	X	X	52
	6	67	56	39	95	X	84

Coherence was assessed subjectively: the descriptions in a group must describe roughly the same type of incident. Clearly we do not understand some of the medical terminology, so in the case of any doubt over the general meaning of the description (which is actually relatively rare) we erred towards placing the groups into the highest plausible group.

'Using the same vocabulary' means that the groups used either identical or very similar language in their incident descriptions, e.g., a group of 20 incidents with 'patient fell' as their description, or a group of 15 incidents all using a Trust's proforma description of a breach of EU blood transfusion guidelines. These types of incident descriptions are deemed to be less interesting than those that use varied vocabulary. The numbers of groups placed into each category for each of the datasets are as shown in Table 5 (where an 'X' is shown, all of the incidents in the sample are classified into the same class). For example, for the Medical Devices dataset, using categorical variable 6 (see Table 2) to calculate linear discriminants, 70 groups were coherent and used varying vocabulary, whilst 67 were clearly incoherent.

Looking at Table 5, we can see that some of the datasets produce results of far higher quality than others. For example, the model produced using categorical variable 1 for the Accident and Emergency dataset contains only one coherent, potentially interesting group; using categorical variable 2 only produces seven of these groups. The number of incoherent groups found for these models are 171 and 85 respectively. However for medical devices using categorical variable 2, 52 out of 91 (57%) of the groups found are coherent and potentially interesting.

The results of Tables 4 and 5 show some similarity: for example the models based on categorical variable 1 produce lower quality results than those based on categorical variable 3. This similarity between the results is reassuring as it provides us with evidence that the assumption that we made for the proxy measure is correct; i.e., that the number of known groups found is related to the overall quality of the model.

An example of one of the groups found in the Surgical Specialty sample is shown in Table 6. This group is a set of incidents where a problem with a lack of notes has had an effect on a medical procedure. It can be seen that the incidents have similar meaning, but for example, the second incident does not include the word 'notes' ('noted' is an entirely separate variable), and the descriptions use different words to refer to the medical procedures: 'surgery', 'theatre', 'surery', 'endoscopic procedure' and so on.

Many of the other groups that are found using our analysis are similar, in that the incident descriptions comprising the groups mean similar things, but use different language to describe them. We have therefore achieved one of the major aims of our project, which is to produce a global, numerical model of the free text incident descriptions, such that text that means the same thing but uses different vocabulary will be in the same part of the low dimensional space.

We have also developed a method by which manageable quantities of high quality information can be passed to the NPSA, but where we can also examine

Table 6. Group Of Incidents Identified From Surgical Specialty Sample

Patient coming for surgery 2 / 2 / 06 . Notes missing . Rang 4987 , last dept to have notes . Notes had been sent in post 31 / 1 / 06 , didn't get to Eye Unit until late afternoon 2 / 2 / 06 . Patient surgery cancelled for 2 / 2 / 06 , no notes .
2 different case sheet numbers for patients with same name and same consultant . Wrong number on list for patient in theatre . Problem noted when getting blood .
Pt planned to go on trauma list , all prepared as per protocol . Pt notes had been requested but not available for anaesthetist . Admissions and medical records aware of urgency for notes . Theatre cancelled in view of medical history and no notes available .
Patient taken to theatre for surery on morning list . ODP noticed that another patient labels were enclosed in the notes .
Patient admitted for surgery 10 / 8 / 06 , preassessed at Hartlepool , notes were then put in for transfer and booked to ward 28 , notes never arrived , unable to find them anywhere , patient surgery cancelled .
Above patient arrived on unit for an endoscopic procedure . When checking notes it was found that the patient had been sent to us with another patients notes . Ward informed and correct notes brought down .
Notes and theatre list for tomorrows list sent from SJUH . On checking list and notes the name and unit number on list did not match one of the sets of notes . On further investigation the name and procedure on the list were correct but the unit number and date of birth were for another patient . The notes sent were for this patient but were not for the named patient on the list and needed a different procedure .

many different types of models using various parameter combinations, in order to optimise our results.

5.3 NPSA Assessment

The groups placed into category A were then sent to the NPSA. Medical experts examined and commented on each of the groups. Two examples are:

- 'all patient to patient aggression'
- 'falls - might be clever little sub-theme of falls out of doors'

This process took several days, and was followed by meetings with the experts. Their feedback is encouraging: around 80% of the groups were found to be coherent, and represent information that had already been extracted from the NRLS by the NPSA. The reviewers state that it would be unlikely that novel group of incidents would be discovered in our samples. This is partly because relatively few truly novel incidents are present in the NRLS, and partly because similar samples from the same areas of the NHS had already been analysed manually. However, some of our groups would not have been known about before the manual analysis work, suggesting that the algorithm would be useful for analysing less well known groups of incidents.

The algorithm is also good at breaking the samples down into themes: the groups describe the main themes found in previous manual thematic analysis of similar data samples. It is possible that our algorithm could be used to avoid much of the laborious manual work involved in this analysis.

Following this pilot study, the NPSA is considering ways in which the algorithm could be used in its everyday work.

6   Discussion and Future Work

We have developed a model that will allow groups of free text descriptions that mean the same thing but use different vocabulary to be found using an anomaly detection algorithm. It appears that using a relatively simple model without sophisticated natural language processing can produce high quality results.

The use of categorical variables to calculate linear discriminants takes advantage of the fact that large numbers of people have entered the data. For example, if it is assumed that it takes 30 seconds to categorise an incident into the six categorical variables described in Table 2, it would take 210 hours of (tedious) work to replicate this information for a sample of 25,000 incidents. However, this information is not ideal for our purposes: the categories tend to be more general than the groups that we are looking for, and groups of interest can cut across locations or specialties. The use of categorical variables to calculate linear discriminants is merely one way to calculate a mapping from the high dimensional vector space model to a lower dimensional representation.

Work that we are carrying out at present aims to produce a low dimensional representation for a training set that reflects the meaning of the incidents, based on elicited information; an optimal mapping between the high and low dimensional spaces can then be calculated and applied to a larger sample. To create the low dimensional representation of the training set we present trios of incident descriptions to a user, who chooses the incident that is the most dissimilar of the three. This process is carried out repeatedly, until sufficient information has been obtained. This information can be used to create a distance matrix between incidents, which can be used to create the low dimensional representation. Alternatively, trios of individual words can be used to create an ontology.

There are many other variations that could be examined: we could use natural language processing techniques to create a more sophisticated model of the text. For example, tagging the words with their parts of speech would disambiguate between homonyms. Given the variable quality of the data, pre-processing using a spellchecking algorithm might improve the data. There are many types of supervised dimensionality reduction, or we could even discard the vector space model and devise an entirely new basic model.

Acknowledgments

James Bentham would like to thank the EPSRC and NPSA for funding his CASE award. The work of David Hand was partially supported by a Royal Society Wolfson Research Merit Award.

References

1. Adams, N., Hand, D., Till, R.: Mining for classes and patterns in behavioural data. Journal of the Operational Research Society 52, 1017–1024 (2001)
2. Aggarwal, C., Hinneburg, A., Keim, D.: On the surprising behavior of distance metrics in high dimensional space. In: Proceedings of the 8th International Conference on Database Theory (2001)
3. Bolton, R., Hand, D., Crowder, M.: Significance tests for unsupervised pattern discovery in large continuous multivariate data sets. Computational Statistics and Data Analysis 46, 57–79 (2004)
4. Manning, C., Raghavan, P., Schütze, H.: An Introduction to Information Retrieval. Cambridge University Press, Cambridge (2008)
5. Salton, G., Buckley, C.: Term-weighting approaches in automatic text retrieval. Information Processing and Management 24(5), 513–523 (1988)
6. http://www.npsa.nhs.uk/corporate/news/npsa-alerts-healthcare-workers-to-new-guidance-for-injecting-adults-and-adolescent-patients-withint/
7. Zhang, Z., Hand, D.: Detecting groups of anomalously similar objects in large data sets. In: Famili, A.F., Kook, J.N., Peña, J.M., Siebes, A., Feelders, A. (eds.) IDA 2005. LNCS, vol. 3646, pp. 509–519. Springer, Heidelberg (2005)

Using Data Mining for Wine Quality Assessment

Paulo Cortez[1], Juliana Teixeira[1], António Cerdeira[2], Fernando Almeida[2],
Telmo Matos[2], and José Reis[1,2]

[1] Dep. of Information Systems/Algoritmi Centre, University of Minho,
4800-058 Guimarães, Portugal
pcortez@dsi.uminho.pt
http://www3.dsi.uminho.pt/pcortez
[2] Viticulture Commission of the Vinho Verde region (CVRVV),
4050-501 Porto, Portugal

Abstract. Certification and quality assessment are crucial issues within
the wine industry. Currently, wine quality is mostly assessed by physicochemical (e.g alcohol levels) and sensory (e.g. human expert evaluation)
tests. In this paper, we propose a data mining approach to predict wine
preferences that is based on easily available analytical tests at the certification step. A large dataset is considered with white *vinho verde* samples
from the Minho region of Portugal. Wine quality is modeled under a regression approach, which preserves the order of the grades. Explanatory
knowledge is given in terms of a sensitivity analysis, which measures the
response changes when a given input variable is varied through its domain. Three regression techniques were applied, under a computationally
efficient procedure that performs simultaneous variable and model selection and that is guided by the sensitivity analysis. The support vector
machine achieved promising results, outperforming the multiple regression and neural network methods. Such model is useful for understanding how physicochemical tests affect the sensory preferences. Moreover,
it can support the wine expert evaluations and ultimately improve the
production.

Keywords: Ordinal Regression, Sensitivity Analysis, Sensory Preferences,
Support Vector Machines, Variable and Model Selection, Wine Science.

1 Introduction

Nowadays wine is increasingly enjoyed by a wider range of consumers. In particular, Portugal is a top ten wine exporting country and exports of its *vinho verde*
wine (from the northwest region) have increased by 36% from 1997 to 2007 [7].
To support this growth, the industry is investing in new technologies for both
wine making and selling processes. Wine certification and quality assessment are
key elements within this context. Certification prevents the illegal adulteration
of wines (to safeguard human health) and assures quality for the wine market.
Quality evaluation is often part of the certification process and can be used to
improve wine making (by identifying the most influential factors) and to stratify
wines such as premium brands (useful for setting prices).

J. Gama et al. (Eds.): DS 2009, LNAI 5808, pp. 66–79, 2009.
© Springer-Verlag Berlin Heidelberg 2009

Wine certification is often assessed by physicochemical and sensory tests [9]. Physicochemical laboratory tests routinely used to characterize wine include determination of density, alcohol or pH values, while sensory tests rely mainly on human experts. It should be stressed that taste is the least understood of the human senses [20], thus wine classification is a difficult task. Moreover, the relationships between the physicochemical and sensory analysis are complex and still not fully understood [16].

On the other hand, advances in information technologies have made it possible to collect, store and process massive, often highly complex datasets. All this data hold valuable information such as trends and patterns, which can be used to improve decision making and optimize chances of success [23]. Data mining (DM) techniques [26] aim at extracting high-level knowledge from raw data. There are several DM algorithms, each one with its own advantages. When modeling continuous data, the linear/multiple regression (MR) is the classic approach. Neural networks (NNs) have become increasingly used since the introduction of the backpropagation algorithm [19]. More recently, support vector machines (SVMs) have also been proposed [3]. Due to their higher flexibility and nonlinear learning capabilities, both NNs and SVMs are gaining an attention within the DM field, often attaining high predictive performances [13]. SVMs present theoretical advantages over NNs, such as the absence of local minima in the learning phase. When applying these methods, performance highly depends on a correct variable and model selection, since simple models may fail in mapping the underlying concept and too complex ones tend to overfit the data [13][12].

The use of decision support systems by the wine industry is mainly focused on the wine production phase [10]. Despite the potential of DM techniques to predict wine quality based on physicochemical data, their use is rather scarce and mostly considers small datasets. For example, in 1991 the famous "Wine" dataset was donated into the UCI repository [2]. The data contain 178 examples with measurements of 13 chemical constituents (e.g. alcohol, Mg) and the goal is to classify three cultivars from Italy. This dataset is very easy to discriminate and has been mainly used as a benchmark for new DM classifiers. In 1997 [22], a NN fed with 15 input variables (e.g. Zn and Mg levels) was used to predict six geographic wine origins. The data included 170 samples from Germany and a 100% predictive rate was reported. In 2001 [24], NNs were used to classify three sensory attributes (e.g. sweetness) of Californian wine, based on grape maturity levels and chemical analysis (e.g. titrable acidity). Only 36 examples were used and a 6% error was achieved. More recently, mineral characterization (e.g. Zn and Mg) was used to discriminate 54 samples into two red wine classes [17]. A probabilistic NN was adopted, attaining 95% accuracy. As a powerful learning tool, SVM has outperformed NN in several applications, such as predicting meat preferences [6]. Yet, in the field of wine quality only one application has been reported, where spectral measurements from 147 bottles were successfully used to predict 3 categories of rice wine age [27].

In this paper, we present a real-world application, where wine taste preferences are modeled by DM algorithms that use analytical data that are easily available at the certification step. In contrast with previous studies, a large dataset is considered with a total of 4898 samples. Wine quality is modeled under a regression approach that preserves the order of the grades. Explanatory knowledge is given by a sensitivity analysis, which measures how the responses are affected when a given input is varied through its domain [14][6]. Variable and model selection are performed simultaneously, in a process that is guided by the sensitivity analysis. Also, we propose a parsimony search method to select the best NN and SVM parameters with a low computational effort. Finally, we show the impact of the obtained models in the wine domain.

2   Materials and Methods

2.1   Wine Data

This study will consider *vinho verde*, a unique product from the Minho (northwest) region of Portugal. Medium in alcohol, is it particularly appreciated due to its freshness (specially in the summer). This wine accounts for 15% of the total Portuguese production [7], and around 10% is exported, mostly white wine. In this work, we will analyze this common variant from the demarcated region of *vinho verde*. The data were collected from May/2004 to February/2007 using only protected designation of origin samples that were tested at the official certification entity (CVRVV). The CVRVV is an inter-professional organization with the goal of improving the quality and marketing of vinho verde. The data were recorded by a computerized system (iLab), which automatically manages the process of wine sample testing from producer requests to laboratory and sensory analysis. Each entry denotes a given test (analytical or sensory) and the final database was exported into a single sheet (.csv).

During the preprocessing stage, the database was transformed in order to include a distinct wine sample (with all tests) per row. To avoid discarding examples, only the most common physicochemical tests were selected. Table 1 presents the physicochemical statistics per dataset. Regarding the preferences, each sample was evaluated by a minimum of three sensory assessors (using blind tastes), which graded the wine in a scale that ranges from 0 (very bad) to 10 (excellent). The final sensory score is given by the median of these evaluations. Fig. 1 plots the histograms of the target variable, denoting a typical normal shape distribution (i.e. with more normal grades that extreme ones).

2.2   Data Mining Approach and Evaluation

We will adopt a regression approach, which preserves the order of the preferences. For instance, if the true grade is 3, then a model that predicts 4 is better than one that predicts 7. A regression dataset D is made up of $k \in \{1, ..., N\}$ examples, each mapping an input vector with I input variables $(x_1^k, \ldots, x_I^k)$ to a given target y_k. The regression performance is commonly measured by an error metric, such as the mean absolute deviation (MAD) [26]:

$$MAD = \sum_{i=1}^{N} |y_i - \widehat{y_i}|/N \qquad (1)$$

Table 1. The physicochemical data statistics

Attribute (units)	Min	Max	Mean
fixed acidity $(g(\text{tartaric acid})/dm^3)$	3.8	14.2	6.9
volatile acidity $(g(\text{acetic acid})/dm^3)$	0.1	1.1	0.3
citric acid (g/dm^3)	0.0	1.0	0.3
residual sugar (g/dm^3)	0.6	65.8	6.4
chlorides $(g(\text{sodium chloride})/dm^3)$	0.01	0.35	0.05
free sulfur dioxide (mg/dm^3)	2	260	35
total sulfur dioxide (mg/dm^3)	9	260	138
density (g/cm^3)	0.987	1.039	0.994
pH	2.7	3.8	3.1
sulphates $(g(\text{potassium sulphate})/dm^3)$	0.2	1.1	0.5
alcohol (% vol.)	8.0	14.2	10.4

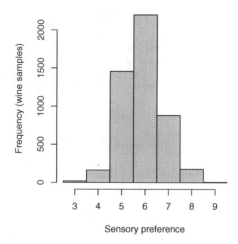

Fig. 1. The histogram for the white wine preferences

where $\widehat{y}_k$ is the predicted value for the k input pattern. The regression error characteristic (REC) curve [1] is also used to compare regression models, with the ideal model presenting an area of 1.0. The curve plots the absolute error tolerance T (x-axis), versus the percentage of points correctly predicted (the accuracy) within the tolerance (y-axis).

The confusion matrix is often used for classification analysis, where a $C \times C$ matrix (C is the number of classes) is created by matching the predicted values (in columns) with the desired classes (in rows). For an ordered output, the predicted class is given by $p_i = y_i$, if $|y_i - \widehat{y}_i| \leq T$, else $p_i = y'_i$, where y'_i denotes the closest class to $\widehat{y}_i$, given that $y'_i \neq y_i$. From the matrix, several metrics can be used to access the overall classification performance, such as the accuracy and precision (i.e. the predicted column accuracies) [26].

The holdout validation is often used to estimate the generalization capability of a model. This method randomly partitions the data into training and test subsets. The former subset is used to fit the model (typically with 2/3 of the data), while the latter (with the remaining 1/3) is used to compute the estimate. A more robust estimation procedure is the k-fold cross-validation [8], where the data is divided into k partitions of equal size. One subset is tested each time and the remaining data are used for fitting the model. The process is repeated sequentially until all subsets have been tested. Therefore, under this scheme, all data are used for training and testing. However, this method requires around k times more computation, since k models are fitted. The validation method will be applied several runs and statistical confidence will be given by the t-student test at the 95% confidence level [11].

2.3   Data Mining Methods

We will adopt the most common NN type, the multilayer perceptron, where neurons are grouped into layers and connected by feedforward links (Fig. 2). Supervised learning is achieved by an iterative adjustment of the network connection weights, called the training procedure, in order to minimize an error function. For regression tasks, this NN architecture is often based on one hidden layer of H hidden nodes with a logistic activation and one output node with a linear function [13]:

$$\widehat{y} = w_{o,0} + \sum_{j=I+1}^{o-1} \frac{1}{1 + \exp(-\sum_{i=1}^{I} x_i w_{j,i} - w_{j,0})} \cdot w_{o,i} \qquad (2)$$

where $w_{i,j}$ denotes the weight of the connection from node j to i and o the output node. The performance is sensitive to the topology choice (H). A NN with $H = 0$ is equivalent to the MR model. By increasing H, more complex mappings can be performed, yet an excess value of H will overfit the data, leading to generalization loss. A computationally efficient method to set H is to search through the range $\{0, 1, 2, 3, \ldots, H_{max}\}$ (i.e. from the simplest NN to more complex ones). For each H value, a NN is trained and its generalization estimate is measured (e.g. over a validation sample). The process is stopped when the generalization decreases or when H reaches the maximum value (H_{max}).

In SVM regression [21], the input $x \in \Re^I$ is transformed into a high m-dimensional feature space, by using a nonlinear mapping (ϕ) that does not need to be explicitly known but that depends of a kernel function (K). The aim of a SVM is to find the best linear separating hyperplane in the feature space:

$$\widehat{y} = w_0 + \sum_{i=1}^{m} w_i \phi_i(x) \qquad (3)$$

To select the best hyperplane, the ϵ-insensitive loss function is often used [21]. This function sets an insensitive tube around the residuals and the tiny errors within the tube are discarded (Fig. 2).

We will adopt the popular gaussian kernel, which presents less parameters than other kernels (e.g. polynomial) [25]: $K(x, x') = exp(-\gamma ||x - x'||^2), \gamma > 0$.

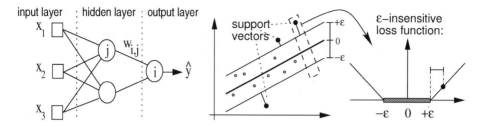

Fig. 2. Example of a multilayer perceptron with 3 inputs, 2 hidden nodes and one output (left) and a linear SVM regression (right, adapted from [21])

Under this setup, the SVM performance is affected by three parameters: γ, ϵ and C (a trade-off between fitting the errors and the flatness of the mapping). To reduce the search space, the first two values will be set using the heuristics [4]: $C = 3$ (for a standardized output) and $\epsilon = \hat{\sigma}/\sqrt{N}$, where $\hat{\sigma} = 1.5/N \times \sum_{i=1}^{N}(y_i - \hat{y}_i)^2$ and $\hat{y}$ is the value predicted by a 3-nearest neighbor algorithm. The kernel parameter (γ) produces the highest impact in the SVM performance, with values that are too large or too small leading to poor predictions. A practical method to set γ is to start the search from one of the extremes and then search towards the middle of the range while the predictive estimate increases [25].

2.4   Input Relevance and Variable/Model Selection

Sensitivity analysis [14] is a simple procedure that is applied after the training phase and analyzes the model responses when the inputs are changed. Originally proposed for NNs, this sensitivity method can also be applied to other algorithms, such as SVM [6]. Let $\hat{y}_{a_j}$ denote the output obtained by holding all input variables at their average values except x_a, which varies through its entire range with $j \in \{1, \ldots, L\}$ levels. If a given input variable ($x_a \in \{x_1, \ldots, x_I\}$) is relevant then it should produce a high variance (V_a). Thus, its relative importance (R_a) can be given by:

$$
\begin{aligned}
V_a &= \sum_{j=1}^{L}(\hat{y}_{a_j} - \overline{\hat{y}_{a_j}})^2/(L-1) \\
R_a &= V_a/\sum_{i=1}^{I} V_i \times 100\,(\%)
\end{aligned}
\tag{4}
$$

The R_a values will be used to measure the relevance of the inputs. For a more detailed input influence analysis, in this work we propose the Variable Effect Characteristic (VEC) curve. For a given a attribute, the VEC plots the x_{a_j} values (x-axis) versus the $\hat{y}_{a_j}$ predictions (y-axis) (see Section 3.3).

The sensitivity analysis will be also used to discard irrelevant inputs, guiding the variable selection algorithm. We will adopt a backward selection scheme, which starts with all variables and iteratively deletes one input until a stopping criterion is met [12]. The difference, when compared to the standard backward selection, is that we guide the variable deletion (at each step) by the sensitivity analysis, in a variant that allows a reduction of the computational effort by a factor of I and that in [14] has outperformed other methods (e.g. backward and genetic algorithms). Similarly to [28], the variable and model selection will be performed simultaneously, i.e. in each backward iteration several models are searched, with the one that presents the best generalization estimate selected. For a given DM method, the overall procedure is:

1. Start with all $F = \{x_1, \ldots, x_I\}$ input variables.
2. If there is a hyperparameter $P \in \{P_1, \ldots, P_k\}$ to tune (e.g. NN or SVM), start with P_1 and go through the remaining range until the generalization estimate decreases. Compute the generalization estimate of the model by using an internal validation method. For instance, if the holdout method is used, the available data are further split into training (to fit the model) and validation sets (to get the predictive estimate).
3. After fitting the model, compute the relative importances (R_i) of all $x_i \in F$ variables and delete from F the least relevant input. Go to step 4 if the stopping criterion is met, otherwise return to step 2.
4. Select the best F (and P in case of NN or SVM) values, i.e., the input variables and model that provide the best predictive estimates. Finally, retrain this configuration with all available data.

3   Empirical Results

3.1   Experimental Setup

All experiments reported in this work were written in **R** [18] and conducted in a Linux server, with an Intel dual core processor. **R** is an open source, multiple platform (e.g. Windows, Linux) and high-level matrix programming language for statistical and data analysis. In particular, we adopted the **RMiner** [5], a library for the **R** tool that facilitates the use of DM techniques in classification and regression tasks.

Before fitting the models, the data was first standardized to a zero mean and one standard deviation [13]. **RMiner** uses the efficient BFGS algorithm to train the NNs (**nnet R** package), while the SVM fit is based on the Sequential Minimal Optimization implementation provided by LIBSVM (**kernlab** package). The the hyperparameters (H and γ) will be set using the procedure described in the previous section and with the search ranges of $H \in \{0, 1, \ldots, 11\}$ [28] and $\gamma \in \{2^3, 2^1, \ldots, 2^{-15}\}$ [25]. While the maximum number of searches is 12/10, in practice the parsimony approach (step 2 of Section 2.4) will reduce this number substantially.

Regarding the variable selection, we set the estimation metric to the MAD value (Eq. 1), as advised in [25]. To reduce the computational effort, we adopted the simpler 2/3 and 1/3 holdout split as the internal validation method. The sensitivity analysis parameter was set to $L = 6$, i.e. $x_a \in \{-1.0, -0.6, \ldots, 1.0\}$ for a standardized input. As a reasonable balance between the pressure towards simpler models and the increase of computational search, the stopping criterion was set to 2 iterations without any improvement or when only one input is available.

3.2   Predictive Knowledge

To evaluate the selected models, we adopted 20 runs of the more robust 5-fold cross-validation, in a total of 20×5=100 experiments for each tested configuration. The results are summarized in Table 2. The test set errors are shown in terms of the mean and 95% confidence intervals. Three metrics are present: MAD, the classification accuracy for different tolerances (i.e. $T = 0.25$, 0.5 and 1.0) and Kappa ($T = 0.5$). The selected models are described in terms of the average number of inputs ($\bar{I}$) and hyperparameter value ($\bar{H}$ or $\bar{\gamma}$). The last row shows the total computational time required in seconds.

For all error metrics, the SVM is the best choice. The differences are higher for small tolerances (e.g. for $T = 0.25$, the SVM accuracy is almost two times better when compared to other methods). This effect is clearly visible when plotting the full REC curves (Fig. 3). The Kappa statistic [26] measures the accuracy when compared with a random classifier (which presents a Kappa value of 0%). The higher the statistic, the more accurate the result. The most practical tolerance values are $T = 0.5$ and $T = 1.0$. The former tolerance rounds the regression response into the nearest class, while the latter accepts a response that is correct within one of the two closest classes (e.g. a 3.1 value can be interpreted as grade 3 or 4 but not 2 or 5). For $T = 0.5$, the SVM accuracy improvement is 11.7 pp (19.9 pp for Kappa). The NN model slightly outperforms the MR results. Regarding the variable selection, the average number of deleted inputs ranges from 1.0 to 1.7, showing that most of the physicochemical tests used are relevant. In terms of computational effort, the SVM is the most expensive method.

A detailed analysis of the SVM classification results is presented by the average confusion matrix for $T = 0.5$ (Table 3). To simplify the visualization, the 3 and 9 grade predictions were omitted, since these were always empty. Most of the values are close to the diagonals (in bold), denoting a good fit by the model. The true predictive accuracy for each class is given by the precision metric (e.g. for the grade 4, $precision_{T=0.5}=18/(18+6+4)=64.3\%$). This statistic is important in practice, since in a real deployment setting the actual values are unknown and all predictions within a given column would be treated the same. For a tolerance

Table 2. The wine modeling results (test set errors and selected models; best values are in **bold**; <u>underline</u> denotes a statistical significance when compared with MR and NN)

	MR	NN	SVM
MAD	0.59±0.00	0.58±0.00	**0.45±0.00**
Accuracy$_{T=0.25}$ (%)	25.6±0.1	26.5±0.3	**50.2±1.1**
Accuracy$_{T=0.50}$ (%)	51.7±0.1	52.6±0.3	**64.3±0.4**
Accuracy$_{T=1.00}$ (%)	84.3±0.1	84.7±0.1	**86.8±0.2**
Kappa$_{T=0.5}$ (%)	20.9±0.1	23.5±0.6	**43.4±0.4**
Inputs ($\bar{I}$)	9.6	9.3	10.0
Model	–	$\bar{H} = 2.1$	$\bar{\gamma} = 2^{0.7}$
Time (s)	**551**	1339	34644

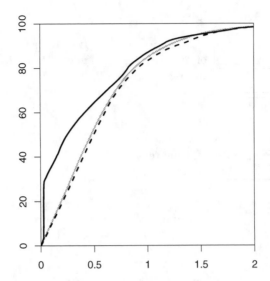

Fig. 3. The average test set REC curves (SVM – solid line, NN - gray line and MR – dashed line)

Table 3. The average confusion matrix ($T = 0.5$) and precision values ($T = 0.5$ and 1.0) for the SVM model (**bold** denotes accurate predictions)

Actual	White wine predictions				
Class	4	5	6	7	8
3	0	3	17	1	0
4	**18**	53	91	1	0
5	6	**832**	598	21	0
6	4	241	**1806**	144	3
7	0	20	418	**436**	6
8	0	2	71	45	**58**
9	0	0	2	2	0
Precision$_{T=0.5}$	64.3%	72.3%	60.1%	67.1%	86.6%
Precision$_{T=1.0}$	89.7%	93.4%	82.0%	90.1%	96.2%

of 0.5, the accuracies are 60.1/64.3% for classes 6 and 4, 67.1/72.3% for grades 7 and 5, and a surprising 86.6% for the class 8 (the exception are the 3 and 9 extremes with 0%, not shown in the table). When the tolerance is increased ($T = 1.0$), high accuracies are obtained, ranging from 82.0 to 96.2%.

3.3   Explanatory Knowledge

The relative importances of the SVM input variables, given in terms of the mean and 95% confidence intervals of the R_a values, are shown in Fig. 4. It should be noted that the whole 11 inputs are shown, since in each simulation different sets

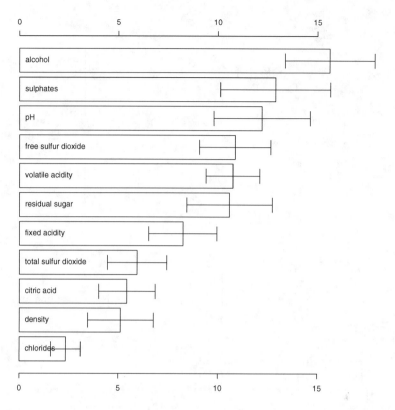

Fig. 4. The relative input importances for the SVM model (in %; bars denote the average value while the whiskers show the 95% confidence intervals)

of variables can be selected. A more detailed analysis will be given to sixth most relevant analytical tests (Fig. 5). For a given input, each plot shows the histogram (frequency values are shown at the right of the y-axis) and the VEC curves ($\widehat{y}_{a_j}$ values, shown at the left of the y-axis) when the analytical test values (x-axis) are changed through their domain. For a given test, we built a VEC curve with $L = 6$ points (the sensitivity levels). Since 100 experiments we performed, we performed a vertical averaging (with the respective 95% confidence intervals) of the 100 curves.

In several cases, the obtained results confirm the oenological theory. For instance, an increase in the alcohol (the most relevant factor) tends to result in a higher quality wine. Fig. 5 shows that this is true between the range from 9 to 13 % (which is related to most samples). In addition, the volatile acidity has a negative impact within the range that corresponds to the majority of the examples. This outcome was expected, since acetic acid is the key ingredient in vinegar. Moreover, residual sugar levels are important in white wine, where the equilibrium between the freshness and sweet taste is more appreciated. The most intriguing result is the high importance of sulphates, ranked second. Oenologically this result could be very interesting. An increase in sulphates might be

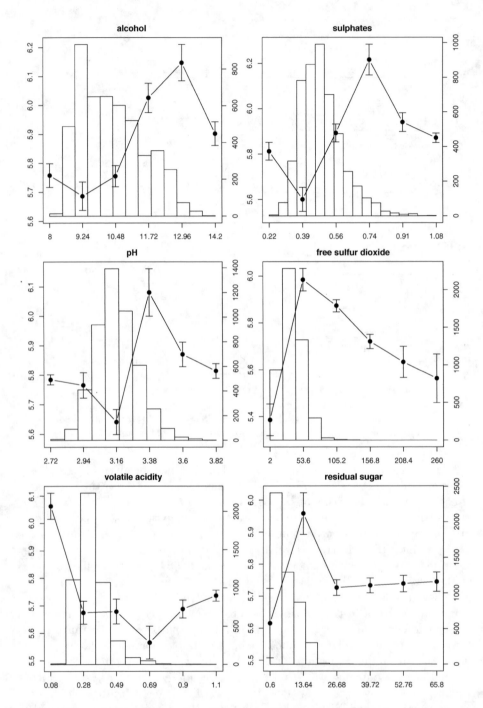

Fig. 5. The vertical averaging of the VEC curves (points and whiskers) and histogram (in bars) for the SVM model and the sixth most relevant physicochemical tests

related to the fermenting nutrition, which is very important to improve the wine aroma, in an effect that occurs within the range 0.4 to 0.7 that contains most of the samples.

4   Conclusions

Due to the increase in the interest in wine, companies are investing in new technologies to improve their production and selling processes. Quality certification is a crucial step for both processes and is currently dependent on wine tasting by human experts. This work aims at the prediction of wine preferences from objective analytical tests that are available at the certification step. A large dataset (with 4898 entries) was considered, including white *vinho verde* samples from the northwest region of Portugal. This case study was addressed by a regression tasks, where wine preference is modeled in a continuous scale, from 0 (very bad) to 10 (excellent). This approach preserves the order of the classes, allowing the evaluation of distinct accuracies, according to the degree of error tolerance (T) that is accepted.

Due to advances in the data mining (DM) field, it is possible to extract knowledge from raw data. Indeed, powerful techniques such as neural networks (NNs) and more recently support vector machines (SVMs) are emerging. While being more flexible models (i.e. no *a priori* restriction is imposed), the performance depends on a correct setting of hyperparameters (e.g. SVM kernel parameter) and the input variables used by the model. In this study, we present an integrated and computationally efficient approach that simultaneously addresses both issues. Sensitivity analysis is used to extract knowledge from the NN/SVM models, given in terms of the effect on the responses when one input is varied, leading to the proposed Variable Effect Characteristic (VEC) curves, and relative importance of the inputs (measured by the variance of the response changes). The the variable selection is guided by sensitivity analysis and the model selection is based on parsimony search that starts from a reasonable value and is stopped when the generalization estimate decreases.

Encouraging results were achieved, with the SVM model providing the best performances, outperforming the NN and MR techniques. The overall accuracies are 64.3% ($T = 0.5$) and 86.8% ($T = 1.0$). It should be noted that the datasets contain six/seven classes (from 3 to 8/9) and these accuracies are much better than the ones expected by a random classifier. While requiring more computation, the SVM fitting can still be achieved within a reasonable time with current processors. For example, one run of the 5-fold cross-validation testing takes around 26 minutes.

The result of this research is relevant to the wine science domain, helping in the understanding of how physicochemical characterization affects the final quality. In addition, this work can have an impact in the wine industry. At the certification phase and by Portuguese law, the sensory analysis has to be performed by human tasters. Yet, the evaluations are based in the experience and knowledge of the experts, which are prone to subjective factors. The proposed data-driven approach is based on objective tests and thus it can be integrated into a decision support system, aiding the speed and quality of the oenologist performance. For instance, the expert could repeat the tasting only if her/his grade is far from the one predicted by the DM model. In effect, within this domain the $T = 1.0$ distance is accepted as a good quality control process and, as shown in this study, high accuracies were achieved for this tolerance. The model could also be used to improve the training of oenology students. Furthermore, the relative importance of the inputs brought interesting insights regarding the impact of the analytical tests. Since some variables can be controlled in the production process this information can be used to improve the wine quality. For instance, alcohol concentration can be increased or decreased by monitoring the grape sugar concentration prior to the harvest. Also, the residual sugar in wine could be raised by suspending the sugar fermentation carried out by yeasts. In future work, we intend to model preferences from niche and/or profitable markets (e.g. for a particular country by providing free wine tastings at supermarkets), aiming at the design of brands that match these market needs. We will also test other DM algorithms that specifically build rankers, such as regression trees [15].

References

1. Bi, J., Bennett, K.: Regression Error Characteristic curves. In: Proceedings of 20th Int. Conf. on Machine Learning (ICML), Washington DC, USA (2003)
2. Blake, C., Merz, C.: UCI Repository of Machine Learning Databases (1998)
3. Boser, B., Guyon, I., Vapnik, V.: A training algorithm for optimal margin classifiers. In: COLT 1992: Proceedings of the Fifth Annual Workshop on Computational Learning Theory, pp. 144–152. ACM Press, New York (1992)
4. Cherkassy, V., Ma, Y.: Practical Selection of SVM Parameters and Noise Estimation for SVM Regression. Neural Networks 17(1), 113–126 (2004)
5. Cortez, P.: RMiner: Data Mining with Neural Networks and Support Vector Machines using R. In: Rajesh, R. (ed.) Introduction to Advanced Scientific Softwares and Toolboxes (in press)
6. Cortez, P., Portelinha, M., Rodrigues, S., Cadavez, V., Teixeira, A.: Lamb Meat Quality Assessment by Support Vector Machines. Neural Processing Letters 24(1), 41–51 (2006)
7. CVRVV. Portuguese Wine - Vinho Verde. Comissão de Viticultura da Região dos Vinhos Verdes (CVRVV) (July 2008), http://www.vinhoverde.pt
8. Dietterich, T.: Approximate Statistical Tests for Comparing Supervised Classification Learning Algorithms. Neural Computation 10(7), 1895–1923 (1998)
9. Ebeler, S.: Linking flavour chemistry to sensory analysis of wine. In: Flavor Chemistry - Thirty Years of Progress, pp. 409–422. Kluwer Academic Publishers, Dordrecht (1999)
10. Ferrer, J., MacCawley, A., Maturana, S., Toloza, S., Vera, J.: An optimization approach for scheduling wine grape harvest operations. Production Economics, pp. 985–999 (2008)
11. Flexer, A.: Statistical evaluation of neural networks experiments: Minimum requirements and current practice. In: Proceedings of the 13th European Meeting on Cybernetics and Systems Research, Vienna, Austria, vol. 2, pp. 1005–1008 (1996)

12. Guyon, I., Elisseeff, A.: An introduction to variable and feature selection. Journal of Machine Learning Research 3, 1157–1182 (2003)
13. Hastie, T., Tibshirani, R., Friedman, J.: The Elements of Statistical Learning: Data Mining, Inference, and Prediction. Springer, NY (2001)
14. Kewley, R., Embrechts, M., Breneman, C.: Data Strip Mining for the Virtual Design of Pharmaceuticals with Neural Networks. IEEE Transactions on Neural Networks 11(3), 668–679 (2000)
15. Kramer, S., Widmer, G., Pfahringer, B., De Groeve, M.: Prediction of Ordinal Classes Using Regression Trees. Fundamenta Informaticae 47(1), 1–13 (2001)
16. Legin, A., Rudnitskaya, A., Luvova, L., Vlasov, Y., Natale, C., D'Amico, A.: Evaluation of Italian wine by the electronic tongue: recognition, quantitative analysis and correlation with human sensory perception. Analytica Chimica Acta, 33–34 (2003)
17. Moreno, I., González-Weller, D., Gutierrez, V., Marino, M., Cameán, A., González, A., Hardisson, A.: Differentiation of two Canary DO red wines according to their metal content from inductively coupled plasma optical emission spectrometry and graphite furnace atomic absorption spectrometry by using Probabilistic Neural Networks. Talanta 72, 263–268 (2007)
18. R Development Core Team: R: A language and environment for statistical computing. R Foundation for Statistical Computing, Vienna, Austria (2008), http://www.R-project.org, ISBN 3-900051-00-3
19. Rumelhart, D., Hinton, G., Williams, R.: Learning Internal Representations by Error Propagation. In: Rulmelhart, D., McClelland, J. (eds.) Parallel Distributed Processing: Explorations in the Microstructures of Cognition, pp. 318–362. MIT Press, Cambridge (1986)
20. Smith, D., Margolskee, R.: Making sense of taste. Scientific American 284, 26–33 (2001)
21. Smola, A., Scholkopf, B.: A tutorial on support vector regression. Statistics and Computing 14, 199–222 (2004)
22. Sun, L., Danzer, K., Thiel, G.: Classification of wine samples by means of artificial neural networks and discrimination analytical methods. Fresenius' Journal of Analytical Chemistry 359, 143–149 (1997)
23. Turban, E., Sharda, R., Aronson, J., King, D.: Business Intelligence, A Managerial Approach. Prentice-Hall, Englewood Cliffs (2007)
24. Vlassides, S., Ferrier, J., Block, D.: Using Historical Data for Bioprocess Optimization: Modeling Wine Characteristics Using Artificial Neural Networks and Archived Process Information. Biotechnology and Bioengineering, 73(1) (2001)
25. Wang, W., Xu, Z., Lu, W., Zhang, X.: Determination of the spread parameter in the Gaussian kernel for classification and regression. Neurocomputing 55, 643–663 (2003)
26. Witten, I.H., Frank, E.: Data Mining: Practical Machine Learning Tools and Techniques with Java Implementations. Morgan Kaufmann, San Francisco (2005)
27. Yu, H., Lin, H., Xu, H., Ying, Y., Li, B., Pan, X.: Prediction of Enological Parameters and Discrimination of Rice Wine Age Using Least-Squares Support Vector Machines and Near Infrared Spectroscopy. Agricultural and Food Chemistry 56, 307–313 (2008)
28. Yu, M., Shanker, M., Zhang, G., Hung, M.: Modeling consumer situational choice of long distance communication with neural networks. Decision Support Systems 44, 899–908 (2008)

MICCLLR: Multiple-Instance Learning Using Class Conditional Log Likelihood Ratio

Yasser EL-Manzalawy[1,2] and Vasant Honavar[1]

[1] Department of Computer Science
Iowa State University
Ames, IA 50011-1040, USA
[2] Systems and Computer Engineering
Al-Azhar University
Cairo, Egypt
{yasser,honavar}@cs.iastate.edu

Abstract. Multiple-instance learning (MIL) is a generalization of the supervised learning problem where each training observation is a labeled bag of unlabeled instances. Several supervised learning algorithms have been successfully adapted for the multiple-instance learning settings. We explore the adaptation of the Naive Bayes (NB) classifier and the utilization of its sufficient statistics for developing novel multiple-instance learning methods. Specifically, we introduce MICCLLR (multiple-instance class conditional log likelihood ratio), a method for mapping each bag of instances as a single meta-instance using class conditional log likelihood ratio statistics such that any supervised base classifier can be applied to the meta-data. The results of our experiments with MICCLLR using different base classifiers suggest that no single base classifier consistently outperforms other base classifiers on all data sets. We show that a substantial improvement in performance is obtained using an ensemble of MICCLLR classifiers trained using different base learners. We also show that an extra gain in classification accuracy is obtained by applying AdaBoost.M1 to weak MICCLLR classifiers. Overall, our results suggest that the predictive performance of the three proposed variants of MICCLLR are competitive to some of the state-of-the-art MIL methods.

Keywords: multiple-instance learning, image retrieval, drug activity prediction, ensemble of multiple-instance learning classifiers, boosted multiple-instance learning.

1 Introduction

Dietterich et al. [1] introduced the multiple-instance learning (MIL) problem motivated by his work on classifying aromatic molecules according to whether or not they are "musky". In this classification task, each molecule can adopt multiple shapes as a consequence of rotation of some internal bonds. Dietterich et al. [1] suggested representing each molecule by multiple conformations (instances) representing possible shapes or conformations that the molecule can

J. Gama et al. (Eds.): DS 2009, LNAI 5808, pp. 80–91, 2009.
© Springer-Verlag Berlin Heidelberg 2009

assume. The multiple conformations yield a multiset (bag) of instances (where each instance corresponds to a conformation) and the task of the classifier is to assign a class label to such a bag. Dieterich's proposed solution to the MIL problem is based on the standard multiple-instance assumption, that all the instances in a bag, in order for it be labeled negative, must contain no positively labeled instance, and a positive bag must have at least one positive instance. The resulting classification task finds application in drug discovery [1], identifying Thioredoxin-fold proteins [2], content-based image retrieval (CBIR) [3,4,5], and computer aided diagnosis (CAD) [6].

Several approaches to MIL have been investigated in the literature including a MIL variant of the backpropagation algorithm [7], variants of the k-nearest neighbor (k-NN) algorithm [8], the Diverse Density (DD) method [9] and EM-DD [10] which improves on DD by using Expectation Maximization (EM), DD-SVM [11] which trains a support vector machine (SVM) classifier in a feature space constructed from a mapping defined by the local maximizers and minimizers of the DD function, and MI logistic regression (MILR) [12]. Most of these methods search for a single instance contributing the positive bag label. Alternatively, a number of MIL methods [13,14,15] have a generalized view of the MIL problem where all the instances in a bag are assumed to participate in determining the bag label.

Two basic approaches for solving the MIL problem have been proposed in the literature: i) adapting supervised learning algorithms for MIL settings. Zhou [16] showed that standard single-instance supervised algorithms can be adapted for MI learning by shifting their focuses from discrimination on the instances to the discrimination on the bags. Many MIL methods can be viewed as single-instance learning methods adapted for the MIL settings. For example, MI-SVM [17], Citation-kNN [8], DD [9], and RBF-MIP [18]; ii) adapting MIL representation for supervised algorithms. The basic idea is to convert each bag of instances into a single feature vector such that supervised classifiers can be trained to discriminate between positive and negative bags by discriminating between their corresponding feature vectors. Several techniques for mapping bags into single feature vectors are discussed in the next section.

Naive Bayes (NB) has proven effective in many practical applications, including text classification, medical diagnosis, and systems performance management [19,20,21]. In this work, we showed that NB classifier can be adapted for MIL setting. However, this adaptation imposes strong and unrealistic independence assumptions (instances within a bag are independent given the bag label and instance attributes are independent given the label of the instance). Alternatively, we propose MICCLLR, a generalized MIL algorithm that uses the class conditional log likelihood ratio statistics to map each bag into a single meta-instance. MICCLLR allows for any supervised learning algorithm to be the base classifier for classifying the meta-instance data. Our results evaluating MICCLLR using different base classifiers suggest that no single base classifier consistently outperforms other base classifiers on all data sets. Consequently, we show that a substantial improvement in performance is obtained using an ensemble of MICCLLR classifiers trained using different base learners. Additional gain in classification accuracy is obtained by applying AdaBoost.M1 [22] to weak MIC-CLLR learners. Overall, our results suggest that the predictive performance of the three proposed variants of MICCLLR are competitive to some of the state-of-the-art MIL methods on five widely used MIL data sets for drug activity prediction [1] and image retrieval [17] domains.

The rest of this paper is organized as follows: Section 2 summarizes the formulations of the MIL problem and overviews a number of related MIL methods that follow the same approach of adapting MIL representation for single-instance learning algorithms. Section 3 introduces our method. Section 4 gives our experimental results. Section 5 concludes with a brief summary and discussion.

2   Preliminaries

2.1   Multiple-Instance Learning Problem

In the standard (single-instance) supervised classifier learning scenario, each instance (input to the classifier) is typically represented by an ordered tuple of attribute values. The instance space $I = D_1 \times D_2 \times ... \times D_n$ where D_i is the domain of the ith attribute. The output of the classifier is a class label drawn from a set C of mutually exclusive classes. A training example is a labeled instance in the form $\langle X_i, c(X_i) \rangle$ where $X_i \in I$ and $c : I \rightarrow C$ is unknown function that assigns to an instance X_i its corresponding class label $c(X_i)$. For simplicity we consider only the binary classification problem in which $C = \{-1, 1\}$. Given a collection of training examples, $E = \{\langle X_1, c(X_1) \rangle, \langle X_2, c(X_2) \rangle, ..., \langle X_n, c(X_n) \rangle\}$, the goal of the (single-instance) learner is to learn a function c^* that approximates c as well as possible.

In the multiple-instance supervised classifier learning scenario, the goal is to train a (multiple-instance) classifier to label a *bag* of instances. Under standard MIL assumption, a bag is labeled negative if and only if all of its instances are negatively labeled and a bag is labeled positive if at least one of its instances is labeled positive. More precisely, Let B_i denotes the ith bag in a set of bags B. Let $X_{ij} \in I$ denotes the jth instance in the bag B_i and X_{ijk} be the value of the kth feature in the instance X_{ij}. The set of MI training examples, E_{MI}, is a collection of ordered pairs $\langle B_i, f(B_i) \rangle$ where f is unknown function that assigns to each bag B_i a class label $f(B_i) \in \{-1, 1\}$. Under the standard MIL assumption [1], $f(B_i) = -1$ iff $\forall_{X_{ij} \in B_i} c(X_{ij}) = -1$; and $f(B_i) = 1$ iff $\exists_{X_{ij} \in B_i}$, such that $c(X_{ij}) = 1$. Given E_{MI}, a collection of MI training examples, the goal of the multiple-instance learner is to learn a good approximation function of f. It should be noted that the function f is defined in terms of a function $c : I \rightarrow C$. However, learning c from the MI training data is challenging since we have labels only associated with bags and we do not have labels for each instance.

A generalization of the MIL problem has been considered by Weidmann et al. [13] and Tao et al. [14]. In this setting, all the bag instances contribute the label assigned to the bag and negative bags may contain some positive instances. Instead of a single concept, the generalized MIL problem considers a set of underlying concepts and requires a positive bag to have a certain number of instances in each of them.

2.2 Adapting MI Representation for Single-Instance Learning Algorithms

A number of existing methods for solving the MIL are based on the idea of adapting the MI representation for single-instance learning algorithms. Gartner et al. [23] mapped each bag into a single meta-instance using an aggregation function (e.g. mean, median, minimum, maximum, etc.) applied to each instance attribute. The resulting labeled meta-instances data set is then used to train an SVM classifier. Weidmann et al. [13] proposed a two-level classifier (TLC) trained from the data at two different levels of abstraction. The first classifier is trained from the MI data at the instance level by assigning the label of each bag to its instances and assigning a weight to each instance such that bags of different size will end up with the same weight. Then, the trained classifier is used to map MI data into a set of meta-instances and a second level supervised classifier is trained. In their experiments, Weidmann et al, [13] used a pruned decision tree and a Logit-boosted decision stumps (DS)[24] with 10 boosting iterations as the first and second level classifier, respectively.

Chen et al. [11] mapped each bag into a meta-instance in a feature space defined by a set of instance prototypes. An instance prototype is an instance that is close as possible to at least one instance in each positive bag and as far as possible from instances in negative bags. The algorithm, named DD-SVM, proceeds in two steps. First a collection of instance prototypes are learned such that each prototype is a local maximizer of the DD function. Second, each bag is mapped into a feature vector where the ith feature is defined by the minimum distance between the ith prototype and each instance in the bag. Finally, a standard SVM classifier is trained in the new feature space.

Recently, Chen et al. [15] proposed multiple-instance learning via embedded instance selection (MILES) which can be viewed as a variant of DD-SVM where the new feature space is defined by the set of all training instances instead of the set of prototypes used with DD-SVM. This feature mapping often provides a large number of irrelevant features. Therefore, 1-norm SVM is applied to select important features and construct classifiers simultaneously.

Zhou and Zhang [25] proposed constructive clustering-based ensemble (CCE) method where all the training instances are clustered into d groups. Then, each bag is mapped into a d binary vector, where the value of the ith feature is set to one if the concerned bag has instances falling into the ith group and zero otherwise. The above procedure is repeated for different values of d. For each value of d, a meta-instance representation of each bag is generated and an SVM classifier is trained. All the classifiers are then combined in an ensemble for prediction.

3   The Algorithm

We motivate our method by first introducing MI Naive Bayes (MINB), an adaptation of Naive Bayes (NB) classifier to MIL settings. The NB classification rule is defined by Eq. 1, where $Pr(c_j)$ is the a priori probability of class c_j and

$Pr(a_k|c_j)$ is the probability that the kth attribute of the instance X takes the value a_k given the class c_j.

$$c(X) = arg \max_{c_j \in C} Pr(c_j) \prod_{k=1}^{n} Pr(a_k|c_j) \qquad (1)$$

These probabilities, which completely specify a NB classifier, can be estimated from the training data using standard probability estimation methods based on relative frequencies of the corresponding classes and attribute value and class label cooccurrences observed in the data [20]. These relative frequencies summarize all the information relevant for constructing a NB classifier from a training data set, and hence constitute sufficient statistics for NB Classifier.

When the class labels are binary, that is, $C = \{-1, 1\}$, the NB classifier can be viewed as a linear discriminant by considering the logarithm of posterior odds as defined by Equations 2 and 3.

$$\phi(X) = ln\frac{Pr(c=1)}{Pr(c=-1)} + ln\frac{Pr(a_1|c=1)}{Pr(a_1|c=-1)} + \ldots + ln\frac{Pr(a_n|c=1)}{Pr(a_n|c=-1)} \qquad (2)$$

$$c(X) = \begin{cases} 1, & \phi(X) > 0 \\ -1, & otherwise \end{cases} \qquad (3)$$

Similarly, given unlabeled bag B_i with m_i instances, MINB assigns a label to B_i as follows:

$$\begin{aligned} c(B_i) &= arg \max_{c_j \in C} Pr(c_j|B_i) \\ &= arg \max_{c_j \in C} Pr(B_i|c_j)Pr(c_j) \\ &= arg \max_{c_j \in C} Pr(X_{i1}, X_{i2}, \ldots, X_{im_i}|c_j)Pr(c_j) \\ &= arg \max_{c_j \in C} Pr(c_j) \prod_{l=1}^{m_i} Pr(X_{il}|c_j) \end{aligned} \qquad (4)$$

The prior probabilities of labels, $Pr(c_j)$, can be easily estimated by counting the number of negative and positive bags. Recalling that instances within a bag are not labeled, estimating $Pr(X_{il}|c_j)$ is not trivial. In order to approximate $Pr(X_{il}|c_j)$, we first need to assign a label to each instance. Then, assuming independence between attributes given the instance class dramatically simplifies the computation of $Pr(X_{il}|c_j)$. That is, $Pr(X_{il}|c_j) = \prod_k Pr(X_{ilk}|c_j)$. Following the approach in [13], we construct a single instance training data set from the set of all instances contained in all bags, labeled with their bag's class label. Instances in a bag B_i are assigned a weight equal $\frac{1}{|B_i|} \cdot \frac{N}{M}$, where $N = \sum_i^m |B_i|$ and M denotes the number of bags in the training data set.

Based on these assumptions, the MINB classification rule can be defined as in Eq. 5, where $X_{ilk} = a_k$ denotes the value of the kth attribute in the lth instance in bag B_i with m_i instances where each instance is represented by an ordered tuple of n attribute values.

$$c(B_i) = arg \max_{c_j \in C} Pr(c_j) \prod_{l=1}^{m_i} \prod_{k=1}^{n} Pr(X_{ilk}|c_j) \qquad (5)$$

Alternatively, we can rewrite the MINB classifier as a linear discriminant:

$$\phi(B_i) = ln \frac{Pr(c=1)}{Pr(c=-1)} + ln \frac{Pr(X_{11}|c=1)}{Pr(X_{i1}|c=-1)} + \ldots + ln \frac{Pr(X_{im_i}|c=1)}{Pr(X_{im_i}|c=-1)} \qquad (6)$$

$$c(B_i) = \begin{cases} 1, & \phi(B_i) > 0 \\ -1, & otherwise \end{cases} \qquad (7)$$

Unfortunately, MINB has strong independence assumptions and its observed cross-validation performance on Musk data sets [1] is not competitive with the performance of the state-of-the art MIL methods (See Table 1). Instead of adapting NB for MIL setting, we propose to use NB to map the MI representation into a single meta-instance representation such that any standard supervised classification algorithm is applicable.

We now proceed to describe, MICCLLR, a MIL algorithm that uses class conditional log likelihood ratio (CCLLR) statistics estimated from the MI training data to map each bag into a single meta-instance. The pseudo code for MICCLLR is described in Algorithm 1. The input to the algorithm is a set of binary labeled bags E_{MI} and a base learner h. First, MICCLLR assigns the label of each bag to its instances and associate a weight with each instance to compensate for the fact that different bags may be of different sizes (i.e, different number of instances). Second, MICCLLR estimates the probability of each possible value for each attribute given the instance label. Under Naive Bayes assumption, the posterior probability of each attribute is independent of other attributes given the instance label. Therefore, the posterior probability of each attribute can be easily estimated from the training data using standard probability methods based on relative frequencies of each attribute value and class label occurrences observed in the labeled training instances [20]. Third, the algorithm uses the collected statistics to map each bag into a single meta-instance. Let $B_i = \{X_{i1}, \ldots, X_{im_i}\}$ be a bag of m_i instances. Each instance is represented by an ordered tuple of n attribute values. We define a function s that maps B_i into a single meta-instance of n real value attributes as; $s(B_i) = \{s_1, s_2, \ldots, s_n\}$ where each meta-instance attribute is computed using Eq. 8.

$$s_q = \frac{1}{m_i} ln \sum_{l=1}^{m_i} \frac{Pr(X_{ilq} = a_q|c=1)}{Pr(X_{ilq} = a_q)|c=-1)} \qquad (8)$$

Once bags in a multiple-instance data set have been transformed into meta-instances, the base learner h is trained on the transformed data set of labeled meta-instances. During the classification phase, each bag to be classified is first transformed into a meta-instance in a similar fashion before being fed to the base classifier h.

Algorithm 1. Training MICCLLR

1: **Input** : $E_{MI} = \{\langle B_1, y_1 \rangle, \ldots \langle B_m, y_m \rangle\}$ set of training bags and h base learner
2: Use E_{MI} to construct the collection of all instances E_{AV} by labeling each instance with its bag's class label and assign to instances in a bag B_i a weight equal to $\frac{1}{|B_i|} \cdot \frac{N}{M}$, where $N = \sum_i |B_i|$ and M denotes the number of bags in the training data set.
3: Estimate the posterior probabilities of each attribute, $Pr(a_q|c_j)$, from E_{AV}.
4: Convert each bag in E_{MI} to a single meta-instance $\{s_1, s_2, ..., s_n\}$ using Eq. 8.
5: Train the base learner h using the meta-instance data.

---

4 Experiments and Results

In our experiments, we implemented MINB and MICCLLR using Java and WEKA API [26]. The rest of classification algorithms considered in our experiments were used as implemented in WEKA. The default parameters for all WEKA classifiers were used unless otherwise specified. As a measure of the predictive performance of the MIL algorithms, we used the classification accuracy obtained by averaging the results of 10 different runs of 10-fold cross-validation tests. We conducted our experiments using five widely used MIL data sets from drug activity prediction [1] and content-based image retrieval (CBIR) [17] application domains.

Recently, Demšar [27] has suggested that non-parametric tests should be preferred over parametric tests for comparing machine learning algorithms because the non-parametric tests, unlike parametric tests, do not assume normal distribution of the samples (e.g., the data sets). Demšar suggested a three-step procedure for performing multiple hypothesis comparisons using non-parametric tests. Unfortunately, this procedure can not be applied to our experimental results because it requires the number of data sets to be greater than 10 and the number of methods to be greater than 5 [27]. However, as noted by Demšar [27], the average ranks by themselves provide a reasonably fair comparison of classifiers. Hence, the classifiers being compared are ranked on the basis of their observed performance on each data set. Then, the average rank of each classifier on all data sets is used to compare the overall performance of different MIL methods.

4.1 Comparison of Base Learners for MICCLLR

As mentioned above, MICCLLR uses class conditional log likelihood ratio statistics collected from the training data for mapping each bag of instances into a single-meta instance such any supervised base classifier becomes applicable. In our experiments, we evaluated MICLLR using a representative set of base classifiers. Specifically, we used Logistic Regression (LR) [28], C4.5 [29], Alternating Decision Trees (ADTree) [30], and 2-norm SVM (SMO) [31] classifiers SVML, SVMP, and SVMR evaluated using three kernels (linear, puk [32], and radial-bias function (RBF) kernel) (respectively) as base classifiers for MICCLLR. Table 1

Table 1. Comparisons of prediction accuracy of MINB with five MICCLLR classifiers evaluated using different base learners on Musk and CBIR data sets. Last row is the performance of an ensemble of the five MICCLLR classifiers constructed using WEKA's Vote method. For each data set, the rank of each classifier is shown in parentheses. Last column is the average performance and rank for each method over the five data sets.

Method	musk1	musk2	elephant	fox	tiger	avg.
MINB	77.06(8)	77.90(6)	81.70(1.5)	56.80(3)	72.00(6)	72.75(4.9)
LR	80.86(7)	85.28(2)	74.35(6.5)	55.8(5)	67.45(8)	72.75(5.7)
J48	87.71(3)	72.05(8)	74.35(6.5)	60.25(1)	73.2(5)	73.51(4.7)
ADTree	84.89(5)	75.35(7)	75.05(5)	59.25(2)	76.85(3)	74.28(4.4)
SVML	86.02(4)	82.20(3)	80.65(3)	52.10(7)	79.10(1)	76.01(3.6)
SVMP	88.39(2)	81.55(4)	75.90(4)	55.00(6)	76.50(4)	75.47(4)
SVMP	82.40(6)	80.53(5)	70.10(8)	50.25(8)	70.25(7)	70.71(6.8)
Vote	91.64(1)	86.12(1)	81.70(1.5)	56.15(4)	78.50(2)	78.82(1.9)

compares the classification accuracy of MINB and six MICCLLR classifiers evaluated using different base learners on Musk and CBIR data sets. Interestingly, MICCLLR classifiers with SVML, SVMP, ADTree, and J48 as base learners have better average ranks than MINB. The results also suggest that MICCLLR performance seems to be sensitive to the choice of the base classifier. However, none of the base classifiers produces a MICCLLR classifier with consistently superior performance on the five data sets. An ensemble of the five reported MICCLLR classifiers developed using WEKA implementation of majority voting, Vote classifier, outperforms any individual classifier on three out of five data sets. The ensemble of MICCLLR classifiers has the best average rank (1.6) followed by MICCLLR classifier using SVML and SVMP with average ranks 3.2 and 3.6, respectively. The predictive performance of the ensemble of MICCLLR classifiers, MICCLLR_Vote, could be further improved by: i) adding more MICCLLR classifiers utilizing other base learners to the ensemble; ii) using more sophisticated methods for constructing the ensemble (e.g., stacking [33]).

4.2   Boosting Weak MICCLLR Classifiers

Several methods for adapting boosting algorithms for the MIL settings have been proposed in the literature [34,35,36,37]. Xu and Frank [36] have noted that supervised learning boosting algorithms (e.g., AdaBoost.M1 [22]) can be applied directly to weak MIL learners. However, they did not compare their proposed MI boosting method with this basic approach. Here, we explored the utility of directly applying AdaBoost.M1 to MICCLLR and MIWrapper [38] weak learners. We compared the performance of the two MIL boosting algorithms implemented in WEKA, MIBoost [36] and MIOpimalBall [35], with AdaBoosted MICCLLR and MIWrapper classifiers. For MIOptimalBall, the weak learner constructs a ball such that at least one instance from each positive bag is included in the ball and all negative instances lie outside the ball. For MIBoost, we used

Table 2. Comparisons of classification accuracy of two MIBoosting algorithms (MIOptimalBall and MIBoost) with boosted MIWrapper and boosted MICCLLR. For MIOptimallBall the weak learner is a ball while for other methods C4.5 and DS were used as weak learners.

Data	MIOptimalBall	MIBoost		boosted MIWrapper		boosted MICCLLR	
		C4.5	DS	C4.5	DS	C4.5	DS
musk1	70.37(7)	84.07(3)	76.98(6)	85.53(2)	78.13(5)	88.57(1)	82.49(4)
musk2	80.80(2)	80.55(3)	74.68(6)	81.61(1)	72.53(7)	79.18(4)	78.03(5)
elephant	72.00(7)	82.05(3)	81.65(4)	86.15(1)	80.75(5)	82.8(2)	78.5(6)
fox	54.60(7)	64.85(1)	62.95(2)	62.8(3)	62.65(4)	62.15(5)	58.8(6)
tiger	65.15(7)	78.95(6)	80.25(4)	81.2(3)	79.25(5)	82.5(1)	81.6(2)
Avg.	68.58(6)	78.09(3.2)	75.3(4.4)	79.46(2)	74.66(5.2)	79.04(2.6)	75.88(4.6)

decision stumps (DS) [24] and C4.5 as the week learners with 25 iterations. For AdaBoost.M1, MICCLLR and MIWrapper weak learners were obtained using DS and C4.5 as the base classifiers and the number of boosting iterations was set to 25.

Table 2 shows that boosted MIWrapper and boosted MICCLLR with C4.5 have the best average ranks of 2 and 2.6, respectively. The results show that boosted MIWrapper and boosted MICCLLR classifiers are competitive with (if not outperforming) the two MIL boosting methods, MIOptimalBall and MIBoost. Interestingly, we observed that MIBoost, boosted MIWrapper, and boosted MICCLLR with C4.5 as the weak learner generally outperform their counterpart classifiers with DS as the weak learner. Among the classifiers using C4.5 as the weak learner, boosted MIWrapper has the best average rank while boosted MICCLLR has the best average rank if we limit our comparison to methods with DS as the weak learner.

4.3   Comparison of MICCLLR to Other MIL Methods

The classification accuracy of the best performing three MICCLLR classifiers, MICCLLR with SVM base learner trained using linear kernel (`MICCLLR_SVML`), ensemble of MICCLLR classifiers (`MICCLLR_Vote`), and boosted `MICCLLR_C4.5` was compared to existing MIL with reported performance on the five data sets considered in this study (See Table 3). The average ranks for the three MICCLLR classifiers are boosted `MICCLLR_C4.5` (4), `MICCLLR_Vote` (5.4), and `MICCLLR_SVML` (8.4). Hence, boosted `MICCLLR_C4.5` improves the predictive performance over the majority vote ensemble of MICCLLR classifiers and the single MICCLLR classifier with SVM as the base learner. The results suggest that boosted `MICCLLR_C4.5` is also competitive in performance with the state-of-the-art MIL methods on Musk and CBIR data sets. The best performing three methods, as measured by the average rank of the classifier, are CH-FD [6], RW-SVM [39], and boosted `MICCLLR_C4.5` with average ranks 3.4, 3.6, and 4, respectively.

Table 3. Comparison of the classification accuracy of three MICCLLR classifiers with different MIL methods on Musk and CBIR data sets

Method	musk1	musk2	elephant	fox	tiger	avg.
EM-DD [17]	84.50(9)	84.90(6)	78.30(10)	56.10(9)	72.10(10)	75.18(8.8)
mi-SVM [17]	87.40(7)	83.60(8)	82.20(4)	58.20(6)	78.9(8)	78.06(6.6)
MI-SVM [17]	77.90(10)	84.30(7)	81.40(7)	59.40(4)	84.00(1)	77.40(5.8)
MICA [40]	88.40(5)	90.50(1)	80.50(9)	58.70(5)	82.60(2)	80.14(4.4)
CH-FD [6]	88.80(3)	85.70(5)	82.40(3)	60.40(2)	82.20(4)	79.90(3.4)
I-DD [41]	90.80(2)	86.40(3)	81.50(6)	57.30(7)	80.70(5)	79.34(4.6)
RW-SVM [39]	87.60(6)	87.10(2)	83.30(1)	60.00(3)	79.50(6)	79.50(3.6)
MICCLLR_SVML	86.02(8)	82.20(9)	80.65(8)	52.10(10)	79.10(7)	76.01(8.4)
MICCLLR_Vote	91.64(1)	86.12(4)	81.70(5)	56.15(8)	78.50(9)	78.82(5.4)
boosted MICCLLR_C4.5	88.57(4)	79.18(10)	82.80(2)	62.15(1)	82.50(3)	79.04(4.0)

5   Conclusions

We introduced MINB, an adaptation of Naive Bayes for the MIL settings. We showed that the proposed MINB algorithm imposes strong and unrealistic independence assumptions (instances within a bag are independent given the bag label and instance attributes are independent given the label of the instance). We empirically showed that class conditional log likelihood ratio statistics estimated from the training data provide useful single feature representation of bags that allows the applicability of standard supervised learning methods (base learners) for predicting labels of MIL bags given their single feature vector representation as an input. The performance of our proposed method, MICCLLR, has been evaluated using different base learners. Moreover, we empirically showed that further improvements in MICCLLR performance is obtained using ensemble of MICCLLR classifiers utilizing different base learners. Finally, we demonstrated that an additional gain in classification accuracy is obtained when AdaBoost.M1 is applied directly to weak MIL learner derived from MIWrapper and MICCLLR using C4.5 as the base learner. Our results suggest that integrating AdaBoost.M1 with MIWrapper and MICCLLR weak learners is a promising approach for developing MIL methods with improved prediction performance.

References

1. Dietterich, T., Lathrop, R., Lozano-Pérez, T.: Solving the multiple instance problem with axis-parallel rectangles. Artif. Intell. 89, 31–71 (1997)
2. Wang, C., Scott, S., Zhang, J., Tao, Q., Fomenko, D., Gladyshev, V.: A Study in Modeling Low-Conservation Protein Superfamilies. Technical Report TR-UNL-CSE-2004-3, Dept. of Computer Science, University of Nebraska (2004)
3. Maron, O., Ratan, A.: Multiple-instance learning for natural scene classification. In: Proceedings of the Fifteenth International Conference on Machine Learning table of contents, pp. 341–349 (1998)

4. Zhang, Q., Goldman, S., Yu, W., Fritts, J.: Content-based image retrieval using multiple-instance learning. In: Proceedings of the Nineteenth International Conference on Machine Learning, pp. 682–689 (2002)
5. Bi, J., Chen, Y., Wang, J.: A sparse support vector machine approach to region-based image categorization. In: 2005 IEEE Computer Society Conference on Computer Vision and Pattern Recognition, pp. 1121–1128 (2005)
6. Fung, G., Dundar, M., Krishnapuram, B., Rao, R.: Multiple instance learning for computer aided diagnosis. In: Advances in Neural Information Processing Systems: Proceedings of the 2006 Conference, pp. 425–432. MIT Press, Cambridge (2007)
7. Ramon, J., De Raedt, L.: Multi instance neural networks. In: Proceedings of the ICML 2000 Workshop on Attribute-Value and Relational Learning (2000)
8. Wang, J., Zucker, J.D.: Solving the multiple-instance problem: a lazy learning approach. In: Proceedings 17th International Conference on Machine Learning, pp. 1119–1125 (2000)
9. Maron, O., Lozano-Perez, T.: A framework for multiple-instance learning. Adv. Neural. Inf. Process. Syst. 10, 570–576 (1998)
10. Zhang, Q., Goldman, S.A.: Em-dd: An improved multiple-instance learning technique. Neural. Inf. Process. Syst. 14 (2001)
11. Chen, Y., Wang, J.: Image categorization by learning and reasoning with regions. J. Mach. Learn. Res. 5, 913–939 (2004)
12. Ray, S., Craven, M.: Supervised versus multiple instance learning: An empirical comparison. In: Proceedings of the Twentieth-Second International Conference on Machine Learning, pp. 697–704 (2005)
13. Weidmann, N., Frank, E., Pfahringer, B.: A two-level learning method for generalized multi-instance problems. In: Lavrač, N., Gamberger, D., Todorovski, L., Blockeel, H. (eds.) ECML 2003. LNCS (LNAI), vol. 2837, pp. 468–479. Springer, Heidelberg (2003)
14. Tao, Q., Scott, S., Vinodchandran, N., Osugi, T.: SVM-based generalized multiple-instance learning via approximate box counting. In: Proceedings of the twenty-first international conference on Machine learning. ACM Press, New York (2004)
15. Chen, Y., Bi, J., Wang, J.: MILES: multiple-instance learning via embedded instance selection. IEEE Trans. Pattern Anal. Mach. Intell. 28, 1931–1947 (2006)
16. Zhou, Z.: Multi-instance learning from supervised view. Journal of Computer Science and Technology 21, 800–809 (2006)
17. Andrews, S., Tsochantaridis, I., Hofmann, T.: Support vector machines for multiple-instance learning. In: Adv. Neural. Inf. Process. Syst., vol. 15, pp. 561–568. MIT Press, Cambridge (2003)
18. Zhang, M., Zhou, Z.: Adapting RBF neural networks to multi-instance learning. Neural Process. Lett. 23, 1–26 (2006)
19. Domingos, P., Pazzani, M.: On the optimality of the simple Bayesian classifier under zero-one loss. Machine learning 29, 103–130 (1997)
20. Mitchell, T.: Machine Learning. McGraw-Hill, New York (1997)
21. Hellerstein, J., Jayram, T., Rish, I.: Recognizing End-User Transactions in Performance Management. In: Proceedings of the Seventeenth National Conference on Artificial Intelligence and Twelfth Conference on Innovative Applications of Artificial Intelligence, pp. 596–602. AAAI Press/The MIT Press (2000)
22. Freund, Y., Schapire, R.: Experiments with a new boosting algorithm. In: Proceedings of 13th International Conference in Machine Learning, pp. 148–156 (1996)
23. Gartner, T., Flach, P., Kowalczyk, A., Smola, A.: Multi-instance kernels. In: Proceedings of the 19th International Conference on Machine Learning, pp. 179–186 (2002)

24. Friedman, J., Hastie, T., Tibshirani, R.: Special invited paper. additive logistic regression: A statistical view of boosting. Annals of statistics, 337–374 (2000)
25. Zhou, Z., Zhang, M.: Solving multi-instance problems with classifier ensemble based on constructive clustering. Knowl. Inf. Syst. 11, 155–170 (2007)
26. Witten, I., Frank, E.: Data mining: Practical machine learning tools and techniques, 2nd edn. Morgan Kaufmann, San Francisco (2005)
27. Demšar, J.: Statistical comparisons of classifiers over multiple data sets. J. Mach. Learn. Res. 7, 1–30 (2006)
28. Le Cessie, S., Van Houwelingrn, J.: Ridge estimators in logistic regression. Appl. Stat. 41, 191–201 (1992)
29. Quinlan, J.: C4. 5: programs for machine learning. Morgan Kaufmann, San Francisco (1993)
30. Freund, Y., Mason, L.: The alternating decision tree learning algorithm. In: Proceedings of the Sixteenth International Conference on Machine Learning table of contents, pp. 124–133. Morgan Kaufmann, San Francisco (1999)
31. Platt, J.: Fast training of support vector machines using sequential minimal optimization. MIT Press, Cambridge (1998)
32. Ustün, B., Melssen, W., Buydens, L.: Facilitating the application of Support Vector Regression by using a universal Pearson VII function based kernel. Chemometrics and Intelligent Laboratory Systems 81, 29–40 (2006)
33. Wolpert, D.: Stacked generalization. Neural Networks 5, 241–259 (1992)
34. Andrews, S., Hofmann, T.: Multiple instance learning via disjunctive programming boosting. In: 2003 Conference on Advances in Neural Information Processing Systems, pp. 65–72. Bradford Book (2004)
35. Auer, P., Ortner, R.: A Boosting Approach to Multiple Instance Learning. In: Boulicaut, J.-F., Esposito, F., Giannotti, F., Pedreschi, D. (eds.) ECML 2004. LNCS (LNAI), vol. 3201, pp. 63–74. Springer, Heidelberg (2004)
36. Xu, X., Frank, E.: Logistic regression and boosting for labeled bags of instances. In: Dai, H., Srikant, R., Zhang, C. (eds.) PAKDD 2004. LNCS (LNAI), vol. 3056, pp. 272–281. Springer, Heidelberg (2004)
37. Viola, P., Platt, J., Zhang, C.: Multiple Instance Boosting for Object Detection. In: Advances in Neural Information Processing Systems, pp. 1417–1424 (2006)
38. Frank, E., Xu, X.: Applying propositional learning algorithms to multi-instance data. Technical report, Dept. of Computer Science, University of Waikato (2003)
39. Wang, D., Li, J., Zhang, B.: Multiple-instance learning via random walk. In: Fürnkranz, J., Scheffer, T., Spiliopoulou, M. (eds.) ECML 2006. LNCS (LNAI), vol. 4212, p. 473. Springer, Heidelberg (2006)
40. Mangasarian, O., Wild, E.: Multiple instance classification via successive linear programming. Data Mining Institute Technical Report 05-02 (2005)
41. Han, F., Wang, D., Liao, X.: An Improved Multiple-Instance Learning Algorithm. In: 4th international symposium on Neural Networks, pp. 1104–1109. Springer, Heidelberg (2007)

On the Complexity of Constraint-Based Theory Extraction

Mario Boley and Thomas Gärtner

Fraunhofer IAIS, Schloss Birlinghoven, Sankt Augustin, Germany
{mario.boley,thomas.gaertner}@iais.fraunhofer.de

Abstract. In this paper we rule out output polynomial listing algorithms for the general problem of discovering theories for a conjunction of monotone and anti-monotone constraints as well as for the particular subproblem in which all constraints are frequency-based. For the general problem we prove a concrete exponential lower time bound that holds for any correct algorithm and even in cases in which the size of the theory as well as the only previous bound are constant. For the case of frequency-based constraints our result holds unless **P** = **NP**. These findings motivate further research to identify tractable subproblems and justify approaches with exponential worst case complexity.

1 Introduction

Many problems in knowledge discovery in databases can be formulated as theory extraction task, i.e., as the problem of extracting all sentences of a pattern language that satisfy some interestingness constraints relative to a database [11]. Many instances of this problem have been considered in literature and examples include subgroup discovery, finding consistent hypotheses, finding minimal keys, as well as different pattern mining problems such as frequent set or subgraph mining. Most research in the area of theory extraction has considered a single, either monotone or anti-monotone, constraint such as frequency in a database. Motivated by important real-world applications such as drug discovery, however, more complex interestingness constraints recently became relevant [4,17]. An often considered case is theory extraction for conjunctions of monotone and anti-monotone constraints. Practical approaches to this problem include the levelwise version space algorithm [18], DualMiner [6], ExAMiner [2], ExAnte [3], and BifoldLeap [8].

In contrast to this considerable number of algorithmic contributions, the problem's computational complexity has, so far, only received little attention. Negative complexity results are important as guidance for the development of new theory extraction algorithms and to provide a theoretical justification for the absence of good worst-case guarantees of existing approaches. Bucila et al. [6] have shown that the number of constraint evaluations needed by any correct general algorithm is lower bounded by the size of the border intersection, i.e., all patterns that are contained in the theory of one constraint as well as in the border of the other. As this quantity can grow exponentially with the problem size and with the size of the theory, this implies that there is no correct general algorithm with runtime polynomial in the input and output. These results

J. Gama et al. (Eds.): DS 2009, LNAI 5808, pp. 92–106, 2009.

do, however, not exclude a correct general algorithm with runtime polynomial in the input, output, and the size of the border intersection. Yet, no such algorithm is known[1], leaving a large gap between the runtime of the practical algorithms and the theoretical lower bound. The *first contribution* of this paper is to significantly reduce this gap for the case of general algorithms. We construct instances of this problem for which both, the size of the theory and the size of the border intersection, are constant. Yet, we show that no correct general algorithm can extract the theory for these instances with a number of constraint evaluations bound by a polynomial in the input. While this reduces the gap between theory and practice for the general problem, there might still be tractable subclasses of the general problem. In particular, specialized algorithms may be more efficient by exploiting the structure of the subclass they consider.

The most commonly used constraints in knowledge discovery are frequency-based, i.e., patterns are required to either be frequent or infrequent with respect to the database. Such constraints are important for instance in the case that the extracted theories are input to learning algorithms as defining the feature set (see, e.g., [20]), as they generalize the problem of consistent hypothesis enumeration for which efficient algorithms are known [16], and as they can be used for explanatory data mining [18]. Despite the importance of such constraints, the complexity of theory extraction for conjunctions of frequency and infrequency constraints has not been investigated. However, it can be shown that this problem is strictly easier than the general problem and hence negative results do not carry over. This observation highlights yet another gap in the complexity analysis of constraint-based theory extraction. The *second contribution* of this paper is to close this gap by a hardness result. In particular, we proof that there is no algorithm extracting all itemsets frequent (with an arbitrary threshold) in one database but not frequent (with another arbitrary threshold) in another database in time polynomial in the input and the output, unless $\mathbf{P} = \mathbf{NP}$.

In contrast to some other complexity results in pattern mining that rule out efficient algorithms for optimization or counting variants for specific pattern classes (e.g., [22,23]), we directly focus on the primary data mining task, which is *listing* all patterns of interest. For listing problems that involve a potentially exponential number of result patterns one usually aims for algorithms having a good time complexity *per pattern* or at least a polynomial time complexity in the combined size of the input and the output, i.e., an output polynomial algorithm. Thus, instead of **NP**-hardness or **#P**-hardness results, which only rule out (input) polynomial algorithms for listing, we are interested in statements with negative implications for output polynomial listing.

2 Preliminaries

This section recalls the basic definition of data mining as theory extraction, frequent set mining, as well as listing algorithms and their notions of efficiency.

Theory Extraction The general formulation of theory extraction can for instance be found in [11]. In this work we use the following conventions: (i) we only consider *finite*

---

[1] This statement refers to the general case. Note that there are theoretically efficient algorithms for instance for the special case of string patterns (see, e.g., [9]).

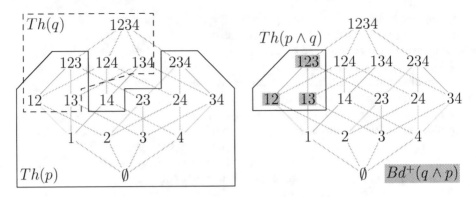

Fig. 1. illustration of Example 1

theory extraction problems that in addition are *representable as sets*, and (ii) we do not explicitly formalize the connection of constraints to a given input database. One can think of this connection as "encapsulated" in the constraints (see the definitions of the constraints frq and ifr in the paragraph about frequent set mining below as an example). This allows us to proceed with simplified definitions. The *pattern language* is always the power set $\mathcal{P}(E)$ of some finite ground set E, and without losing generality we assume that $E = \{1, \ldots, n\}$ for some positive integer n. A *constraint* is then a boolean function $q : \mathcal{P}(E) \rightarrow \{0, 1\}$ and the *theory* of q, denoted as $\mathrm{Th}(q)$, is the set of all elements of $\mathcal{P}(E)$ that satisfy q, i.e., $\mathrm{Th}(q) = \{F \subseteq E : q(F) = 1\}$. A constraint q is called *anti-monotone* if $F' \subseteq F \subseteq E$ and $q(F) = 1$ together imply $q(F') = 1$. Similar, q is called *monotone* if $F' \subseteq F \subseteq E$ and $q(F') = 1$ together imply $q(F) = 1$. The *positive border* of q, denoted $\mathrm{Bd}^+(q)$, is the family of maximal and minimal sets of $\mathrm{Th}(q)$, i.e., $\mathrm{Bd}^+(q) = \mathrm{Up}^+(q) \cup \mathrm{Lw}^+(q)$ with

$$\mathrm{Up}^+(q) = \{F \in \mathcal{P}(E) : q(F) = 1 \wedge \forall e \in E \setminus F, q(F \cup \{e\}) = 0\}$$
$$\mathrm{Lw}^+(q) = \{F \in \mathcal{P}(E) \setminus \{\emptyset\} : q(F) = 1 \wedge \forall e \in F, q(F \setminus \{e\}) = 0\} .$$

Conversely, the *negative border* of q, denoted $\mathrm{Bd}^-(q)$, is the family of maximal and minimal sets of $\mathcal{P}(E) \setminus \mathrm{Th}(q)$, i.e., $\mathrm{Bd}^-(q) = \mathrm{Up}^-(q) \cup \mathrm{Lw}^-(q)$ with

$$\mathrm{Up}^-(q) = \{F \in \mathcal{P}(E) : q(F) = 0 \wedge \forall e \in F, q(F \setminus \{e\}) = 1\}$$
$$\mathrm{Lw}^-(q) = \{F \in \mathcal{P}(E) \setminus \{E\} : q(F) = 0 \wedge \forall e \in E \setminus F, q(F \cup \{e\}) = 1\} .$$

The *border* of q is then defined as $\mathrm{Bd}(q) = \mathrm{Bd}^+(q) \cup \mathrm{Bd}^-(q)$. Note that in contrast to the definitions of [11] and [6] our notion of border is similar for anti-monotone and monotone constraints as well as for constraints that do not satisfy any monotonicity condition. It is, however, equivalent to the traditional definition for the cases of anti-monotone and monotone constraints. For conjunctions $c = p \wedge q$ of a monotone constraint p and an anti-monotone constraints q, the positive border $\mathrm{Bd}^+(c)$ concisely represents the theory of c because for all $F \subseteq E$ it holds that

$$F \in \mathrm{Th}(c) \iff (\exists L \in \mathrm{Lw}^+(c), L \subseteq F) \wedge (\exists U \in \mathrm{Up}^+(c), U \subseteq F) .$$

Frequent Set Mining. A *dataset* over *items* E is a finite multiset $\mathcal{D}$ containing subsets of E, i.e., $\mathcal{D} \subseteq \mathcal{P}(E)$ is a set with associated multiplicities $\lambda : \mathcal{D} \to \mathbb{N}$. The elements of $\mathcal{D}$ are usually called *transactions*. The cardinality of a submultiset $\mathcal{D}' \subseteq \mathcal{D}$ is defined as $|\mathcal{D}'| = \sum_{D \in \mathcal{D}'} \lambda(D)$. Let $F \subseteq E$ be a set of items and $t \in \mathbb{N}$ a *frequency threshold*. The *support set* of F in a dataset $\mathcal{D}$, denoted as $\mathcal{D}[F]$, is the submultiset of $\mathcal{D}$ containing all transactions that are a superset of F, i.e., $\mathcal{D}[F] = \{D \in \mathcal{D} : D \supseteq F\}$. The set F is called t-*frequent* in $\mathcal{D}$ if $|\mathcal{D}[F]| \geq t$, and it is called t-*infrequent* in $\mathcal{D}$ if it is not t-frequent in $\mathcal{D}$. Clearly, the constraints $\mathrm{frq}_{\mathcal{D},t}, \mathrm{ifr}_{\mathcal{D},t} : \mathcal{P}(E) \to \{0,1\}$ defined by

$$\mathrm{frq}_{\mathcal{D},t}(F) = 1 \text{ iff } F \text{ is } t\text{-frequent in } \mathcal{D}$$

$$\mathrm{ifr}_{\mathcal{D},t}(F) = 1 \text{ iff } F \text{ is } t\text{-infrequent in } \mathcal{D}$$

are anti-monotone respectively monotone. As the *size* of a dataset $\mathcal{D}$ we regard the sum of its transaction sizes, i.e., $\mathrm{size}(\mathcal{D}) = \sum_{D \in \mathcal{D}} \lambda(D) |D|$, corresponding to an incidence list representation of $\mathcal{D}$.

Listing Algorithms. A *listing problem* can be formalized as follows: given an instance $x \in X$ where X is the set of all problem instances, list some associated finite set $S(x)$. Naturally, we say that an algorithm $\mathcal{A}$ is a *listing algorithm* for that problem (solves the associated listing problem) if for all $x \in X$ the algorithm's output $\mathcal{A}(x)$ is equal to $S(x)$. Two important notions of efficiency for listing algorithms are:

- *output polynomial time*, i.e., there is a polynomial p such that for all $x \in X$ the number of computational steps that $\mathcal{A}$ performs on input x, denoted as $\mathrm{time}_{\mathcal{A}}(x)$, is bounded by $p(\mathrm{size}(x) + |S(x)|)$,
- *polynomial delay*, i.e., there is a polynomial p such that for all inputs $x \in X$ the number of computational steps that $\mathcal{A}$ performs before printing the first element, after printing the last element, and between printing two consecutive elements is bounded by $p(\mathrm{size}(x))$.

Clearly, polynomial delay is a strictly stronger condition than output polynomial time. For instance dfs-algorithms for frequent set mining like Fpgrowth [12] run with polynomial delay, while bfs-algorithms like Apriori [1] run in output polynomial time but do not guarantee polynomial delay. We note that there are also other notions of efficiency for listing algorithms—in particular incremental polynomial time [13] and cumulative polynomial delay [10]—both of which are stronger than output polynomial time but weaker than polynomial delay.

For theory extraction algorithms that expect constraints as input we are also interested in the *communication complexity*, denoted as $\mathrm{comm}_{\mathcal{A}}(n)$, i.e., the maximum number of evaluations of the constraints that $\mathcal{A}$ performs when given an input of size n. Clearly, the communication complexity is a lower bound to an algorithm's time complexity. Moreover, for many tasks in data mining and database system the communication complexity is the actual quantity of interest because the cost of constraint evaluations dominates the running time.

We close this section with an example illustrating some of the introduced notions. Figure 2 depicts the involved theories.

Example 1. *Consider the datasets $\mathcal{D}_1$ and $\mathcal{D}_2$ over items $E = \{1, \ldots, 4\}$ given by*

$$\mathcal{D}_1 = \{123, 234, 1234\}$$
$$\mathcal{D}_2 = \{124, 134, 234, 234\}$$

with the constraints $p = frq_{\mathcal{D}_1,2}$ and $q = ifr_{\mathcal{D}_2,2}$. The resulting positive and negative borders are $Bd^+(q) = \{12, 13\}$, $Bd^-(q) = \{14, 234\}$, $Bd^+(p) = \{123, 234\}$, and $Bd^-(p) = \{14\}$. The theory of the conjunction $Th(p \wedge q)$ is $\{12, 13, 123\}$. Consequently, an output polynomial theory extraction algorithm with respect to a polynomial p would terminate after a number of steps not greater than

$$p(\text{size}(\mathcal{D}_1, \mathcal{D}_2) + |Th(p \wedge q)|) = p(22 + 3) \ .$$

3   General Problem Lower Bound

In this section we prove an exponential lower bound on the worst-case number of constraint evaluations, i.e., the communication complexity, that is needed to solve the general conjunctive constraint problem:

Problem 1 (LIST-CONJUNCTIVE-THEORY). Given a finite set E with an anti-monotone constraint p and a monotone constraint q each defined on $\mathcal{P}(E)$, list $\text{Th}(p \wedge q)$.

More specifically, we consider only deterministic algorithms for that problem, i.e., those whose behavior is completely specified by the sequence of queries and their corresponding evaluations. Moreover, we are aiming for a lower bound that also holds when Problem 1 is restricted to instances that satisfy

$$k \geq |\text{Th}(p \wedge q)| \text{ and} \tag{1}$$
$$k' \geq |\text{Th}(p) \cap \text{Bd}(q)| + |\text{Th}(q) \cap \text{Bd}(p)| \tag{2}$$

for some constants k, k'. An exponential lower bound that holds for the thus defined subproblem would rule out algorithms having a communication complexity that is polynomially bounded in the right-hand side of Equations 1 and 2, respectively. For Condition 1 this is motivated by the fact that the communication complexity of an algorithm is a lower bound to its time complexity, and consequently this would also rule out an output polynomial algorithm for Problem 1. For Condition 2 this is motivated by the fact that complexity result and algorithm of [6] leave it open whether Problem 1 can be solved efficiently in the right-hand side of (2).

The idea of the proof is simple: we consider an algorithm that does not meet the lower bound and then construct two problem instances that have a different theory but "cannot be distinguished" by the algorithm. Consequently, it has to output an incorrect theory for at least one of the two problem instances.

A first approach for such a construction could be the following. For a ground set $E = \{1, \ldots, n\}$ with an even n consider the constraints

$$q(F) = 1 \text{ iff } |F| \leq n/2 - 1$$
$$p(F) = 1 \text{ iff } |F| \geq n/2 + 1 \ .$$

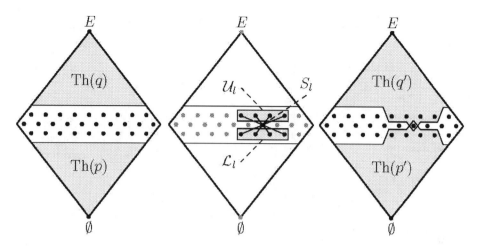

Fig. 2. construction used in proof of Theorem 1

Then, for any algorithm $\mathcal{A}$ invoking less than $\binom{n}{n/2}$ queries there is a subset $S \subseteq E$ of size $n/2$ that $\mathcal{A}$ neither queries by p nor by q. With this we can construct a second problem instance

$$p'(F) = 1 \text{ iff } p(F) \vee (F = S)$$
$$q'(F) = 1 \text{ iff } q(F) \vee (F = S) \ .$$

As the only set in which the two instances differ is not queried by $\mathcal{A}$, the algorithm's behavior and in particular its output must be identical in both cases. But

$$\text{Th}(p \wedge q) = \{\} \neq \{S\} = \text{Th}(p' \wedge q')$$

as required to show that $\mathcal{A}$ is incorrect.

The problem with the construction above is that the attained lower bound on the communication complexity is equal to the border intersection, i.e., $\binom{n}{n/2}$. Thereby it does not rule out an algorithm that lists $\text{Th}(p \wedge q)$ efficiently in that quantity. For that purpose problem instances are needed that have small—ideally constant— border intersections while they retain the property of proving the incorrectness of $\mathcal{A}$.

For that reason we modify the construction such that not only one but three levels of the first problem instance are left unoccupied. The resulting theories then have an empty border intersection but, in turn, this requires the addition of more than one set to $\text{Th}(p')$ and $\text{Th}(q')$ as to not violate the monotonicity requirements. The formal theorem statement and proof are given below. Figure 3 illustrates the construction.

Theorem 1. *Let $\mathcal{A}$ be a deterministic algorithm that correctly solves* LIST-CONJUNCTIVE-THEORY. *Then for problem instances with $|E| = n > 4$ it holds for the communication complexity of $\mathcal{A}$ that*

$$\text{comm}_{\mathcal{A}}(n) \geq 2^{n/4} \ .$$

This lower bound even holds when the problem is further restricted to instances satisfying

$$1 \geq |Th(p \wedge q)| \quad and \tag{3}$$

$$1 \geq |Th(p) \cap Bd(q)| + |Th(q) \cap Bd(p)| \quad . \tag{4}$$

Proof. Assume for a contradiction that there is a deterministic algorithm $\mathcal{A}$ that correctly solves the restricted problem defined by Equations 3 and 4 such that the worst-case number of constraint queries satisfies

$$\mathrm{comm}_{\mathcal{A}}(n) < 2^{n/4} < \binom{n/2}{n/4}$$

We show that $\mathcal{A}$ is incorrect by constructing two inputs such that $\mathcal{A}$ fails to compute the correct theory for at least one of them.

Without loss of generality assume that n is divisible by 4. Then consider the "uniform" predicates p and q defined on the power set $\mathcal{P}(E)$ of the ground set $E = \{1, \dots, n\}$ by

$$q(F) = 1 \quad \text{iff} \quad |F| \leq n/2 - 2$$
$$p(F) = 1 \quad \text{iff} \quad |F| \geq n/2 + 2 \quad .$$

Clearly p is anti-monotone and q is monotone. Moreover, their borders are simply

$$Bd(p) = \{F \subseteq E \colon |F| \in \{n/2 - 1, n/2 - 2\}\}$$
$$Bd(q) = \{F \subseteq E \colon |F| \in \{n/2 + 1, n/2 + 2\}\}$$

and in particular the theory of their conjunction is empty, i.e., $Th(p \wedge q) = \emptyset$. Thus, these predicates satisfy Equations 3 and 4.

Next, we construct $k = \binom{n/2}{n/4}$ subsets of cardinality $n/2$ having a pairwise symmetric difference of at least 4. Such a family of subsets is for instance given by the family $\mathcal{S}$ defined by

$$\mathcal{S} = \{F \cup F^- \colon F \subseteq \{2, 4, \dots, n\}, |F| = n/4\}$$

where F^- denotes the set $\{e - 1 \colon e \in F\}$. Note that we have $|\mathcal{S}| = k$ and for distinct sets $S, S' \in \mathcal{S}$ it holds for their symmetric difference that $|S \Delta S'| \geq 4$ as desired. Then let $\{S_1, \dots, S_k\} = \mathcal{S}$ be some ordering of $\mathcal{S}$. For each S_i we define the family of its immediate predecessors respectively immediate successors as follows:

$$\mathcal{L}_i = \{S_i \setminus \{s\} \colon s \in S_i\}$$
$$\mathcal{U}_i = \{S_i \cup \{e\} \colon e \in E \setminus S_i\} \quad .$$

Let $L_i \in \mathcal{L}_i$ and $L_j \in \mathcal{L}_j$ with $i \neq j$. Then

$$|L_i \Delta L_j| \geq |S_i \Delta S_j| - 2 \geq 2 \quad .$$

This implies $L_i \neq L_j$ and consequently $\mathcal{L}_i \cap \mathcal{L}_j = \emptyset$. Similar we know that the families $\mathcal{U}_i$ are pairwise disjoint and, consequently, that the combined families

$$\mathcal{C}_i = \{S_i\} \cup \mathcal{L}_i \cup \mathcal{U}_i$$

do not intersect. As there are $k > \mathrm{comm}_{\mathcal{A}}(n)$ such families and the algorithm $\mathcal{A}$ invokes the membership oracles only $\mathrm{comm}_{\mathcal{A}}(n)$ times, it follows then that there is an index $l \in \{1, \ldots, k\}$ such that $\mathcal{A}$ does not access any element of the family $\mathcal{C}_l$.

We use this family $\mathcal{C}_l$ to construct a second problem instance p', q' with the same ground set E that for $\mathcal{A}$ "cannot distinguish" from the first one.

$$p'(F) = 1 \text{ iff } p(F) \vee (F \in \{S_l\} \cup \mathcal{L}_l)$$
$$q'(F) = 1 \text{ iff } q(F) \vee (F \in \{S_l\} \cup \mathcal{U}_l) .$$

Observe that p' is anti-monotone: S_l is the only element of $\mathrm{Th}(p')$ with cardinality $n/2$ and all of its $n/2 - 1$ subsets are in $\mathcal{L}_l \subset \mathrm{Th}(p')$. Moreover all subsets $F \subseteq E$ of cardinality $n/2 - 2$ and less are in $\mathrm{Th}(p) \subset \mathrm{Th}(p')$. The anti-monotonicity of p' follows, and, similar, one can check that q' is monotone.

Moreover, it is straightforward to check that S_l is the only element of $\mathcal{P}(E) \setminus \mathrm{Bd}(p)$ respectively $\mathcal{P}(E) \setminus \mathrm{Bd}(q)$ that satisfies the defining formula of being a border element of p' respectively q'. It follows that $\mathrm{Bd}(p') = \{S_l\} \cup \mathcal{B}_1$ and $\mathrm{Bd}(q') = \{S_l\} \cup \mathcal{B}_2$, where $\mathcal{B}_1$ contains only elements of cardinality less than $n/2$, and $\mathcal{B}_2$ contains only elements of cardinality greater than $n/2$. Consequently,

$$\mathrm{Bd}(p') \cap \mathrm{Th}(q') = \mathrm{Bd}(q') \cap \mathrm{Th}(p') = \mathrm{Th}(p' \wedge q') = \{S_l\}$$

and in particular p', q' is an instance of the restricted problem defined by Equations 3 and 4.

Let $A_1(F_1), \ldots, A_v(F_v)$ be the sequence of queries performed by $\mathcal{A}$ on the first instance defined by p and q. That is, $A_i(F_i)$ is either $p(F_i)$ or $q(F_i)$ for some $F_i \in \mathcal{P}(E) \setminus \mathcal{C}_l$ for every $i = 1, \ldots, v$. Since $\mathcal{A}$ is deterministic, $F_i \notin \mathcal{C}_l$, and $p(F) = p'(F)$ and $q(F) = q'(F)$ for every $F \in \mathcal{P}(E) \setminus \mathcal{C}_l$, $\mathcal{A}$ will perform the same sequence of queries for the second instance defined by p' and q' and generate the same output. But this implies that $\mathcal{A}$ is incorrect on at least one of the two instances, as their results are distinct, i.e.,

$$\mathrm{Th}(p \wedge q) = \{\} \neq \{S_l\} = \mathrm{Th}(p' \wedge q') .$$

As desired this theorem rules out not only an output polynomial algorithm for Problem 1, i.e., one that lists the result with a communication complexity polynomial in $|\mathrm{Th}(p \wedge q)|$, but also an algorithm with a communication complexity that is polynomial in the border intersection (Equation 2). Moreover, since we have chosen the constants k of Equation 1 to be 1, it follows that the lower bound also applies to the problem of constructing and sampling one element of $\mathrm{Th}(p \wedge q)$. Similar, the theorem rules out efficient algorithms for listing any non-empty subset or a polynomially bounded superset of the theory.

Finally, as we restricted ourselves to pattern languages representable as sets, we note that the concrete exponential lower bound is not caused by non-uniformity of the pattern language but by its size alone.

4 Frequent and Infrequent Sets

Next, we leave the general theory extraction problem and turn to the subproblem of anti-monotone and monotone constraints that are frequency-induced. More specifically, the problem is to list sets that are frequent in one dataset while they are infrequent in another or formally:

Problem 2 (LIST-FREQUENCY-THEORY). *Given* two datasets $\mathcal{D}_1, \mathcal{D}_2$ over common elements E and frequency thresholds $t_1, t_2 \in \mathbb{N}$, *list* the family $\mathcal{F}$ of all sets $F \subseteq E$ that are t_1-frequent in $\mathcal{D}_1$ and t_2-infrequent in $\mathcal{D}_2$.

This is a variant of the problem of listing *emerging patterns* [7] that fits into the description of LIST-CONJUNCTIVE-THEORY: implicitly the task of Problem 2 is to list the conjunctive theory $\text{Th}(\text{frq}_{\mathcal{D}_1, t_1} \wedge \text{ifr}_{\mathcal{D}_2, t_2})$ of the anti-monotone frequency constraint $\text{frq}_{\mathcal{D}_1, t_1}$ and the monotone infrequency constraint $\text{ifr}_{\mathcal{D}_2, t_2}$. Thus, Problem 2 is a subproblem of Problem 1.

For this concrete task, however, we are interested in ruling out efficient algorithm that may exploit the structure of $\text{frq}_{\mathcal{D}_1, t_1}$ and $\text{ifr}_{\mathcal{D}_2, t_2}$, i.e., algorithms that directly access the underlying datasets $\mathcal{D}_1$ and $\mathcal{D}_2$. Showing intractability for this problem, hence, means to rule out algorithms that have an output polynomial time complexity with respect to the combined input size $\text{size}(\mathcal{D}_1, \mathcal{D}_2) = \text{size}(\mathcal{D}_1) + \text{size}(\mathcal{D}_2)$.

It was shown in [21] that there are anti-monotone families that cannot be concisely represented as the family of frequent sets of some dataset. Consequently, Problem 2 is strictly easier than Problem 1. Still, as we will show below, there is no efficient, i.e., output polynomial time, listing algorithm for that task. In order to achieve this result, we first prove a stronger claim, namely the **NP**-hardness of deciding the existence of a set that satisfies the constraints. Hardness of efficient listing follows then as a corollary among other related hardness results.

The idea of the construction is that an infrequency constraint can be used to encode a minimum size constraint. In addition the proof builds on the **NP**-hardness of the problem FREQUENT-SET [11], i.e., *given* a dataset $\mathcal{D}$, a frequency threshold t, and a positive integer k, *decide* whether there is a t-frequent set of size k in $\mathcal{D}$.

Theorem 2. *The following problem is **NP**-complete:*
Given *an instance of Problem 2,* decide *whether the result set $\mathcal{F}$ is non-empty.*

Proof. Elements of $\mathcal{F}$ can serve as polynomial time checkable certificates for yes-instances. This shows membership in **NP**.

For the hardness we give a polynomial reduction from FREQUENT-SET. Given an instance of this problem, i.e., a dataset $\mathcal{D}$ over E, a frequency threshold t, and a positive integer k, we define a second dataset $\mathcal{D}'$ over E by

$$\mathcal{D}' = \{E \setminus \{e\} \colon e \in E\}$$

with simple multiplicities $\lambda \equiv 1$. Since any element of E is contained in all but one element of $\mathcal{D}'$ and $|\mathcal{D}'| = |E|$ it holds that for all $F \subseteq E$ that $|\mathcal{D}'[F]| = |E| - |F|$. With this we can deduce:

$$F \text{ is } (|E| - k + 1)\text{-infrequent in } \mathcal{D}' \Leftrightarrow |\mathcal{D}'[F]| \le |E| - k$$
$$\Leftrightarrow |E| - |F| \le |E| - k$$
$$\Leftrightarrow |F| \ge k \ .$$

Thus, for $\mathcal{D}_1 = \mathcal{D}$ and $\mathcal{D}_2 = \mathcal{D}'$ as constructed above and $t_1 = t$ and $t_2 = |V| - k + 1$ it holds for the resulting solution $\mathcal{F}$ of Problem 2 that $\mathcal{F} \neq \emptyset$ if and only if $\mathcal{D}$ contains a t-frequent set of size k. Since the above construction can be performed in polynomial time, we have a polynomial reduction as required.   □

Theorem 2 directly implies the hardness of several related problems, namely of constructing an element of $\mathcal{F}$, sampling from $\mathcal{F}$ according to any distribution, as well as listing $\mathcal{F}$ with polynomial delay. The last statement follows because a polynomial delay algorithm for listing $\mathcal{F}$ does after a time bounded polynomially in $\text{size}(\mathcal{D}_1, \mathcal{D}_2)$ either lists an element of $\mathcal{F}$ in case $\mathcal{F} \neq \emptyset$ or terminates otherwise.

For ruling out output polynomial listing one can use the following idea: if the number of steps of an output polynomial algorithm exceeds some input polynomial threshold then the algorithm has to produce at least one element. This is exploited in the proof of Corollary 3 below.

Corollary 3 *There is no output polynomial listing algorithm for* LIST-FREQUENCY-THEORY *(unless* **P** $=$ **NP***).*

Proof. Assume there is an algorithm $\mathcal{A}$ and a (without loss of generality) monotone polynomial p such that $\mathcal{A}$ lists $\mathcal{F}$ in time $p(\text{size}(\mathcal{D}_1, \mathcal{D}_2) + |\mathcal{F}|)$. Then one can construct another algorithm $\mathcal{A}'$ that decides whether $\mathcal{F}$ is empty in polynomial time. The claim then follows from Theorem 2. The construction of $\mathcal{A}'$ is as follows: Simulate $\mathcal{A}$ for $t = p(\text{size}(\mathcal{D}_1, \mathcal{D}_2))$ steps. If $\mathcal{A}$ has not stopped after t steps then

$$p(\text{size}(\mathcal{D}_1, \mathcal{D}_2)) < \# \text{ steps } \mathcal{A} \text{ performs on input } \mathcal{D}_1, \mathcal{D}_2$$
$$\leq p(\text{size}(\mathcal{D}_1, \mathcal{D}_2) + |\mathcal{F}|) \ .$$

By monotonicity of p this implies $0 < |\mathcal{F}|$, i.e., $\mathcal{F}$ is non-empty. If $\mathcal{A}$ terminates after t or less steps then the length of the output $\mathcal{F}$ is bounded by t and can be examined explicitly in that time to decide whether $\mathcal{F}$ is empty.

Note that for the reasoning in the proof of Corollary 3 the statement of Theorem 2 is in fact stronger than required: It would have been sufficient to know that it is **NP**-hard to decide whether $|\mathcal{F}| \leq k$ where k is polynomially bounded in $\text{size}(\mathcal{D}_1, \mathcal{D}_2)$. We call a construction that uses this idea a *polynomially cardinality reduction*, as it reduces an **NP**-hard decision to the question of whether the size of some set is smaller than or equal to some polynomial threshold. In fact, the construction in [5] used to prove intractability of listing all maximal frequent sets as well as the one in [14] to prove hardness of listing all bases of an independence system both can be regarded as such a polynomially cardinality reduction.

5   Conclusion

Summary In this paper we investigated the complexity of listing the theory of constraints that are formed as the conjunction of anti-monotone and monotone constraints. We showed that no general output polynomial algorithm for that task exists, justifying the exponential behavior of systems that operate on such a general level. In particular we gave a lower bound that is exponential even in cases when the only known prior bound is constant. Then, we showed that listing all patterns that are frequent in one database and infrequent in another database cannot be done efficiently. It is important to emphasize that our proofs also rule out efficient algorithms listing any non-empty subset or any small superset of the interesting patterns for the considered tasks. Similar, our theorems imply hardness of the corresponding construction and sampling problems.

Discussion and Related Work. The lower bound for the LIST-CONJUNCTIVE-THEORY problem stands in line with the already mentioned complexity result of Bucila et al. [6] for the same problem and the results of Mannila and Toivonen [15], which says that the communication complexity of listing $\text{Th}(p)$ for a monotone constraint p is lower bounded by $\text{Bd}(p)$. All these results have in common that they do not require any complexity assumption such as $\mathbf{P} \neq \mathbf{NP}$. This is due to the restricted computational model they are based on.

In contrast, the intractability result of Section 4 for LIST-FREQUENCY-THEORY requires the assumption that $\mathbf{P} \neq \mathbf{NP}$. This is similar to the result Boros et al. [5] that rules out output polynomial time algorithms for the task of listing the maximal frequent sets for a given dataset (the minimal infrequent sets can at least be listed in output quasi-polynomial time). A noteworthy difference between maximal frequent sets and LIST-FREQUENCY-THEORY is that for the first problem the construction problem is in $\mathbf{P}$ while it is $\mathbf{NP}$-hard for the latter problem.

Although the proof of Theorem 2 did not use a reduction to the problem of listing all maximal frequent sets, the intractability of that problem might suggest that there is connection between the complexity of listing the border of the involved theories to listing their intersection, i.e., the theory of the conjunction. But at least in one direction, namely from the tractability of the border enumeration to the tractability of the conjunction theory such an implication does not hold. As example consider the following result.

Theorem 4. *Let $G = (V, E)$ be an undirected graph. Define the constraints p and q on $\mathcal{P}(E)$ as follows:*

$$p(F) = 1 \;\; \text{iff} \;\; \forall v \in V, \; \left| \delta_{(V,F)}(v) \right| \leq 2$$
$$q(F) = 1 \;\; \text{iff} \;\; (V, F) \; \text{is connected} \;,$$

where $\delta_{(V,F)}(v)$ denotes the set of edges that are incident to v in the graph (V, F).

Then $Bd^+(p)$, $Bd^+(q)$, $Bd^-(p)$, and $Bd^-(q)$ can each be listed with polynomial delay given an input graph G. But it is $\mathbf{NP}$-hard to decide whether $Th(p \wedge q) = \emptyset$, and consequently $Th(p \wedge q)$ cannot be listed in output polynomial time (unless $\mathbf{P} = \mathbf{NP}$).

The hardness follows because $\text{Th}(p \wedge q)$ is the family of edgesets that form Hamiltonian paths and cycles in G. The complete proof can be found in the appendix.

Future Work. In the light of the results presented in this paper there are several important directions for future work on the efficiency of theory extraction methods. On the one hand it is important to identify classes of constraints that allow for efficient algorithms. A natural candidate here may be convex constraints as they are a subset of the conjunctions of monotone and anti-monotone constraints. On the other hand it is important to investigate the complexity of alternative algorithms and try to devise e.g., randomized listing algorithms and/or try to find fixed-parameter tractable subclasses.

References

1. Agrawal, R., Srikant, R.: Fast algorithms for mining association rules in large databases. In: VLDB, pp. 487–499 (1994)
2. Bonchi, F., Giannotti, F., Mazzanti, A., Pedreschi, D.: ExAMiner: Optimized level-wise frequent pattern mining with monotone constraint. In: ICDM, pp. 11–18. IEEE Computer Society Press, Los Alamitos (2003)
3. Bonchi, F., Giannotti, F., Mazzanti, A., Pedreschi, D.: Exante: A preprocessing method for frequent-pattern mining. IEEE Intelligent Systems 20(3), 25–31 (2005)
4. Bonchi, F., Lucchese, C.: Pushing tougher constraints in frequent pattern mining. In: Ho, T.-B., Cheung, D., Liu, H. (eds.) PAKDD 2005. LNCS (LNAI), vol. 3518, pp. 114–124. Springer, Heidelberg (2005)
5. Boros, E., Gurvich, V., Khachiyan, L., Makino, K.: On the complexity of generating maximal frequent and minimal infrequent sets. In: Alt, H., Ferreira, A. (eds.) STACS 2002. LNCS, vol. 2285, p. 733. Springer, Heidelberg (2002)
6. Bucila, C., Gehrke, J., Kifer, D., White, W.: DualMiner: A dual-pruning algorithm for itemsets with constraints. Data Mining and Knowledge Discovery 7(3), 241–272 (2003)
7. Dong, G., Li, J.: Efficient mining of emerging patterns: discovering trends and differences. In: KDD 1999: Proceedings of the fifth ACM SIGKDD international conference on Knowledge discovery and data mining, pp. 43–52. ACM, New York (1999)
8. El-Hajj, M., Zaïane, O.R., Nalos, P.: Bifold constraint-based mining by simultaneous monotone and anti-monotone checking. In: ICDM, pp. 146–153. IEEE Computer Society Press, Los Alamitos (2005)
9. Fischer, J., Heun, V., Kramer, S.: Optimal string mining under frequency constraints. In: Fürnkranz, J., Scheffer, T., Spiliopoulou, M. (eds.) PKDD 2006. LNCS (LNAI), vol. 4213, pp. 139–150. Springer, Heidelberg (2006)
10. Goldberg, L.A.: Efficient algorithms for listing combinatorial structures. Cambridge University Press, New York (1993)
11. Gunopulos, D., Khardon, R., Mannila, H., Saluja, S., Toivonen, H., Sharma, R.S.: Discovering all most specific sentences. ACM Trans. Database Syst. 28(2), 140–174 (2003)
12. Han, J., Pei, J., Yin, Y., Mao, R.: Mining frequent patterns without candidate generation: A frequent-pattern tree approach. Data Mining and Knowledge Discovery 8(1), 53–87 (2004)
13. Johnson, D.S., Papadimitriou, C.H.: On generating all maximal independent sets. Inf. Process. Lett. 27(3), 119–123 (1988)
14. Lawler, E.L., Lenstra, J.K., Kan, A.H.G.R.: Generating all maximal independent sets: Np-hardness and polynomial-time algorithms. SIAM J. Comput. 9(3), 558–565 (1980)
15. Mannila, H., Toivonen, H.: Levelwise search and borders of theories in knowledge discovery. Data Mining and Knowledge Discovery 1(3), 241–258 (1997)
16. Mitchell, T.M.: Generalization as search. Artificial Intelligence 18, 203 (1982)
17. Pei, J., Han, J.: Can we push more constraints into frequent pattern mining? In: KDD, pp. 350–354 (2000)
18. De Raedt, L., Kramer, S.: The levelwise version space algorithm and its application to molecular fragment finding. In: Nebel, B. (ed.) Proceedings of the Seventeenth International Joint Conference on Artificial Intelligence, IJCAI 2001, Seattle, Washington, USA, August 4-10, 2001, pp. 853–862. Morgan Kaufmann, San Francisco (2001)

19. Read, R.C., Tarjan, R.E.: Bounds on backtrack algorithms for listing cycles, paths, and spanning trees. Networks 5, 237–252 (1975)
20. Saigo, H., Nowozin, S., Kadowaki, T., Kudo, T., Tsuda, K.: gBoost: A mathematical programming approach to graph classification and regression. Machine Learning (2009)
21. Sloan, R.H., Takata, K., Turán, G.: On frequent sets of boolean matrices. Annals of Mathematics and Artificial Intelligence 24(1-4), 193–209 (1998)
22. Wang, L., Zhao, H., Dong, G., Li, J.: On the complexity of finding emerging patterns. Theoretical Computer Science 335(1), 15–27 (2005); Pattern Discovery in the Post Genome
23. Yang, G.: Computational aspects of mining maximal frequent patterns. Theoretical Computer Science 362(1-3), 63–85 (2006)

A   Proofs

In this section we prove Theorem 4. This involves two positive listing results that may be interesting in their own right. In order to formulate these proofs some more graph terminology is needed.

Let $G = (V, E)$ be an (undirected) graph. The set of edges incident to a vertex v is denoted $\delta(v)$. The maximum vertex degree of G, $\max\{|\delta(v)| : v \in V\}$ is denoted $\Delta(G)$. For a set of vertices $X \subseteq V$ we mean by its neighbors $\Gamma(X) = \bigcup\{\Gamma(x) : x \in X\} \setminus X$ and by its incident edges $\delta(X) = \{(v, w) \in E : v \in X, w \notin X\}$, i.e., the *cut* induced by X. In the context of more than one graph we use an index to clarify the above notations, e.g., $\delta_G(v)$. A subgraph of G is a pair (V', E') with $V' \subseteq V$, $E' \subseteq E$ and $e \subseteq V'$ for all $e \in E'$. A *component* of G is a maximal connected subgraph. We denote the subgraph induced by a set $U \subseteq V$ as $G[U] = (U, E \cap (2^U))$. For convenience we often identify a component C with the set of vertices U inducing it ($G[U] = C$). If $G[V \setminus \{v\}]$ has more components than G then v is called an *articulation*.

Proof (of Theorem 4). Anti-monotonicity of p respectively monotonicity of q are clear. The family $\mathrm{Bd}^-(q)$ contains the complements of the minimal cuts of G and can be listed with polynomial delay (Lemma 5) and $\mathrm{Bd}^+(q)$ contains the spanning trees of G, which can also be listed with polynomial delay with the algorithm from [19]. Polynomial delay for $\mathrm{Bd}^+(p)$ follows from Lemma 6 with $k = 2$. This also holds for $\mathrm{Bd}^-(p)$, which contains exactly the subsets of $\{e \in E : v \in e\}$ with cardinality 3 for all vertices $v \in V$.

Lemma 5 *For all graphs $G = (V, E)$, the set of minimal cuts of G can be listed with delay $O(|V|^3|E|)$.*

Proof. As the set of minimal cuts of a graph is the union of the minimal cuts of its components, we can wlog assume that G is connected. Moreover, a cut $\delta(X)$ is minimal if and only if each part of its inducing bipartition X and $V \setminus X$ is connected. Thus, the task is equivalent to the enumeration of a subset $\mathcal{F}'$ of the set system

$$\mathcal{F} = \{X \subseteq V : G[X] \text{ connected and } G[V \setminus X] \text{ connected }\}$$

with the property that for all $X \in \mathcal{F}$ exactly either $X \in \mathcal{F}'$ or $(V \setminus X) \in \mathcal{F}'$. For that we construct a polynomial mapping $\rho : \mathcal{F} \to \mathcal{P}(\mathcal{F})$ such that all $X \in \mathcal{F}$ can be reached from $\emptyset$ by applying ρ at most $|X|$ times. Formally this means that for all $X \in \mathcal{F}$ there is an $i \leq |X|$ with $X \in \rho^i(\emptyset)$ where ρ^i is defined recursively by $\rho^1(X) = \rho(X)$ and $\rho^i(X) = \bigcup\{\rho(Y): Y \in \rho^{i-1}(X)\}$. For $X \subseteq V$ inducing a minimal cut and $v \in \Gamma(X)$ define

$$\rho_X(v) = \{V \setminus C: C \text{ a component of } G[V \setminus (X \cup \{v\})]\}$$

and based on that $\rho(X) = \bigcup_{v \in \Gamma(X)} \rho_X(v)$. Note that if v is no articulation in $G[V \setminus X]$ then $X \cup \{v\}$ is the only element of $\rho_X(v)$. It is easy to check that $\rho(X) \subseteq \mathcal{F}$ for all $X \in \mathcal{F}$.

We now show that ρ can be used to generate $\mathcal{F}$ completely. Let $X, Y \in \mathcal{F}$ with $X \subset Y$. Since Y is connected, there is a $y \in \Gamma(X) \cap Y$. Since $V \setminus Y$ is also connected, there is a component C of $G[V \setminus (X \cup \{y\})]$ such that $V \setminus Y \subseteq C$. But then $Y \supseteq (V \setminus C) \in \rho(X)$ implying that for all $X, Y \in \mathcal{F}$ with $X \subset Y$ there is an $X' \in \rho(X)$ with $X' \in \mathcal{F}$ and $X \subset X' \subseteq Y$. It follows by induction that all elements of $\mathcal{F}$ get visited by a depth-first-search traversal of the directed enumeration graph $T = (\mathcal{F}, \{(X, X'): X' \in \rho(X)\})$ starting in $\emptyset$. To distinguish between the vertices of T and the underlying graph G we subsequently call the vertices of T 'nodes'.

It remains to show that this search can be implemented in such a way that it visits a node corresponding to a new minimal cut (or terminates) with polynomial delay. An $X \in \mathcal{F}$ can have at most $|V|$ neighbors in G and for every neighbor $v \in \Gamma(X)$ the complement graph $G[V \setminus (X \cup \{v\})]$ can at most have $|V|$ components. It follows that $|\rho(X)| \leq |V|^2$ and each element of $\rho(X)$ can trivially be found within polynomial time. In particular depth-first-search can encounter not more than $|V|^2$ already visited children in a given node before it tracks back. Moreover, it can track back at most $|V|$ times, because $Y \supset X$ for all $Y \in \rho(X)$ and $X \in \mathcal{F}$. It follows that depth-first-search finds a new node or terminates after $O(t|V|^3)$ where t is the time needed to check whether some node has already been visited.

So the remaining problems are to check in polynomial time whether an $X \subseteq V$ has already been visited and to avoid double enumeration of $\delta(X) = \delta(V \setminus X)$. Both can be taken care of by storing $\delta(X)$ for each node X using a string representation induced by some fixed order of the edges E. Storing all enumerated edge sets in a prefix tree it is possible to check whether an $F \subseteq E$ has already been visited in time $O(|F|)$. $\square$

Lemma 6 *For all graphs $G = (V, E)$ and all positive integers $k \leq \Delta(G)$, the family of all maximal edge sets with degree bounded by k, i.e., the maximal elements of $\mathcal{F} = \{F \subseteq E: \Delta(V, F) \leq k\}$, can be listed with delay $O(k^4|E|^3)$.*

Proof. Just as for Lemma 5 we prove the theorem by constructing a polynomially bounded operator ρ and use it for depth-first-search. Let $F \in \max \mathcal{F}$ be a maximal degree bounded edge set. For $e \in E \setminus F$ define

$$\mathcal{D}_{F,e} = \min\{F' \in \mathcal{P}(F): (F \setminus F') \cup \{e\} \in \mathcal{F}\}$$

the family of all minimal sets of edges one has to delete from F such that it can be legally augmented with e. Note that $(F \setminus F') \cup \{e\}$ is in general not a maximal element of $\mathcal{F}$ for $F' \in \mathcal{D}_{F,e}$. Denote the family of all maximal extensions of an $X \in \mathcal{F}$ by $\mathcal{B}(X) = \{B \in \max \mathcal{F}: B \supset X\}$. With this we can define $\rho_F(e) = \bigcup_{F' \in \mathcal{D}_{F,e}} \mathcal{B}((F \setminus F') \cup \{e\})$

the family of all maximal extensions of $(F \setminus F') \cup \{e\}$ for all deletions $F' \in \mathcal{D}_{F,e}$ and finally the mapping $\rho : \max \mathcal{F} \to \mathcal{P}(\max \mathcal{F})$ by $\rho : F \mapsto \bigcup_{e \in E \setminus F} \rho_F(e)$.

Now we show that for any two distinct elements X, Y of $\max \mathcal{F}$, Y can be reached from X by applying ρ at most $|E|$ times. Let $e \in Y \setminus X$. Then, for $v \in e$ and $|\delta(v) \cap (X \cup \{e\})| > k$ there must be an edge $f_v \in (\delta(v) \cap X) \setminus Y$. Hence, $F' = \{f_v : v \in e\} \in \mathcal{D}_{X,e}$ and $F' \cap Y = \emptyset$. This implies for the symmetric difference $X \triangle Y$:

$$|Y \triangle ((X \setminus F') \cup \{e\})| = |Y \triangle X| - (|F'| + 1) . \tag{5}$$

Furthermore, for each element of F' the edge set $(X \setminus F') \cup \{e\}$ has at most one unsaturated vertex, i.e., one to which an edge can be added without leaving $\mathcal{F}$. So for an $B \in \max \mathcal{F}$ with $B \supset (X \setminus F') \cup \{e\}$ we have

$$|B \setminus ((X \setminus F') \cup \{e\})| \le |F'| . \tag{6}$$

(5) and (6) together imply that

$$|Y \triangle B| \le |Y \triangle ((X \setminus F') \cup \{e\})| + |B \setminus ((X \setminus F') \cup \{e\})|$$
$$\le |Y \triangle X| - 1 < |Y \triangle X| .$$

Since $B \in \rho_X(e)$, we conclude that for all $X, Y \in \max \mathcal{F}$ with $X \neq Y$ there is an element $X' \in \rho(X)$ with $|X' \triangle Y| < |X \triangle Y|$.

Let $F \in \max \mathcal{F}$ and $e \in E \setminus F$. Since $(V, F \cup \{e\})$ has at most two vertices with degree greater than k, all deletions $F' \in \mathcal{D}_{F,e}$ also have at most two elements. Equation (6) implies the same for $\{B \in \max \mathcal{F} : B \supseteq (F \setminus F') \cup \{e\}\}$. These two elements can be chosen from at most k edges. Hence, it holds for $|\rho(F)|$ that

$$|\rho(F)| \le |E| \max_{e \in E, \, F' \in \mathcal{D}_{F,e}} |\mathcal{D}_{F,e}| \, |\mathcal{B}((F \setminus F') \cup \{e\})| \le k^4 |E| .$$

Again, it is no problem to enumerate the elements of $\rho(F)$ with polynomial delay and ρ can be computed in time polynomial in the size of G. Using the same techniques as in the proof of Lemma 5 a depth-first-search traversal of $\max \mathcal{F}$ using ρ can be performed. This time we can start in any one element $X \in \max \mathcal{F}$. Such an element can be obtained from the empty set by greedy augmentations. A prefix tree look-up can be performed in $O(|E|)$ and the maximal recursion depth is $\max_{Y \in \max \mathcal{F}} |X \triangle Y| = O(|E|)$.   □

Algorithm and Feature Selection for VegOut: A Vegetation Condition Prediction Tool

Sherri Harms[1,*], Tsegaye Tadesse[2], and Brian Wardlow[2]

[1] Department of Computer Science and Information Systems, University of Nebraska at Kearney, Kearney, NE, USA
[2] National Drought Mitigation Center, University of Nebraska - Lincoln, Lincoln, NE, USA
harmssk@unk.edu, ttadesse2@unl.edu, bwardlow2@unl.edu

Abstract. The goal of the VegOut tool is to provide accurate early warning drought prediction. VegOut integrates climate, oceanic, and satellite-based vegetation indicators to identify historical patterns between drought and vegetation conditions indices and predict future vegetation conditions based on these patterns at multiple time steps (2, 4 and 6-week outlooks). This paper evaluates different sets of data mining techniques and various climatic indices for providing the improved prediction accuracy to the VegOut tool.

Keywords: Drought Risk Management, Algorithm Selection, Feature Selection, Vegetation Outlook, VegOut.

1 Introduction

The Vegetation Outlook (VegOut) tool, introduced by Tadesse and Wardlow in [1], provides a spatio-temporal prediction outlook of general vegetation conditions using a regression tree data mining technique. VegOut integrates climate, oceanic, and satellite-based vegetation indicators and general biophysical characteristics of the environment to identify historical patterns between drought intensity and vegetation conditions and to predict future vegetation conditions based on these patterns at multiple time steps. Cross-validation (withholding years) revealed that the seasonal VegOut model had relatively high prediction accuracy across the growing season.

Presently, the VegOut tool uses the rule-based regression tree algorithm, Cubist [2], to predict satellite-observed seasonal greenness (SSG) at 2-, 4- and 6-weeks out during the growing season in fifteen Midwest and Great Plains states. The VegOut models are applied to geospatial data to produce vegetation condition maps [1]. Even though this approach had relatively high prediction accuracy, no alternative modeling methods have been investigated for use in the VegOut tool. This paper examines which climatic index gives the best prediction, as well as the prediction accuracy of

* This research is partially funded by the United States Department of Agriculture (USDA) Risk Management Association (RMA) partnership (02-IE-0831-0228) with the National Drought Mitigation Center (NDMC), University of Nebraska-Lincoln, and a subcontract with the University of Nebraska at Kearney.

J. Gama et al. (Eds.): DS 2009, LNAI 5808, pp. 107–120, 2009.
© Springer-Verlag Berlin Heidelberg 2009

several data mining algorithms compared with the current regression tree-based model used in VegOut for various temporal prediction periods. The goal is to select the algorithm that generalizes well, has the best prediction accuracy for this problem domain, and is able to run in real-time.

2   The Algorithm Selection Problem

The algorithm selection problem introduced by Rice [3] in 1976 seeks to answer the question: Which algorithm is likely to perform best for my problem? Although some algorithms are better than others on average, there is rarely a best algorithm for a given problem. By far, the most common approach to algorithm selection has been to measure different algorithms' performance on a given problem, and then to use the algorithm that has the best predictive value [4]. This is the planned approach for the VegOut domain.

There are four essential components of the algorithm selection model [3]:

1. The *problem space P* which represents the set of instances of a problem class;
2. The *feature space F* which contains measurable characteristics of the instances generated by a computational feature extraction process applied to P;
3. The *algorithm space A* which is the set of all considered algorithms for tackling the problem; and
4. The *performance space Y* which represents the mapping of each algorithm to a set of performance metrics.

The algorithm selection problem can be formally stated as follows [3]:

> *For a given problem instance $x \in P$, with features $f(x) \in F$, find the selection mapping $S(f(x))$ into algorithm space A, such that the selected algorithm $\alpha \in A$ maximizes the performance mapping $y(\alpha(x)) \in Y$.*

2.1   Problem Space Definition

For the VegOut tool, the goal is to predict the general vegetation condition and provide accurate early warning in drought years. Drought is responsible for major crop and other agricultural losses in the U.S. each year. Recent extended periods of drought in the U.S. emphasized this vulnerability of the agricultural sector and the need for a more proactive, risk management approach to drought-induced water shortages that negatively affect vegetation conditions [5]. Preparedness and mitigation reduces drought vulnerability and its devastating impacts. A critical component of planning for drought is the provision of timely and reliable information that aids decision makers at all levels in making critical management decisions [6].

Predicting drought and its impact on vegetation is challenging because there is inevitable uncertainty in predicting precipitation. Moreover, even if it is possible to get accurate forecasts, the complex spatial and temporal relationships between climate and vegetation makes the prediction of vegetation condition difficult [1]. Recent improvements in meteorological observations and forecasts have greatly enhanced the capability to monitor vegetation conditions [7]. In addition, remote sensing observations from satellite-based platforms provide spatially continuous and repeat measurements of general vegetation conditions over large areas where no weather observing stations are present. Satellite observations have proven useful over the past two decades for monitoring vegetation conditions at regional to global scales [8]. Most remote sensing studies have analyzed normalized difference vegetation index (NDVI) data, which provides a multi-spectral-based measure (combined spectral data from the visible red and near infrared wavelength regions) of plant vigor and general vegetation condition. Even though this information is valuable for monitoring the vegetation condition, drought-affected areas cannot be distinguished from locations affected by other environmental stressors (e.g., pest infestation) solely from satellite-based observations [1].

2.2 Feature Space Definition

As discussed in [4], features must be chosen so that the varying complexities of the problem instances are exposed, any known structural properties of the problems are captured, and any known advantages and limitations of the different algorithms are related to features. Clearly, it is not straightforward to design suitable features for a given problem domain. Rice [3] concludes that "the determination of the proper nonlinear form is still somewhat of an art and there is no algorithm for making the choice".

For the VegOut feature space, recent studies that investigated ocean-atmosphere relationships found significant improvements in seasonal climate predictions (e.g., precipitation and temperature) with the inclusion of oceanic indicators [9]. The integration of climate, satellite, and ocean data, with other biophysical information such as available soil water capacity, ecoregion, and land cover type holds considerable potential for enhancing our drought monitoring and prediction capabilities [1]. The feature space used in this study is the same as that used in the initial VegOut study.

Selected Weather Stations. More than 3000 weather stations were available for the 15 states to build the historical database. However, only 1402 stations were selected to be used in the VegOut model, as shown in Figure 1. Stations that did not have a long historical climate records (i.e., > 30 years of precipitation data and > 20 years of temperature data), incomplete data record (< 10% missing observations over the historical record), or were not currently in operation were excluded. Stations that were predominately surrounded by either an urban area or water (i.e., > 50% of the surrounding 3 km x 3 km area) were also eliminated because they would not be representative of vegetation conditions. For VegOut model development, a training database was built to extract historical climate and satellite information, as well as the biophysical parameters (considered static over the 18-year record) at the 1402 weather station locations across the 15-state study area.

Climate-based Data. Two commonly used climate-based drought indices, the Palmer Drought Severity Index (PDSI) [10][11] and the Standardized Precipitation Index (SPI) [12], were used to represent the climatic variability that affects the vegetation

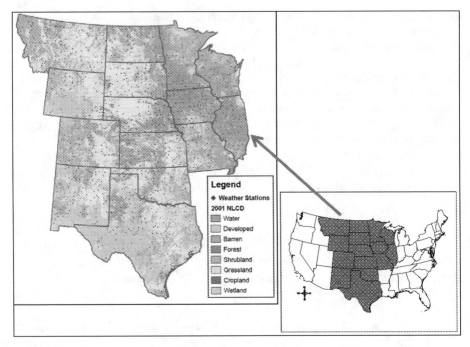

Fig. 1. Land cover and the 1402 weather station locations for the 15-state study area

condition in the VegOut models. The SPI is based on precipitation data and has the flexibility to detect both short- and long-term drought. The PDSI is calculated from a soil water balance model that considers precipitation, temperature, and available soil water capacity observations at the station for VegOut model development. The historical time series of bi-weekly climate (i.e. SPI and PDSI) values for each of the 1402 stations was then generated for an 18-year period from (1989 to 2006).

For this study, one goal was to decide which climate-based index to use for the VegOut model. We experimented with 19 different indices: 1 week SPI, 2 week SPI, 4 week SPI, 8 week SPI, 12 week SPI, 16 week SPI, 20 week SPI, 24 week SPI, 28 week SPI, 32 week SPI, 36 week SPI, 40 week SPI, 44 week SPI, 48 week SPI, 52 week SPI, 1 week PDSI, 2 week PDSI, 1 week PalmerZ and 2 week PalmerZ.

Satellite Data. The Standardized Seasonal Greenness (SSG) metric, which represents the general condition of vegetation, was calculated from 1-km2 resolution NDVI data over the study area. The SSG is calculated from the Seasonal Greenness (SG) measure, which represents the accumulated NDVI through time from the start of the growing season (as defined from satellite) [13]. From the SG data, the SSG is calculated at 2-week time steps throughout the growing season using a standardization formula (i.e., the current SG minus the average SG divided by the standard deviation). Since the satellite data are in gridded, raster format, they require a geographic summarization of each variable across a window of grid cells (or pixels) that surrounded each station's location. A 3-by-3 window centered on the station's location was implemented for this study whereby the mean of the 9 grid cells values was calculated for continuous variables.

The Oceanic Indices. Eight oceanic indices are integrated into the predictions to account for the temporal and spatial relationships between ocean-atmosphere dynamics and climate-vegetation interactions (i.e., teleconnection patterns) that have been observed over the central United States ([14,15,and 16]). The indices include the Southern Oscillation Index (SOI), Multivariate El Niño and Southern Oscillation Index (MEI), Pacific Decadal Oscillation (PDO), Atlantic Multi-decadal Oscillation (AMO), Pacific-North American index (PNA), North Atlantic Oscillation index (NAO), Madden-Julian Oscillation (MJO), and Sea Surface Temperature anomalies (SST). For each oceanic index, the same value for a given bi-week was assigned to all stations. Furthermore, for oceanic indices reported on a monthly time step, the same monthly value was assigned to all bi-weeks where the majority of the 14-day windows occurred in that specific month.

Biophysical Data. The biophysical parameters used in this study included land cover type, available soil water capacity, percent of irrigated land and ecosystem type. The dominant (or majority) value within a 9-km^2 window (i.e., 3-by-3 window of 1km pixels) surrounding each weather station was calculated from the 1-km2 images for each biophysical variable and used for VegOut model development.

For the dynamic climate, oceanic (2-week values extrapolated from monthly data), and satellite variables (the bi-weekly historical records from 1989 to 2006); and single, static values for the biophysical variables were extracted for each weather station and organized into a database which would be used in the analysis.

Seasonal Periods. The VegOut product is generated bi-weekly from the start of the growing season. However, as in [1], only 3 VegOut models developed for spring (April to June), mid-summer (Jun to August), and fall (August to October) dates were presented to illustrate the seasonal predictive ability of the VegOut approach across the growing season. The biweekly periods selected to test the VegOut models were Period 10 (the first two weeks of May), Period 15 (the last two weeks of July), and Period 17 (the last 2 weeks of August) representing the spring, mid-summer and fall season, respectively.

2.3 Algorithmic Space Definition

For the VegOut algorithmic space, recent studies have shown that data mining techniques are an effective means to integrate this diverse collection of data sets and to identify hidden and complex spatio-temporal patterns within the data that are related to drought [7].

Due to the need in the problem space for drought experts to be able to examine the model and study the parameters used in determining the predicted value, only "white box" methods were used. White box methods include decision trees, rules and regression algorithms, as the user can examine the formulas in these models. (As a side note, we did compare these results with a few "black box" methods, such as multilayer perceptron (a neural network), to see if we were losing any prediction value, and found the neural network results to be both slower and less accurate.)

Because of the need for predicting numerical values, rather than predicting a classification label, only methods that can handle numerical prediction were used.

Additionally, recent algorithmic comparisons across problem sets indicate that combinations of algorithms are likely to increase prediction accuracy over a single model. These combined methods share the disadvantage of being difficult to analyze: they can comprise dozens or even hundreds of individual models, and although they perform well it is not easy to understand in intuitive terms what factors are contributing to the improved decisions [17]. Because of our problem domain's need to examine the model, we will examine a few combinations of algorithms that have comprehensible models.

The data mining methods used in this study were linear regression, least median square regression, M5P model tree, M5P rules, support vector machines, bagging with linear regression, bagging with least median square, and bagging with M5P. The data mining methods used were all part of the Waikato Environment for Knowledge Analysis (Weka) workbench. The descriptions of the Weka algorithms are based on the documentation of the Weka system [17].

Linear Regression. This is the traditional linear regression statistical model. The Weka version uses the Akaike criterion for model selection, and is able to deal with weighted instances.

Least Median Square Regression. This is a least median squared linear regression algorithm that utilizes the existing Weka linear regression class to form predictions. The least squared regression functions are generated from random subsamples of the data. The least squared regression with the lowest median squared error is chosen as the final model. The basis of the algorithm is from Rousseeuw and Leroy [18].

M5P Model Tree. Model trees have a conventional decision tree structure but use linear functions at the leaves instead of discrete class labels. The original algorithm M5 was invented by Quinlan [19] and Wang and Witten [20] made improvements called M5' (referred to as M5P in Weka).

M5 Rules/Cubist. Cubist is a commercially available rule-based regression tree method [2]. The technique is generally referred to as regression tree modeling. In Weka, the non-commercial version of Cubist is the M5P Rules algorithm.

The M5 rules algorithm generates a decision list for regression problems using separate-and-conquer [21]. A separate and conquer technique identifies a rule that covers instances in the class (and excludes ones not in the class), separates them out, and continues on those that are left. The M5 algorithm combines regression trees and classification to make Model Trees (MT)s. The input space is partitioned into subsets based on entropy measures, and a regression equation is then fit to each subset. Each leaf in the decision tree is a multivariate regression model. Linear models that are collections of rules, each of which is associated with a linear expression for computing a target value, are generated.

The continuous splitting often results in a too complex tree that needs to be pruned (reduced) to a simpler tree to improve the generalization capacity. Finally, the value predicted by the model at the appropriate leaf is adjusted by the smoothing operation to reflect the predicted values at the nodes along the path from the root to that leaf.

The overall global model, which is the collection of these linear (and locally accurate) models, brings the required non-linearity in dealing with the problem. The difference from pure linear regression is that the necessary (sub)optimal splitting of input space is performed automatically. Model Trees can learn efficiently and can tackle tasks with high dimensionality which can be up to hundreds of attributes. The resulting MTs are transparent and simple – this makes them potentially more successful in the eyes of decision makers. The disadvantage, as with any rule-based algorithm, is that it tends to over-fit the training data and does not generalize well to independent test sets, particularly on noisy data.

Bagging. Bagging is an ensemble method that averages the prediction of a group of the same type of models to reduce variance and improve overall prediction accuracy [22]. For each model in the ensemble, bagging simulates the process of getting fresh new data by randomly deleting some instances and replicating others. Once the individual models have finished, the prediction for each model receives equal weight and an average prediction is computed. It can be shown theoretically that averaging over multiple models built from independent training sets always reduces the expected value of the mean-squared error, which is the key indicator of the quality of the prediction of a model [17]. For our experiments, bagging was used with linear regression, least median square, and the M5P model tree.

Because bagging averages the prediction of the same type of models, the resulting model is still a comprehensible model, which is a requirement for the VegOut problem domain. Bagging is also quite efficient and generalizes well.

Support Vector Machine Regression. SVMreg implements the support vector machine for regression. Support vector machines (SVMs) are a set of related supervised learning methods used for classification and regression. Viewing input data as two sets of vectors in an n-dimensional space, an SVM will construct a separating hyperplane in that space, one which maximizes the margin between the two data sets. To calculate the margin, two parallel hyperplanes are constructed, one on each side of the separating hyperplane, which are "pushed up against" the two data sets. Intuitively, a good separation is achieved by the hyperplane that has the largest distance to the neighboring data points of both classes, since in general the larger the margin the better the generalization error of the classifier. The parameters can be learned using various algorithms. The algorithm is selected by setting the RegOptimizer. The most popular algorithm (RegSMOImproved) is due to Shevade, et al, [23] and used as the default RegOptimizer.

The advantage of SVM regression models are their excellent overall prediction accuracy. However, this algorithm has extremely slow training times. It took almost a week to run each SVM experiment. Thus, it was not included in the results section, as the speed performance is not acceptable for use with the VegOut model, which is intended to be an operational tool.

2.4 Performance Space

For our experiments on the VegOut problem, we used the Mean Absolute Difference (MAE) as the performance metric. Due to the yearly variations that occur in weather data, instead of using cross validation for testing, we held out one year's worth of data at a time (a total of 16 hold out year iterations were used in this study).   For each year, we built the model without that year's data, and then tested the model using that year's data. The MAE was then averaged over all of the years.

3 Results/Discussion

The first set of comprehensive experiments we conducted was used to determine which climate-based index to use in the VegOut tool. For these experiments, we predict the SSG for the two-week time periods 10, 15 and 17 (i.e., 7 to 20 May, 14 to 29 Jul and 27 Aug to 9 Sept) using a 2, 4 and 6-week prediction outlook with one previous time period data for the data described above. We compared the average MAE for each of the 19 climate-based indices by averaging the algorithmic results for each climatic index. A second set of experiments was conducted to determine which algorithm had the best overall prediction accuracy. Finally, we compared the results from these experiments with the earlier results in [1].

3.1 Feature Selection Results

One goal of this study was to decide which climate-based index to use for the Ve-gOut model. Table 1 shows the results from the experimentation over the climatic indices for the 2-week outlook. This table shows the average MAE and the 99% confidence interval for the average MAE.   Confidence interval values are used instead of p-values (which were used in [1]), since they provide a plausible range for the true value of the population mean, whereas p-values do not. The difference between the results in this paper and those reported in [1] is that the results in [1] use 2000 as the test year, whereas the results in this paper are the overall average from holding out each year individually. These averages are shown instead of an individual year, to examine how well these algorithms and climatic indices generalize to unknown data.

As shown in Table 1, for the 2-week outlook, the 8 week SPI gave the best overall average MAE on the periods 10 and 15 data whereas the 4 week SPI gave the best overall average MAE on the period 17 data. Also shown in Table 1, the 4 week SPI gave the best overall average MAE on the period 17 data for the 2-week outlook.

The results support our expectation that a short prediction outlook favors shorter climatic indices, because the shorter climatic indices incorporate smaller variations in weather, such as a two-week rainy period or a two-week drought, than the longer climatic indices.

Table 1. Average results over all algorithms by climatic index using a 2-week outlook

Climatic	Period 10		Period 15		Period 17	
Index	Avg MAE	99% CI	Avg MAE	99% CI	Avg MAE	99% CI
1wkSPI	29.69	0.46	11.38	2.13	9.60	0.69
2wkSPI	29.38	0.57	11.33	2.18	9.37	0.60
4wkSPI	29.75	0.39	11.35	2.24	*9.06*	*0.48*
8wkSPI	*29.20*	*0.42*	*10.44*	*0.35*	9.26	0.63
12wkSPI	29.51	0.29	10.55	0.42	9.31	0.60
16wkSPI	29.50	0.24	10.55	0.41	9.46	0.64
20wkSPI	29.49	0.28	10.58	0.40	9.47	0.64
24wkSPI	29.47	0.21	10.58	0.41	9.58	0.69
28wkSPI	29.97	0.44	10.55	0.36	9.58	0.68
32wkSPI	29.73	0.27	10.57	0.36	9.56	0.66
36wkSPI	29.80	0.40	10.54	0.38	9.56	0.66
40wkSPI	29.91	0.79	10.63	0.41	9.57	0.70
44wkSPI	29.54	0.54	10.60	0.40	9.58	0.70
48wkSPI	29.70	0.74	10.61	0.39	9.56	0.69
52wkSPI	29.58	0.62	10.61	0.40	9.61	0.69
2wkPDSI	29.65	0.66	10.54	0.39	9.44	0.62
1wkPDSI	29.69	0.63	10.59	0.41	9.44	0.65
1wkPalmerZ	29.62	0.41	10.52	0.38	9.64	0.71
2wkPalmerZ	29.44	0.51	10.46	0.36	9.46	0.64
AVERAGE	**29.61**	**0.47**	**10.68**	**0.67**	**9.48**	**0.65**
RANGE	**0.77**	**0.59**	**0.95**	**1.89**	**0.58**	**0.23**

Similar to [1], the prediction accuracy was slightly lower for the spring phase compared with the mid-summer (peak growing season) and late summer (senescence) periods of the growing seasons. This was expected, as spring vegetation is much more variable than mid and late summer vegetation. A late frost can delay the green-up process, while an early warm-up with above average rainfall can speed-up the green-up process.

Even though the 4 week and 8 week SPI indices gave the best results over all the time periods tested for the 2-week outlook, the variability between the results from the different precipitation indices was not statistically significant. The range of the average MAE was only 0.77 for period 10, .95 for period 15, and .58 for period 17. The range of the 99% confidence intervals was quite small for these results, with a range of .59 for period 10, 1.89 for period 15, and .23 for period 17. Individual 99%

confidence interval values were also quite small, indicating that the true means for these experiments are likely to be close to the reported values.

For the 4-week outlook (not shown), the 28 week SPI gave the best average MAE over all the algorithms on the period 10 data (76.16) and period 17 data (84.66), while the 36week SPI gave the best average MAE on the period 15 data (77.12). For the 6-week outlook (also not shown), the 20 week SPI gave the best overall average MAE over all of the algorithms on the period 15 data (79.77), while the 24 week SPI gave the best overall average MAE on the period 17 data (82.30). These results support our expectation that a longer prediction outlook favors longer climatic indices.

Note that the prediction accuracy decreased as the prediction period increased from two to four or six weeks, similar to the results in [1].

The variability between the results obtained from different precipitation indices using the same algorithm for the 4-week outlook was slightly greater than the variability in 2-week outlook results, with a range of the average MAE of 7.62 on the spring phase data, 15.38 on the mid-summer data, and 5.50 on the late summer data, but was still not statistically significant. The variability in the 6-week outlook results was also insignificant, but larger than the variability in the 2-week outlook results.

Both the decrease in prediction accuracy and the increase in variability in results for the 4-and 6-week outlooks compared with the 2-week outlook are reasonable, since it is harder to predict vegetation conditions four and six weeks in the future than two weeks. Increasing the length of time between the current time period and the predicted time period allows for greater differences in weather conditions. Four weeks is a large timeframe when predicting vegetation conditions. The weather can range from four weeks without any significant rainfall to four weeks with extremely high rainfall amounts. Interestingly, the results from the 6-week outlook were similar to the results from the 4-week outlook, and in the case of period 17, the 6-week outlook prediction accuracy was slightly better than the 4-week outlook over the same algorithms.

Using t-tests, we tested for differences between the average MAEs for each climatic index, using period 10, 15, and 17 data and at the 2-, 4- and 6-week outlooks. We found no statistical significant difference in these mean values at the $\alpha=.01$ level, for any climatic index, during any period, with any outlook. This indicates that any of the climatic indices could be used with the VegOut tool while compromising little prediction accuracy. Even so, we would still like to determine the best climatic index feature to select for use in the VegOut tool. The 28 week SPI had the lowest overall average MAE of 56.58 with an averaged 99% confidence interval of 4.75, and would make a good choice for the climatic index feature to use for the VegOut model. The 24 week SPI had an overall average MAE of 56.79 and an averaged 99% confidence interval of 4.05, making its range of values for the average MAE smaller than that of the 28 week SPI values. Either of these climatic indices would make a good choice.

3.2 Algorithm Selection Results

The second goal of this study was to decide which algorithm to use for the VegOut model. Table 2 shows the results when averaging the MAE over all climatic indices by algorithm, using the 2-, 4-, and 6-week outlooks for periods 10, 15 and 17.

Table 2. Average MAE results over all climatic indices by algorithm

Algorithm	Period 10		Period 15			Period 17		
	2 wk	4 wk	2 wk	4 wk	6 wk	2 wk	4 wk	6 wk
Linear Regression	29.01	79.66	10.87	86.76	95.51	9.53	97.76	93.08
LeastMedSq	*29.00*	77.91	10.58	85.56	90.39	9.58	94.06	90.73
M5Rules	30.04	83.25	11.07	90.47	84.87	10.47	85.50	84.74
M5P	30.11	82.67	11.04	78.26	88.93	10.46	*81.98*	85.71
Bagging	29.71	*76.71*	*10.11*	72.57	73.02	*8.77*	82.51	*77.94*
AVERAGE	**29.57**	**80.04**	**10.73**	**82.73**	**86.54**	**9.76**	**88.36**	**86.44**
RANGE	**1.11**	**6.54**	**0.96**	**17.9**	**22.49**	**1.70**	**15.78**	**15.14**

No experiments were conducted for predicting period 10 with a six week outlook. Period 10 is the time period from 7 to 20 May, when the vegetation is just starting to green up. For the 15 states used in these experiments, there is little vegetation four weeks prior to period 10, and almost no vegetation six weeks prior to period 10, especially on agricultural land.

As shown in Table 2, the bagging[1] algorithm gave the best overall average MAE on most experiments, including the 4-week outlook for period 10, the 2-, 4- and 6-week outlook for period 15 data, and the 2- and 6-week outlook for the period 17 data, when averaged over all of the climatic indices.

The variability in these algorithms' means on the 2-week outlook over the precipitation indices was quite small for all three periods, and not statistically significant. For the 4-week outlook on periods 15 and17 data, we found statistically meaningful differences from the bagging, M5P and M5Rules algorithms' means, for the linear regression and least median square algorithms using a t-test, with α=.05. We also found statistically meaningful differences from the bagging mean, for the linear regression, least median square, M5Rules, and M5P algorithms for the 6-week outlook on period 15 data and for the linear regression, least median square, and M5P algorithms on the period 17 data.

Bagging is the only algorithm from the algorithms we experimented with that had the statistically lowest average MAE value in all experiments. As stated earlier, we used the bagging algorithm with three base algorithms: linear regression, least median square, and M5P. However, since the speed of the bagging with linear regression is the fastest and the comprehension of linear regression is quite good, bagging with linear regression would probably be the best overall choice of algorithm for VegOut.

When comparing the averages between the results obtained from the different algorithms over the precipitation indices, with the results obtained from the different climatic indices over the various algorithms, there was great statistical significance

---

[1] Bagging was used independently with three algorithms: Linear Regression, Least Median Square, and M5P. Similar results were found for all three bagging approaches.

between the algorithms, and no statistical significance between the climatic indices. This suggests that it is more important to choose the most appropriate algorithm than to have the best precipitation index in the feature set.

3.3 Comparison with Earlier Results

A third goal of this study was to compare the results from these experiments with the earlier results for the VegOut model. Table 3 shows the best algorithm and climatic index for each period-outlook combination. Table 3 also shows the average Mean Absolute Difference (MAD) expressed in SSG data units, for comparison with the results in [1], as well as the level at which these experimental results were statically significantly better than the results in [1].

Table 3. Best results over all precipitation indices and seven different algorithms for periods 10, 15, and 17

Period	Outlook	Best		Test Results		Results from [1]	Significant level
		Classifier	Climatic Index	Avg MAE	MAD	MAD	
10	2 week	LeastMedSq	8 week SPI	28.48	0.05	0.16	α=.01
	4 week	Bagging	20 week SPI	73.95	0.12	0.23	α=.01
15	2 week	Bagging	8 week SPI	10.01	0.02	0.09	α=.01
	4 week	Bagging	36 week SPI	70.32	0.11	0.14	α=.10
	6 week	Bagging	28 week SPI	71.43	0.11	0.18	α=.01
17	2 week	Bagging	4 week SPI	8.53	0.01	0.07	α=.01
	4 week	M5Rules	32 week SPI	76.77	0.12	0.11	--
	6 week	Bagging	40 week SPI	76.00	0.12	0.15	--

The MAD results improved over the results in [1], particularly for the 2-week outlooks, the 4-week outlook on period ten, and slightly on the 4- and 6-week outlooks for periods 15 and 17. As shown in Table 3, most of these results were significantly better than the earlier results at the α=.01 level. Only two of these results were not significantly better than the earlier results (period 17 with 4- and 6-week outlooks), but those results were statistically equivalent to the earlier results. These results again support that algorithm selection is important to improving VegOut prediction capability.

4   Future Work

The prediction accuracy of all 4- and 6-week prediction outlooks needs further examination. When looking at Table 3, the best average MAE of 70.32, using a 4-week outlook to predict period 15, is poor compared to the 2-week prediction outlook average MAE of 8.12. In fact, all of the 4- and 6-week prediction outlook results could be improved. Models need to be investigated to see if more accurate prediction on the longer outlooks can be achieved. This is critical for allowing practitioners to make more accurate decisions, such as irrigation quotas and water releases from dams – decisions that must be made several weeks in advance. Ensembles besides bagging need to be explored to see if they could be used to improve prediction accuracy without compromising the ability for practitioners to comprehend the model. New inputs, such as sun spot data, need to be investigated for inclusion in the VegOut model for improving overall prediction accuracy. Finally, VegOut currently is focused on 2-, 4- and 6-week outlooks in fifteen U.S. states in the Midwest and Great Plains regions, but expansion to other areas of the U.S. is planned in the near future.

5 Conclusion

As with any scientific discovery model, it is important to evaluate the algorithm and features selected for use with the VegOut drought monitoring tool. The current version of the VegOut tool uses the rule-based regression tree algorithm, Cubist. This study experimented with several data mining algorithms, as well as various climatic indices for the purpose of improving the overall accuracy of the VegOut prediction at the 2-, 4-, and 6-week outlooks.

Overall, the 28-week SPI climatic index had the lowest overall average MAE of all the climatic indices, for each of the time periods in the experiments and all of the prediction outlooks, and would likely make a good choice for the climatic index feature to use for the VegOut model. However, since the variability between the various precipitation indices for any of the algorithms was not statistically significant, any of the precipitation indices could be chosen for the VegOut application without compromising much prediction accuracy.

Overall, the bagging algorithm produced the lowest overall average MAE for the time periods and prediction outlooks studied. The variability between the various algorithms was statistically significant. The bagging algorithm was the only algorithm that had the smallest average MAE over all of the experiments and should be used for this application.

The best results from these experiments compared well with the best results from the experiments in earlier studies. In fact, these results were either statistically equivalent or statistically better than the earlier results.

The experimental results from this study support the results from [1] that the use of data mining in monitoring drought and its impact on vegetation conditions *ahead* of time over large geographic areas is possible. Further progress and improvements to VegOut are expected in the future as the research initiatives outlined in section 4 are undertaken.

References

1. Tadesse, T., Wardlow, B.: The Vegetation Outlook (VegOut): A New Tool for Providing Outlooks of General Vegetation Conditions Using Data Mining Techniques. In: 7th IEEE International Conference on Data Mining Workshops, pp. 667–672. IEEE Press, Washington (2007)
2. Rulequest Research. An overview of Cubist,
   http://rulequest.com/cubist-win.html
3. Rice, J.R.: The algorithm selection problem. Advances in Computers 15, 65–118 (1976)

4. Smith-Miles, K.A.: Cross-Disciplinary perspectives on meta-learning for algorithm selection. ACM Computing Surveys 41, 1–25 (2008)
5. Wilhite, D.: Preparing for drought: a methodology. In: Wilhite, D.A. (ed.) Drought: A Global Assessment. Routledge Hazards and Disaster Series, vol. II, pp. 89–104 (2002)
6. Wilhite, D., Svoboda, M.: Drought Early Warning Systems in the Context of Drought Preparedness and Mitigation. Preparedness and Mitigation Proceedings of an Expert Group, Lisbon, Portugal (2000)
7. Tadesse, T., Brown, J.F., Hayes, M.J.: A new approach for predicting drought-related vegetation stress: Integrating satellite, climate, and biophysical data over the U.S. central plains. ISPRS Journal of Photogrammetry and Remote Sensing 59(4), 244–253 (2005)
8. Bayarjargala, Y., Karnieli, A., Bayasgalan, M., Khudulmurb, S., Gandush, C., Tucker, C.J.: A comparative study of NOAA–AVHRR derived drought indices using change vector analysis. Remote Sensing of Environment 105(1), 9–22 (2006)
9. Tadesse, T., Wilhite, D.A., Harms, S.K., Hayes, M.J., Goddard, S.: Drought monitoring using data mining techniques: A case study for Nebraska, U.S.A. Natural Hazards Journal 33, 137–159 (2004)
10. Palmer, W.C.: Meteorological Drought. Research Paper No. 45, U.S. Department of Commerce Weather Bureau, Washington, D.C. p. 58 (1965)
11. Wells, N., Goddard, S., Hayes, M.J.: A Self-Calibrating Palmer Drought Severity Index. Journal of Climate 17(12), 2335–2351 (2004)
12. McKee, T.B., Doesken, N.J., Kleist, J.: Drought Monitoring with Multiple Time Scales. In: Preprints, 9th Conference on Applied Climatology, Dallas, Texas, January 15–20, pp. 233–236 (1995)
13. Reed, B.C., Brown, J.F., VanderZee, D., Loveland, T.R., Merchant, J.W., Ohlen, D.O.: Measuring phonological variability from satellite imagery. Journal of Vegetation Science 5, 703–714 (1994)
14. Barnston, A.G., Kumar, A., Goddard, L., Hoerling, M.P.: Improving seasonal prediction practices through attribution of climate variability. Bull. Amer. Meteor. Soc. 86, 59–72 (2005)
15. Tadesse, T., Wilhite, D.A., Hayes, M.J., Harms, S.K., Goddard, S.: Discovering associations between climatic and oceanic parameters to monitor drought in Nebraska using data-mining techniques. Journal of Climate 18(10), 1541–1550 (2005)
16. Los, S.O., Collatz, G.J., Bounoua, L., Sellers, P.J., Tucker, C.J.: Global Interannual Variations in Sea Surface Temperature and Land Surface Vegetation, Air Temperature, and Precipitation. Journal of Climate, 1535–1549 (2001)
17. Witten, I., Frank, E.: Data Mining: Practical Machine Learning Tools and Techniques, 2nd edn. Morgan Kaufmann, San Francisco (2005)
18. Rousseeuw, P.J., Leroy, A.M.: Robust Regression and Outlier Detection. John Wiley & Sons, Inc., Chichester (1987)
19. Quinlan, R.J.: Learning with Continuous Classes. In: 5th Australian Joint Conference on Artificial Intelligence, Singapore, pp. 343–348 (1992)
20. Wang, Y., Witten, I.H.: Induction of model trees for predicting continuous classes. In: Poster papers of the 9th European Conference on Machine Learning (1997)
21. Holmes, G., Hall, M., Frank, E.: Generating Rule Sets from Model Trees. In: Twelfth Australian Joint Conference on Artificial Intelligence, pp. 1–12 (1999)
22. Breiman, L.: Bagging predictors. Machine Learning. 24(2), 123-140 (1996)
23. Shevade, S.K., Keerthi, S.S., Bhattacharyya, C., Murthy, K.R.K.: Improvements to the SMO Algorithm for SVM Regression. IEEE Transactions on Neural Networks (1999)

Regression Trees from Data Streams with Drift Detection

Elena Ikonomovska[1], João Gama[2,3], Raquel Sebastião[2,4], and Dejan Gjorgjevik[1]

[1] FEEIT – Ss. Cyril and Methodius University, Karpos II bb, 1000 Skopje, Macedonia
[2] LIAAD/INESC – University of Porto, Rua de Ceuta, 118 – 6, 4050-190 Porto, Portugal
[3] Faculty of Economics – University of Porto, Rua Roberto Frias, 4200 Porto, Portugal
[4] Faculty of Science – University of Porto, R. Campo Alegre 823, 4100 Porto, Portugal

Abstract. The problem of extracting meaningful patterns from time changing data streams is of increasing importance for the machine learning and data mining communities. We present an algorithm which is able to learn regression trees from fast and unbounded data streams in the presence of concept drifts. To our best knowledge there is no other algorithm for incremental learning regression trees equipped with change detection abilities. The FIRT-DD algorithm has mechanisms for drift detection and model adaptation, which enable to maintain accurate and updated regression models at any time. The drift detection mechanism is based on sequential statistical tests that track the evolution of the local error, at each node of the tree, and inform the learning process for the detected changes. As a response to a local drift, the algorithm is able to adapt the model only locally, avoiding the necessity of a global model adaptation. The adaptation strategy consists of building a new tree whenever a change is suspected in the region and replacing the old ones when the new trees become more accurate. This enables smooth and granular adaptation of the global model. The results from the empirical evaluation performed over several different types of drift show that the algorithm has good capability of consistent detection and proper adaptation to concept drifts.

Keywords: data stream, regression trees, concept drift, change detection, stream data mining.

1 Introduction

Our environment is naturally dynamic, constantly changing in time. Huge amounts of data are being constantly generated by various dynamic systems or applications. Real-time surveillance systems, telecommunication systems, sensor networks and other dynamic environments are such examples. Learning algorithms that model the underlying processes must be able to track this behavior and adapt the decision models accordingly. The problem takes the form of changes in the target function or the data distribution over time, and is known as concept drift. Examples of real world problems where drift detection is relevant include user modeling, real-time monitoring industrial processes, fault detection, fraud detection, spam, safety of complex systems, and many others [1]. In all these dynamic processes, the new concepts replace the old concepts, while the interest of the final user is always to have available model that

J. Gama et al. (Eds.): DS 2009, LNAI 5808, pp. 121–135, 2009.
© Springer-Verlag Berlin Heidelberg 2009

will describe or accurately predict the state of the underlying process at every time. Therefore, the importance of drift detection when learning from data streams is evident and must be taken into consideration. Most machine-learning algorithms, including the FIMT algorithm [2] make an assumption that the training data is generated by a single concept from a stationary distribution, and are designed for static environments. However, when learning from data streams dynamic distributions are rule and not an exception. To meet the challenges posed by the dynamic environment, they must be able to detect changes and react properly on time. This is the challenge we address in this work: how to embed change detection mechanisms inside a regression tree learning algorithm and adapt the model properly.

Having in mind the importance of the concept drifting problem when learning from data streams, we have studied the effect of local and global drift over the accuracy and the structure of the learned regression tree. We propose the FIRT-DD (Fast and Incremental Regression Tree with Drift Detection) algorithm which is able to learn regression trees from possibly unbounded, high-speed and time-changing data streams. FIRT-DD algorithm has mechanisms for drift detection and model adaptation, which enable to maintain an accurate and updated model at any time. The drift detection mechanism is consisted of distributed statistical tests that track the evolution of the error at every region of the instance space, and inform the algorithm about significant changes that have affected the learned model locally or globally. If the drift is local (affects only some regions of the instance space) the algorithm will be able to localize the change and update only those parts of the model that cover the influenced regions.

The paper is organized as follows. In the next section we present the related work in the field of drift detection when learning in dynamic environments. Section 3 describes our new algorithm FIRT-DD. Section 4 describes the experimental evaluation and presents a discussion of the obtained results. We conclude in section 5 and give further directions.

2   Learning with Drift Detection

The nature of change is diverse. Changes may occur in the context of learning, due to changes in hidden variables or in the intrinsic properties of the observed variables. Often these changes make the model built on old data inconsistent with the new data, and regular updating of the model is necessary. In this work we look for changes in the joint probability $P(X, Y)$, in particular for changes in the Y values given the attribute values X, that is $P(Y|X)$. This is usually called concept drift. There are two types of drift that are commonly distinguished in the literature: abrupt (sudden, instantaneous) and gradual concept drift. We can also make a distinction between local and global drift. The local type of drift affects only some parts of the instance space, while global concept drift affects the whole instance space. Hidden changes in the joint probability may also cause a change in the underlying data distribution, which is usually referred to as virtual concept drift (sampling shift). A good review of the types of concept drift and the existing approaches to the problem is given in [3, 4]. We distinguish three main approaches for dealing with concept drift:

1. Methods based on data management. These include weighting of examples, or example selection using time-windows with fixed or adaptive size. Relevant work is [5].
2. Methods that explicitly detect a change point or a small window where change occurs. They may follow two different approaches: (1) monitoring the evolution of some drift indicators [4], or (2) monitoring the data distribution over two different time-windows. Examples of the former are the FLORA family of algorithms [6], and the works of Klinkenberg presented in [7, 8]. Examples of the latter are the algorithms presented in [9, 10].
3. Methods based on managing ensembles of decision models. The key idea is to maintain several different decision models that correspond to different data distributions and manage an ensemble of decision models according to the changes in the performance. All ensemble based methods use some criteria to dynamically delete, reactivate or create new ensemble members, which are normally based on the model's consistency with the current data [3]. Such examples are [11, 12].

The adaptation strategies are usually divided on blind and informed methods. The latter adapt the model without any explicit detection of changes. These are usually used with the data management methods (time-windows). The former methods adapt the model only after a change has been explicitly detected. These are usually used with the drift detection methods and decision model management methods.

The motivation of our approach is behind the advantages of explicit detection and informed adaptation methods, because they include information about process dynamics: meaningful description of the change and quantification of the changes. Another important aspect of drift management methods that we adopt and stress is the ability to detect local drift influence and adapt only parts of the learned decision model. In the case of local concept drift, many global models are discarded simply because their accuracy on the current data chunks falls, although they could be good experts for the stable parts of the instance space. Therefore, the ability to incrementally update local parts of the model when required is very important. Example of a system that possesses this capability is the CVFDT system [13]. CVFDT algorithm performs regular periodic validation of its splitting decisions by maintaining the necessary statistics at each node over a window of examples. Every time a split is discovered as invalid it starts growing new decision tree rooted at the corresponding node. The new sub-trees grown in parallel are aimed to replace the old ones since they are generated using data which corresponds to the now concepts. To smooth the process of adaptation, CVFDT keeps the old tree rooted at the influenced node until one of the new ones becomes more accurate. Maintaining the necessary counts for class distributions at each node requires significant amount of additional memory and computations, especially when the tree becomes large. We address this problem with the utilization of a light weight detection units positioned in each node of the tree, which evaluate the goodness of the split continuously for every region of the space using only few incrementally maintained variables. This approach does not require significant additional amount of memory or time and is therefore suitable for the streaming scenario, while at the same time enables drift localization.

3 The FIRT-DD Algorithm

The FIRT-DD algorithm is an adaptation of the FIMT algorithm [2] to dynamic environments and time-changing distributions. FIMT is an incremental any-time algorithm for learning model trees from data streams. FIMT builds trees following the top-down approach where each splitting decision is based on a set of examples corresponding to a certain time period or a sequence of the data stream. Decisions made in upper nodes of the tree are therefore based on older examples, while the leaves receive the most current set of examples. Each node has a prediction obtained during its growing phase. The FIMT algorithm can guarantee high asymptotic similarity of the incrementally learned tree to the one learned in a batch manner if provided with enough data. This is done by determining a bound on the probability of selecting the best splitting attribute. The probability bound provides statistical support and therefore stability to the splitting decision as long as the distribution of the data is stationary. However, when the distribution is not stationary and the data contains concept drifts, some of the splits become invalid. We approach this problem using statistical process control methodologies for change detection, which are particularly suitable for data streams. Statistical process control (SPC) methodologies [14] are capable to handle large volume of data and have been widely used to monitor, control and improve industrial processes quality. In recent years some SPC techniques were developed to accommodate auto-correlated data, such as process residual charts.

3.1 A Fully Distributed Change Detection Method Based on Statistical Tests

Although the regression tree is a global model it can be decomposed according to the partitions of the instance space obtained with the recursive splitting. Namely, each node with the sub-tree below covers a region (a hyper-rectangle) of the instance space. The root node covers the whole space, while the descendant nodes cover subspaces of the space covered by their parent node. When a concept drift occurs locally in some parts of the instance space, it is much less costly to make adaptations only to the models that correspond to that region of the instance space. The perception of the possible advantages of localization of drift has led us to the idea of a fully distributed change detection system.

In order to detect where in the instance space drift has occurred, we bound each node of the tree with a change detection unit. The change detection units bounded with each node perform local, simultaneous and separate monitoring of every region of the instance space. If a change has been detected, it suggests that the instance space to which the node belongs has been influenced by a concept drift. The adaptation will be made only at that sub-tree of the regression tree. This strategy has a major advantage over global change detection methods, because the costs of updating the decision model are significantly lower. Further, it can detect and adapt to changes in different parts of the instance space at the same time, which speeds up the adaptation process.

The change detection units are designed following the approach of monitoring the evolution of one or several concept drift indicators. The authors in [8] describe as common indicators: performance measures, properties of the decision model, and properties of the data. In the current implementation of our algorithm, we use the absolute loss as performance measure[1]. When a change occurred in the target concept the actual model does not correspond to the current status of nature and the absolute loss will increase. This observation suggests a simple method to detect changes: monitor the evolution of the loss. The absolute loss at each node is the absolute value of the difference between the prediction (i.e. average of the target variable) from the period the node was a leaf and the prediction now. If it starts increasing it may be a sign of change in the examples' distribution or in the target concept. To confidently tell that there is a significant increase of the error which is due to a concept drift, we propose continuous performing statistical tests at every internal node of the tree. The test would monitor the absolute loss at the node, tracking major statistical increase which will be a sure sign that a change occurred. The alarm of the statistical test will trigger mechanisms for adaptation of the model.

In this work, we have considered two methods for monitoring the evolution of the loss. We have studied the CUSUM algorithm [15] and Page-Hinkley method [16], both from the same author. The CUSUM charts were proposed by Page [15] to detect small to moderate shifts in the process mean. Since we are only interested in detecting increases of the error the CUSUM algorithm is suitable to use. However, when compared with the second option, the Page-Hinkley (PH) test [16], the results attested stability and more consistent detection in favor of the PH test. Therefore in the following sections we have only analyzed and evaluated the second option. The PH test is a sequential adaptation of the detection of an abrupt change in the average of a Gaussian signal [1]. This test considers two variables: a cumulative variable m_T, and its minimum value M_T. The first variable m_T is defined as the cumulated difference between the observed values and their mean till the current moment, where T is the number of all examples seen so far and x_t is the variable's value at time t:

$$m_T = \sum_{t=1}^{T} (x_t - \overline{x}_t - \alpha),$$

(1)

$$\text{where } \overline{x}_t = \frac{1}{t} \sum_{l=1}^{t} x_l,$$

(2)

Its minimum value after seeing T examples is computed using the following formula:

$$M_T = \min\{m_T, t = 1...T\}.$$

(3)

At every moment the PH test monitors the difference between the minimum M_T and m_T: $PH_T = m_T - M_T$. When this difference is greater than a given threshold (λ) it alarms a change in the distribution (i.e. $PH_T > \lambda$). The threshold parameter λ depends on the admissible false alarm rate. Increasing λ will entail fewer false alarms, but might miss some changes. The parameter α corresponds to the magnitude of changes that are allowed.

---

[1] Other measures like square loss could be used. For the purposes of change detection both metrics provide similar results.

Each example from the data stream is traversing the tree until reaching one of the leaves where the necessary statistic for building the tree is maintained. On its path from the root till the leaf it will traverse several internal nodes, each of them bounded with a drift detection unit which continuously performs the PH test, monitoring the evolution of the absolute loss after every consecutive example passing through the node. When the difference between the observed absolute loss and its average is increasing fast and continuously, eventually exceeding a user predefined limit (λ), we can confidently conclude that the predictions used in the computation of the absolute loss are continuously and significantly wrong. This is an indication of a non-valid split at the corresponding node.

The computation of the absolute loss can be done using the prediction from the node where the PH test is performed or from the parent of the leaf node where the example will fall (the leaf node is not used because it is still in its growing phase). As a consequence of this we will consider two methods for change detection and localization. The prediction (whether it is from the node where the PH test is or from the parent node of the leaf) can be: 1) the mean of y-values of the examples seen at that node during its growing phase or; 2) the perceptron's output for a given example which is trained incrementally at the node. Preliminary experiments have shown that the detection of the changes is more consistent and more reliable when using the first option, that is the average of y values. If the loss is computed using the prediction from the current node the computation can be performed while the example is passing the node on its path to the leaf. Therefore, the loss will be monitored first at the root node and after at its descendant nodes. Because the direction of monitoring the loss is from the top towards the "bottom" of the tree, this method will be referred to as Top-Down (TD) method. If the loss is computed using the prediction from the parent of the leaf node, the example must first reach the leaf and then the computed difference at the leaf will be back-propagated towards the root node. While back-propagating the loss (using the same path the example reached the leaf) the PH tests located in the internal nodes will monitor the evolution. This reversed monitoring gives the name of the second method for change detection which will be referred to as Bottom-Up (BU) method. The idea for the BU method came from the following observation: with the TD method the loss is being monitored from the root towards the leaves, starting with the whole instance space and moving towards smaller sub-regions. While moving towards the leaves, predictions in the internal nodes become more accurate which reduces the probability of false alarms. This was additionally confirmed with the empirical evaluation: most of the false alarms were triggered at the root node and its immediate descendants. Besides this, the predictions from the leaves which are more precise than those at the internal nodes (as a consequence of splitting) emphasize the loss in case of drift. Therefore, using them in the process of change detection would shorten the delays and reduce the number of false alarms. However, there is one negative side of this approach, and it concerns the problem of gradual and slow concept drift. Namely, the tree grows in parallel with the process of monitoring the loss. If drift is present, consecutive splitting enables new and better predictions according to the new concepts, which would disable proper change detections. The slower the drift is the probability of non-detection gets higher.

3.2   Adapting to Concept Drift

The natural way of responding to concept drift for the incremental algorithm (if no adaptation strategy is employed) would be to grow and append new sub-trees to the old structure, which would eventually give correct predictions. Although the predictions might be good, the structure of the whole tree would be completely wrong and misleading. Therefore, proper adaptation of the structure is necessary. The FIRT-DD without the change detection abilities is the incremental algorithm FIRT [2]. For comparison FIRT is also used in the evaluation and is referred to as "No detection".

Our adaptation mechanism falls in the group of informed adaptation methods: methods that modify the decision model only after a change was detected. Common adaptation to concept drift is forgetting the outdated and un-appropriate models, and re-learning new ones that will reflect the current concept. The most straightforward adaptation strategy consists of pruning the sub-models that correspond to the parts of the instance space influenced by the drift. If the change is properly detected the current model will be most consistent with the data and the concept which is being modeled. Depending on whether the TD or the BU detection method is used, in the empirical evaluation this strategy will be referred to as "TD Prune" and "BU Prune" correspondingly. However when the change has been detected relatively high in the tree, pruning the sub-tree will decrease significantly the number of predicting nodes – leaves which will lead to unavoidable drastic short-time degradation in accuracy. In these circumstances, an outdated sub-tree may still be better than a single leaf. Instead of pruning when a change is detected we can start building an alternate tree (a new tree) from the examples that will reach the node. A similar strategy is used in CVFDT [13] where on one node can be grown several alternate trees at the same time.

This is the general idea of the adaptation method proposed in the FIRT-DD algorithm. When a change is detected the change detection unit will trigger the adaptation mechanism for the node where the drift was detected. The node will be marked for regrowing and new memory will be allocated for maintaining the necessary statistic used for growing a leaf. Examples which will traverse a marked node will be used for updating its statistic, as well as the statistic at the leaf node where they will eventually fall. The regrowing process will initiate a new alternate tree rooted at the node which will grow in parallel with the old one. Every example that will reach a node with an alternate tree will be used for growing both of the trees. The nodes in the alternate tree won't perform change detection till the moment when the new tree will replace the old one. The old tree will be kept and grown in parallel until the new alternate tree becomes more accurate.

However, if the detected change was a false alarm or the alternate tree cannot achieve better accuracy, replacing the old tree might never happen. If the alternate tree shows slow progress or starts to degrade in performance this should be considered as a sign that growing should be stopped and the alternate tree should be removed. In order to prevent reactions to false alarms the node monitors the evolution of the alternate tree and compares its accuracy with the accuracy of the original sub-tree. This is performed by monitoring the average difference in accuracy with every example reaching the node. The process of monitoring the average difference starts after the node has received twice of the growing process chunk size (e.g. 400 examples) of examples, which should be enough to determine the first split and to grow an alternate tree with at least three nodes. When this number is reached the nodes starts to maintain the mean squared error for the old and the alternate tree simultaneously. On a user predetermined evaluation interval (e.g. 5000 examples) the difference of the mean squared error between the old and the new tree is evaluated, and if it is positive and at least 10% greater than the MSE of the old tree the new alternate tree will replace the old one. The old tree will be removed, or as an alternative it can be stored for possible reuse in case of reoccurring concepts. If the MSE difference does not fulfill the previous conditions its average is incrementally updated. Additionally in the evaluation process a separate parameter can be used to specify a time period which determines how much time a tree is allowed to grow in order to become more accurate than the old one. When the time period for growing an alternate tree has passed or the average of the difference started to increase instead to decrease, the alternate tree will be removed from the node together with the maintained statistic and the memory will be correspondingly released. In order to prevent from premature discarding of the alternate tree the average of the MSE difference is being evaluated only after several evaluation intervals have passed. The strategy of growing alternate trees will be referred to as "TD AltTree" and "BU AltTree" depending on which change detection method is used (TD/BU).

4   Experimental Evaluation

To provide empirical support to FIRT-DD we have performed an evaluation over several versions of an artificial "benchmark" dataset (simulating several different types of drift) and over a real-world dataset from the DataExpo09 [18] competition. Using artificial datasets allows us to control relevant parameters and to empirically evaluate the drift detection and adaptation mechanisms. The real-problem dataset enables us to evaluate the merit of the method for real-life problems. Empirical evaluation showed that the FIRT-DD algorithm possesses satisfactory capability of detection and adaptation to changes and is able to maintain an accurate and well structured decision model.

4.1   Experimental Setup and Metrics

In the typical streaming scenario data comes in sequential order, reflecting the current state of the physical process that generates the examples. Traditional evaluation methodologies common for static distributions are not adequate for the streaming setting because they don't take into consideration the importance of the order of examples (i.e. their dependency on the time factor), as well as the evolution of the model in time. One convenient methodology is to use the predictive sequential or *prequential* error which can be computed as a cumulative sum of a loss function (error obtained after every consecutive example) typical for the regression domain (mean squared or root relative mean squared error). The *prequential* approach uses all the available data for training and testing, and draws a pessimistic learning curve that traces the evolution of the error. Its main disadvantage is that it accumulates the errors from the first examples of the data stream and therefore hinders precise on-line evaluation of real performances. Current improvements cannot be easily seen due to past degradation in accuracy accumulated in the prequential error.

More adequate methodology in evaluating the performance of an incremental algorithm is to use an "exponential decay"/"fading factor" evaluation or a sliding window evaluation. Using the "exponential decay"/"fading factor" method we can diminish the influence of earlier errors by multiplying the cumulated loss with an $e^{-\delta t}$ function of the time t or a fading factor (constant value less than one, e.g. 0.99) before summing the most current error. This method requires setting only the parameter δ or the fading factor, but since it still includes all the previous information of the error it gives slightly smoothed learning curve. With the sliding window method for evaluation we can obtain detailed and precise evaluation over the whole period of training/learning without the influence of earlier errors. With this method we evaluate the model over a test set determined by a window of examples which the algorithm has not used for training. The window of examples manipulates the data like a FIFO (first-in-first-out) queue. The examples which have been used for testing are given to the algorithm one by one for the purpose of training. The size of the sliding window determines the level of aggregation and it can be adjusted to the quantity a user is willing to have. The sliding step determines the level of details or the smoothness of the learning curve and can be also adjusted. Using the sliding window test set we measure accuracy in terms of the mean squared error (MSE) or root relative squared error (RRSE) and the current dimensions of the learned model.

We have performed a sensitivity analysis on the values of the parameters α and λ which resulted in the pairs of values (0.1, 200), (0.05, 500) and (0.01, 1000), correspondingly. Smaller values for α would increase the sensibility, while smaller values for λ would shorten the delays in the change detection. However, we should have in mind that smaller λ values increase the probability of detecting false alarms. In the empirical evaluation we have used $\alpha = 0.05$ and $\lambda = 500$ for all the simulations of drift over the Fried artificial dataset.

4.2   The Datasets

For simulation of the different types of drift we have used the Fried dataset used by Friedman in [18]. This is an artificial dataset containing 10 continuous predictor attributes with independent values uniformly distributed in the interval [0, 1]. From those 10 predictor attributes only five attributes are relevant, while the rest are redundant. The original function for computing the predicted variable y is:

$$y = 10sin(\pi x_1 x_2) + 20(x_3 - 0.5)^2 + 10x_4 + 5x_5 + \sigma(0, 1),$$

where $\sigma(0,1)$ is a random number generated from a normal distribution with mean 0 and variance 1. The second dataset used is from the Data Expo competition [18] and contains large amount of records containing flight arrival and departure details for all commercial flights within the USA, from October 1987 to April 2008. This is a large dataset since there are nearly 120 million records. The *Depdelay* dataset we have used in the evaluation contains around 14 million records starting from January 2006 to April 2008. The dataset is cleaned and records are sorted according to the departure date (year, month, day) and time (converted in seconds). The target variable is the departure delay in seconds.

4.3 Results over Artificial Datasets

Using the artificial dataset we performed a set of controlled experiments. We have studied several scenarios related with different types of change:

1. **Local abrupt drift**. The first type of simulated drift is local and abrupt. We have introduced concept drift in two distinct regions of the instance space. The first region where the drift occurs is defined by the inequalities: $x_2 < 0.3$ and $x_3 < 0.3$ and $x_4 > 0.7$ and $x_5 < 0.3$. The second region is defined by: $x_2 > 0.7$ and $x_3 > 0.7$ and $x_4 < 0.3$ and $x_5 > 0.7$. We have introduced three points of abrupt concept drift in the dataset, the first one at one quarter of examples, the second one at one half of examples and the third at three quarters of examples. For all the examples falling in the first region ($x_2 < 0.3$ and $x_3 < 0.3$ and $x_4 > 0.7$ and $x_5 < 0.3$) the new function for computing the predicted variable y is: $y = 10x_1x_2 + 20(x_3 - 0.5) + 10x_4 + 5x_5 + \sigma(0, 1)$. For the second region ($x_2 > 0.7$ and $x_3 > 0.7$ and $x_4 < 0.3$ and $x_5 > 0.7$) the new function for computing the predicted variable y is: $y = 10cos(x_1x_2) + 20(x_3 - 0.5) + e^{x_4} + 5x_5^2 + \sigma(0, 1)$. At every consecutive change the region of drift is expanded. This is done by reducing one of the inequalities at a time. More precisely, at the second point of change the first inequality $x_2 < 0.3$ ($x_2 > 0.7$) is removed, while at the third point of change two of the inequalities are removed: $x_2 < 0.3$ and $x_3 < 0.3$ ($x_2 > 0.7$ and $x_3 > 0.7$).
2. **Global abrupt drift**. The second type of simulated drift is global and abrupt. The concept drift is performed with a change in the original function over the whole instance space, which is consisted of misplacing the variables from their original position. We have introduced two points of concept drift, first at one half of examples when the function for computing the predicted variable becomes: $y = 10sin(\pi x_4x_5) + 20(x_2 - 0.5)^2 + 10x_1 + 5x_3 + \sigma(0,1)$, and the second point at three quarters of examples, when the old function is returned (reoccurrence).
3. **Global gradual drift**. The third type of simulated drift is global and gradual. The gradual concept drift is initiated the first time at one half of examples. Starting from this point examples from the new concept: $y = 10sin(\pi x_4x_5) + 20(x_2 - 0.5)^2 + 10x_1 + 5x_3 + \sigma(0,1)$ are being gradually introduced among the examples from the first concept. On every 1000 examples the probability of generating an example using the new function is incremented. This way after 100000 examples only the new concept is present. At three quarters of examples a second gradual concept drift is initiated on the same way. The new concept function is: $y = 10sin(\pi x_2x_5) + 20(x_4 - 0.5)^2 + 10x_3 + 5x_1 + \sigma(0,1)$. Examples from the new concept will gradually replace the ones from the last concept like before. The gradual drift ends again after 100000 examples. From this point only the last concept is present in the data.

The first part of the experimental evaluation was focused on analyzing and comparing the effectiveness of the change detection methods proposed. The comparison was performed only over the artificial datasets (each with size of 1 million examples) because the drift is known and controllable and because they enable precise measurement of delays, false alarms and miss detections. The Table 1 presents the averaged results over 10 experiments for each of the artificial datasets. We have measured the number of false alarms for each point of drift, the Page-Hinkley test delay (number of examples monitored by the PH test before the detection) and the global delay (number of examples processed in the tree from the point of the concept drift till the moment of the detection). The delay of the Page-Hinkley test measures how fast the algorithm will be able to start the adaptation strategy at the local level, while the global delay measures how fast the change was detected globally. The "Num. of change" column specifies the number of the change point.

Table 1. Averaged results from the evaluation of change detection over 10 experiments

Data set	Num. of change	Top – Down (TD)			Bottom – Up (BU)		
		FA's	PH test delay	Global delay	FA's	PH test delay	Global delay
Local abrupt drift	1	0.5	1896.8	5111.7	0	698.6	14799.8
	2	1.7	1128.2	2551.1	0	494.1	3928.6
	3	0.8	3107.1	5734.5	0	461.1	5502.4
Global abrupt drift	1	1.5	284.5	1325.7	0	260.4	260.4
	2	0	492.2	3586.3	0	319.9	319.9
Global gradual drift	1	1	2619.7	16692.9	0	1094.2	14726.3
	2	2.8	4377.5	10846.5	0	644.7	11838.2
No drift	0	0.9	-	-	0	-	-

Results in Table 1 show in general that the TD method for change detection triggers significant number of false alarms as compared with the BU method (which never detects false alarms). Both of the methods detect all the simulated changes for all the types of drift. The detailed analysis for the *Local abrupt drift* dataset has shown that most of the false alarms with TD were triggered at the root node. The true positives with TD are also detected higher in the tree, while with the BU method changes are detected typically in lower nodes, but precisely at the lowest node whose region is closest to the region with local drift. On this way the BU method performs most precise localization of the drift. With the TD method the global delay for the first and the second point of drift is smaller (because changes are detected higher), but with the BU method the PH test delays are much smaller. This enables faster local adaptation. The analysis from Table 1 over the *Global abrupt drift* dataset has shown clear advantage of the BU method over the TD method. The BU method does not detect false alarms, and both of the delays are smaller especially the global delay. With BU changes are detected first at the root node (which covers the whole region where the global drift occurs) and only after in the nodes below. This is not the case with the TD method when changes are detected first in the lower nodes and after at their parents (moving towards to root). False alarms are triggered usually for the root node. For the *Global gradual drift* dataset the detailed analysis showed that with both of the methods changes were detected similarly, starting at the lower parts of the tree and moving towards the root. The main difference is that the BU method detects changes significantly faster, having smaller delays. However, once a change is detected at the highest point, the adaptation strategy (Prune or AltTree) prevents from detecting more changes, although the drift might still be present and increasing. The fast detection of the drift in this case is a disadvantage for the BU method, rather than an advantage. The TD method whose delays are bigger detects changes during the whole period of gradual increase of the drift. This enables to perform the adaptation right on time when all of the examples belong to the new concept. The last dataset is the original Fried dataset without any drift. From the table it can be noted again that the BU method doesn't trigger false alarms, while the TD method detected at least one in nine of ten generations of the same dataset.

Table 2. Performance results over the last 10000 examples of the data stream averaged over 10 experiments for each type of simulated drift

Data set	Measures	No detection	Top – Down (TD)		Bottom – Up (BU)	
			Prune	AltTrees	Prune	AltTrees
Local abrupt drift	MSE/Std. dev.	4.947±0.14	6.768±0.13	3.82±0.08	4.969±0.1	3.696±0.08
	RRSE	0.4114	0.4845	0.363	0.4149	0.3571
	Growing nodes	1292.8	145.5	818.7	220.1	817.3
Global abrupt drift	MSE/Std. dev.	5.496±0.11	4.758±0.09	4.664±0.09	4.459±0.08	4.465±0.08
	RRSE	0.46	0.4364	0.4316	0.4221	0.4223
	Growing nodes	977.8	195.4	310.1	229.7	228.8
Global gradual drift	MSE/Std. dev.	16.629±0.32	4.682±0.08	3.85±0.07	5.149±0.11	7.461±0.16
	RRSE	0.7284	0.3917	0.3541	0.4081	0.4892
	Growing nodes	963.4	125.8	180.4	179.6	178.5

In Table 2 are given performance results over the last 10000 examples. Results in this table enable to evaluate the adaptation of model for the different types of drift, when learning has ended. For the *Local abrupt drift* dataset it is evident that the BU Prune strategy gives better results than the TD Prune strategy. This is easy to explain having in mind the comments from the last paragraph. Namely, in the case of TD detection much bigger portions of the tree are pruned than necessary because the drift is detected inadequately higher. The tree ends up smaller and even has lower accuracy as compared to the tree grown without change detection. The BU method performs precise drift localization, which enables pruning just the right parts of the tree and therefore achieving the same performance results as the "No detection" strategy but with a significantly lower number of rules. With the AltTree adaptation strategy reacting to false alarms is avoided. According to that performance results for TD AltTree are much better and even similar with the BU AltTree. For the *Global abrupt drift* dataset in general the BU approach gives slightly better results. It is interesting to notice that both of the adaptation strategies are equally good. Since drift is global they perform the same thing, regrowing the whole tree using examples from the new concept. However, the TD approach is not very adequate because the change is not detected first at the root node but somewhere lower. Because of that, neither of the adaptation strategies will enable proper correction of structure, although accuracy might still be good. The "No detection" strategy gives the worst results. For the *Global gradual drift* dataset the performance results are in favor of the TD method. Trees obtained are smaller and with better accuracy because of the on-time adaptation.

On Fig. 2 are given the learning curves obtained with the sliding window evaluation only for the *Local abrupt drift* and the *Global gradual drift* datasets due to

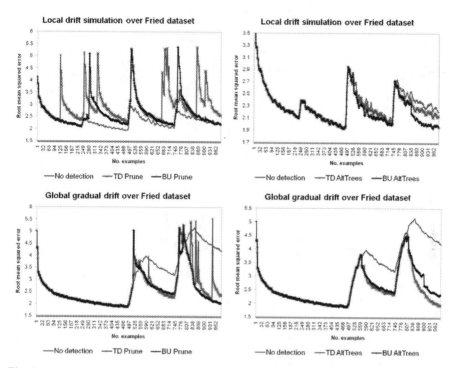

Fig. 1. Local abrupt and global abrupt/gradual drift simulation over Fried dataset using sliding window evaluation over a window of size 10000 and sliding step 1000 examples

lack of space. On the top left figure are evident many peaks corresponding to drastic degradation in accuracy when pruning huge parts of the tree or as a reaction false alarms (before the first point of drift). On the top right figure are shown the effects of smooth adaptation using the AltTree strategy. Obtained trees are smaller and continuously more accurate. Similar conclusions can also be obtained from the lower figures, but here more interesting is the advantage of the TD method, which is especially evident for the second point of drift. Comments on this type of drift are given below Table 1, but the general conclusion is that the tree obtained using the BU method shows worst results mainly because it has been grown during the presence of the two different concepts. Therefore, many of its splitting decisions are invalid.

4.4   Results over a Real-World Dataset

The Depdelay dataset represents a highly variable concept which depends on many time-changing variables. Performance results in Table 3 were obtained using the slid-ing window validation over the last 100000 examples. The results show significant improvement of the accuracy when change detection and adaptation is performed. The size of the model is also substantially smaller (in an order of magnitude). Stan-dard deviation of the error for TD/BU methods is bigger compared to the "No detec-tion" situation, but detailed results show that this is due to a sudden increase over the last 100000 examples (3 to 7 times). This can be seen on Fig.3. Both TD/BU AltTrees methods perform continuously better compared the "No detection" situation.

Table 3. Performance results over the last 100000 examples of the Depdelay dataset

Measures	No detection	Top – Down (TD)		Bottom – Up (BU)	
		Prune	AltTrees	Prune	AltTrees
MSE/Std. dev.	738.995±13.6	175.877±26.11	150.072±21.63	181.884±23.54	136.35±20.06
RRSE	0.396951	0.200305	0.185353	0.20379	0.181785
Growing nodes	4531	121	365	103	309

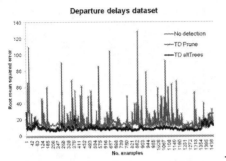
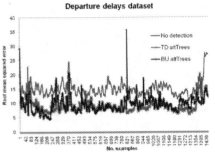

Fig. 2. Departure delays dataset

On Fig. 3 it can be also seen that when growing alternate trees the accuracy of the model is stable, persistent and continuously better than the accuracy of the model when no drift detection is performed. This is the evidence that data contains drifts and that the FIRT-DD algorithm is able to detect and adapt the model properly.

5 Conclusion

This paper presents a new algorithm for learning regression trees from time-changing data streams. To our best knowledge, FIRT-DD is the first algorithm for inducing regression trees for time-changing data streams. It is equipped with drift detection mechanism that exploits the structure of the regression tree. It is based on change-detection units installed in each internal node that monitor the growing process. The tree structure is being monitored at every moment and every part of the instance space. The change-detection units use only small constant amount of memory per node and small, constant amount of time for each example. FIRT-DD algorithm is able to cope with different types of drift including: abrupt or gradual, and local or global concept drift. It effectively maintains its model up-to-date with the continuous flow of data even when concept drifts occur. The algorithm enables local adaptation when required, reducing the costs of updating the whole decision model and perform-ing faster and better adaptation to the changes in data. Using an adaptation strategy based on growing alternate trees FIRT-DD avoids short-term significant performance degradation adapting the model smoothly. The model maintained with the FIRT-DD algorithm continuously exhibits better accuracy than the model grown without any change detection and proper adaptation. Preliminary application of FIRT-DD to a real-world domain shows promising results. Our future work will be focused on im-plementing these ideas in the domain of classification trees.

Acknowledgments. Thanks to the financial support by FCT under the PhD Grant SFRH/BD/41569/2007.

References

1. Basseville, M., Nikiforov, I.: Detection of Abrupt Changes: Theory and Applications. Prentice-Hall Inc., Englewood Cliffs (1987)
2. Ikonomovska, E., Gama, J.: Learning Model Trees from Data Streams. In: Boulicaut, J.-F., Berthold, M.R., Horváth, T. (eds.) DS 2008. LNCS (LNAI), vol. 5255, pp. 5–63. Springer, Heidelberg (2008)
3. Tsymbal, A.: The problem of concept drift: definitions and related work. Technical Report, TCD-CS-2004-15, Department of Computer Science, Trinity College Dublin, Ireland (2004)
4. Gama, J., Castillo, G.: Learning with Local Drift Detection. In: Bazzan, A.L.C., Labidi, S. (eds.) SBIA 2004. LNCS (LNAI), vol. 3171, pp. 286–295. Springer, Heidelberg (2004)
5. Klinkenberg, R.: Learning drifting concepts: Example selection vs. example weighting. J. Intelligent Data Analysis (IDA), Special Issue on Incremental Learning Systems Capable of Dealing with Concept Drift 8(3), 281–300 (2004)
6. Widmer, G., Kubat, M.: Learning in the presence of concept drifts and hidden contexts. J. Machine Learning 23, 69–101 (1996)
7. Klinkenberg, R., Joachims, T.: Detecting concept drift with support vector machines. In: Langley, P. (ed.) 17th International Conference on Machine Learning, pp. 487–494. Morgan Kaufmann, San Francisco (2000)
8. Klinkenberg, R., Renz, I.: Adaptive information filtering: Learning in the presence of concept drifts. In: Learning for Text Categorization, pp. 33–40. AAAI Press, Menlo Park (1998)
9. Kifer, D., Ben-David, S., Gehrke, J.: Detecting change in data streams. In: 30th International Conference on Very Large Data Bases, pp. 180–191. Morgan Kaufmann, San Francisco (2004)
10. Gama, J., Fernandes, R., Rocha, R.: Decision trees for mining data streams. J. Intelligent Data Analysis 10(1), 23–46 (2006)
11. Kolter, J.Z., Maloof, M.: Using additive expert ensembles to cope with concept drift. In: 22nd International Conference on Machine Learning, pp. 449–456. ACM, New York (2005)
12. Kolter, J.Z., Maloof, M.: Dynamic weighted majority: A new ensemble method for tracking concept drift. In: 3rd International Conference on Data Mining, pp. 123–130. IEEE Computer Society Press, Los Alamitos (2003)
13. Hulten, G., Spencer, L., Domingos, P.: Mining time-changing data streams. In: 7th ACM SIGKDD International Conference on Knowledge Discovery and Data Mining, pp. 97–106. ACM Press, Menlo Park (2001)
14. Grant, L., Leavenworth, S.: Statistical Quality Control. McGraw-Hill, United States (1996)
15. Page, E.S.: Continuous Inspection Schemes. J. Biometrika 41, 100–115 (1954)
16. Mouss, H., Mouss, D., Mouss, N., Sefouhi, L.: Test of Page-Hinkley, an Approach for Fault Detection in an Agro-Alimentary Production System. In: 5th Asian Control Conference, vol. 2, pp. 815–818. IEEE Computer Society Press, Los Alamitos (2004)
17. Friedman, J.H.: Multivariate Adaptive Regression Splines. J. The Annals of Statistics 19, 1–141 (1991)
18. ASA Sections on Statistical Computing and Statistical Graphics, Data Expo (2009), http://stat-computing.org/dataexpo/2009/

Mining Frequent Bipartite Episode from Event Sequences[*]

Takashi Katoh[1], Hiroki Arimura[1], and Kouichi Hirata[2]

[1] Graduate School of Information Science and Technology, Hokkaido University
Kita 14-jo Nishi 9-chome, Sapporo 060-0814, Japan
Tel.: +81-11-706-7678, Fax: +81-11-706-7890
{t-katou,arim}@ist.hokudai.ac.jp
[2] Department of Artificial Intelligence, Kyushu Institute of Technology
Kawazu 680-4, Iizuka 820-8502, Japan
Tel.: +81-948-29-7622, Fax: +81-948-29-7601
hirata@ai.kyutech.ac.jp

Abstract. In this paper, first we introduce a *bipartite episode* of the form $A \mapsto B$ for two sets A and B of events, which means that every event of A is followed by every event of B. Then, we present an algorithm that finds all frequent bipartite episodes from an input sequence without duplication in $O(|\Sigma| \cdot N)$ time per an episode and in $O(|\Sigma|^2 n)$ space, where Σ is an alphabet, N is total input size of $\mathcal{S}$, and n is the length of S. Finally, we give experimental results on artificial and real sequences to evaluate the efficiency of the algorithm.

1   Introduction

It is one of the important tasks in data mining to discover frequent patterns from time-related data. For such a task, Mannila *et al.* [10] have introduced *episode mining* to discover frequent *episodes* in an event sequence. Here, an episode is formulated as an acyclic labeled digraphs in which labels correspond to events and arcs represent a temporal precedent-subsequent relation in an event sequence. Then, the episode is a richer representation of temporal relationship than a subsequence, which represents just a linearly ordered relation in sequential pattern mining (*cf.*, [3,12]). Furthermore, since the frequency of the episode is formulated by a window that is a subsequence of an event sequence under a fixed time span, the episode mining is more appropriate than the sequential pattern mining when considering the time span.

For subclasses of episodes [8,9,5,10], a number of efficient algorithms have been developed so far (in Fig. 1). Mannila *et al.* [10] presented efficient mining algorithm for subclasses of episodes, called *parallel episodes* and *serial episodes* Mannila *et al.* [10] have designed an algorithm to construct general class of episodes from serial and parallel episodes, which is general but inefficient. On the other hand, in order to capture the direct relationship between premises

---

[*] This work is partially supported by Grand-in-Aid for JSPS Fellows (20·3406).

J. Gama et al. (Eds.): DS 2009, LNAI 5808, pp. 136–151, 2009.
© Springer-Verlag Berlin Heidelberg 2009

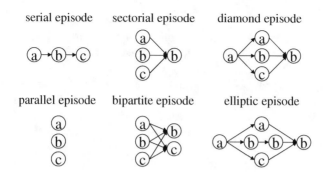

Fig. 1. Examples of subclasses of episode. serial episode (Mannila *et al.* [10]), parallel episode (Mannila *et al.* [10]), sectorial episode (Katoh *et al.* [8]), bipartite episode (this paper), diamond episode (Katoh *et al.* [9]), and elliptic episode (Katoh *et al.* [5]).

and consequences, Katoh *et al.* have introduced *sectorial episodes* [8], *diamond episodes* [9], and *elliptic episodes* [5]. Both episodes have the special events, a source as a premise and a sink as a consequence. In particular, from bacterial culture data [5,9], they have succeeded to find frequent diamond and elliptic episodes concerned with the replacement of bacteria and the changes for drug resistance from the medical viewpoint. Here, the source and the sink are set to the bacteria and another bacteria for the former episodes, and the sensitivity of antibiotic and the resistant of the same antibiotic for the latter episodes.

On the other hand, since both diamond and elliptic episodes have just a single source and a single sink, it is insufficient to represent the relationship including plural premises and plural consequences that simultaneously occur, for example, the replacement of *the families* of bacteria or the changes for *the families* of drug resistance. As the simplest forms of episodes to represent such a relationship, in this paper, we newly introduce *bipartite episodes* of the form $A \mapsto B$, where A and B are sets of events. The bipartite episode $A \mapsto B$ means that every event of A is followed by every event of B, so A and B are regarded as the sets of sources and sinks, respectively, and the graph representation of it forms a directed bipartite graph from A to B.

By paying our attention to enumeration methods, Katoh *et al.* [5,9] have designed so called level-wise algorithms for enumerating all of the frequent diamond or elliptic episodes, based on the frequent itemset mining algorithm APRIORI-TID [1]. While the level-wise algorithms are sufficient to find frequent episodes efficiently in practice, it is difficult to give theoretical guarantee of the efficiency. Recently, in order to give such theoretical guarantee, Katoh *et al.* [7] have developed the enumeration algorithm for frequent diamond episodes in polynomial delay and in polynomial space, based on depth-first search.

In this paper, we design the algorithm BIPAR to enumerate frequent bipartite episodes efficiently based on depth-first search. Then, it finds all of the frequent bipartite episodes in an input sequence S without duplication in $O(|\Sigma|N)$ time per an episode and in $O(|\Sigma|^2 n)$ space, where $|\Sigma|$, n, and N are an alphabet size, the length of $\mathcal{S}$, and the total size of $\mathcal{S}$, respectively. Hence, we can enumerate frequent bipartite episodes in polynomial delay and in polynomial space. We also give the incremental computation for occurrences, and practical speed-up by dynamic programming and prefix-based classes.

This paper is organized as follows. In Section 2, we introduce bipartite episodes and other notions necessary to the later discussion. In Section 3, we discuss several properties of bipartite episodes. In Section 4, we present the algorithm BIPAR and show its correctness and complexity. In Section 5, we give some experimental results from randomly generated event sequences to evaluate the practical performance of the algorithms. In Section 6, we conclude this paper and discuss the future works.

2   Preliminaries

In this section, we introduce the frequent episode mining problem and the related notions necessary to later discussion. We denote the sets of all integers and all natural numbers by $\mathbf{Z}$ and $\mathbf{N}$, respectively. For a set S, we denote the cardinality of S by $|S|$. A *digraph* is a graph with directed edges (*arcs*). A *directed acyclic graph* (*dag*, for short) is a digraph without cycles.

2.1   An Input Event Sequence and Its Windows

Let $\Sigma = \{1, \ldots, m\}$ ($m \geq 1$) be a finite alphabet with the total order $\leq$ over $\mathbf{N}$. Each element $e \in \Sigma$ is called an *event*[1]. Let *null* be the special, smallest event, called the *null event*, such that $a < null$ for all $a \in \Sigma$. Then, we define $\max \emptyset = null$. An *input event sequence* (*input sequence*, for short) $\mathcal{S}$ on Σ is a finite sequence $\langle S_1, \ldots, S_n \rangle \in (2^\Sigma)^*$ of events ($n \geq 0$), where $S_i \subseteq \Sigma$ is called the i-th *event set* for every $1 \leq i \leq n$. For any $i < 0$ or $i > n$, we define $S_i = \emptyset$. Then, we define n the *length* of $\mathcal{S}$ by $|\mathcal{S}| = n$ and define the *total size* of $\mathcal{S}$ by $||\mathcal{S}|| = \sum_{i=1}^{n} |S_i|$. Clearly, $||\mathcal{S}|| = O(|\Sigma|n)$, but the converse is not always true, that is, $O(||\mathcal{S}||) \neq |\Sigma|n$.

2.2   Episodes

Mannila *et al.* [10] defined an episode as a partially ordered set of labeled nodes.

Definition 1 (Mannila *et al.* [10]). A labeled acyclic digraph $X = (V, E, g)$ is an *episode* over Σ where V is a set of nodes, $E \subseteq V \times V$ is a set of arcs and $g : V \to \Sigma$ is a mapping associating each vertices with an event.

An episode is an acyclic digraph in the above definition, while it is define as a partial order in Mannila *et al.* [10]. It is not hard to see that two definitions are essentially same each other. For an arc set E on a vertex set V, let E^+ be the *transitive closure* of E such that $E^+ = \{ (u, v) \mid \text{there is some directed path from } u \text{ to } v \}$.

---

[1] Mannila *et al.* [10] originally referred to each element $e \in \Sigma$ itself as an *event type* and an occurrence of e as an *event*. However, we simply call both of them as *events*.

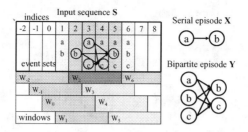

Fig. 2. (Left) An input sequence $\mathcal{S} = (S_1, \ldots, S_6)$ of length $n = 6$ over $\Sigma = \{a, b, c\}$ and their k-windows. (Right) Serial episode $X = a \mapsto b$ and a bipartite episode $Y = (\{a, b, c\} \mapsto \{b, c\})$. In the sequence $\mathcal{S}$, we indicate an occurrence (embedding) of Y in the second window W_2 in circles and arrows. See Example 1 and 2 for details.

Definition 2 (embedding). For episodes $X_i = (V_i, E_i, g_i)$ ($i = 1, 2$), X_1 is *embedded in* X_2, denoted by $X_1 \sqsubseteq X_2$, if there exists some mapping $f : V_1 \to V_2$ such that (i) f preserves vertex labels, i.e., for all $v \in V_1$, $g_1(v) = g_2(f(v))$, and (ii) f preserves precedence relation, i.e., for all $u, v \in V$ with $u \neq v$, if $(u, v) \in E_1$ then $(f(u), f(v)) \in (E_2)^+$. The mapping f is called an *embedding* from X_1 to X_2.

Given an input sequence $\mathcal{S} = \langle S_1, \ldots, S_n \rangle \in (2^{\Sigma})^*$, an *window* in $\mathcal{S}$ is a contiguous subsequence $W = \langle S_i \cdots S_{i+k-1} \rangle \in (2^{\Sigma})^*$ of $\mathcal{S}$ for some i, where $k \geq 0$ is the *width* of W.

Definition 3 (occurrence for an episode). An episode $X = (V, E, g)$ *occurs in* an window $W = \langle S_1 \cdots S_k \rangle \in (2^{\Sigma})^*$, denoted by $X \sqsubseteq W$, if there exists some mapping $h : V \to \{1, \ldots, k\}$ such that (i) h preserves vertex labels, i.e., for all $v \in V$, $g(v) \in S_{h(x)}$, and (ii) h preserves precedence relation, i.e., for all $u, v \in V$ with $u \neq v$, if $(u, v) \in E$ then $h(u) < h(v)$. The mapping h in the above definition is called an *embedding* of X into W.

An *window width* is a fixed positive integer $1 \leq k \leq n$. For any $-k + 1 \leq i \leq n$, we say that an episode X *occurs at* position i in $\mathcal{S}$ if $X \sqsubseteq W_i$, where $W_i = \langle S_i, \ldots, S_{i+k-1} \rangle$ is the i-th window of width k in $\mathcal{S}$. Then, we call i an *occurrence* or *label* of X in $\mathcal{S}$. In what follows, we denote the i-th window W_i by $\mathbf{W}_i^{\mathcal{S}, k}$. Let $\mathbf{W}_{\mathcal{S}, k} = \{ i \mid -k + 1 \leq i \leq \}$ be the domain of occurrences. For an episode X, we define the *occurrence list* for X in $\mathcal{S}$ by $\mathbf{W}_{\mathcal{S}, k}(X) = \{ -k + 1 \leq i \leq n \mid X \sqsubseteq W_i \}$, the set of occurrences of X in an input $\mathcal{S}$.

Example 1. Consider an alphabet $\Sigma = \{a, b, c\}$ and an input event sequence $\mathcal{S} = \langle \{a, b\}, \{b\}, \{a, c\}, \{a, c\}, \{a, b, c\}, \{a, b, c\} \rangle$ in Figure 2. Then, if the window width k is 4, has nine 4-windows from W_{-2} to W_6 for all $-2 \leq i \leq 6$, i.e., $\mathbf{W}_{\mathcal{S}, 5} = \{ W_i \mid -2 \leq i \leq 6 \}$.

Let $\mathcal{C}$ be a subclass of episodes, $\mathcal{S}$ be an input sequence, and $k \geq 1$ a window width. Let $X \in \mathcal{C}$ be an episode in the class $\mathcal{C}$. The *frequency* of X in $\mathcal{S}$ is defined by the number of k-windows $freq_{S,k}(X) = |\mathbf{W}_{S,k}(X)| = O(n)$. A *minimum frequency threshold* is any positive integer $\sigma \geq 1$. Without loss of generality, we can assume that $\sigma \leq |\mathbf{W}_{S,k}|$ for the length n of S. Then, the episode X is σ-*frequent in* S if $freq_{S,k}(X) \geq \sigma$. We denote by $\mathcal{F}_{S,k,\sigma}$ be the set of all σ-frequent episodes occurring in S. Let $\mathcal{C}$ be a subclass of episodes we consider.

Definition 4. FREQUENT EPISODE MINING PROBLEM FOR $\mathcal{C}$:
Given an input sequence $S \in (2^\Sigma)^*$, an window width $k \geq 1$, and a minimum frequency threshold $\sigma \geq 1$, the task is to find all σ-frequent episodes X within class $\mathcal{C}$ that occur in S with window width k without duplicates.

Our goal is to design an efficient algorithm for the frequent episode mining problem in the framework of enumeration algorithms [2,4]. Let N be the total input size and M the number of all solutions. An enumeration algorithm $\mathcal{A}$ is of *output-polynomial time*, if $\mathcal{A}$ finds all solutions $S \in \mathcal{S}$ in total polynomial time both in N and M. Also $\mathcal{A}$ is of *polynomial delay*, if the *delay*, which is the maximum computation time between two consecutive outputs, is bounded by a polynomial in N alone.

3   Bipartite Episodes

In this section, we introduce the class of bipartite episodes and other notions and discuss their properties.

3.1   Definition

Definition 5. For $m \geq 1$, m-*serial episode* (or *serial episode*) over Σ is a sequence $P = (a_1 \mapsto \cdots \mapsto a_m)$ of events $a_1, \ldots, a_m \in \Sigma$. This P represents an episode $X = (V, E, g)$, where $V = \{v_1 \ldots v_m\}$, $E = \{(v_i, v_{i+1}) \mid 1 \leq i < m\}$, and $g(i) = a_i$ for every $i = 1, \ldots, m$.

Definition 6. An episode $X = (V, E, g)$ is a *partial bipartite episode (or partial bi-episode)* if the underlying acyclic digraph X is bipartite, i.e., (i) $V = V_1 \cup V_2$ for mutually disjoint sets V_1, V_2, (ii) for every arc $(x, y) \in E$, $(x, y) \in V_1 \times V_2$. Then, we call V_1 and V_2 the source and sink sets.

Definition 7. A *bipartite episode (bi-episode, for short)* is an episode $X = (V, E, g)$ that satisfies the following conditions (i) – (iii):

(i) X is a partial bipartite episode with $V = V_1 \cup V_2$.
(ii) X is *complete*, i.e., $E = V_1 \times V_2$ holds.
(iii) X is *partwise-linear*, that is, for every $i = 1, 2$, the set V_i contains no distinct vertices with the same labeling by g.

In what follows, we represent a bipartite episode $X = (V_1 \cup V_2, E, g)$ by a pair $(A, B) \in 2^\Sigma \times 2^\Sigma$ of two subsets $A, B \subset \Sigma$ of events, or equivalently, an expression $(A \mapsto B)$, where $A = g(V_1)$ and $B = g(V_2)$ are the images of V_1 and V_2 by label mapping g. We also write (a, b) or $(a \mapsto b)$ for a 2-serial episode. In what follows, then, we define the *size* of a bipartite episode X by $||X|| = |A| + |B| = |V_1| + |V_2|$.

Example 2. In Figure 2, we show examples of an input event sequence $S = \langle \{a, b\}, \{b\}, \{a, c\}, \{a, c\}, \{a, b, c\}, \{a, b, c\} \rangle$ of length $n = 6$, a serial episode $X = a \mapsto b$ and a bipartite episode $Y = (\{a, b, c\} \mapsto \{b, c\})$ on an alphabet of events $\Sigma = \{a, b, c\}$. Then, the window list for a bipartite episode $Y = (\{a, b, c\} \mapsto \{b, c\})$ is $\mathbf{W}(Y) = \{W_2, W_3, W_4\}$.

In what follows, we denote by $\mathcal{SE}_k$, $\mathcal{SE} = \cup_{k \geq 1} \mathcal{SE}_k$, $\mathcal{PE}$, $\mathcal{SEC}$, $\mathcal{BE}$, $\mathcal{DE}$, and $\mathcal{EE}$, respectively, the classes of k-serial, serial, parallel, sectorial, bipartite, diamond, and elliptic episodes over Σ. For subclasses of episodes, the following inclusion relation hold: (i) $\mathcal{SE}_2 \subseteq \mathcal{SEC} \subseteq \mathcal{BE}$ and (ii) $\mathcal{PE} \subseteq \mathcal{BE}$.

3.2 Serial Constructibility

In this section, we introduce properties of bipartite episode that are necessary to devise an efficient algorithm for the frequent bipartite episode mining problem. We define the set of all serial episodes embedded in episode X by $Ser(X) = \{ S \in \mathcal{SE} \mid S \sqsubseteq X \}$. An episode X is said to be *serially constructible* on Σ if for any input event sequence S on Σ and for any window W of S, $X \sqsubseteq W$ holds iff for every serial episode $S \in Ser(X)$, $S \sqsubseteq W$ holds.

Katoh and Hirata [6] gave a necessary and sufficient condition for serially constructibility, called the parallel-freeness.

Definition 8 (Katoh and Hirata [6]). An episode $X = (V, E, g)$ is *parallel-free* if any pair of vertices labeled by the same event are reachable, that is, for any pair of mutually distinct vertices $u, v \in V$ $(u \neq v)$, if $g(u) = g(v)$ then there exists a directed path from u to v or v to u in X.

Theorem 1 (Katoh and Hirata [6]). *Let Σ be any alphabet. X is parallel-free iff X is serially constructible.*

Theorem 2. *Let X be partial bi-episode. If X is bipartite then X is parallel-free.*

Corollary 1. *Any bipartite episode is serially constructible.*

Let $X_i = (A_i \mapsto B_i)$ be a bipartite episode for every $i = 1, 2$. We define that $X_1 \subseteq X_2$ if $A_1 \subseteq A_2$ and $B_1 \subseteq B_2$.

Lemma 1 (anti-monotonicity of frequency). *Let σ be any frequency threshold and $k \geq 1$ be a window width. Let $X_i = (A_i \mapsto B_i)$ $(i = 1, 2)$ be a bipartite episode. If $X_1 \subseteq X_2$ then $freq_{S,k}(X_1) \geq freq_{S,k}(X_2)$.*

Proof. Let W be any window in S. Suppose that $X_2 \sqsubseteq W$. By Corollary 1, $S \sqsubseteq W$ for all serial episodes $S \in Ser(X_2)$. Since $X_1 \subseteq X_2$, we can show that $Ser(X_1) \subseteq Ser(X_2)$. For all serial episodes $S \in Ser(X_1)$, it holds that $S \sqsubseteq W$. By Corollary 1, $X_1 \sqsubseteq W$. From the above, if $X_2 \sqsubseteq W$ then $X_1 \sqsubseteq W$ for any W. Therefore, $\mathbf{W}_{S,k}(W_1) \supseteq \mathbf{W}_{S,k}(W_2)$. Then, $freq(X_1) \geq freq(X_2)$. □

Now, we have shown the serial constructibility for bipartite episodes. In the following, however, we further make detailed analysis on the serial constructibility for bipartite episodes by giving a simpler proof of Corollary 1 that does not use Theorem 1. For a window W and an event $e \in \Sigma$, we denote by $st(e, W)$ and $et(e, W)$, respectively, the first and the last positions in W at which e occurs.

Lemma 2 (characterization of the occurrences for a bipartite episode).
Let $X = (U, V, A, g)$ be any bipartite episode and W any window in $\boldsymbol{W}_{S,k}$. Then, $X \sqsubseteq W$ iff $(\max_{u \in U} st(g(u), W)) < (\min_{v \in V} et(g(v), W))$ holds.

Lemma 3. *For any bipartite episode $X = (A \mapsto B)$, $Ser(X) = A \cup B \cup \{ (a \mapsto b) \mid (a, b) \in A \times B \}$*

Theorem 3 (a detailed version of serial construction). *Let X be a bipartite episode and $W = \langle S_1, \ldots, S_k \rangle$ a window in $\boldsymbol{W}_{S,k}$. Let, $A, B \subseteq \Sigma$ be non-empty sets. Then,*

(1) If $X = (A \mapsto B)$ then, $X \sqsubseteq W$ iff $\forall (a, b) \in A \times B$, $(a \mapsto b) \sqsubseteq W$.
(2) if $X = (A \mapsto \emptyset)$ or $X = (\emptyset \mapsto B)$, $X \sqsubseteq W$ iff $\forall a \in A \cup B$, $a \sqsubseteq W$.

We define the *merge* of two bipartite episodes $X_i = (A_i \mapsto B_i)$ $(i = 1, 2)$ by $X_1 \cup X_2 = (A_1 \cup A_2 \mapsto B_1 \cup B_2)$, such that the edge set is the set unions of their edge sets. The *downward closure property* for a class $\mathcal{C}$ of episodes says that for any episodes $X_1, X_2 \in \mathcal{C}$, the condition $\boldsymbol{W}_{S,k}(X_1 \cup X_2) = \boldsymbol{W}_{S,k}(X_1) \cap \boldsymbol{W}_{S,k}(X_2)$ holds. Unfortunately, the class of bipartite episodes does not satisfy this property in general. The next lemma is essential to fast incremental computation of occurrence lists for the class of bipartite episodes in the next section.

Theorem 4 (downward closure property). *Let $X_i = (A_i \mapsto B_i)$ $(i = 1, 2)$. For any input sequence S and any $k \geq 1$, if $A_1 = A_2$ then $\boldsymbol{W}_{S,k}(X_1 \cup X_2) = \boldsymbol{W}_{S,k}(X_1) \cap \boldsymbol{W}_{S,k}(X_2)$.*

4   A Polynomial-Delay and Polynomial-Space Algorithm

4.1   The Outline of the Algorithm

In this section, we present a polynomial-delay and polynomial-space algorithm BIPAR for extracting all frequent bipartite episodes in a given input sequence. Let $S = (S_1, \ldots, S_n) \in (2^\Sigma)^*$ be an input sequence of length n and total input size $N = ||S||$, $k \geq 1$ be the window width, and $\sigma \geq 1$ be the minimum frequency threshold.

4.2   Enumeration of Bipartite Episodes

The main idea of our algorithm is to enumerate all frequent bipartite episodes by searching the whole search space from general to specific using depth-first search. For the search space, we define the parent-child relationships for bipartite episodes.

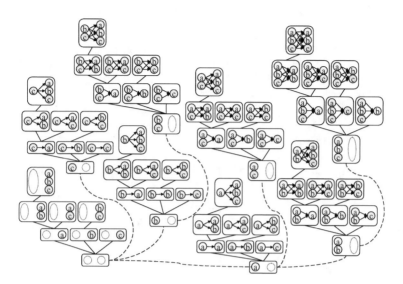

Fig. 3. The parent-child relationships on the alphabet $\Sigma = \{a, b, c\}$, where each white empty circle indicates the empty set

Definition 9. The bi-episode $\perp = (\emptyset \mapsto \emptyset)$ is the *root*. Then, the parent of a non-root bipartite episode $X = A \mapsto B$ is defined by

$$parent(A \mapsto B) = \begin{cases} (A - \{\max(A)\}) \mapsto B & (\text{if } B = \emptyset) \\ A \mapsto (B - \{\max(B)\}) & (\text{otherwise}) \end{cases}$$

We define the set of all children of X by $Children(X) = \{Y \mid parent(Y) = X\}$. Then, we define the *family tree* for $\mathcal{BE}$ by the rooted digraph $\mathcal{T}(\mathcal{BE}) = (V, E, \perp)$ with the root $\perp$, the vertex set $V = (\mathcal{BE})$, and the edge set $E = \{(X, Y) \mid X$ is the parent of $Y, Y \neq \perp\}$. As shown in Fig. 3, we can show that the family tree $\mathcal{T}(\mathcal{BE})$ forms the spanning tree for all bi-episodes of $\mathcal{BE}$.

In Fig. 4, we show the basic version of our polynomial-delay and polynomial-space algorithm BIPAR and its subprocedure FREQBIPARREC for extracting frequent bipartite episodes from input sequence $\mathcal{S}$. The algorithm is a backtracking algorithm that traverses the spanning tree $\mathcal{T}(\mathcal{BE})$ based on depth-first search starting from the root $\perp$ using the parent-child relationships over $\mathcal{BE}$.

The subprocedure BIPAROCC is a straightforward algorithm that computes the occurrence list $\mathbf{W}_{\mathcal{S},k}(X)$ for bi-episode X by testing the embedding $X \sqsubseteq \mathbf{W}_i^{\mathcal{S},k}$ for each position i while scanning the input sequence. Its definition is omitted here. We can show that BIPAROCC computes $\mathbf{W}_{\mathcal{S},k}(X)$ from X of size $m = ||X||$ and an input sequence $\mathcal{S}$ of length n in $O(|\Sigma|kmn)$ time.

Lemma 4. *Let $\mathcal{S}$ be any input sequence of length n. For any window width $k \geq 1$ and minimum frequency threshold $\sigma \geq 1$, the algorithm BIPAR in Fig. 4 with BIPAROCC finds all σ-frequent bipartite episodes occurring in $\mathcal{S}$ without duplicates in $O(|\Sigma|^4 kn)$ delay and $O(|\Sigma|^2 + n)$ space.*

---

algorithm $\text{BIPAR}(\mathcal{S}, k, \Sigma, \sigma)$
input: input event sequence $\mathcal{S} = \langle S_1, \ldots S_n \rangle \in (2^\Sigma)^*$ of length $n \geq 0$,
window width $k > 0$, alphabet of events Σ, the minimum frequency $1 \leq \sigma \leq n + k$;
output: the set of all σ-frequent bipartite episodes in $\mathcal{S}$ with window width k;
method:
1   $\bot := (\emptyset \mapsto \emptyset)$; // The root bipartite episode $\bot$;
2   $\text{FREQBIPARREC}(\bot, \mathcal{S}, k, \Sigma, \sigma)$;

procedure $\text{FREQBIPARREC}(X, \mathcal{S}, k, \Sigma, \sigma)$
input: bipartite episode $X = (A \mapsto B)$ and $\mathcal{S}$, k, Σ, and k are same as in BIPAR.
output: the set of all σ-frequent bipartite episodes in $\mathcal{S}$ that are descendants of X;
method:
1   **if** ($|\mathbf{W}_{\mathcal{S},k}(X)| \geq \sigma$) **then**
2   **output** X;
3   // Execute in the special case that the right hand side of X is empty.
4   **if** ($B = \emptyset$) **then**
5   **foreach** $e \in \Sigma$ **such that** $e > \max(A)$ **do**
6   // Expand the left hand side.
7               $\text{FREQBIPARREC}(((A \cup \{e\}) \mapsto B), \mathcal{S}, k, \Sigma, \sigma)$;
8   **end if**
9   // Execute always.
10   **foreach** ($e \in \Sigma$ ($e > \max(B)$)) **do**
11   // Expand the right hand side.
12          $\text{FREQBIPARREC}((A \mapsto (B \cup \{e\})), \mathcal{S}, k, \Sigma, \sigma)$;
13 **end if**

Fig. 4. The main algorithm BIPAR and a recursive subprocedure FREQBIPARREC for mining frequent bipartite episodes in a sequence

4.3   Incremental Computation of Occurrences

The algorithm BIPAROCCINC in Fig.5 computes the occurrence list $W = \mathbf{W}_{\mathcal{S},k}(Y)$ for the newly created child episode $Y = (A \mapsto B \cup \{b\})$ from the list $\mathbf{W}_{\mathcal{S},k}(X))$ for its parent $X = (A \mapsto B)$ by calling the subprocedure SERIALOCC. The next lemma is derived from Theorem 3 on the downward closure property for $\mathcal{BE}$.

Lemma 5 (correctness of BIPAROCCINC). *Let $X = (A \mapsto B)$ be a bipartite episode and $e \in \Sigma$ be an event. Then, we have the next (1) and (2):*

(1) If $A = \emptyset$ then $\mathbf{W}(A \mapsto (B \cup \{e\})) = \mathbf{W}(X) \cap \mathbf{W}(\emptyset \mapsto \{e\})$.
(2) if $A \neq \emptyset$ then $\mathbf{W}(A \mapsto (B \cup \{e\})) = \mathbf{W}(X) \cap \bigcap_{a \in A} \mathbf{W}(a \mapsto \{e\})$.

The algorithm BIPAROCCINC uses the subprocedure SERIALOCC for computing the occurrence list for a 2-serial episode. This algorithm is a modification of FASTSERIALOCC for 3-serial episodes in [7] and its definition is omitted here. We can show that SERIALOCC can be implemented to run in $O(N) = ||\mathcal{S}||$ time in the total input size $N = ||\mathcal{S}||$ regardless window width k.

---

procedure BIPAROCCINC$(X, X_0, W_0, k, \mathcal{S})$
input: bipartite episodes $X = (A \mapsto (B_0 \cup \{e\}))$ and $X_0 = (A \mapsto B_0) = parent(X)$,
the occurrence list W_0 for X_0, window width $k > 0$
, an input sequence $\mathcal{S} = \langle S_1, \ldots S_n \rangle$;
output: the occurrence list W for X;
method:
1    $W := W_0$;
2   **if** ($A = \emptyset$) **then** $W := W \cap$ SERIALOCC$((\emptyset \mapsto \{e\}), k, \mathcal{S})$;
3   **else**
4   **foreach** $(a \in A)$ $W := W \cap$ SERIALOCC$((a \mapsto e), k, \mathcal{S})$;
5   **return** W;

Fig. 5. An improved algorithm BIPAROCCINC for computing the occurrence list of a bipartite episode

Lemma 6. *The algorithm* BIPAROCCINC *in Fig.5 computes the new occurrence list* $W = \boldsymbol{W}_{\mathcal{S},k}(Y)$ *for the child episode* $Y = (A \mapsto B \cup \{b\})$ *in* $O(N|A|) = O(|\Sigma|^2 n)$ *time from a bi-episode* $X = (A \mapsto B)$, $\boldsymbol{W}_{\mathcal{S},k}(X)$, *any event* $b \in \Sigma$, *and* k, *where* $n = |\mathcal{S}|$ *and* $N = ||\mathcal{S}||$.

4.4 Practical Improvement by Dynamic Programming

We can further improve the computation of occurrence list by BIPAROCCINC using dynamic programming technique as follows.

During the execution of the algorithm FREQBIPARREC the subprocedure SE-RIALOCC for $\mathcal{SE}$ are called many times inside BIPAROCCINC with the same arguments $(a \mapsto b, k, \mathcal{S})$ $(a, b \in \Sigma)$. Fig. 6 shows the algorithm LOOKUPSERIALOCC that is a modification version of SERIALOCC using dynamic programming. This algorithm uses a hash table $TABLE$ in Fig. 6 that stores pairs $\langle X, \boldsymbol{W}(X) \rangle$ of a 2-serial episode $X = (a \mapsto b)$ and its occurrence list $\boldsymbol{W}(X)$.

We modify the main algorithm BIPAR and BIPAROCCINC such that after initializating the hash table, we call LOOKUPSERIALOCC instead of SERIALOCC. This modification does not change the behavior, while it reduces the total number of the calls for SERIALOCC from at most $|\Sigma||F|$ to at most $|\Sigma|^2$, where $\mathcal{F} \subseteq \mathcal{BE}$ is the set of solutions.

Lemma 7. *After initializating the hash table* $TABLE$, *the algorithm* LOOKUPSE-RIALOCC *calls* SERIALOCC *at most* $O(|\Sigma|^2)$ *times during the execution of the main procedure* BIPAR *using* $O(|\Sigma|^2 n)$ *memory.*

4.5 Reducing the Number of Scan on the an Input Sequence by Prefix-Based Classes

We can improve the computation of occurrence list by BIPAROCCINC using the idea of prefix-based classes, which is originally invented by Zaki [14,15].

---

global variable: a hash table $TABLE : \Sigma^2 \to 2^{\{-k+1,\dots,n\}}$;
initialization: $TABLE := \emptyset$;

procedure LOOKUPSERIALOCC($X, k \in \mathbf{N}, \mathcal{S}$)
input: serial episode $X = (a \mapsto b)$, window width $k > 0$,
an input sequence $\mathcal{S} = \langle S_1, \dots S_n \rangle$;
output: the occurrence list W for X;
method:
1   **if** $(TABLE[(a,b)] = UNDEF)$ **then**
2       $W :=$ SERIALOCC($(a \mapsto b), k, \mathcal{S}$);
3       $TABLE := TABLE \cup \{\langle (a,b), W \rangle \}$;
4   **end if**
5   **return** $TABLE[(a,b)]$;

Fig. 6. Practical speed-up for computing occurrence lists of serial episodes using dynamic programming

For a bipartite episode $P = (A, B)$, called a *common prefix*, we define the *prefix-based class* related to P by the set of bi-episodes

$$\mathcal{C}_P = \{ X = (A \mapsto B \cup \{b\}) \mid P = (A \mapsto B), b \in \Sigma, \max B < b \}.$$

In our modified algorithm BIPARFAST, we enumerate each prefix-based classes for $\mathcal{BE}$ instead of each episode in $\mathcal{BE}$. We start with defining enumeration procedure of bi-episodes using prefix-based classes induced in a new class of family trees for $\mathcal{BE}$. We define the parent function $parent : \mathcal{BE} \backslash \{\bot\} \to \mathcal{BE}$.

Definition 10. For any non-root bipartite episode $X = A \mapsto B$,

$$parent(A \mapsto B) = \begin{cases} ((B - \{\max B\}) \mapsto \emptyset) & \text{if } A = \emptyset, B \neq \emptyset \\ ((A - \{\max A\}) \mapsto B) & \text{if } A \neq \emptyset, |B| = 0 \text{ or } |B| = 1 \\ (A \mapsto (B - \{\max B\})) & \text{if } A \neq \emptyset, |B| \geq 2 \end{cases}$$

By using the parent function above, we can define the family tree $\mathcal{T} = (V, E, \bot)$ in a similar way as in Section 4.2.

Next, we give a procedure to enumerate all bi-partite episodes based on depth-first search on $\mathcal{T}$. Starting with $\bot = (\emptyset, \emptyset)$, we enumerate bi-episodes in $\mathcal{BE}$ by the following rules.

Lemma 8. *For any bi-episodes $X, Y \in \mathcal{BE}$, Y is a child of X if and only if Y is obtained from X by applying one of the following rules to X. The new occurrence list $\mathbf{W}(Y)$ is also obtained by the corresponding rule.*

(i) *If $X = (A \mapsto \emptyset)$, then for any $e \in \Sigma$, $Y = (\emptyset \mapsto A)$ and*
    $\mathbf{W}(Y) = \mathbf{W}(X)$.
(ii) *If $X = (A \mapsto \emptyset)$, then for any $e \in \Sigma$, $Y = (A \cup \{e\} \mapsto \emptyset)$ and*
    $\mathbf{W}(Y) = \mathbf{W}(X) \cap \mathbf{W}(e)$.

(iii) If $X = (A \mapsto \{b\})$, then for any $e \in \Sigma$, $Y = (A \cup \{e\} \mapsto \{b\})$ and
$\mathbf{W}(Y) = \mathbf{W}(X) \cap \mathbf{W}((e \mapsto b))$.

(iv) If $X = (A \mapsto C \cup \{a\}) \in \mathcal{C}_P$, then for any $Z = (A \mapsto C \cup \{b\}) \in \mathcal{C}_P$ such that
$a < b$, $Y = (A \mapsto C \cup \{a, b\})$, where $\mathcal{C}_P$ is the unique prefix-based class to
which X belongs, and $\mathbf{W}(Y) = \mathbf{W}(X) \cap \mathbf{W}(Z)$.

Proof. The statements (i) – (iii) are easily proved by construction of the parents.
In statements (iv), it follows from the condition $\max B < a, b$ and $a < b$ that
the parent for Y is uniquely determined to be X. The proof for the property
$\mathbf{W}(Y) = \mathbf{W}(X) \cap \mathbf{W}(Z)$ follows from Theorem 5 on downward closure property
for $\mathcal{BE}$. □

By the above lemma, provided that each prefix-based class $\mathcal{C}_P$ is available, we
do not need to compute the new occurrence lists $\mathbf{W}(Y)$ for each child in the
cases of (i) and (iv). We have to explicitly compute the occurrence lists only for
a single event in case (i) and for a 2-serial episodes in case (iii).

We can apply further improvements to our algorithm BIPARFAST as shown in
[7]. We can improve the delay of the algorithm BIPARFAST by the factor of the
height of the search tree $\mathcal{T}$ using *alternating output* technique of [13]. We can
also reduce the space complexity of the algorithm BIPARFAST from $O(|\Sigma|^3 n)$ to
$O(|\Sigma|^2 n)$ space by using the *diffset* technique, of Zaki [15] for itemset mining.

Combining all improvements discussed above, we can modify the basic version
of our backtracking algorithm BIPAR. In what follows, we call this modified
algorithm by BIPARFAST. Now, we have the main theorem of this paper on the
delay and the space complexities of the modified algorithm BIPARFAST.

Theorem 5. *Let $\mathcal{S}$ be any input sequence of length n on event alphabet Σ.
For any window width $k \geq 1$ and minimum frequency threshold $\sigma \geq 1$, the
algorithm BIPARFAST can be implemented to find all σ-frequent bipartite episodes
occurring in $\mathcal{S}$ without duplicates in $O(|\Sigma| N)$ delay (time per frequent episode)
and $O(|\Sigma|^2 n)$ space, where $N = ||\mathcal{S}||$ is the total size of input.*

5   Experimental Results

In this section, we give the experimental results for the following combinations of
the algorithms given in Section 4, by applying to the randomly generated event
sequences and the real event sequence.

Data. As randomly generated data, we adopt an event sequence $\mathcal{S} = (S_1, \ldots, S_n)$
over an alphabet $\Sigma = \{1, \ldots, s\}$ from four parameters (n, s, p, r), by generating
each event set S_i $(i = 1, \ldots, n)$ under the probability $P(e \in S_i) = p(e/s)^r$ for
each $e \in \Sigma$. On the other hand, as the real event sequence, we adopt bacterial
culture data provided from Osaka Prefectural General Medical Center from 2000
to 2005. In particular, we adopt an event sequence obtained by regarding a
detected bacterium as an event type, fixing the sample of sputum and connecting
data of every patient with same span.

Method. We implemented the following three depth-first search (DFS) algorithms given in Section 4:

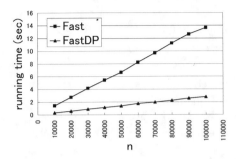

Fig. 7. Running time for the input length n, where $s = 10$, $p = 0.1$, $r = 0.0$, $k = 10$, and $\sigma = 0.1n$

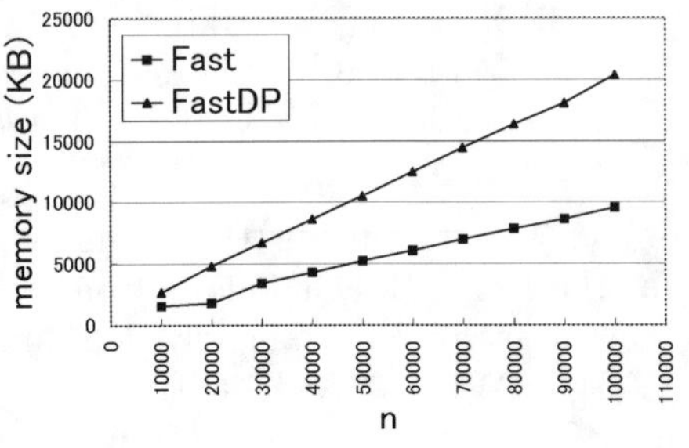

Fig. 8. Memory size for the input length n, where $s = 10$, $p = 0.1$, $r = 0.0$, $k = 10$, and $\sigma = 0.1n$

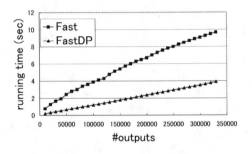

Fig. 9. Running time for the number of outputs, where $n = 10,000$, $s = 10$, $p = 0.1$, $r = 0.0$, $k = 10$, and $\sigma = 0.001n$

Basic : the basic DFS algorithm BIPARBASIC with OIPAROCC
   in Fig. 4 (sec. 4.1).
Fast : the modified DFS algorithm BIPARFAST with SEREALOCC (sec. 4.5).
FastDP : the modified DFS algorithm BIPARFAST with LOOKUPSEREALOCC
   based on dynamic programming (sec. 4.4).

All experiments were run in a PC (AMD Mobile Athlon64 Processor 3000+, 1.81GHz, 2.00GB memory, Window XP, Visual C++) with window width $K \geq 1$ and minimum frequency threshold $\sigma \geq 1$.

Experiment A. Fig. 7 and Fig. 8 show the running time and the size of virtual memory usage of the algorithms Fast and FastDP for the randomly generated event sequences from the parameter ($10000 \leq n \leq 100000, s = 10, p = 0.1, r = 0.0$), where $k = 10$ and $\sigma = 0.1n$. Then, both time and space complexity of these algorithms seem to be linear in the input size and thus expected to scales well on large datasets. Furthermore, FastDP is five hundred times as faster as Fast. On the other hand, FastDP tends to occupy more memory than Fast.

Table 1. Parameter settings for Experiment C, where rand and bact indicates a randomly generated data and a bacterial culture data, respectively. The first, second, third, and fourth rows show the name of setting, the data, the parameters, and the number of output episodes, respectively.

exp	exp1	exp2	exp3	exp4	exp5	exp6	exp7
type	rand	rand	rand	rand	rand	bact	bact
n	1,000	10,000	10,000	1,000	100,000	70,606	70,606
s	10	10	1,000	10	10	174	174
p	0.1	0.1	0.1	0.1	0.001		
r	0.0	0.0	10.0	0.0	0.0		
k	10	10	10	100	1,000	15	15
σ	$0.1n$	$0.001n$	$0.25n$	$0.1n$	$0.1n$	$0.01n$	1
#outputs	2,330	334,461	3,512	1,048,576	1,780	162	177,216

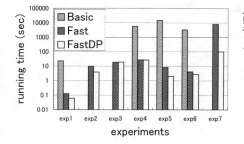

Fig. 10. Running time for exp1 to exp7

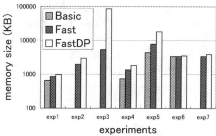

Fig. 11. Memory size for exp1 to exp7

Experiment B. Fig. 9 shows the running time of the algorithms Fast and FastDP for the number of outputs for the randomly generated event sequences from the parameter $(n = 10,000, s = 10, p = 0.1, r = 0.0)$, where $k = 10$ and $\sigma = 0.001n$. Then, the slopes are almost constant and thus the delays are just determined by the input size as indicated by Theorem 5.

Experiment C. Table 1 gives the seven experiments for the algorithms Basic, Fast, and FastDP under the various parameter settings par1 – par7. The input data from exp1 to exp5 are randomly generated event sequences with parameters (n, s, p, r), and ones of exp6 and exp7 are the bacterial culture data.

Fig. 10 and Fig. 11 show the running time and the virtual memory size of the algorithms. Here, for exp2, exp3, and exp7, we give no results for Basic, because the running time is over 20,000 (sec). Then, for exp1, exp4, exp5, and exp6, Fast and FastDP are faster than Basic. Especially, for exp5, FastDP was 7300 times faster than Basic. On the other hand, Fast and FastDP occupy more memory space than Basic. From exp1 to exp7, the algorithm FastDP occupy more memory space than Fast. Then, the algorithm FastDP is the fastest algorithm except exp3. For exp3 with large alphabet size $|\Sigma| = 1000$, FastDP occupies sixteen (16) times large memory space than Fast, and Fast is faster than FastDP.

$$X = \{\texttt{Serratia-marcescens, Staphylococcus-aureus}\}$$
$$\mapsto \{\texttt{yeast, Stenotrophomonas-maltophilia}\}$$

Fig. 12. An example of bipartite episode extracted from a bacterial culture data

Fig. 12 shows an example of the bipartite episode X with frequency 21 extracted from the bacterial culture data for exp7, where an event in type writer fonts denotes the names of bacteria. This episode represents that the bacteria of `Serratia-marcescens` and `Staphylococcus-aureus` are followed by another bacteria of `yeast` and `Stenotrophomonas-maltophilia` within fifteen days.

6   Conclusion

This paper studied the problem of frequent bipartite episode mining, and presented an efficient algorithm BIPAR that finds all frequent bipartite episodes in an input sequence in polynomial delay and polynomial space in the input size. We have further studied several techniques for reducing the time and the space complexities of the algorithm. Possible future problems are extension of BIPAR for general fragments of DAGs [10,11]. Also, we plan to apply the proposed algorithm to bacterial culture data [5,9].

References

1. Agrawal, R., Srikant, R.: Fast algorithms for mining association rules in large databases. In: Proc. 20th VLDB, pp. 487–499 (1994)
2. Arimura, H.: Efficient algorithms for mining frequent and closed patterns from semi-structured data. In: Washio, T., Suzuki, E., Ting, K.M., Inokuchi, A. (eds.) PAKDD 2008. LNCS (LNAI), vol. 5012, pp. 2–13. Springer, Heidelberg (2008)
3. Arimura, H., Uno, T.: A polynomial space and polynomial delay algorithm for enumeration of maximal motifs in a sequence. In: Deng, X., Du, D.-Z. (eds.) ISAAC 2005. LNCS, vol. 3827, pp. 724–737. Springer, Heidelberg (2005)
4. Avis, D., Fukuda, K.: Reverse search for enumeration. Discrete Applied Mathematics 65, 21–46 (1996)
5. Katoh, T., Hirata, K.: Mining frequent elliptic episodes from event sequences. In: Proc. 5th LLLL, pp. 46–52 (2007)
6. Katoh, T., Hirata, K.: A simple characterization on serially constructible episodes. In: Washio, T., Suzuki, E., Ting, K.M., Inokuchi, A. (eds.) PAKDD 2008. LNCS (LNAI), vol. 5012, pp. 600–607. Springer, Heidelberg (2008)
7. Katoh, T., Arimura, H., Hirata, K.: A Polynomial-Delay Polynomial-Space Algorithm for Extracting Frequent Diamond Episodes from Event Sequences. In: Theeramunkong, T., et al. (eds.) PAKDD 2009. LNCS (LNAI), vol. 5476, pp. 172–183. Springer, Heidelberg (2009)
8. Katoh, T., Hirata, K., Harao, M.: Mining sectorial episodes from event sequences. In: Todorovski, L., Lavrač, N., Jantke, K.P. (eds.) DS 2006. LNCS (LNAI), vol. 4265, pp. 137–148. Springer, Heidelberg (2006)

9. Katoh, T., Hirata, K., Harao, M.: Mining frequent diamond episodes from event sequences. In: Torra, V., Narukawa, Y., Yoshida, Y. (eds.) MDAI 2007. LNCS (LNAI), vol. 4617, pp. 477–488. Springer, Heidelberg (2007)

10. Mannila, H., Toivonen, H., Verkamo, A.I.: Discovery of frequent episodes in event sequences. Data Mining and Knowledge Discovery 1, 259–289 (1997)

11. Pei, J., Wang, H., Liu, J., Wang, K., Wang, J., Yu, P.S.: Discovering frequent closed partial orders from strings. IEEE TKDE 18, 1467–1481 (2006)

12. Pei, J., Han, J., Mortazavi-Asi, B., Wang, J., Pinto, H., Chen, Q., Dayal, U., Hsu, M.-C.: Mining sequential patterns by pattern-growth: The PrefixSpan approach. IEEE Trans. Knowledge and Data Engineering. 16, 1–17 (2004)

13. Uno, T.: Two general methods to reduce delay and change of enumeration algorithms, NII Technical Report, NII-2003-004E (April 2003)

14. Zaki, M.J.: Scalable Algorithms for Association Mining. IEEE TKDE 12, 372–390 (2000)

15. Zaki, M.J., Hsiao, C.-J.: CHARM: An efficient algorithm for closed itemset mining. In: Proc. 2nd SDM, pp. 457–478. SIAM, Philadelphia (2002)

CHRONICLE: A Two-Stage Density-Based Clustering Algorithm for Dynamic Networks

Min-Soo Kim and Jiawei Han

Department of Computer Science, University of Illinois at Urbana-Champaign
{msk,hanj}@cs.uiuc.edu

Abstract. Information networks, such as social networks and that extracted from bibliographic data, are changing dynamically over time. It is crucial to discover time-evolving communities in dynamic networks. In this paper, we study the problem of finding time-evolving communities such that each community freely forms, evolves, and dissolves for any time period. Although the previous *t*-partite graph based methods are quite effective for discovering such communities from large-scale dynamic networks, they have some weak points such as finding only stable clusters of *single path* type and not being scalable *w.r.t.* the time period. We propose *CHRONICLE*, an efficient clustering algorithm that discovers not only clusters of single path type but also clusters of *path group* type. In order to find clusters of both types and also control the dynamicity of clusters, CHRONICLE performs the *two-stage* density-based clustering, which performs the 2nd-stage density-based clustering for the *t*-partite graph constructed from the 1st-stage density-based clustering result for each timestamp network. For a given data set, CHRONICLE finds all clusters in a fixed time by using a fixed amount of memory, regardless of the number of clusters and the length of clusters. Experimental results using real data sets show that CHRONICLE finds a wider range of clusters in a shorter time with a much smaller amount of memory than the previous method.

1   Introduction

Recently, there is an increasing interest to mining dynamics of information networks that evolve over time. Examples of dynamic networks include network traffic data [12], telephone traffic data[1], bibliographic data[2], dynamic social network data [6,14], and time-series microarray data [16], where a dynamic network is regarded as a sequence of networks with different timestamps (or temporal intervals). A cluster in dynamic network data, which is also called a *community*, typically represents *cohesive subgroup of individuals* within a network that persists for a specific time interval [12,14,2]. That is, a community is cohesive both in a timestamp network and across time. Identifying communities from dynamic networks has been paid much attention lately as an important research topic.

---

[1] http://reality.media.mit.edu/download.php
[2] http://www.informatik.uni-trier.de/~ley/db

J. Gama et al. (Eds.): DS 2009, LNAI 5808, pp. 152–167, 2009.
© Springer-Verlag Berlin Heidelberg 2009

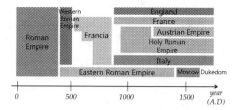

Fig. 1. Some chronicle patterns of the European historical diagram

In this paper, we study the problem of finding time-evolving communities such that each community freely forms, evolves, and dissolves for any time period. Such communities would look like *chronicle patterns* of a historical diagram. Figure 1 shows some chronicle patterns of the European historical diagram of between A.D. 0 and 1,700, where multiple "dynasties" co-existed at the same period of time or started/ended at different times. Mining chronicle patterns over dynamic networks would be able to discover some interesting and important knowledge that could be invisible with the previous clustering methods for static networks. For example, it could find the evolutionary collaborating groups of researchers from the DBLP data and identify how they evolve as time goes on.

There have been several approaches for finding communities in dynamic networks. First, the concept of *evolutionary clustering* has been proposed to capture the evolutionary process of clusters in temporal data. Several evolutionary clustering methods [4,13,10] have been proposed, but they have some drawbacks such as assuming only a fixed number of communities over time and not being scalable with data size. Since the forming of new communities or the dissolving of existing communities is quite natural and common phenomena in real dynamic networks [1], and moreover, the size of real dynamic networks tends to be large, the existing evolutionary clustering methods could not be very useful in real applications. Second, there have been some methods to detect break points through clustering [12,6]. For example, GraphScope [12] identifies the change points where the subgroup structure of hosts (or servers) is largely changed in network communication data. However, they do not allow arbitrary insertion/deletion of nodes or arbitrary start/stop of communities over time, which occurs quite often in real dynamic networks, and thus, they could be less useful for real data. Third, there have been few studies that discover communities by using t-partite graph from temporal data [11,2]. They perform clustering for each network, construct a t-partite graph by connecting between similar local clusters at different times, and find a sequence of local clusters as a community. Especially, the methods proposed by Bansal et al. [2] explicitly handles dynamic network data, and the *BFS method* among them is known to be the most efficient method. It is scalable with the number of nodes, finds a variable number of communities with arbitrary start/stop over time, and allows arbitrary insertion/deletion of nodes.

Although the existing t-partite graph based method, especially the BFS method is quite effective for discovering clusters from large-scale dynamic networks, it has also some weak points: (1) finding only stable clusters of *single path* (*i.e.*, a sequence of local clusters over time); (2) finding a very small number of clusters (*i.e.*, the most stable *top-k* clusters); (3) not being scalable *w.r.t.* the length of dynamic networks; and (4) using a large amount of memory depending on its parameters. All these weak points are caused by the fact that the BFS method is based on a dynamic programming (DP) algorithm. Besides the clusters of single path type, actually, there are many cohesive clusters of non-single path type in t-partite graph. Figure 2 shows an example of t-partite graph over three timestamp networks. Each network has 3~4 local clusters. The numbers on lines between T_1 and T_2 (or T_2 and T_3) indicate that they have a non-zero similarity. When $k = 1$, the BFS method finds a single path cluster $c_{11}c_{21}c_{31}$ because it has the strongest similarities between local clusters. However, if some members in c_{12} transfer to c_{24}, some members in c_{13} to c_{23}, and the members in c_{23} and c_{24} are merged into c_{33}, then there could be another cluster like $(c_{12}c_{13})(c_{23}c_{24})c_{33}$, where the similarity between $(c_{12}c_{13})$ and $(c_{23}c_{24})$ and the similarity between $(c_{23}c_{24})$ and c_{33} might be very high (*i.e.*, very cohesive) although the similarity of each single path (*e.g.*, $c_{13}c_{23}c_{33}$) are not so high. Here, () represents a consolidation of multiple local clusters. We call a cluster of this type as a *path group* cluster since there are multiple paths over time in the cluster. The BFS cannot find the clusters of this type.

In this paper, we propose a density-based clustering algorithm, *CHRONICLE*, that efficiently discovers both single path clusters and path group clusters. For finding clusters of both types, CHRONICLE performs the density-based clustering in *two stages*: the 1st-stage density-based clustering for each timestamp network and the 2nd-stage density-based clustering for the t-partite graph. In case of the previous BFS method, it only performs the 1st-stage clustering and finds single path clusters by using a DP algorithm. A density-based clustering approach has several advantages such as discovering clusters of arbitrary shape, handling noise, and being fast. These features allow us to find a wider range of clusters (*i.e.*, not only single path clusters, but also path group clusters) in an efficient way. As the length of dynamic networks, the number of clusters, or the length of cluster (*i.e.*, path length) increases, the running time and the amount of memory usage of the BFS method largely increase. Using disk for saving

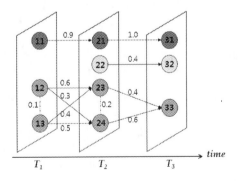

Fig. 2. An example of t-partite graph constructed from a dynamic network

the amount of memory would cause performance degradation. In contrast, for a given data set, CHRONICLE finds all clusters in a fixed time by using a fixed amount of memory, regardless of the number of clusters and the length of clusters. The density parameter ε allows us to control both the number of clusters and the dynamicity of communities. Here, the *dynamicity* indicates how much communities change over time.

Our proposed algorithm, CHRONICLE, has the following key features:

- Discovering a variable number of communities with arbitrary start/stop over time in dynamic networks
- Finding a wide range (*i.e.*, stable and dynamic) communities of both single path type and path group type
- Being scalable *w.r.t.* the length of dynamic networks
- Being fast and using only a small amount of memory irrespective of both the number of clusters and the length of clusters

The rest of this paper is organized as follows. Section 2 presents the most related piece of work, the BFS method [2]. Sections 3 presents our CHRONICLE algorithm, and Section 4 the results of experimental evaluation. Finally, Section 5 concludes the study.

2   Related Work

In this section, we briefly explain the BFS method [2]. Given a dynamic network $\mathcal{G} = \{G_1, \ldots, G_t\}$, the BFS method first finds local clusters for each timestamp network G_i by using the biconnected component algorithm. Actually, biconnected components are not very good clusters for network/graph data. There are many other better methods using well-known measures such as min-max cut, normalized cut, modularity, betweenness, and structural similarity. Bansal et al. mentioned the BFS method used biconnected components as local clusters for simplicity and fast mining. After clustering for each timestamp network, the BFS method constructs a t-partite graph by connecting between similar local clusters at different times. The core and unique part of the BFS method is finding the most stable top-k clusters of length at least l_{min} over the t-partite graph.

For finding top-k paths of length l_{min}, the BFS method needs to maintain up to l_{min} heaps of size k for each node c_{ij} of the t-partite graph. We denote this heap as h_{ij}^x $(1 \leq x \leq l_{min})$, where each h_{ij}^x contains at most k entries. During scanning each node c_{ij} of each partite of the t-partite graph from 1 to t, the BFS method calculates top-k (or fewer) highest weighting subpaths of length x ending at c_{ij} $(1 \leq x \leq l_{min})$. This task is easily performed by merging the subpaths in the heaps of the previous partite, *i.e.*, $h_{i'j'}^x$ $(i' < i)$ with the subpaths in the current heaps, *i.e.*, h_{ij}^x, where $c_{i'j'}$ and c_{ij} are connected in the t-partite graph, and choosing new top-k subpaths. For finding paths of length greater than l_{min}, the BFS method should maintain additional data structure to contain the top scoring paths of all length greater than l_{min} for each c_{ij}.

In fact, the BFS method is a modified version of the Viterbi algorithm, a DP algorithm for finding the most probable state path in Hidden Markov Model (HMM). The original Viterbi algorithm does not need to maintain heaps like h_{ij}^x because it finds only the top-1 state path of the same length with a given sequence. In contrast, the BFS method requires l_{min} heaps of size k because it finds the top-k paths of length at least l_{min}. Although the BFS method has a strong point that finds the exact top-k paths by using a DP algorithm, it consumes a large amount of memory due to a lot of heaps, and moreover, such memory consumption increases as the length of dynamic network n_{ts}, the number of clusters k, or the minimum length of clusters l_{min} gets larger. The excessive computation for heaps also incurs an increase of running time. Using disk swapping might save memory, but it would also cause an additional performance degradation. Besides, the BFS method tends to discover only a small number of very stable clusters that are hardly changed over time. However, many applications might need to find not only a small number of stable clusters, but also a large number of various clusters enough to capture the structure of the entire dynamic network.

3   Chronicle Algorithm

In this section, we present the three parts of our CHRONICLE algorithm, the 1st-stage clustering, constructing t-partite graph, and the 2nd-stage clustering in Sections 3.1, 3.2, and 3.3, respectively.

3.1   The 1st-Stage Clustering

In this section, we briefly present the 1st-stage density-based clustering method, which we denote as CHRONICLE$_{1st}$, to find all local clusters for each timestamp network. We define a dynamic network $\mathcal{G}$ as a sequence of networks $G_i(V_i, E_i)$, i.e., $\mathcal{G} = \{G_1, \ldots, G_t\}$.

For similarity measure, CHRONICLE$_{1st}$ uses the *cosine similarity* (or *structure similarity*) [9,15], which is one of the well-known measures for network data. Definitions 1~2 show the concept of the structural similarity. Intuitively, $\sigma(v, w)$ indicates how many nodes v and w share w.r.t. the overall number of their adjacent nodes including themselves. By definition, $\sigma(v, w)$ becomes non-zero only if v is directly connected to w with an edge. The value of $\sigma(v, w)$ is in the range 0.0~1.0 and especially becomes 1.0 when both v and w are in a clique.

Definition 1. *The neighborhood $N(v)$ of a node $v \in V$ is defined by $N(v) = \{x \in V \mid \langle v, x \rangle \in E\} \cup \{v\}$.*

Definition 2. *The structural similarity $\sigma(v, w)$ of a node pair $(v, w) \in V \times V$ is defined by*

$$\sigma(v, w) = \frac{|N(v) \cap N(w)|}{\sqrt{|N(v)| \times |N(w)|}}.$$

Fig. 3. An example network and four local clusters

CHRONICLE$_{1st}$ is actually equivalent to the SCAN algorithm [15], which is again basically the same with the original density-based clustering algorithm DBSCAN [5], but uses the structural similarity as the distance measure between two nodes instead of Euclidean distance. With the notions of density-based clustering, CHRONICLE$_{1st}$ finds a high density subset of nodes, *i.e.*, a *topologically dense subgraph* like a quasi-clique, as a cluster in each G_i. Figure 3 shows an example network and four local clusters found by CHRONICLE$_{1st}$ over the network. After clustering, there remain some nodes that have relatively low similarity with its adjacent nodes, and thus, do not belong to any cluster. Such nodes are considered *noises* by the notions of density-based clustering. We skip the explanation of how to determine two density parameters μ and ε since it is not a core part of this paper and there are some methods for it.

3.2   Constructing t-Partite Graph

CHRONICLE constructs a t-partite graph from the 1st-stage clustering result. This is performed by connecting between two local clusters c_{ij} and $c_{i'j'}$ $(i' < i)$ that have a non-zero similarity (or affinity). For a similarity measure, there are many candidates, and Jaccard coefficient $Jaccard(c_{ij}, c_{i'j'})$ would be a good candidate. For example, if $|c_{ij}| = 6$, $|c_{i'j'}| = 8$, and c_{ij} and $c_{i'j'}$ have four common nodes, then $Jaccard(c_{ij}, c_{i'j'}) = \frac{4}{6+8-4} = 0.4$. We might be able to connect between two local clusters in non-adjacent timestamp networks $G_{i'}$ and G_i $(|i' - i| > 1)$, where we call $g = |i' - i|$ as a *gap*, but we only deal with the case of $g = 1$ in this paper due to space limit. Each link between a pair of nodes of t-partite graph has its own weight based on the similarity between the corresponding local clusters. Hereafter, we call a connection between two different partites in the t-partite graph as a *link* for discriminating it from normal edges. We also denote the t-partite graph as T.

Besides links, CHRONICLE also connects two local clusters in the same partite that have *inter cluster edges* (simply, *ic-edges*) between them. Let ν_j be the number of edges with one endpoint in c_{ij} and the other endpoint in $\overline{c_{ij}}$, where $\overline{c_{ij}}$ denotes the complement of c_{ij}, and let ω_j be the number of edges with both endpoints in c_{ij}. We denote the number of edges with one endpoint in c_{ij} and the other endpoint in $c_{ij'}$ as $\nu_{jj'}$ and denote $\nu_j + \omega_j = \tau_j$, which is equal to the sum of degrees of nodes in c_{ij}. Then, we define a similarity measure $InterEdge(c_{ij}, c_{ij'}) = \frac{\nu_{jj'}}{\tau_j + \tau_{j'}}$ for two local clusters in the same network. As two local clusters c_{ij} and $c_{ij'}$ are more tightly connected with more ic-edges, the value of $InterEdge(c_{ij}, c_{ij'})$ increases. CHRONICLE uses these ic-edges for the 2nd-stage clustering. In Figure 2, the dashed line between c_{12} and c_{13} and that between c_{23} and c_{24} represent ic-edges. We denote the nodes, links, and ic-edges of T as V_T, L_T, and E_T, respectively.

3.3   The 2nd-Stage Clustering

Similarity measure. In this section, we present the similarity measure for the 2nd-stage clustering method CHRONICLE$_{2nd}$. Different from a timestamp network G_i, the t-partite graph T has weights on each link/ic-edge, and has time semantics from T_1 to T_t. Thus, our similarity measure, *general similarity (GS)*, for CHRONICLE$_{2nd}$ considers those two features of T.

The key concept of GS is the integration of the *structural affinity (SA)* and *weight affinity (WA)* between two nodes so as to discover both single path clusters and path group clusters. GS is defined as in Eq. 1. We just use the structural similarity $\sigma(v, w)$ as $SA(v, w)$. If we only use $SA(v, w)$ instead of GS as a measure for T, the clustering result could be awkward. For example, in Figure 2, $\sigma(v, w)$ would identify $c_{22}c_{32}$ as a very strong cluster since $\sigma(c_{22}, c_{32}) = 1$ even though its weight is just 0.4.

$$GS(v, w) = SA(v, w) \times WA(v, w) \qquad (1)$$

For considering time semantics of t-partite graph, CHRONICLE$_{2nd}$ restricts the scope of measuring similarity to each bipartite within time interval $[i, i + 1]$. Figure 4 shows the general case of the relationship between v in T_i and w in T_{i+1} in T. When measuring the similarity between v and w, CHRONICLE$_{2nd}$ only takes account of nodes and links/ci-edges within time interval $[i, i + 1]$ except the links within time interval $[i - 1, i]$ and those within time interval $[i + 1, i + 2]$. This restriction prevents that the links of such outside timestamps affect the similarity of v and w, and at the same time, allows CHRONICLE$_{2nd}$ to find clusters in an on-line fashion, *i.e.*, perform incremental clustering for a new timestamp network G_{t+1}. Unless having such restriction, the nodes in two extreme partites of T, *i.e.*, T_1 and T_t would always have some distorted similarity values due to the imbalanced number of neighborhood nodes.

The sub-similarity measure SA gives a cohesive path group a chance to be found as a cluster. In Figure 4, every pair of v and w would have zero or more

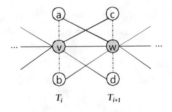

Fig. 4. The general case of the relationship between v of T_i and w of T_{i+1}

neighborhood nodes. If there is no neighborhood, v and w forms a simplest bi-clique, and so, the weight itself between v and w determines the overall similarity between them. However, if there are some neighborhood nodes, we consider that not only the weights on links or ic-edges, but also the structural cohesion between them affects the overall similarity. That is, the more common neighborhood nodes v and w share compared with all their neighborhood nodes, v and w receive a high score of structural affinity. For example, in Figure 2, four links of $c_{12}c_{23}$, $c_{12}c_{24}$, $c_{13}c_{23}$, and $c_{13}c_{24}$ belong to the bi-clique of $\{c_{12}, c_{13}, c_{23}, c_{24}\}$, and so, all their structural similarities become 1.0 just like a single path. This is reasonable since it gives a chance to a group of paths that have weak similarities, but are highly intertwined, to be found as one strong cluster of path group type.

The sub-similarity measure weight affinity WA considers the weights of links/ic-edges between v and w including common neighborhood nodes. WA is defined as in Definitions 3~4. Intuitively, $\Phi(v, w)$ in Definition 3 represents the sum of weights conveyed from v to w through their common neighborhood nodes. WA is a weighted combination of a direct affinity between v and w, i.e., $\phi(v, w)$, and a indirect affinity between v and w, i.e., $\Phi(v, w)$. This combination indicates the expected maximum possible affinity between two consolidations of v and w with their common neighborhood nodes (e.g., affinity between $\{a, b, v\}$ and $\{c, d, w\}$ in Figure 4). The parameter α allows user to control the weights of the direct and indirect affinities.

Definition 3. *Let $\phi(v, w)$ be a similarity weight between v and w, and $\Omega(v, w)$ common neighborhood nodes of v and w, i.e., $\Omega(v, w) = (N(v) \cap N(w) - \{v, w\})$. The common neighborhood weights $\Phi(v, w)$ of between $v \in V_i$ and $w \in V_{i+1}$ is defined by*

$$\Phi(v, w) = \sum_{x \in \Omega(v,w)} (\phi(v, x) + \phi(x, w)).$$

Definition 4. *The weight affinity $WA(v, w)$ of a node pair $(v, w) \in V_i \times V_{i+1}$ is defined by*

$$WA(v, w) = \alpha \cdot \phi(v, w) + (1 - \alpha) \cdot \Phi(v, w), \quad for\ 0 \le \alpha \le 1.$$

Finally, CHRONICLE$_{2nd}$ gives a high general similarity GS score to a pair of nodes whose both SA and WA are high. For example, in Figure 2, $SA(c_{12}, c_{23}) = 1$ since c_{12} and c_{13} are a part of the bi-clique, and, when $\alpha = 0.5$, $WA(c_{12}, c_{23}) = 0.5 \times 0.6 + 0.5 \times (0.5 + 0.5) = 0.8$. Thus, $GS(c_{12}, c_{23}) = 1 \times 0.8 = 0.8$. Likewise, $GS(c_{12}, c_{24}) = 1 \times 0.85 = 0.85$, $GS(c_{13}, c_{23}) = 1 \times 0.9 = 0.9$, and $GS(c_{13}, c_{24}) = 1 \times 0.75 = 0.75$.

Notions of density-based clustering. We summarize the notions of density-based clustering using the GS measure for t-partite graph through Definitions 5~11. The similar notions are used in other density-based clustering methods such as DBSCAN [5], SCAN [15], and TRACLUS [8] for points, static network, and trajectory data, respectively. However, different from those methods, our notions consider time semantics existing in T. The CHRONICLE$_{2nd}$ using this notions takes two density parameters, μ and ε, and discovers all subgraphs with high GS scores as clusters (i.e., communities) on T.

Definition 5. *The ε-neighborhood $N_\varepsilon(v)$ of a node $v \in V_T$ is defined by $N_\varepsilon(v) = \{x \in N(v) \mid \langle v, x \rangle \in L_T \wedge GS(v, x) \geq \varepsilon\}$.*

Definition 6. *A node $v \in V_T$ is called a core node w.r.t. ε and μ if $|N_\varepsilon(v)| \geq \mu$.*

We note that CHRONICLE$_{2nd}$ considers only the nodes connected with v not by *ic-edges* but by *links* as the candidates of $N_\varepsilon(v)$ in order to expand a cluster in the direction of time. We also note that $v \in N(v)$ and $GS(v, v) = 1$ in Definition 5, and so, the value of $|N_\varepsilon(v)|$ is always at least 1 in Definition 6.

Definition 7. *A node $x \in V_T$ is gs-direct reachable from a node $v \in V_T$ w.r.t. ε and μ if (1) v is a core node and (2) $x \in N_\varepsilon(v)$.*

Definition 8. *A node $v_j \in V_T$ is gs-reachable from a node $v_i \in V_T$ w.r.t. ε and μ if there is a chain of nodes $v_i, v_{i+1}, \ldots, v_{j-1}, v_j \in V_T$ such that v_{i+1} is gs-direct reachable from v_i ($i < j$) w.r.t. ε and μ.*

Definition 9. *A node $v \in V_T$ is gs-connected to a node $w \in V_T$ w.r.t. ε and μ if there is a node $x \in V_T$ such that both v and w are gs-reachable from x w.r.t. ε and μ.*

Definition 10. *A non-empty subset $S \subseteq V_T$ is called a gs-connected cluster w.r.t. ε and μ if S satisfies the following two conditions:*
(1) Connectivity: $\forall v, w \in S$, v is gs-connected to w w.r.t. ε and μ
(2) Maximality: $\forall v, w \in V_T$, if $v \in S$ and w is gs-reachable from v w.r.t. ε and μ, then $w \in S$.

We note that the gs-reachability is the transitive closure of direct gs-reachability, and it is asymmetric. It is only symmetric for a pair of core nodes. However, the gs-connectivity is a symmetric relation, which is an important property for incremental clustering because it guarantees the consistency of the clustering result regardless of whether performing batch clustering for the whole T or performing incremental clustering for every bipartite of T, i.e., $\{\langle T_1, T_2 \rangle, \ldots, \langle T_{t-1}, T_t \rangle\}$.

Definition 11. *Let P be a set of gs-connected clusters found by Definition 10. A node $v \in V_T$ is a noise if v is not contained in any cluster of P.*

CHRONICLE$_{2nd}$ algorithm. Algorithm 1 outlines the pseudo-code of the CHRONICLE$_{2nd}$ algorithm for batch clustering. CHRONICLE$_{2nd}$ makes one scan over a t-partite graph $T = \langle V_T, L_T, E_T \rangle$ and finds a set of gs-connected clusters $\mathcal{CR} = \{C_i\}$ w.r.t. ε and μ. Since there could exist many small-sized clusters composed of only two nodes like $c_{22}c_{32}$ in Figure 2, we set $\mu = 2$ in most cases for not missing them.

Algorithm 1. CHRONICLE$_{2nd}$

Input: (1) t-partite graph $T = \langle V_T, L_T, E_T \rangle$,
         (2) Minimum number of nodes μ,
         (3) Similarity threshold ε.
Output: Communities $\mathcal{CR} = \{C_i\}$.
 1: $\mathcal{CR} \leftarrow \emptyset$;
 2: $\forall v \in V : v \leftarrow$ UNCLASSIFIED;
 3: **for each** UNCLASSIFIED node $v \in V$ **do**
 4:      **if** v is a core node **then**
 5:          $\mathcal{CR} \leftarrow \mathcal{CR} \cup findCluster(v)$;
 6:      **else**
 7:          $v \leftarrow$ NON_MEMBER;

 8: **return** $\mathcal{CR}$;

Algorithm 2. FindCluster

Input: (1) A core node v,
         (2) T, μ, ε.
Output: A gs-connected cluster C.
 1: $C \leftarrow \emptyset$;
 2: $Q.push(v)$;   // Q is a queue
 3: **while** $Q.empty() =$ **false do**
 4:      $x \leftarrow Q.front()$;
 5:      $R \leftarrow \{y \in V \mid y$ is direct gs-reachable from $x\}$;
 6:      **for each** $y \in R$ **do**
 7:          **if** y is UNCLASSIFIED **then**
 8:              $C \leftarrow C \cup \{y\}$;
 9:              $Q.push(y)$;
10:          **if** y is NON_MEMBER **then**
11:              $C \leftarrow C \cup \{y\}$;
12:      $Q.pop()$;
13: **return** C;

For each node of T, there are two kinds of labels: UNCLASSIFIED and NON_MEMBER. At first, all nodes are labeled as UNCLASSIFIED (line 2). If there is a node that is not classified yet (line 3), CHRONICLE$_{2nd}$ checks whether the node is a core node (line 4). If the node v is a core node, CHRONICLE$_{2nd}$ finds a gs-connected cluster containing v and adds the cluster to $\mathcal{CR}$ (line 5). Otherwise, CHRONICLE$_{2nd}$ labels the node as NON_MEMBER (line 7). After finding all clusters, the NON_MEMBER nodes can be further classified into *outliers* or *hubs* by whether the node has edges to only one cluster or multiple clusters, respectively, although the corresponding codes are not presented in the algorithm.

Algorithm 2 outlines the pseudo-code of the *FindCluster* algorithm, a subroutine of CHRONICLE$_{2nd}$. FindCluster finds all nodes that are gs-reachable from a given seed node v. It starts with inserting v into a queue Q (line 2).

Then, FindCluster searches those nodes by repeating the following steps until Q is empty: (1) calculating the direct gs-reachable nodes R from the front node x of Q; (2) inserting the part of R into Q; and (3) deleting x from Q (lines $3\sim5, 9, 12$). For each node y of R, FindCluster inserts y into a result cluster C and a queue Q if y is UNCLASSIFIED (lines $8\sim9$), or inserts y only into C if y is NON_MEMBER (line 11). Here, that y is NON_MEMBER means that y is visited before and is not a core node.

The time complexity of the $CHRONICLE_{2nd}$ algorithm is $O(2 \cdot |L_T| + 2 \cdot |E_T|)$. It is because the algorithm visits each node $v \in V_T$ only once and checks the GS scores between v and its neighborhood nodes. We note that this complexity is not affected by the number of clusters or any parameters such as μ, ϵ, and α.

Example 1. Consider performing $CHRONICLE_{2nd}$ for the t-partite graph in Figure 2 with the parameter setting of $\alpha = 0.5$, $\mu = 2$, and $\varepsilon = 0.7$. First, the single path of $c_{11}c_{21}c_{31}$ is easily found as a cluster since those three nodes form a gs-connected cluster, where they all are core nodes satisfying $GS(c_{11}, c_{21}) = 0.9 \geq \varepsilon$ and $GS(c_{21}, c_{31}) = 1.0 \geq \varepsilon$. Next, the path group of $(c_{12}c_{13})(c_{23}c_{24})$ is also found as another cluster since those four nodes form a gs-connected cluster with satisfying $GS(c_{12}, c_{23}) = 0.8 \geq \varepsilon$, $GS(c_{12}, c_{24}) = 0.85 \geq \varepsilon$, $GS(c_{13}, c_{23}) = 0.9 \geq \varepsilon$, and $GS(c_{13}, c_{24}) = 0.75 \geq \varepsilon$. The remaining nodes $\{c_{22}, c_{32}, c_{33}\}$ are identified as noises due to their low GS scores with their neighborhood. However, if we loosen the threshold ε to 0.6, the path group of $(c_{12}c_{13})(c_{23}c_{24})c_{33}$ is found as a cluster instead of $(c_{12}c_{13})(c_{23}c_{24})$ because $GS(c_{23}, c_{23}) = 0.6 \geq \varepsilon$ and $GS(c_{24}, c_{33}) = 0.6 \geq \varepsilon$. Here, the nodes $\{c_{22}, c_{32}\}$ are still identified as noises, which would be found as a cluster when ε decreases into 0.4.   □

Online version of $CHRONICLE_{2nd}$. Although we present $CHRONICLE_{2nd}$ for batch clustering in Algorithm 1, the online version of $CHRONICLE_{2nd}$, i.e., incremental clustering can be easily performed under the same concept. When a new timestamp network G_{t+1} arrives, we perform the 1st-stage clustering $CHRONICLE_{1st}$ for G_{t+1}, and obtain a new partite of T, i.e., T_{t+1} by calculating links and ic-edges within time interval $[t, t + 1]$. Then, we perform $CHRONICLE_{2nd}$ on the bipartite graph $\{T_t, T_{t+1}\}$ while maintaining the community ID of each node in T_t. Since $CHRONICLE_{2nd}$ only takes account of nodes and links/ci-edges within time interval $[i, i + 1]$ for calculating the similarity, and at the same time, the gs-connectivity is a symmetric relation, this incremental clustering does not hurt the consistency of a new clustering result compared with the past clustering result for $\{T_1, \ldots, T_t\}$. In Example 1, suppose that we perform the initial clustering on the first bipartite $\{T_1, T_2\}$, then incremental clustering on the second bipartite $\{T_2, T_3\}$ under the parameter setting of $\alpha = 0.5$, $\mu = 2$, and $\varepsilon = 0.6$. As a result of the first clustering for $\{T_1, T_2\}$, two clusters $c_{11}c_{21}$ and $(c_{12}c_{13})(c_{23}c_{24})$ are found, which we assign community ID 1 and 2, respectively. As a result of the second clustering for $\{T_2, T_3\}$, two clusters $c_{21}c_{23}$ and $(c_{23}c_{24})c_{33}$ are found, and we already know the community IDs of c_{21} and $(c_{23}c_{24})$ are 1 and 2, respectively, and thus, we finally obtain the same clustering result with that in Example 1 by assigning the community IDs 1 and 2 to c_{23} and c_{33}, respectively.

4   Experimental Evaluation

In this section, we evaluate the effectiveness and efficiency of our algorithm
CHRONICLE compared with the previous state-of-art t-partite graph based
method, the BFS method. We describe the experimental data and environment
in Section 4.1, and present the comparison results in Sections 4.2 and 4.3.

4.1   Experimental Setting

We use real dynamic network data set, the DBLP data[3]. We regard authors
as nodes, co-authorships as edges, and years as timestamps. We extract the
bibliographic information of all journals and conferences of the years from 1993
to 2007 (*i.e.*, 15 years). By filtering authors of low number of publications, we
generate four data sets of 15,000 authors, 30,000 authors, 60,000 authors, and
120,000 authors. We call each data sets as DBLP-15K, DBLP-30K, DBLP-60K,
and DBLP-120K, respectively.

In both the CHRONICLE algorithm and the BFS method, the 1st-stage clustering and the 2nd-stage clustering are independent with each other, and the
key part of them is the 2nd-stage clustering. Since the quality of the 1st-stage
clustering results of the BFS method (*i.e.*, biconnected components) are not as
good as those of CHRONICLE$_{1st}$, and at the same time, in order to be fair in
our comparison, we use the t-partite graph constructed by CHRONICLE$_{1st}$ as
an input t-partite graph for the BFS method. For all the 1st-stage clustering,
we use $\mu = 3$ and $\varepsilon = 0.6$.

To compare the efficiency, we measure the elapsed time and memory usage of
both CHRONICLE$_{2nd}$ and the BFS method while varying the data size and the
number of timestamps, n_{ts}. In case of CHRONICLE$_{2nd}$, the elapsed time and
memory usage are not affected by the parameters μ, ε or α. On the contrary, in
case of the BFS method, the performance is largely affected by the parameters k
or l_{min}. Thus, we also measure the elapsed time and memory usage of the BFS
method while varying k and l_{min}. For CHRONICLE$_{2nd}$, we set $\alpha = 0.5$, $\mu = 2$,
and $\varepsilon = 0.5$.

The BFS method and CHRONICLE$_{2nd}$ performs some *different mining task*
with different purpose for t-partite graph. Moreover, there are no explicit ground
truth answers for communities in the DBLP data. Thus, it is very difficult to
evaluate the effectiveness by using common measures such as precision/recall.
Instead, we measure how many clusters of the BFS method are overlapped with
those of CHRONICLE$_{2nd}$, *i.e.*, how much the result of CHRONICLE$_{2nd}$ contains
the result of the BFS method. We note that, while the BFS method discovers
the exact top-k clusters based on a DP algorithm, CHRONICLE$_{2nd}$ discovers
less exact but more various clusters, where the variousness is controlled by ε. Let
$\mathcal{CR}_{BFS}$ be a set of clusters of the BFS method, and $\mathcal{CR}_{common}$ a set of clusters
of the BFS method overlapped with those of CHRONICLE$_{2nd}$. We measure the
ratio between the sizes of two sets, $\frac{|\mathcal{CR}_{common}|}{|\mathcal{CR}_{BFS}|}$ while varying ε and k. In order to

---

[3] http://www.informatik.uni-trier.de/~ley/db

show the effectiveness of path group type clusters of CHRONICLE$_{2nd}$, we also measure the average similarity of clusters across time by using Jaccard coefficient for both single path type and path group type while varying ε.

We conduct all the experiments on a Pentium Core2 Duo 2.0 GHz PC with 2 GBytes of main memory, running on Windows XP. We implement our algorithm andthe BFS method in C++ using Microsoft Visual Studio 2005.

4.2   Results of Efficiency

Figures 5~7 show the result of efficiency evaluation of the BFS method and CHRONICLE$_{2nd}$. In Figure 5, CHRONICLE$_{2nd}$ has the better efficiency than the BFS method as the number of nodes and the number of timestamps

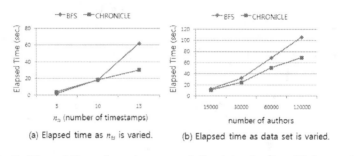

Fig. 5. Elapsed time of varying n_{ts} and the data size ($k = 20$, $l_{min} = 3$)

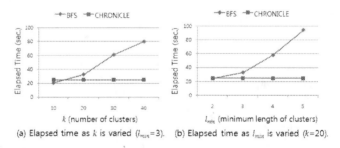

Fig. 6. Elapsed time of varying k and l_{min} (data set = DBLP-30K, $n_{ts} = 10$)

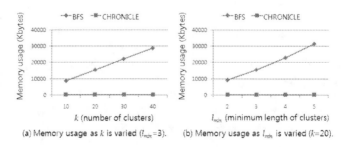

Fig. 7. Memory usage of of varying k and l_{min} (data set = DBLP-30K, $n_{ts} = 10$)

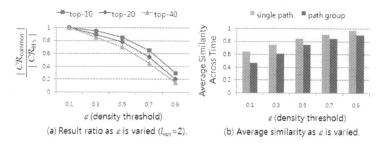

Fig. 8. The result ratio of $\frac{|\mathcal{CR}_{common}|}{|\mathcal{CR}_{BFS}|}$ and the average similarity across time of clusters of CHRONICLE$_{2nd}$ (data set = DBLP-30K, $n_{ts} = 10$)

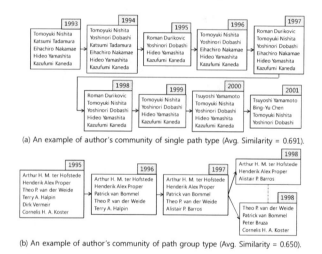

(a) An example of author's community of single path type (Avg. Similarity = 0.691).

(b) An example of author's community of path group type (Avg. Similarity = 0.650).

Fig. 9. Examples of author's communities (data set = DBLP-30K)

increase. Especially, since the BFS method needs to maintain paths of all lengths greater than l_{min}, its performance gets much worse as n_{ts} gets longer as in Figure 5(a). In Figure 6, the performance of the BFS method is much more degraded as k and l_{min} increase due to heavy computation for heaps, whereas that of CHRONICLE$_{2nd}$ is the same regardless of the parameters μ and ε. In Figure 7, the amount of memory usage of the BFS method also largely increases as k and l_{min} get larger due to the increased heap usage, whereas that of CHRONICLE$_{2nd}$ is the same regardless of μ and ε since it uses only two fixed-size data structures: t-partite graph itself and status list for checking UN-CLASSIFIED/NON_MEMBER.

4.3   Results of Effectiveness

Figure 8(a) shows $\frac{|\mathcal{CR}_{common}|}{|\mathcal{CR}_{BFS}|}$ while varying k of the BFS method and ε of CHRONICLE$_{2nd}$. Even though CHRONICLE$_{2nd}$ discovers a wide range clusters instead of only the most stable clusters, its clustering result includes a fairly high percentage (about 0.7~0.9) of the clustering result of the BFS method at the moderate density threshold $\varepsilon = 0.5$. Thus, a user can obtain the approximated top-k clusters by sorting the clusters found by CHRONICLE$_{2nd}$. Figure 8(b) shows that not only the clusters of single path type but also the clusters of path group type have a high similarity across time with the similarity becoming stronger as ε increases. This means that the dynamicity of both type clusters is controlled by ε.

Figure 9 shows examples of author's communities of single path type and path group type. Our CHRONICLE algorithm finds dynamic communities of single path type as in Figure 9(a) as well as communities of path group type as in Figure 9(b), whereas the BFS method cannot find both communities.

5   Conclusions

In this paper, we have proposed a two-stage density-based clustering algorithm, CHRONICLE, that efficiently discovers time-evolving communities over large-scale dynamic networks. By performing the density-based clustering in the 2nd-stage for t-partite graph, CHRONICLE$_{2nd}$ finds both single path clusters and path group clusters, and at the same time, achieves a very high performance compared with the previous BFS method. Through extensive experiments over the DBLP data set, we have shown the effectiveness and efficiency of CHRONICLE, especially, that CHRONICLE$_{2nd}$ is far more scalable than the BFS method with respect to the length of dynamic network and the parameter values.

Acknowledgement

The work was supported in part by the Korea Research Foundation grant funded by the Korean Government (MOEHRD), KRF-2007-357-D00203, by the U.S. National Science Foundation grants IIS-08-42769 and BDI-05-15813, and the Air Force Office of Scientific Research MURI award FA9550-08-1-0265. Any opinions, findings, and conclusions expressed here are those of the authors and do not necessarily reflect the views of the funding agencies.

References

1. Asur, S., Parthasarathy, S., Ucar, D.: An event-based framework for characterizing the evolutionary behavior of interaction graphs. In: Proc. KDD, pp. 913–921 (2007)
2. Bansal, N., Chiang, F., Koudas, N., Tompa, F.W.: Seeking stable clusters in the blogosphere. In: Proc. VLDB 2007, pp. 806–817 (2007)
3. Chakrabarti, D., Kumar, R., Tomkins, A.: Evolutionary clustering. In: Proc. KDD 2006, pp. 554–560 (2006)
4. Chi, Y., Song, X., Zhou, D., Hino, K., Tseng, B.L.: Evolutionary spectral clustering by incorporating temporal smoothness. In: Proc. KDD 2007, pp. 153–162 (2007)
5. Ester, M., Kriegel, H.-P., Sander, J., Xu, X.: A density-based algorithm for discovering clusters in large spatial databases with noise. In: Proc. KDD 1996, pp. 226–231 (1996)

6. Falkowski, T., Bartelheimer, J., Spiliopoulou, M.: Mining and visualizing the evolution of subgroups in social networks. In: Proc. IEEE/WIC/ACM Web Intelligence 2006, pp. 52–58 (2006)
7. Han, J., Kamber, M.: Data Mining: Concepts and Techniques, 2nd edn. Morgan Kaufmann, San Francisco (2006)
8. Lee, J.-G., Han, J., Whang, K.-Y.: Trajectory clustering: A partition-and-group framework. In: Proc. SIGMOD 2007, pp. 593–604 (2007)
9. Leicht, E.A., Holme, P., Newman, M.E.J.: Vertex similarity in networks. Physical Review E73, 026120 (2006)
10. Lin, Y.-R., Chi, Y., Zhu, S., Sundaram, H., Tseng, B.L.: FacetNet: A framework for analyzing communities and their evolutions in dynamic networks. In: Proc. WWW 2008, pp. 685–694 (2008)
11. Mei, Q., Zhai, C.: Discovering evolutionary theme patterns from text - an exploration of temporal text mining. In: Proc. KDD 2005, pp. 198–207 (2005)
12. Sun, J., Faloutsos, C., Papadimitriou, S., Yu, P.S.: GraphScope: Parameter-free mining of large time-evolving graphs. In: Proc. KDD 2007, pp. 687–696 (2007)
13. Tang, L., Liu, H., Zhang, J., Nazeri, Z.: Community evolution in dynamic multi-mode networks. In: Proc. KDD 2008, pp. 677–685 (2008)
14. Tantipathananandh, C., Berger-Wolf, T.Y., Kempe, D.: A framework for community identification in dynamic social networks. In: Proc. KDD 2007, pp. 717–726 (2007)
15. Xu, X., Yuruk, N., Feng, Z., Schweiger, T.A.J.: SCAN: A structural clustering algorithm for networks. In: Proc. KDD 2007, pp. 824–833 (2007)
16. Zhao, L., Zaki, M.J.: Tricluster: An effective algorithm for mining coherent clusters in 3d microarray data. In: Proc. SIGMOD 2005, pp. 694–705 (2005)

Learning Large Margin First Order Decision Lists for Multi-Class Classification

Huma Lodhi[1], Stephen Muggleton[1], and Mike J.E. Sternberg[2]

[1] Department of Computing, Imperial College London, SW7 2AZ
hml@doc.ic.ac.uk, shm@doc.ic.ac.uk
[2] Centre for Bioinformatics, Imperial College London, SW7 2AZ
m.sternberg@imperial.ac.uk

Abstract. Inductive Logic Programming (ILP) systems have been successfully applied to solve binary classification problems. It remains an open question how an accurate solution to a multi-class problem can be obtained by using a logic based learning method. In this paper we present a novel logic based approach to solve challenging multi-class classification problems. Our technique is based on the use of large margin methods in conjunction with the kernels constructed from first order rules induced by an ILP system. The proposed approach learns a multi-class classifier by using a divide and conquer reduction strategy that splits multi-classes into binary groups and solves each individual problem recursively hence generating an underlying decision list structure. We also study the well known one-vs-all scheme in conjunction with logic-based kernel learning. In order to construct a highly informative logical and relational space we introduce a low dimensional embedding method. The technique is amenable to skewed/non-skewed class distribution where multi-class problems such as protein fold recognition are generally characterized by highly uneven class distribution. We performed a series of experiments to evaluate the proposed rule selection and multi-class schemes. The methods were applied to solve challenging problems in computation biology and bioinformatics, namely multi-class protein fold recognition and mutagenicity detection. Experimental comparisons of the performance of large margin first order decision list based multi-class scheme with the standard multi-class ILP algorithm and multi-class Support Vector Machine yielded statistically significant results. The results also demonstrated a favorable comparison between the performances of decision list based scheme and one-vs-all strategy.

1 Introduction

The underlying aim of a multi-class approach is to learn a highly accurate function that categorizes examples into predefined classes. Effective multi-class techniques are crucial to solving the problems ranging from multiple object recognition to multi-class protein fold recognition.

The two areas of machine learning, namely Inductive Logic Programming (ILP) and Kernel based methods (KMs) are well known for their distinguishing

J. Gama et al. (Eds.): DS 2009, LNAI 5808, pp. 168–183, 2009.

features: ILP techniques are characterized by their use of background knowledge and expressive language formalism whereas strong mathematical foundations and high generalization ability are remarkable characteristics of KMs. Recently some logic based techniques (such as Support Vector Inductive Logic Programming (SVILP) [1], kFOIL [2] and RUMBLE [3]) have been designed which use kernels to solving binary classification problems and performing real-valued predictions. In this paper we study multi-class classification in the combined ILP and kernel learning scenario by extending SVILP. We also propose an effective method to constructing highly informative relational and logical low dimensional feature space. The method is designed in a way so as a classifier trained in the feature space is amenable to highly imbalance category distribution. A skewed class distribution is a common phenomenon in multi-class classification tasks.

SVILP solves binary classification problems in a multi-stage learning process. In the first stage, a set of all the first order horn clauses (rules), constructed during the search of the hypothesis space, is obtained from an ILP system. In the next stages similarity between the examples is computed by the use of novel kernel function that captures semantic and structural commonalities between examples. The computed relational and logic based kernel is used in conjunction with a large margin learning algorithm to induce a binary classifier. In this way, SVILP performs classification task by training a large margin first order classifier.

SVILP [1] uses all the clauses with positive compression: an information theoretic measure. The number of positively compressed rules can vary from zero to thousands for the particular task. Furthermore rules with negative compression can contain crucial information to solving the problem at hand. This scenario can cause a decrease in generalization performance of the learning machine. In order to handle such issues we extend SVILP by introducing a novel rule selection method where the selected rules can have very high information content, generalization ability and can handle class imbalance problem.

In order to solve multi-class problems we propose a simple but accurate approach. The method is designed by reducing the multi-class classification task to binary problems. However our approach is distinguished from the existing reduction techniques as it learns the hidden structure and characteristics of the data and hence improves the performance of the classifier. The proposed method is based on divide-and-conquer strategy and it discriminates different classes by using an underlying structure based on decision lists. The multi-class problem is reduced by recursively breaking it down into binary problems where each binary task is solved by invoking an SVILP machine. At each node of the decision list the algorithm induces a classifier and updates the training set by removing the examples of the class chosen at the previous node. A label is assigned to a new example by traversing the list. We also study the well known one-vs-all scheme in conjunction with SVILP.

During recent years, a number of multi-class classification method have been proposed [4,5,6,7,8]. The focus of the methods has been on the construction of different effective multi-class schemes whereas less attention has been paid to manipulating the hidden structures and characteristics of the data by using expressive representations. In ILP, which is well known for its use of expressive language formalism, the standard method to solve multi-class problems is based upon inducing a set of disjunctive rules for each class and a new example is predicted if it satisfies the conditions of the rules. In the case that multiple classes are assigned to an example, that is common in ILP, the method is biased towards majority class. Within ILP algorithms, the use of decision lists [9] was explored by Mooney and Califf [10] for binary concept learning. The method extended FOIL [11] by incorporating intensional background knowledge and it is characterized by it's ability to induce logic programs without explicitly taking negative examples as input. The logic program generated by the technique comprised ordered list of clauses (rules). The method was successfully applied to the complex problem of learning past tense of English verbs.

In order to evaluate the performance of proposed methods, we conducted a series of experiments. We applied the techniques to solving multi-class protein fold recognition problem and binary class mutagenicity detection and identification task. The results show that the techniques yield substantial and significant improvements in performance.

2   Multi-class Inductive Logic Programming (MC_ILP)

ILP systems have been successfully applied to binary classification tasks in computational biology, bioinformatics, and chemoinformatics. There are few ILP systems that can perform multi-class classification tasks [12]. The standard multi-class logic based method, described below, is biased towards the majority class. The method is based on learning theories H_j (first order horn clauses) for each class j. The obtained theories for r classes are merged into a multi-theory H. For each class the number of correctly classified training examples are recorded. A class is assigned to a new example if the example satisfies the conditions of the rules. In the case that an example is predicted to have multiple classes, then the class with the maximum number of predicted training examples is assigned to the example. If an example fails to satisfy the conditions of all rules in H, a default class (majority class) is assigned to it. The method is termed as multi-class ILP (MC_ILP).

3   Support Vector Inductive Logic Programming

Support Vector Inductive Logic Programming [1] is a new machine learning technique that is at the intersection of Inductive Logic Programming and Support Vector Machines [13]. SVILP extends ILP with SVMs where the similarity between the examples is measured by computing an inner product on the subset of rules induced by an ILP system. It can be viewed as a multi-stage learning algorithm. The four stages that comprise SVILP learning are described as follows.

In the first stage a set of rules $\mathcal{H}$ is obtained from an ILP system that takes relationally encoded examples (positive, negative) and background knowledge as input. The set, $\mathcal{H}$, comprises all the rules constructed during the search of the hypothesis space. This stage maps the examples into a logic based relational space. A first order rule, $h \in \mathcal{H}$, can be viewed as a boolean function of the form, $h : D \rightarrow \{0,1\}$.

In the next stage a subset $H \in \mathcal{H}$ is selected by using an information theoretic measure, namely compression, described below. The stage maps the examples into another lower dimensional space containing the information relevant to the task at hand. The compression value of a rule is computed by the expression, $C = \frac{PT*(ps-(ng+cl))}{ps}$, where ps is the number of positive examples correctly deducible from the rule, ng is the number of negative examples that satisfy the conditions of the rules, cl is the length of the rule and PT is the total number of positive examples.

In the third stage a kernel function is defined on the selected set of rules where rules can be weighted/unweighted. The kernel is based on the idea of comparing two examples by means of structural and relational features they contain; the more features in common the more similar they are. The function is given by the inner product between the mapped examples where the mapping ϕ is implied by the set of rules H. The mapping ϕ for an example d is given by,[1]

$$\phi : d \rightarrow \left(\sqrt{\pi(h_1(d))}, \sqrt{\pi(h_2(d))}, \ldots, \sqrt{\pi(h_t(d))} \right)', \text{ where } h_1, \ldots, h_t \text{ are rules}$$

and π is the weight assigned to each rule h_i. The kernel for examples d_i and d_j is given by, $k(d_i, d_j) = \langle \phi(d_i), \phi(d_j) \rangle = \sum_{l=1}^{t} \sqrt{\pi(h_l(d_i))} \sqrt{\pi(h_l(d_j))}$. The kernel specified by an inner product between two mapped examples is a sum over all the common hypothesized rules. Given that ϕ maps the data into feature space spanned by ILP rules, we can construct Gaussian RBF kernels, $k_{RBF}(d_i, d_j) = \exp\left(\frac{-\|(\phi(d_i)-\phi(d_j)\|^2}{2\sigma^2} \right)$, where $\|(\phi(d_i) - \phi(d_j)\| = \sqrt{k(d_i, d_i) - 2k(d_i, d_j) + k(d_j, d_j)}$.

In the final stage learning is performed by using an SVM in conjunction with the kernel. SVILP is flexible to construct any kernel in the space spanned by the rules. However, in the present work we used RBF kernels, k_{RBF}, and linear kernels, k, in conjunction with an SVILP machine.

We now consider an example that shows how SVILP kernel measures similarity between two protein domains, 'd2hbg_' and 'd1alla__' which belong to α structural class and 'Globin-like' fold (SCOP classification scheme). Figures 1, 3, 2 and 4 show the two domains and their relationally encoded features. Here predicates 'len', 'nb_alpha', and 'nb_beta' denote the length of the polypeptide chain, number of α-helices and β strands respectively. The other predicates represent the relationship between the secondary structure elements and their properties (hydrophobicity, the hydrophobic moment, the length of proline and etc.). Figure 5 shows a set of induced rules together with their English conversion. A rule classifies an example positive (1) if it fulfils the conditions of the rule while an example that fails to satisfy the conditions is classified negative (0). The set of equally weighted rules maps the two examples as follows:

---

[1] $'$ specifies column vector.

Fig. 1. Protein domain 'd1all__'

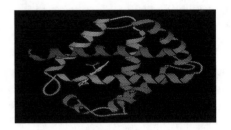

Fig. 2. Protein domain 'd2hbg'

dom_t(d1alla_).
len(d1alla_, 161). nb_alpha(d1alla_,7).
nb_beta(d1alla_,0). has_pro(d1alla_h1).
sec_struc(d1alla_, d1alla_h3).
unit_t(d1alla_h3).
sst(d1alla_h3,4,4,a,104,9,h,0.443,3.003,
116.199, [v,t,p,i,e,e,i,g,v]).
unit_hmom(d1alla_h2, hi).···

Fig. 3. Relationally encoded features of protein domain. 'd1alla_'.

dom_t(d2hbg__).
len(d2hbg__, 147). nb_alpha(d2hbg__,6).
nb_beta(d2hbg__,0).
has_pro(d2hbg__h5).
sec_struc(d2hbg__, d2hbg__h2).
unit_t(d2hbg__h2).
sst(d2hbg__h2,3,3,blank,40,7,h,0.540,
1.812, 213.564, [q,m,a,a,v,f,g]). ···

Fig. 4. Relational encoded features of protein domain 'd2hbg__'

$\phi(d1alla_) = \phi(d1) = \left(1 \times 1 \ \ 1 \times 1 \ \ 1 \times 1\right)'$ and $\phi(d2hbg__) = \phi(d2) = \left(1 \times 1 \ \ 0 \times 1 \ \ 1 \times 1\right)'$. Given that the rules are equally weighted, each entry of the vector is multiplied by 1. The kernel values between the examples are as follows: $k(d1, d2) = k(d2, d1) = 2$, $k(d1, d1) = 3$ and $k(d2, d2) = 2$. In the proceeding sections we present rule selection and multi-class classification schemes for SVILP.

4   Extending Support Vector Inductive Logic Programming

4.1   Low Dimensional Embedding

As described earlier, an SVILP machine obtains a set $\mathcal{H}$ of all the rules, constructed during the search of the hypothesis space, from an ILP algorithm. The number of rules can be very large and the compression value of a rule can be positive or negative. The wide ranging set $\mathcal{H}$ includes rules that are highly relevant to build a classifier with high generalization ability and rules that are highly irrelevant (noise) and can decrease the generalization performance of the classifier. The irrelevant rules establishe a need to present an effective method to selecting relevant rules and hence embedding the data into an informative lower dimensional logical space. In [1] data was embedded into a lower dimensional space H,

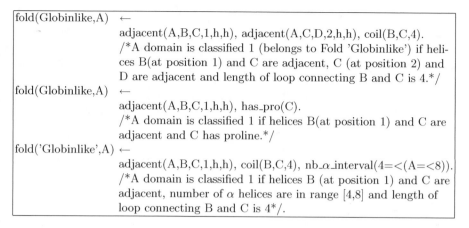

fold(Globinlike,A)   ←
> adjacent(A,B,C,1,h,h), adjacent(A,C,D,2,h,h), coil(B,C,4).
> /*A domain is classified 1 (belongs to Fold 'Globinlike') if heli-
> ces B(at position 1) and C are adjacent, C (at position 2) and
> D are adjacent and length of loop connecting B and C is 4.*/

fold(Globinlike,A)   ←
> adjacent(A,B,C,1,h,h), has_pro(C).
> /*A domain is classified 1 if helices B(at position 1) and C are
> adjacent and C has proline.*/

fold('Globinlike',A) ←
> adjacent(A,B,C,1,h,h), coil(B,C,4), nb_α_interval(4=<(A=<8)).
> /*A domain is classified 1 if helices B (at position 1) and C are
> adjacent, number of α helices are in range [4,8] and length of
> loop connecting B and C is 4*/.

Fig. 5. Rules followed by English conversion for Protein domains in Globin-like fold

where $H \ll \mathcal{H}$, by selecting all the rules with positive compression values. However negatively compressed rules can contain highly relevant information such
as structural and relational features that can be crucial to solving the complex
problem at hand. In this section we present a novel method to embed data into
a lower dimensional space with extra information.

The proposed method is based on the construction of feature space by exploiting the information content and discriminatory power of the rules. The
constructed space is characterized by its amenability to multi-class (/ binary)
classification. We now derive an expression to measure the influence of the rules.
We use P to denote the number of positive example, and N represent number
of negative examples. Similarly, the number of positive examples that fail to
satisfy the conditions of a rule are represented by P^-, where N^+ shows the
number of negative examples that incorrectly fulfils the conditions of the rule.
The expression is given by

$$HD = W_P * P^- + W_N * N^+ \tag{1}$$

where W_P and W_N are the weights assigned to P^-, and N^+ respectively.

The smaller value of HD illustrates the goodness of fit for a rule. The expression can be viewed as weighted sum of hamming distances between two boolean
vectors. Let $\hat{\mathbf{c_P}}$ and $\hat{\mathbf{c_N}}$ denote vectors of positive (1) and negative (0) examples respectively. We use $\hat{\mathbf{f_P}}$ to represent vector of the predictions on positive
examples by a rule. Similarly, $\hat{\mathbf{f_N}}$ denotes vector of the predictions on negative examples by the rule. The distance between $\hat{\mathbf{c_P}}$ and $\hat{\mathbf{f_P}}$ can be computed
by counting the number of entries which differ in both the vectors. Formally,
$HD_P(\hat{\mathbf{c_P}}, \hat{\mathbf{f_P}}) = \sum_{i=1}^{P} |c_{P_i} - f_{P_i}|.$ $\left(\text{For labels } \{+1, -1\} \text{ the distance can be}\right.$
computed by $\sum_{i=1}^{P} \frac{|c_{P_i} - f_{P_i}|}{2}\Big).$ Similarly, $HD_N(\hat{\mathbf{c_N}}, \hat{\mathbf{f_N}}) = \sum_{i=1}^{N} |c_{N_i} - f_{N_i}|.$ The
weighted sum of the distances is given by $HD = W_P * HD_P(\hat{\mathbf{c_P}}, \hat{\mathbf{f_P}}) + W_N *$
$HD_N(\hat{\mathbf{c_N}}, \hat{\mathbf{f_N}}).$ That is like computing the expression HD given in 1.

We now describe how we utilize the expression 1 to obtain a lower dimensional logical and relational feature space with extra information. A set of rules, $\mathcal{H}$, is obtained by an ILP system. In order to measure the score (influence) of rules a validation set is used. For each rule the values of P^- and N^+ are counted and the goodness of fit is measured by expression, $HD = W_P * P^- + W_N * N^+$. The calculated scores are recorded in a list. Once a list is created, the next step involves sorting it in ascending order. The first t rules with lowest HD values are selected.

The idea behind the use of weights in the expression 1 is to give equal importance to all the classes in a dataset that is characterized by uneven class distribution. We now describe a heuristic method to assign weights. We assume a scenario where a set of examples belong to two classes (positive, negative) and the examples belonging to the negative class make the majority class. In this scenario W_P is set to $\frac{N}{P}$ and W_N is set to 1. We used this approach to compute W_P and W_N for the experiments reported in section 5.

4.2   Multi-Class Classification

We now propose a novel logic based method to solving multi-class classification problems. We apply inductive learning in which an algorithm is provided with a set of examples, D, of the form $D = \{(d_1, c_1), (d_2, c2), \ldots, (d_n, c_n)\}$ where d_i are training examples and $c_i \in \{1, 2, \ldots, r\}$ are classes (labels). The goal of the classification algorithm is to generate a function $f : d \rightarrow \{1, 2, \ldots, r\}$ that assigns a new example d to the class with low error probability.

In order to solve multi-class problems we apply powerful but simple divide and conquer strategy. The complex multi-class classification task is divided into binary problems and each problem is solved recursively. The method constructs a decision list as shown in figure 6. Here each non-leaf node has two children.

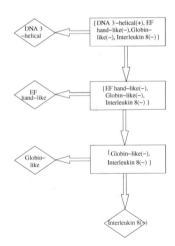

Fig. 6. A decision list, learned by the large margin first order rule learner, for multi-class classification

---

Algorithm 1. Support Vector Inductive Logic Programming (DL_SVILP) for multi-class classification

---

Input: A set of training examples $\{(d_1, c_1), (d_2, c_2), \ldots, (d_n, c_n)\}$, where $d_i \in D$ and $c_i \in \{1, 2, \ldots, r\}$ and a vector *index* that represents learned structure of the list.

    for $j = 1$ to $r - 1$ **do**

        /* Select a class p from r classes */

        $p = index[j]$

        /* Formulate the binary class problem by assigning label '1' to examples of class p and '-1' to examples of remaining classes */

        $D_i = \{(d_1, c_1), (d_2, c_2), \ldots, (d_n, c_n)\}$, where $d_i \in D$ and $c_i \in \{1, -1\}$

        /* Induce a binary classification function f_i by applying SVILP to set D_i */

        $f_i : D_i \rightarrow \{1, -1\}$

        /* Reduce the size of set D_i by removing the examples belonging to class p */

        $D_{i+1} = D_i \setminus D_p$

    end for

    return f_i for $i = 1, \ldots, r - 1$

Classes are represented by non leaf nodes where edges are labeled by the binary classifier's output. We term the technique as decision list based SVILP (DL_SVILP). The method is shown as Algorithm 1. The technique reduces multi-class classification problem to $r - 1$ binary problems, where r is the total number of classes. The algorithm can be viewed as comprising $r - 1$ iterations. In each iteration a class is selected as the positive class and the remaining classes are reduced to the negative class. The binary problem is solved by using a large margin first order rule learner. The training set is updated by removing the examples of the chosen class. In this way the root node contains all the classes whereas the node at depth $r - 1$ contains two classes. The size of the training set used at depth $r - 1$ is (much) smaller than the size of the training set for the root node. DL_SVILP assigns a class j to a new example d as follows:

1. Begin at the root node
2. Apply the classifier associated with the node to example d
3. Travel down the edge labeled by the classifier's output
4. If the edge is labeled positive output the class associated with the leaf. If the edge is labeled negative repeat steps 2 and 3 until the last positive edge is reached. Output the label given by the node.

We now describe how the underlying structure of the list is constructed. The method is dynamic and adaptive to the learning process. At each node the selection of the positive class is made in way so as the classifier can have high generalization ability. The method is presented as Algorithm 2. For each class j a binary class problem is formulated by assigning label '1' to examples of chosen class and '-1' to examples of remaining classes. The classifier, induced from the dataset, is evaluated on a validation set. The performance of the classifier is measured by the expression 1 and the values are recorded in a list. In short, r one-vs-all classifiers are trained and a list of scores that represent the performance of

---

Algorithm 2. Learning underlying structure for DL_SVILP

Input: Training set, $d_1, d_2, \ldots, d_n$, validation set, $d'_1, d'_2, \ldots, d'_s$, r classes and a large
margin first order rule learner (such as SVILP)

    for $j = 1$ to r **do**

        /* Formulate the binary class problem by assigning label '1' to examples of class
j and '-1' to examples of remaining classes */

        /* Induce a binary classification function by applying SVILP to training data,
$d_1, d_2, \ldots, d_n$ */

        /* Apply the learned function to validation set, $d'_1, d'_2, \ldots, d'_s$ */

        /* Measure performance of classifier by using expression 1 */

        $S[j]' = W_P * P^- + W_N * N^+$

        where P = total number of positive example, N = total number of negative
examples, P^- = number of misclassified positive examples, N^- = number of
misclassified negative examples, $W_P = \frac{N}{P}$ and $W_N = 1$

        $index[j]' = j$

    end for

    /* Sort list S' in ascending order and reorder list $index'$ accordingly */

    $S = sort(S')$

    $index = reorder(index')$

    return $index$ and S

---

the classifiers, is obtained. Finally the list is sorted and this ranked list defines
the underlying structure.

4.3   One-vs-All

One-vs-all is a well known multi-class classification strategy. The recent research
[7] showed that the solution obtained by the scheme is accurate. We now describe
how we design one-vs-all based SVILP multi-class classifier that we term one-vs-
all Support Vector Inductive Logic Programming (OVA_SVILP). We construct
OVA_SVILP by learning r binary classifiers by using SVILP. A new example is
classified by applying all the classifiers to it. The example is assigned a label by
the classifier that outputs the largest value(margin).

5   Experiments and Results

We conducted a series of experiments to evaluate the performance of the proposed methods for selecting informative rules and solving multi-class classification problems. We applied the methods to complex tasks, such as mutagenicity
detection and protein fold recognition.

    For multi-class classification problems we used accuracy and positive predic-
tive value (precision rate) as evaluation measures. Let P_j denote the number of
examples belonging to class j, $P = \sum_{j=1}^{j=k} P_j$ represent total number of examples
belonging to k classes, and TP_j denote the number of correctly classified examples belonging to class j. The accuracy for each class j is given by $\frac{TP_j}{P_j}$ whereas

Table 1. Cross-validated accuracy for mutagenesis

kFOIL	nFOIL	c-ARMR+SVM	RUMBLE	PROGOL	SVILP$_C$	SVILP$_{HD}$
81.3	75.4	73.9	84.0	78.7	85.6	**87.2**

the overall accuracy is defined by the expression $\frac{\sum_{j=1}^{j=k} TP_j}{P}$. We used two-sample t-test to assess the significance of our results. The performance of the methods was also analyzed in relation to their average positive predictive values (PPVs) that is given by $\frac{TP_j}{TP_j+FP_j}$ for each class j. In the expression FP_j denotes the numbers of examples that are incorrectly classified in class j.

In order to construct underlying binary SVILP classifiers we used CProgol5 (PROGOL) [14] and SVM^{light} [15]. We refer SVILP to SVILP$_C$ for compression based rule selection whereas SVILP is termed as SVILP$_{HD}$ for the proposed rule selection method. For multi-class classification OVA_SVILP$_C$, DL_SVILP$_C$, OVA_SVILP$_{HD}$, DL_SVILP$_{HD}$ represent compression based and HD (hamming distance) based schemes respectively.

Mutagen Classification. In drug design and development, toxicity classification including mutagen detection and identification is a key task. Mutagenic compounds produce mutations in DNA. In order to validate the use of SVILP as a binary classifier, we applied the algorithm to the mutagen classification problem. For comparison with related techniques, we conducted experiments on a benchmark machine learning dataset, namely mutagenesis [16] that has been widely used for the evaluation of new techniques [17]. We used regression friendly subset comprising 188 molecules and atom and bond background information so that we could compare the performance of SVILP with closely related methods kFOIL and RUMBLE. 10-fold cross validation was used as experimental methodology. At each cross-validation iteration, a classifier was trained on 8 folds, 1 fold was used as the validation set while the remaining 1 fold comprised the test set. We tuned the free parameters clause length and noise of PROGOL, the regularization parameter C of SVMs and width parameter γ of RBF kernels by using the validation set. The set of values for clause length is {2,4}, noise is {5,10,20}, C is {1, 10, 100} and γ is {0.001, 0.01,0.1, 1}. Optimal number of rules were selected from the set {25, 100, 200, 400}. Table 1 shows the results of kFOIL, nFOIL, c-ARMR+SVM, PROGOL, SVILP$_C$ and SVILP$_{HD}$. The reported results of kFOIL, nFOIL, c-ARMR+SVM and RUMBLE appeared in [2] and [3]. The results show that SVILP compares favorably with related approaches. The results also validate the efficacy of the proposed rule selection methodology.

Protein Fold Classification. The recognition of proteins having similar structure is a challenging and complex task in computational biology and bioinformatics. It has key importance in studying protein structure and function and can provide answers to biological problems. In fold recognition, labels are assigned to proteins from a set of predefined annotations (labels, folds). In this way protein fold recognition can be viewed as the multi-class classification task where

Table 2. 5-fold cross-validated over all accuracy (OA) ± standard deviation for protein fold dataset for MC_ILP, OVA_SVILP$_C$, OVA_SVILP$_{HD}$, DL_SVILP$_C$, DL_SVILP$_{HD}$ and MC_SVM. We also report cross-validated accuracy ± standard deviation for 20 folds. The higher values (shown in bold) demonstrate the advantage of the methods.

Fold	MC_ILP	OVA_SVILP$_C$	OVA_SVILP$_{HD}$	DL_SVILP$_C$	DL_SVILP$_{HD}$	MC_SVM
α						
1	43.3 ± 9.0	76.7 ± 7.7	76.7 ± 7.7	73.3 ± 8.1	66.7 ± 8.6	43.3 ± 9.0
2	28.6 ± 12.1	28.6 ± 12.1	**64.3 ± 12.8**	21.4 ± 11.0	57.1 ± 13.2	14.3 ± 9.4
3	46.2 ± 13.8	**92.3 ± 7.4**	69.2 ± 12.8	61.5 ± 13.5	53.9 ± 13.8	53.8 ± 13.8
4	10.0 ± 9.5	10.0 ± 9.5	30.0 ± 14.5	**40.0 ± 15.5**	30.0 ± 14.5	0.0 ± 0.0
5	40.0 ± 15.5	30.0 ± 14.5	**50.0 ± 15.8**	40.0 ± 15.5	40.0 ± 15.5	20.0 ± 12.6
OA	36.4 ± 5.5	55.8 ± 5.7	**63.6 ± 5.5**	53.3 ± 5.7	54.6 ± 5.7	31.2 ± 5.3
β						
6	73.3 ± 6.6	88.9 ± 4.7	75.6 ± 6.4	**91.1 ± 4.2**	88.9 ± 4.7	71.1 ± 6.8
7	57.1 ± 10.8	90.5 ± 6.4	95.2 ± 4.7	95.2 ± 4.7	90.5 ± 6.4	66.7 ± 10.3
8	0.0 ± 0.0	10.0 ± 6.7	15.0 ± 8.0	15.0 ± 8.0	**35.0 ± 10.7**	15.0 ± 8.0
9	43.8 ± 12.4	68.8 ± 11.6	75.0 ± 10.8	75.0 ± 10.8	75.0 ± 10.8	68.8 ± 11.6
10	64.3 ± 12.8	85.7 ± 9.4	**92.9 ± 6.9**	71.4 ± 12.1	71.4 ± 12.1	64.3 ± 12.8
OA	52.6 ± 4.6	72.4 ± 4.2	70.7 ± 4.2	74.1 ± 4.1	**75.9 ± 4.0**	59.5 ± 4.6
α/β						
11	85.5 ± 4.8	85.5 ± 4.8	**87.3 ± 4.5**	67.3 ± 6.3	76.4 ± 5.7	58.2 ± 6.7
12	52.4 ± 10.9	81.0 ± 8.6	61.9 ± 10.6	76.2 ± 9.3	**90.5 ± 6.4**	28.6 ± 9.9
13	28.6 ± 12.1	35.7 ± 12.8	50.0 ± 13.4	50.0 ± 13.4	50.0 ± 13.4	7.1 ± 6.9
14	7.7 ± 7.4	7.7 ± 7.4	15.4 ± 10.0	30.8 ± 12.8	**38.5 ± 13.5**	0.0 ± 0.0
15	0.0 ± 0.0	0.0 ± 0.0	0.0 ± 0.0	8.3 ± 8.0	8.3 ± 8.0	**16.7 ± 10.8**
OA	54.8 ± 4.6	60.9 ± 4.6	60.9 ± 4.6	56.5 ± 4.6	**64.4 ± 4.5**	35.7 ± 4.5
α $+\beta$						
16	53.8 ± 9.8	69.2 ± 9.1	**73.1 ± 8.7**	69.2 ± 9.1	69.2 ± 9.1	23.1 ± 8.3
17	15.4 ± 10.0	30.8 ± 12.8	38.5 ± 13.5	53.9 ± 13.8	53.9 ± 13.8	30.8 ± 12.8
18	7.7 ± 7.4	53.8 ± 13.8	**61.5 ± 13.5**	46.2 ± 13.8	46.2 ± 13.8	30.8 ± 12.8
19	0.0 ± 0.0	8.3 ± 8.0	8.3 ± 8.0	8.3 ± 8.0	25.0 ± 12.5	25.0 ± 12.5
20	77.8 ± 13.9	77.8 ± 13.9	77.8 ± 13.9	66.7 ± 15.7	66.7 ± 15.7	22.2 ± 13.9
OA	32.9 ± 5.8	50.7 ± 5.7	54.8 ± 5.7	52.1 ± 5.8	54.8 ± 5.6	26.0 ± 5.6
OA	46.2 ± 2.6	61.4 ± 2.5	63.3 ± 2.5	60.4 ± 2.5	**64.0 ± 2.5**	40.2 ± 2.5

the problem is characterized by highly skewed class distribution. The aim of a protein fold classification system is to assign proteins to one of many folds with high accuracy. Machine learning methods have been applied to investigate the problem. The studies reported in [18,4] applied SVMs [13] to solving multi-class protein fold classification problem. Chen and Kurgan [19] and Shen and Chou [20] studied ensemble methods to assign 27 folds, from SCOP, to proteins. [21].

Dataset1. We solved protein fold classification problem by applying the proposed multi-class methods to the dataset presented in [22]. In order to compare the

Fold	#Exm	Fold	#Exm
α		α/β	
1	30	11	55
2	14	12	21
3	13	13	14
4	10	14	13
5	10	15	12
β		$\alpha+\beta$	
6	45	16	26
7	21	17	13
8	20	18	13
9	16	19	12
10	14	20	9

Fig. 7. Class distribution for 20 protein folds of dataset1

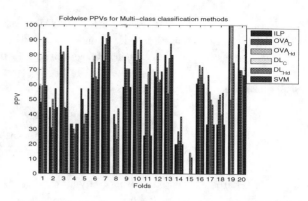

Fig. 8. Fold-wise positive predictive values (PPVs) for MC_ILP (MC), OVA_SVILP$_C$ (OVA$_C$), OVA_SVILP$_{HD}$ (OVA$_{HD}$), DL_SVILP$_C$ (DL$_C$), DL_SVILP$_{HD}$ (DL$_{HD}$) and MC_SVM (SVM).

performance of SVILP based multi-class classification schemes with non-SVILP based methods we used multi-class SVM (MC_SVM) and MC_ILP. MC_SVM was trained by using SVM^{light} [15] where the method was presented in [23]. For MC_SVM, we represented protein domains by using non-relational features namely, total number of residues, α-helices and β-strands. Previous research demonstrated the effectiveness of these features for protein fold classification task. For MC_ILP and SVILP based techniques we used relational fold discriminatory features described in [22]. These features are polypeptide chain length, number of α-helices and β-strands, adjacent secondary structure elements, properties of the secondary structure such as the hydrophobicity, the hydrophobic moment, the length of proline (number of proline residues) and the length of the loop.

The dataset comprises 381 protein domains. They belong to 20 folds of SCOP that have been categorized into 4 structural classes, namely α, β, α/β and $\alpha+\beta$. The indices 1 to 20 shown in Table 2 represent SCOP folds DNA 3-helical, EF hand-like, Globin-like, 4-Helical cytokines, Lambda repressor, Ig beta-sandwich, Tryp ser proteases, OB-fold, SH3-like barrel, Lipocalins, α/β (TIM)-barrel, Rossmann-fold, P-loop, Periplasmic II, α/β-Hydrolases, Ferredoxin-like, Zincin-like, SH2-like, β-Grasp, and Interleukin respectively. The dataset is characterized by uneven class distribution as shown in figure 7.

We randomly divided the dataset into 5 equal-sized folds and followed the experimental methodology as follows. At each cross validation round 3-folds were used for training the classifiers where the remaining two folds were used as validation set and test set. The free parameter of SVM_MC ($\mathcal{C}$, width of the Gaussian kernel), SVILP_OVA$_C$ ($\mathcal{C}$, width of the Gaussian kernel), SVILP_OVA$_{HD}$ (number of rules, $\mathcal{C}$, width of the Gaussian kernel), SVILP_DL$_C$ ($\mathcal{C}$, width of the Gaussian kernel), and SVILP_DL$_{HD}$ (number of rules, $\mathcal{C}$, width of the Gaussian kernel) were tuned by using the validation set. Table 2 lists the cross-validated accuracy for each protein fold for the multi-class classification methods. Overall accuracy over 20 folds is also given. From the results it is clear the DL_SVILP$_{HD}$

outperforms all other methods in the study. We first focus on the performance of SVILP based methods. In order to assess the effect of low dimensional embedding methods (compression based rule selection, HD based rule selection) on the quality of the trained multi-class classifiers, the performance of DL_SVILP$_{HD}$ was compared with DL_SVILP$_C$. DL_SVILP$_{HD}$ improved the performance over DL_SVILP$_C$ and two sample t-test verified the significance of the gain in accuracy (with p $\ll$ 0.1). Comparison of the performances of OVA_SVILP$_{HD}$ with OVA_SVILP$_C$ demonstrated that OVA_SVILP$_{HD}$ also yielded substantial (but not statistically significant) gain in accuracy. In summary, the results validate the efficacy of HD based rule selection method where the gain in performance is generally substantial and statistically significant.

We now analyze the performance of large margin first order decision list based learner, DL_SVILP$_{HD}$, for multi-class classification. Table 2 shows that the accuracy values of DL_SVILP$_{HD}$ are higher than the other methods. It yielded higher over all accuracy than OVA_SVILP$_C$, OVA_SVILP$_{HD}$ and DL_SVILP$_C$ for folds of β and α/β structural classes. The significance of the results was checked by two sample t-test. The classifiers trained by DL_SVILP$_{HD}$ are statistically significantly better than OVA_SVILP$_C$ (with p $\ll$ 0.1) and DL_SVILP$_C$ (with p=0.11). We also compared the performance of DL_SVILP$_{HD}$ with MC_ILP and MC_SVMs. Table 2 shows the effectiveness of DL_SVILP$_{HD}$ where there is a substantial gain in accuracy values. Again, we used two sample t-test to confirm the statistical significance of the results. The performance of DL_SVILP$_{HD}$ is highly significantly better (with p $\ll$ 0.001) than the performance of MC_ILP and MC_SVM.

The performance of the techniques were also analyzed in terms of average positive predictive values. The values are depicted in figure 8 for 20 folds. The figure demonstrates that SVILP based techniques capture structural and relational similarities between proteins and hence learn accurate classifiers.

Dataset2. We further studied the performance of new logic based multi-class classification strategy, DL_SVILP$_{HD}$, by conducting experiments on the protein folds dataset described in [24]. For this set of experiments we only focused on MC_ILP and DL_SVILP$_{HD}$. In the original study protein fold classification problem was solved by viewing it as a binary problem. The dataset comprises 45 protein folds and 441 protein domains that belong to 4 structural classes. The background knowledge comprised structural information for each protein domain that was derived from known secondary structure and multiple structure alignment information. We performed experiments by using the train/test split as described in [24]. As there was no validation set, we, therefore, did not tune the parameters of DL_SVILP$_{HD}$ and MC_ILP. Alternatively, we set PROGOL's clause length and noise parameters to 10 and 20 respectively. The regularization parameter C was set to 1. A linear kernel was used and the number of rules for DL_SVILP$_{HD}$ was set to 100. The performance of DL_SVILP$_{HD}$ was compared to MC_ILP. Table 9 and figure 10 show the results that confirm the usefulness of DL_SVILP$_{HD}$ to solving multi-class classification problems. For the sake of space we only report over all accuracy values for α, β, α/β and $\alpha+\beta$ structural classes.

Fold	MC_ILP	DL_SVILP$_{HD}$
α	57.78 ± 5.21	**62.22 ± 5.11**
β	33.64 ± 4.57	**45.79 ± 4.82**
α/β	56.45 ± 4.45	**62.90 ± 4.33**
$\alpha + \beta$	66.67 ± 5.41	**72.62 ± 5.27**
All	52.84 ± 2.48	**60.25 ± 2.43**

Fig. 9. Accuracy ± standard deviation for protein fold dataset for MC_ILP and DL_SVILP$_{HD}$. The results are averaged over 5 runs of the techniques.

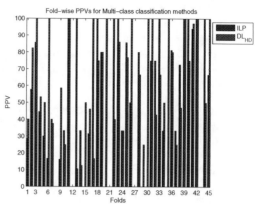

Fig. 10. Fold-wise positive predictive values (PPVs) for MC_ILP (ILP), DL_SVILP$_{HD}$ (DL$_{HD}$)

The results show that DL_SVILP$_{HD}$ yielded higher over all accuracy values for all the structural classes. According to the two sample t-test, the performance of DL_SVILP$_{HD}$ is statistically significantly (with p $\ll$ .01) better than the performance of MC_ILP. Figure 10 depicts average positive predictive values for 45 protein folds that also confirm the efficacy of DL_SVILP$_{HD}$ to solving protein fold recognition problem.

6   Conclusion

In this paper we proposed a novel logic based multi-class classification method. Furthermore we designed an effective low dimensional embedding technique. The efficacy of the proposed methods was evaluated by applying the techniques to mutagen detection and identification and multi-class protein fold recognition problems. The experimental results demonstrated the efficacy of proposed techniques in selecting highly informative rules and producing accurate solutions to complex (binary) multi-class problems. The method, DL_SVILP, captured structural and relational similarities between examples. The results show that the proposed approach can provide an effective alternative to solving multi-class problems.

Acknowledgements

The authors would like to acknowledge the support of the BBSRC project "Protein Function Prediction using Machine Learning by Enhanced Novel Support Vector Logic-based Approach", Grant Reference BB/E000940/1.

References

1. Muggleton, S., Lodhi, H., Amini, A., Sternberg, M.J.E.: Support Vector Inductive Logic Programming. In: Hoffmann, A., Motoda, H., Scheffer, T. (eds.) DS 2005. LNCS (LNAI), vol. 3735, pp. 163–175. Springer, Heidelberg (2005)
2. Landwehr, N., Passerini, A., Raedt, L., Frasconi, P.: kFOIL: Learning simple relational kernels. In: Proceedings of the National Conference on Artificial Intelligence (AAAI), vol. 21, pp. 389–394 (2006)
3. Ruckert, U., Kramer, S.: Margin-base first-order rule learning. Machine Learning 70(2-3), 189–206 (2008)
4. Ding, C.H., Dubchak, I.: Multi-class protein fold recognition using support vector machines and neural networks. Bioinformatics 17, 349–358 (2001)
5. Allwein, E.L., Schapire, R.E., Singer, Y.: Reducing multiclass to binary: a unifying approach for margin classifiers. Journal of Machine Learning Research 1, 113–141 (2000)
6. Platt, J.C., Cristianini, N., Shawe-Taylor, J.: Large margin dags for multiclass classification. In: Advances in Neural Information Processing Systems, vol. 12, pp. 457–553. MIT Press, Cambridge (2000)
7. Rifkin, R., Klautau, A.: In defense of one-vs-all classification. Journal of Machine Learning Research 5, 101–141 (2004)
8. Krebel, U.: Pairwise classification and support vector machines. In: Advances in Kernel Methods: Support Vector Learning, pp. 255–268. MIT Press, Cambridge (1999)
9. Riverst, R.: Learning decision list. Machine Learning 2(3), 229–246 (1987)
10. Mooney, R.J., Califf, M.E.: Induction of first-order decision lits: results on learning the past tense of english verbs. Journal of Artificial Intelligence Research 3, 1–24 (1995)
11. Quinlan, J.: Learning logical definitions from relations. Machine Learning 5(3), 239–266 (1990)
12. Laer, W.V., de Raedt, L., Dzeroski, S.: On multi-class problems and discretization in Iductive Logic Programming. In: Proceedings of the 10th International Symposium on Foundations of Intelligent Systems, pp. 277–286 (1997)
13. Vapnik, V.: The Nature of Statistical Learning Theory. Springer, New York (1995)
14. Muggleton, S.: Inverse entailment and progol. New Generation Computing 13, 245–286 (1995)
15. Joachims, T.: Making large–scale SVM learning practical. In: Schölkopf, B., Burges, C.J.C., Smola, A.J. (eds.) Advances in Kernel Methods — Support Vector Learning, pp. 169–184. MIT Press, Cambridge (1999)
16. Debnath, A.K., de Compadre, R.L.L., Debnath, G., Schusterman, A.J., Hansch, C.: Structure-activity relationship of mutagenic aromatic and heteroaromatics nito compounds. correlation with molecular orbital energies and hydrophobicity. Journal of Medicinal Chemistry 34(2), 786–797 (1991)
17. Lodhi, H., Muggleton, S.: Is mutagenesis still challenging? In: International Conference on Inductive Logic Programming (ILP - Late-Breaking Papers), pp. 35–40 (2005)
18. Shamim, M., Anwaruddin, M., Nagarajaram, H.A.N.J.: Support Vector Machine based classification of protein folds using the structural properties of amino acid residue pairs. Bioinformatocs (2006)

19. Chen, K., Kurgan, L.: PFRES: Protein fold classification by using evolutionary information and predicted secondary structure. Bioinformatics 23, 2843–2850 (2007)
20. Shen, H.B., Chou, C.K.: Ensemble classifier for protein fold recognition. Bioinformatics 22, 1717–1722 (2006)
21. Murzin, A.G., Brenner, S.E., Hubbard, T., Chothia, C.: SCOP: a structural classification of proteins database for the investigation of sequences and structures. J. Mol. Biol. 247, 536–540 (1995)
22. Turcotte, M., Muggleton, S., Sternberg, J.E.: Automated discovery of structural signatures of protein fold and function. J. Mol. Biol. 306, 591–605 (2001)
23. Crammer, K., Singer, Y.: On the algorithmic implementation of multi-class svms. In: JMLR (2001)
24. Cootes, A.P., Muggleton, S., Sternberg, M.J.: The automatic discovery of structural principles describing protein fold space. Journal of Molecular Biology 330(4), 839–850 (2003)

Centrality Measures from Complex Networks in Active Learning

Robson Motta, Alneu de Andrade Lopes, and Maria Cristina F. de Oliveira

Instituto de Ciências Matemáticas e de Computação (ICMC)
University of São Paulo, P.O. Box 668, 13560-970, São Carlos, SP, Brazil

Abstract. In this paper, we present some preliminary results indicating that Complex Network properties may be useful to improve performance of Active Learning algorithms. In fact, centrality measures derived from networks generated from the data allow ranking the instances to find out the best ones to be presented to a human expert for manual classification. We discuss how to rank the instances based on the network vertex properties of closeness and betweenness. Such measures, used in isolation or combined, enable identifying regions in the data space that characterize prototypical or critical examples in terms of the classification task. Results obtained on different data sets indicate that, as compared to random selection of training instances, the approach reduces error rate and variance, as well as the number of instances required to reach representatives of all classes.

Keywords: Complex networks, Active learning, Text mining.

1   Introduction

Text Mining [1] addresses the development of techniques and tools to help humans in tasks that require discriminating potentially useful content from irrelevant material. It encompasses a wide range of techniques in information retrieval, information and topic extraction from texts, automatic text clustering and classification and also strategies supported by visual interfaces [2]. Yet, identifying and selecting relevant information in large repositories of textual documents may still be very difficult, despite the wide availability of text mining techniques.

The problem of automatic text classification requires a set of examples, or instances from the problem domain. A set of labeled (i.e., already classified) instances is input to train a classifier algorithm that will (hopefully) be able to predict the label of new (non-classified) examples, within certain precision. Obtaining training sets for text classification tasks is particularly critical, as labeling an even moderately large set of textual documents demands considerable human effort and time [3].

Active Learning handles this problem departing from the assumption that even when faced with a reduced training set, a learning algorithm may still achieve good precision rates as long as training instances are carefully selected [3]. An active learner may pose to an 'oracle' (e.g., a human expert) queries relative

J. Gama et al. (Eds.): DS 2009, LNAI 5808, pp. 184–196, 2009.

to critical examples, such as those located in class borders. This approach is strongly motivated by the scarcity of labeled instances, particularly severe in the case of text data.

We present some preliminary results indicating that Complex Network properties may be useful to improve performance of active learning algorithms. In fact, centrality measures derived from networks generated from the data allow ranking the training instances to find out the best examples to be presented to a human expert for manual classification.

The rest of the paper is organized as follows. In Section 2 we briefly discuss related work on active learning and also introduce a few concepts in complex networks. In Section 3 we describe how to generate a similarity network from examples on a particular domain – a corpus of scientific papers. We also discuss how centrality measures obtained from such network derived from a corpus of scientific papers may help to select training instances for a paper classification task. In Section 4 we evaluate the proposed approach on different data sets, and finally Section 5 includes some final remarks and a brief comment on further work.

2   Background

2.1   Active Learning

The amount of labeled examples available for training is an important parameter for inductive learning algorithms. They may adopt a supervised learning process, if a large enough set of labeled examples exists, or a semi-supervised learning approach, if otherwise few labeled examples are available. The low number of training examples poses additional challenges in semi-supervised learning.

A popular algorithm for semi-supervised learning is *Co-training* [4], which employs two independent views of data to induce two different hypotheses, adopting either one or two distinct supervised learning algorithms. The presence of two views of each example suggests iterative strategies in which models are induced separately on each view. Then, predictions of one algorithm on new unlabeled instances are employed to expand the training set of the other.

Labeled examples are hardly available in many practical situations involving real data. Nonetheless, a classifier may still be trained with the assistance of a human expert. This is the underlying principle of the active learning paradigm, which addresses the construction of reduced training sets capable of ensuring good precision performance [5]. The rationale is to carefully select the training instances, so that the training set includes those examples that are most likely to strongly impact classifier precision.

Given a set of non-labeled instances, highly representative examples located in the decision boundaries are selected and presented to the expert for labeling. The resulting labeled set is then employed to train the classifier. The problem, of course, is how to identify the representative examples.

Cohn [5] proposes a statistical approach, considering that three factors affect classification error: (a) noise, which is inherent to the data and independent of the classifier; (b) bias, due to the overall strategy and choices of the induction algorithm; and (c) variance, which measures the variation of the correctness rates obtained by the induced models. Representativeness of a particular example is therefore measured by how much it reduces the errors due to bias and classifier variance. The author formulates techniques to select examples that reduce variance working with Gaussian mixtures and locally weighted regression.

Two strategies may be considered to reduce error due to variance. Committee-based methods [6] employ different classifiers to predict the class of an example, querying the expert whenever there is a conflict. Uncertainty-based methods [7] require expert intervention if the classifier prediction has low confidence.

Addressing the problem of text classification, Tong and Koller [8] introduced an active learning algorithm with *Support Vector Machines*. They attempt to select the instance that comes closest to the hyperplanes separating the data. Also handling texts, Hoi et al. [9] employ the *Fisher information* to select a subset of non-labeled examples at each iteration, while reducing redundancy among selected examples. The Fisher information represents a global uncertainty of each example in the classification model. The authors report experiments showing high efficiency in text classification tasks.

We suggest using vertex centrality and community measures from complex networks to assist example selection in active learning. The goal is to minimize the manual classification effort and, simultaneously, improve precision of automatic classifiers. In the following we introduce the relevant concepts in complex networks.

2.2   Complex Networks

Complex Networks [10] are large scale graphs that model phenomena described by a large number of interacting objects. Objects are represented as graph vertices, and relationships are indicated by edges – which may be directed and/or weighted, depending on the nature of the problem. Objects and relationships are usually dynamic and determine the network behavior.

Several models of network behavior have been identified and extensively discussed in the literature – a detailed description and discussion may be found elsewhere [11]. We shall restrict ourselves to briefly introducing a few properties of networks and their vertices that are directly related to the approach described in Section 3.

Vertex Degree: As in ordinary graphs, the degree of a vertex is given by the number of its adjacent edges.

Connectivity Distribution: Defined as the probability of a randomly selected vertex having degree k. The connectivity distribution of a network is defined by a histogram of the degrees of its vertices.

Closeness: Central vertices are strategic elements in network topology. A measure of centrality referred to as vertex *closeness* [12] is obtained by computing the inverse of the average shortest-path length from the vertex to all the other vertices in the network. The higher its closeness, the closer the vertex is, in average, to the remaining network vertices.

Betweenness: This is another centrality measure. Equation 1 describes the *betweenness* of a vertex v_i, contained in a set of vertices V. $amt\_shortest\_paths_{ab}$ is the number of shortest paths between an arbitrary pair of vertices v_a e v_b and $amt\_shortest\_paths_{ab}(i)$ is the number of such paths that include vertex v_i. A high value of betweenness is typical of vertices that link groups of highly connected vertices.

$$b_i = \sum_{a \neq b \neq i \in V} \frac{amt\_shortest\_paths_{ab}(i)}{amt\_shortest\_paths_{ab}} \tag{1}$$

Community Structure: Newman [10] defines community structure as a property of networks that are organized in groups of vertices that are strongly connected amongst themselves, (i.e., internally, within the group), and weakly connected with elements in other (external) groups. This is typical of networks organized into a modular structure. Several strategies and algorithms have been proposed to identify community structures in networks. Newman [13], for example, introduced an agglomerative hierarchical algorithm that has the advantage of not requiring the number of communities as an input parameter [13].

3   Centrality Measures in Active Learning

Given an unlabeled data set and a measure of similarity between any two examples, it is possible to derive a network from the examples. Data examples are represented as network vertices, which will be connected by edges based on a certain criterion. In this case, the chosen criterion is a measure of similarity between the examples represented by the vertices.

The goal is to derive a hierarchical similarity-based network that attempts to (i) capture the community structure of the data; (ii) prioritize links among highly similar data instances; and (iii) search for a network with a desired average degree. The rationale is that instance data similarity structure will be expressed in the topology of networks constructed employing these criteria. The following procedure has been adopted to create a hierarchical similarity-based network.

An initial network includes all available examples as vertices, and has no edges, so that each vertex constitutes a single component. An iterative hierarchical agglomerative process starts that gradually connects vertex pairs, based on a given similarity threshold. Assuming the similarity measure takes values in the range $[0, 1]$, where 1 indicates highly similar examples, the similarity threshold is initialized with a high value, close to one.

An outer loop (the second While in Algorithm 1) inspects all vertex pairs whose similarity measure is above the current similarity threshold. It is responsible for identifying the potential edges to be added, i.e., those pairs whose similarity is above the threshold, and calls an inner loop, shown in Algorithm 2, to select the vertices or components to be actually joined. The similarity threshold is updated at each external iteration step in order to ensure that only the

5% most similar vertex pairs still unconnected are considered for potential connection. The whole process stops when all vertices have been connected into a single component forming a connected network.

Edge inclusion in a component stops when the component reaches a user defined average degree. In the inner loop depicted in Algorithm 2, an edge inclusion that joins components is performed if (i) the edge will link two highly similar vertices, and (ii) the two components share a high number of potential edges (i.e., highly similar vertices). A subset of the potential edges is actually added at each iteration, until the average degree of each component reaches a desired (user defined, for each component) value. The edges effectively added are those that, if included – thus causing their respective components C_i and C_j to be joined – will maximize the measure given by Equation 2.

$$interconnectivity(C_i, C_j) = \frac{1}{\#C_i + \#C_j} \sum_{\substack{x \in C_i, y \in C_j, \\ \exists edge(x,y)}} sim(x,y) \qquad (2)$$

In the above equation, C_i and C_j denote components, $\#C$ represents the number of vertices in component C, $sim(x,y)$ denotes the similarity between vertices $x \in C_i$, $y \in C_j$. The equation is computed for all pairs of components defined in the current iteration, seeking the set of edges that maximizes its result. This approach ensures that components resulting from this iteration have maximum intra-component connectivity and minimum inter-component connectivity.

Algorithm 1. Construction of the Hierarchical Similarity Based Network
___

Input:
   Set of vertices: $V = v_1,...,v_n$
   Average Degree: $averageDegree$
   Data similarity matrix: $similarity$
Output:
   Network, given by a set of vertices and a set of edges: (V,E)

Components $C \leftarrow V$
Edges $E \leftarrow \varnothing$
$minSim \leftarrow$ similarity threshold to obtain the 5% most similar pairs of vertices
While $(\#C > 1)$
   While $(\exists$ pair $(x,y) \mid$ similarity$(x,y) \geq minSim,$ $x \in C_i, C_i \in C, y \in C\text{-}C_i)$
      Components Coalescing$(C,E,averageDegree,minSim)$ % selection of components to be joined
   $minSim \leftarrow$ similarity to add the 5% most similar pairs of vertices

Returns (V,E)
___

A characteristic of such a network is that similar examples – typically expected to be associated to the same class – are likely to define communities. In other words, they form groups of vertices that are densely connected among themselves, with few connections to external groups. Thus, one would expect

Algorithm 2. Components Coalition

Input:
    Set of components: C
    Set of Edges: E
    Average Degree: $averageDegree$
    Similarity Threshold: $minSim$
Output:
    Set of components: C
    Set of Edges: E

preSelectedEdges $\leftarrow \emptyset$
For each component C_i of C
    $numberOfEdges \leftarrow$ ($averageDegree$ * #C_i / 2) - #$E(C_i)$
    If (numberOfEdges ≤ 0)
        numberOfEdges $\leftarrow 1$)
    For all pair$(i,j) \mid i \in C_i$ and $j \in C$ - C_i % Edges (i,j) taken from a priority queue
        If (similarity$(i,j) \geq minSim$)
            preSelectedEdges$(C_i,C_j) \leftarrow$ preSelectedEdges$(C_i,C_j) \cup (i,j)$
            $numberOfEdges--$
            If $(numberOfEdges == 0)$
                $break$
    $(C_a,C_b) \leftarrow$ max(interconnectivity$(C_i,C_j))$ % selected components
    $C_a \leftarrow C_a \cup C_b$ %join components
    $E(C_a) \leftarrow E(C_a) \cup E(C_b) \cup$ preSelectedEdges(C_a,C_b) %add edges of the joined components
    $C \leftarrow C$ - C_b %removes joined component
    $A \leftarrow E$ - $E(C_b)$ %remove from E edges of the removed component

Returns (C,E)

the community structure of the network to reflect the underlying data similarity structure. The network would ideally have few connections between vertices in different communities, and the communities would be formed by groups of similar examples likely to belong to the same class. A probabilistic version of this algorithm has been fully described by Motta et al. [14].

The schema shown in Figure 1 illustrates our working hypothesis, that centrality measures bear a strong relation with the role of examples in a classification process. To illustrate this point, let us consider that two communities have been identified in a hypothetical network generated by the above process. Notice that:

1. *Boundary vertices*, located in the borders of the communities, labeled regions R1 in the figure, typically have low values for closeness, computed for the vertex relative to its component.
2. *Inner vertices*, located in the regions labeled R2 in the figure, are those well identified with a specific community. They typically have high values for closeness, again computed relative to the vertex component.

3. *Critical vertices*, located in the region labeled R3, are those placed across different communities. They typically have low values for closeness computed relative to their own community. Moreover, their betweenness values, computed in the context of the network as a whole, are high, as they are elements linking different communities.

We argue that these regions identified in the above network reflect, to some extent, the topology of the data examples relative to their possible classes. Thus, vertex centrality measures help identify the examples representative of the different regions. Hence, by analyzing such measures, or a combination of them, one may identify the interesting examples in an active learning process.

Boundary vertices correspond to examples that, although not prototypical of a class, have good chances of being properly classified. *Inner vertices* correspond to the prototypical examples that would be easily classified, whereas *Critical vertices* are likely to represent the problematic examples, those in the borders between classes. Thus, we suggest an active learning approach that adopts the following steps:

1. given the examples, generate the hierarchical similarity-based network;
2. partition this network into communities;
3. compute the closeness for all vertices in each community;
4. compute the betweenness for all vertices in the whole network;
5. rank vertices based on (a combination of) selected centrality measures to select examples

As we assume that the three types of region identifiable in the network communities define data set topology, a representative training sample should therefore include examples taken from the three regions from each major community. As measures for betweenness and closeness have different ranges, values are normalized in the interval [0,1]. To rank Critical vertices (in regions R3) one may consider the difference between the normalized measures of betweenness and closeness. A value closer to 1 is indicative of vertices in the critical region.

We shall first illustrate the approach on a data set of scientific papers. Figure 2(a) illustrates a similarity network obtained with the above algorithms from a corpus of nearly 600 papers in the subjects *Case-Based Reasoning*, *Inductive Logic Programming* e *Information Retrieval* (named corpus CBR-ILP-IR). Examples have been labeled by a human expert, based on their source vehicle, so that each subject is taken as a class.

To construct the network, the papers are described by their vector representation, according to the well known bag-of-words model [15]. A data similarity matrix has been computed with the cosine distance over the vector space model, for all document pairs. The average degree threshold has been set to five (see Section 4 for a discussion on this choice).

In the network graph shown in Figure 2(a) each vertex represents a paper, whose known class is indicated by the vertex shade. The same figure, in (b), shows the network after applying the community detection algorithm by Newman [13], mentioned in Section 2. One observes from the figure that most communities have a clear predominance of elements from a single class. This suggests

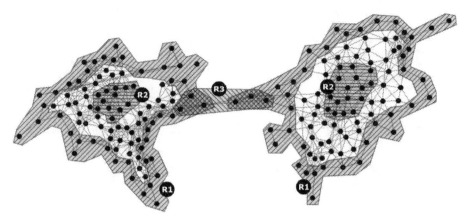

Fig. 1. Schema illustrating values of centrality measures and their relationship to the location of examples relative to their possible classes

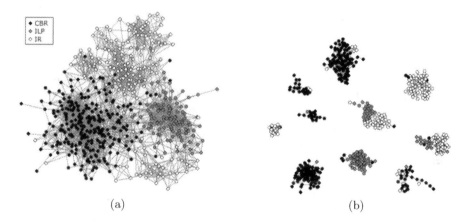

(a)   (b)

Fig. 2. (a) Hierarchical similarity network for the corpus of scientific papers (CBR, ILP e IR) and (b) communities identified in the network

that the community structure identified by the algorithm, without using any class information, reflects reasonably well the known class information.

Figure 3 (a) depict information about vertex closeness, computed for the larger community structure identified in the network. Vertex size maps measure values: greater values are shown as larger glyphs. As expected, one observes vertices with lower closeness in the boundary of the community.

Figure 3 (b) shows another community from the network, with vertex shade mapping the class of the corresponding example. Notice this community includes a single example from the (CBR) class, shown in black. In the same figure the 10% vertices of this community with higher values computed for betweenness are shown as larger glyphs. Inspection shows that these higher values are either in or close to the borders between classes.

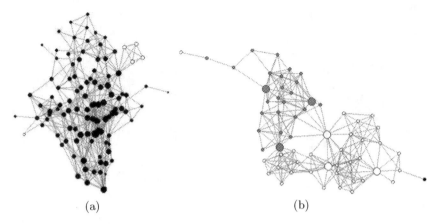

(a)   (b)

Fig. 3. (a): vertex size maps the value of vertex closeness, for a particular community. (b) shows class distribution in another community that has a predominance of two classes (shown in white and gray). Also in this figure, the larger vertices shown correspond to the 10% vertices from this community with higher values for betweenness.

4   Results

The evaluation process was carried out on 10 data sets from UCI repository and on the CBR-ILP-IR corpus already mentioned. We conducted two kinds of experiments on each data set: the first one to verify how many instances should be labeled in order to ensure the identification of all classes present in the training data set; and a second to measure the variation of the classifier error relative to the number of examples selected for labeling.

In both cases, results reported were averaged on 100 runs for data sets with cardinality lower than 500, and with 30 runs otherwise. Selection of training instances with the proposed approach was compared with a random selection. Similar results were obtained with Naïve Bayes and K-Nearest Neighbor (KNN) (with $K = 1$) classifiers. Figure 4 shows the classification errors for the KNN classifier on the CBR-ILP-IR data, considering different policies to select the training data, as described next.

First, the similarity based network is created for the data set. All networks were created by setting the desired average degree for the communities equal to 5. Various experiments conducted on these and other data sets have shown that when lower average degrees (up to 3) are set community structures are not well defined. By 'well defined' we mean communities densely connected internally and sparsely connect with other communities. On the other hand, when setting values above 5 the number of edges increased unnecessarily for the purpose of obtaining clearly defined community structures in the data.

The similarity matrices were computed with the cosine distance for the CBR-ILP-IR data and with the Euclidean distance for the numerical (normalized) data sets. The community structure was then extracted from the resulting network using the aforementioned Newman's algorithm. The following procedure has been adopted to select instances.

Communities are first sorted in decreasing order of their size (number of vertices). Then, the examples in each community are ranked by decreasing values of their closeness (in the community), and all the network examples are ranked in decreasing value of their betweenness.

The overall process is conducted by iteratively scanning the ordered list of communities, selecting one sample from each community at each iteration. Several policies may be adopted to select instances, taking examples from the different regions (inner, boundary or critical). The following alternative policies have been considered to create test data sets in this study:

- Policy $R1$: in forming the test data set, priority is given to instances from the *boundary* region. At each iteration, the bottom ranked instance from each community is selected, i.e., the instance with the lowest closeness;
- Policy $R2$: priority is given to vertices from the *inner* region. The top ranked instance from each community is selected, i.e., the instance with the highest closeness;
- Policy $R3$: priority is given to vertices from the *critical* region. The instance with the highest critical value is selected from each community, computed from the difference *betweenness − closeness*;
- Policy $R2 + R3$: the test data set is formed selecting vertices based on a combination of *critical* and *inner* regions. At successive iterations, alternate between selecting from each community an instance from the Inner region (highest closeness) and an instance from the Critical region (highest critical value);

In all cases, once an instance is selected for the training data it is removed from further consideration. The iterative selection process stops when the desired number of training instances is reached. In these studies, this number has been set as equal to the number of communities identified in the network. We also generated training data sets of the corresponding size by selecting instances *randomly* and, for comparison purposes, employed a training data set including 100% of the labelled instances.

In all experiments we observed that selection of examples that correspond to vertices in Inner and Critical regions resulted in greater error reduction in the classifier output. Considered in isolation, Inner vertices provide the best examples for error reduction, followed by Critical vertices and then Boundary vertices.

Notice that policy $R3$ (selecting from critical region alone) does not produce good results, although this region is associated with higher classification uncertainty. In fact, the experiment indicates that the human expert must also classify some 'prototypical' examples (those in region $R2$). In these experiments an equal number of elements was selected from each region – we did not investigate whether assigning more weight to either regions $R2$ or $R3$ would improve the results.

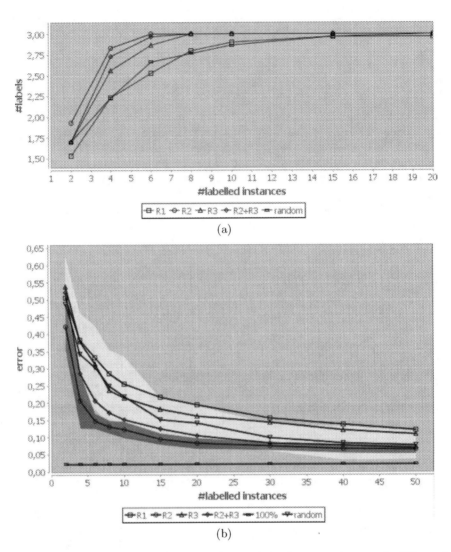

Fig. 4. Comparing different policies to select training instances (average in 100 runs). (a) Number of examples required to identify all classes and (b) KNN classifier error for the CBR-ILP-IR corpus, considering examples selected with the proposed approach under different policies, and at random. For the sake of clarity, variance is represented in the figure only for the random selection policy (lighter region in the figure) and for the $R2$ policy (darker region).

In Figure 4 (a) one observes that the number of examples required to identify all classes is considerably reduced when examples are selected with either policy $R2$ or policy $R2 + R3$. Figure 4(b) shows the error of the KNN classifier as the number of labeled examples in the training set increases. The Naïve Bayes classifier presented a similar behavior. Errors have been computed for training

Table 1. Comparing the classification errors for different data sets. For each test data set we inform the number of known examples, the number of classes, the number of instances desired in the training set, equal to the average number of communities in each run, and the errors for the different instance selection policies. In the table, results from the policy that presented the lowest error (ignoring the error obtained with the training set that includes all examples) are highlighted in bold, whereas the second lowest error policy results are highlighted in italics.

dataset	#instances	#classes	#labelled instances	R1	R2	R3	R2+R3	random	100%
balance	625	3	14	0.312	**0.297**	0.311	*0.306*	0.316	0.223
ecoli	336	8	11	0.327	**0.263**	0.333	*0.29*	0.300	0.193
glass	214	7	10	*0.345*	**0.283**	0.451	0.361	0.352	0.090
ionosphere	351	2	12	0.276	*0.038*	0.082	**0.025**	0.172	0.011
iris	150	3	8	0.199	**0.057**	0.178	*0.111*	0.153	0.047
sonar	208	2	11	0.416	**0.331**	0.382	*0.342*	0.390	0.136
wdbc	569	2	14	0.155	**0.084**	0.108	*0.094*	0.095	0.049
wine	178	3	8	0.234	**0.087**	0.209	*0.091*	0.169	0.050
yeast	1484	10	20	0.654	**0.577**	0.627	0.594	*0.590*	0.473
zoo	101	7	7	0.288	**0.157**	0.352	*0.210*	0.299	0.040
CBR-ILP-IR	574	3	14	0.225	**0.103**	0.181	*0.130*	0.182	0.022

sets obtained with each of the above policies, and also a training set consisting of all known examples.

In Table 1 we synthesize the classification errors obtained for multiple data sets. Notice that all data sets, except the *glass* and the *yeast*, present similar behavior: the lowest error rates and lowest variances are obtained by selecting the training instances using policies $R2$ (inner vertices) and $R2 + R3$ (inner and critical vertices).

5   Conclusions and Further Work

We investigate a novel approach to support active learning in classification tasks, based on properties computed from hierarchical similarity networks derived from the known data instances. A discussion has been presented on how vertex centrality measures obtained from network communities enable identifying regions in the data space that characterize critical examples. Such measures may guide the selection of examples to be presented to a human expert in an active learning approach for classification tasks.

Results of applying the proposed approach on several data sets have been presented and discussed, including an illustrative example of its application on a corpus of scientific papers. These results indicate the potential of this approach for the problem. As further work, it would be desirable to investigate further the potential role of these and other network measures, under different classification algorithms (e.g., Support Vector Machines) and for distinct data set domains. How different network construction procedures may affect the instance selection process is also a topic that deserves further investigation.

Acknowledgments

The authors acknowledge the financial support of FAPESP (Grants 2008/04622-8 and 2009/03306-8) and CNPq (Grant 305861/2006-9).

References

1. Berry, M.: Survey of Text Mining: Clustering, Classification, and Retrieval. Springer, Heidelberg (2003)
2. Minghim, R., Levkowitz, H.: Visual mining of text collections, Tutorial Notes. In: Proc. of EUROGRAPHICS 2007, Computer Graphics Forum, pp. 929–1021 (2007)
3. Settles, B.: Active learning literature survey. Computer Sciences Technical Report 1648, University of Wisconsin–Madison (2009)
4. Blum, A., Mitchell, T.: Combining labeled and unlabeled data with co-training. In: COLT 1998: Proceedings of the eleventh annual conference on Computational learning theory, pp. 92–100. ACM, New York (1998)
5. Cohn, D.A., Ghahramani, Z., Jordan, M.I.: Active learning with statistical models. Journal of Artificial Intelligence Research 4, 129–145 (1996)
6. Engelson, S., Dagan, I.: Minimizing manual annotation cost in supervised training from corpora. In: Proceedings of the 34th annual meeting on Association for Computational Linguistics, Morristown, NJ, USA, pp. 319–326. Association for Computational Linguistics (1996)
7. Hwa, R.: Sample selection for statistical grammar induction. In: Proceedings of EMNLP/VLC 2000, pp. 45–52 (2000)
8. Tong, S., Koller, D.: Support vector machine active learning with applications to text classification. Journal of Machine Learning Research 2, 45–66 (2002)
9. Hoi, S., Jin, R., Lyu, M.: Large-scale text categorization by batch mode active learning. In: WWW 2006: Proceedings of the 15th international conference on World Wide Web, pp. 633–642. ACM, New York (2006)
10. Newman, M.: The structure and function of complex networks. SIAM Review 45(2), 167–256 (2003)
11. Costa, L.F., Rodrigues, F.A., Travieso, G., Boas, P.V.: Characterization of complex networks: A survey of measurements. Advances In Physics 56, 167 (2007)
12. Wasserman, S., Faust, K.: Social Network Analysis: Methods and Applications. Cambridge University Press, Cambridge (1994)
13. Newman, M.: Fast algorithm for detecting community structure in networks. Physical Review E 69, 066133 (2004)
14. Motta, R., Almeida, L.J., Lopes, A.A.: Probabilistic similarity based networks for exploring communities. In: I Workshop on Web and Text Intelligence (SBIA-WTI 2008), Salvador, Brasil, pp. 1–8 (2008) (in Portuguese)
15. Baeza-Yates, R.A., Ribeiro-Neto, B.A.: Modern Information Retrieval. ACM Press/Addison-Wesley (1999)

Player Modeling for Intelligent Difficulty Adjustment

Olana Missura and Thomas Gärtner

Fraunhofer Institute Intelligent Analysis and Information Systems IAIS,
Schloss Birlinghoven, D-53754 Sankt Augustin, Germany
firstname.lastname@iais.fraunhofer.de

Abstract. In this paper we aim at automatically adjusting the difficulty of computer games by clustering players into different types and supervised prediction of the type from short traces of gameplay. An important ingredient of video games is to challenge players by providing them with tasks of appropriate and increasing difficulty. How this difficulty should be chosen and increase over time strongly depends on the ability, experience, perception and learning curve of each individual player. It is a subjective parameter that is very difficult to set. Wrong choices can easily lead to players stopping to play the game as they get bored (if underburdened) or frustrated (if overburdened). An ideal game should be able to adjust its difficulty dynamically governed by the player's performance. Modern video games utilise a game-testing process to investigate among other factors the perceived difficulty for a multitude of players. In this paper, we investigate how machine learning techniques can be used for automatic difficulty adjustment. Our experiments confirm the potential of machine learning in this application.

Keywords: clustering, supervised learning, player modeling, difficulty adjustment.

1 Introduction

We aim at developing games that provide challenges of the "right" difficulty, i.e., such that players are stimulated but not overburdened. Naturally, what is the right difficulty depends on many factors and can not be fixed once and for all players. For that, we investigate how general machine learning techniques can be employed to automatically adjust the difficulty of games. A general technique for this problem has natural applications in the huge markets of computer and video games but can also be used to improve the learning rates when applied to serious games.

The traditional way in which games are adjusted to different users is by providing them with a way of controlling the difficulty level of the game. To this end, typical levels would be 'beginner', 'medium', and 'hard'. Such a strategy has many problems. On the one hand, if the number of levels is small, it may be easy to choose the right level but it is unlikely that the difficulty is then set in a very satisfying way. On the other hand, if the number of levels is large, it is more

J. Gama et al. (Eds.): DS 2009, LNAI 5808, pp. 197–211, 2009.

likely that a satisfying setting is available but finding it becomes more difficult. Furthermore, choosing the game setting for each of these levels is a difficult and time-consuming task.

In this paper we investigate the use of supervised learning for dynamical difficulty adjustment. Our aim is to devise a difficulty adjustment algorithm that does not bother the actual players. For that, we assume there is a phase of the game development in which the game is played and the difficulty is manually adjusted to be just right. From the data collected in this way, we induce a difficulty model and build it into the game. The actual players do not notice any of this and are always challenged at the difficulty that is estimated to be just right for them.

Our approach to building a difficulty model consists of three steps: (i) cluster the recorded game traces, (ii) average the supervision over each cluster, and (iii) learn to predict the right cluster from a short period of gameplay. In order to validate this approach, we use a leave-one-player-out strategy on data collected from a simple game and compare our approach to less sophisticated, yet realistic, baselines. All approaches are chosen such that the players are not bothered. In particular, we want to compare the performance of dynamic difficulty versus constant difficulty as well as the performance of cluster prediction versus no-cluster. Our experimental results confirm that dynamic adjustment and cluster prediction together outperform the alternatives significantly.

2   Motivation and Context

A game and its player are two interacting entities. A typical player plays to have fun, while a typical game wants its players to have fun. What constitutes the *fun* when playing a game?

One theory is that our brains are physiologically driven by a desire to learn something new: new skills, new patterns, new ideas [1]. We have an instinct to play because during our evolution as a species playing generally provided a safe way of learning new things that were potentially beneficial for our life. Daniel Cook [3] created a psychological model of a player as an entity that is driven to learn new skills that are high in perceived value. This drive works because we are rewarded for each new mastered skill or gained knowledge: The moment of mastery provides us with the feeling of joy. The games create additional rewards for their players such as new items available, new areas to explore. At the same time there are new challenges to overcome, new goals to achieve, and new skills to learn, which creates a loop of learning-mastery-reward and keeps the player involved and engaged.

Thus, an important ingredient of the games that are fun to play is providing the players with the challenges corresponding to their skills. It appears that an inherent property of any challenge (and of the learning required to master it) is its difficulty level. Here the difficulty is a subjective factor that stems from the interaction between the player and the challenge. The perceived difficulty is also not a static property: It changes with the time that the player spends learning a skill.

To complicate things further, not only the perceived difficulty depends on the current state of the player's skills and her learning process, the dependency is actually bidirectional: The ability to learn the skill and the speed of the learning process are also controlled by how difficult the player perceives the task. If the bar is set too high and the task appears too difficult, the player will end up frustrated and will give up on the process in favour of something more rewarding. Then again if the challenge turns out to be too easy (meaning that the player already possesses the skill necessary to deal with it) then there is no learning involved, which makes the game appear boring.

It becomes obvious that the game should provide the challenges for the player of the "right" difficulty level: The one that stimulates the learning without pushing the players too far or not enough. Ideally then, the difficulty of any particular instance of the game should be determined by who is playing it at this moment.

Game development process usually includes multiple testing stages, where a multitude of players is requested to play the game to provide data and feedback. This data is analysed to tweak the games parameters in an attempt to provide a fair challenge for as many players as possible. The question we investigate in this work is how the data from the α/β tests can be used for the intelligent difficulty settings with the help of machine learning.

We proceed as follows: After reviewing related work in Section 3, we describe the algorithm for the dynamic difficulty adjustment in general terms in Section 4. In Sections 5 and 6 we present the experimental setup and the results of the evaluation before concluding in Section 7.

3   Related Work

In the games existing today we can see two general approaches to the question of difficulty adjustment. The traditional way is to provide a player with a way to set up the difficulty level for herself. Unfortunately, this method is rarely satisfactory. For game developers it is not an easy task to map a complex gameworld into a single parameter. When constructed, such a mapping requires additional extensive testing, creating time and money costs. Consider also the fact that generally games require several different skills to play them. The necessity of going back and forth between the gameplay and the settings when the tasks become too difficult or too easy disrupts the flow component of the game.

An alternative way is to implement a mechanism for dynamic difficult adjustment (DDA). One quite popular approach to DDA is a so called *Rubber Band AI*, which basically means that the player and her opponents are virtually held together by a rubber band: If the player is "pulling" in one direction (playing better or worse than her opponents), the rubber band makes sure that her opponents are "pulled" in the same direction (that is they play better or worse respectively). While the idea that the better you play the harder the game should be is sound, the implementation of the Rubber Band AI often suffers from disbalance and exploitability.

There exist a few games with a well designed DDA mechanism, but all of them employ heuristics and as such suffer from the typical disadvantages (being not transferable easily to other games, requiring extensive testing, etc). What we would like to have instead of heuristics is a universal mechanism for DDA: An online algorithm that takes as an input (game-specific) ways to modify difficulty and the current player's in-game history (actions, performance, reactions, . . .) and produces as an output an appropriate difficulty modification.

Both artificial intelligence researchers and the game developers community display an interest in the problem of automatic difficulty scaling. Different approaches can be seen in the work of R. Hunicke and V. Chapman [9], R. Herbich and T. Graepel [8], Danzi et al [5], and others. As can be seen from these examples the problem of dynamic difficulty adjustment in video games was attacked from different angles, but a unifying approach is still missing.

Let us reiterate that as the perceived difficulty and the preferred difficulty are subjective parameters, the DDA algorithm should be able to choose the "right" difficulty level in a comparatively short time for any particular player. It makes sense, therefore, to conduct the learning in the offline manner and to make use of the data created during the test phases to construct the player models. These models can be used afterwards to generalise to the unseen players.

Player modeling in computer games is a relatively new area of interest for the researchers. Nevertheless, existing work [12,11,2] demonstrates the power of utilising the player models to create the games or in-game situations of high interest and satisfaction for the players.

In the following section we present an algorithm that learns a mapping from different player types to the difficulty adjustments and predicts an appropriate one given a new player.

4   Algorithm

To simplify the problem we assume that there exists a finite number of types of players, where by type we mean a certain pattern in behaviour with regard to challenges. That is certainly true, since we have a finite amount of players altogether, possibly times a finite amount of challenges, or timesteps in a game. However, this realistic number is too large to be practical and certainly not fitting the purpose here. Therefore, we discretize the space of all possible players' behaviours to get something more manageable. The simplest such discretization would be into beginners, averagely skilled, and experts (corresponding to easy, average, and difficult settings).

In our experiments we do not predefine the types, but rather infer them using the clustering of the collected data. Instead of attempting to create a universal mechanism for a game to adapt its difficulty to a particular player, we focus on the question of how a game can adapt to a particular player type given two sources of information:

1. the data collected from the alpha/beta-testing stages (offline phase);
2. the data collected from the new player (online phase).

The idea is rather simple. By giving the testers control over the difficulty settings in the offline phase the game can learn a mapping from the set of types into the set of difficulty adjustments. In the online phase, given a new player, the game needs only to determine which type he belongs to and then apply the learned model. Therefore, the algorithm in general consists of the following steps:

1. Given data about the game instances in the form of time sequences

$$T_k = ((t_1, f_1(t_1), \ldots, f_L(t_1)), \ldots, (t_N, f_1(t_N), \ldots, f_L(t_N))),$$

   where t_i are the time steps and $f_i(t_j)$ are the values of corresponding features, cluster it in such a way that instances exhibiting similar player types are in the same cluster.
2. Given a new player, decide on which cluster he belongs to and predict the difficulty adjustment using the corresponding model.

Note that it is desirable to adapt to the new player as quickly as possible. To this purpose we propose to split the time trace of each game instance into two parts:

- a prefix, the relatively short beginning that is used for the training of the predictor in the offline phase and the prediction itself in the online phase;
- a suffix, the rest of the trace that is used for the clustering.

In our experiments we used the K-means algorithm [7] for the clustering step and an SVM with a gaussian kernel function [4] for the prediction step of the algorithm outlined above.

We considered the following approaches to model the adjustment curves in the clusters:

1. The constant model. Given the cluster, this function averages over all instances in the cluster and additionally over the time, resulting in a static difficulty adjustment.
2. The regression model. Given the cluster, we train the regularised least squares regression [10] with the gaussian kernel on its instances.

The results stemming from using these models are described in Section 6.

5 Experimental Setup

To test our approach we implemented a rather simple game using the Microsoft XNA framework[1] and one of the tutorials from the XNA Creators Club community, namely "Beginner's Guide to 2D Games"[2]. The player controls a cannon that can shoot cannonballs. The gameplay consists of shooting down the alien spaceships while they are shooting at the cannon (Figure 1). A total of five

---

[1] http://msdn.microsoft.com/en-us/xna/default.aspx
[2] http://creators.xna.com/en-GB/

Fig. 1. A screenshot showing the gameplay

spaceships can be simultaneously on the screen. They appear on the right side of the game screen and move on a constant height from the right to the left. The spaceships are generated so that they have a random speed within a specific δ-interval from a given average speed. Whenever one of the spaceships is shot down or leaves the game screen, a new one is generated. At the beginning of the game the player's cannon has a certain amount of hitpoints, which is reduced by one every time the cannon is hit. At random timepoints a repair kit appears on the top of the screen, floats down, and disappears again after a few seconds. If the player manages to hit the repair kit, the cannon's hitpoints are increased by one. The game is over if the hitpoints are reduced to zero or a given time limit of 100 seconds is up.

Additionally to the controls that allow the player to rotate the cannon and to shoot, there are also two buttons by pressing which the player can increase or decrease the difficulty at any point in the game. In the current implementation the difficulty is controlled by the average speed of the alien ships. For every destroyed spaceship the player receives a certain amount of score points, which increases quadratically with the difficulty level. During each game all the information concerning the game state (e.g. the amount of hitpoints, the positions of the aliens, the buttons pressed, etc) is logged together with a timestamp. At the current state of our work we held one gaming session with 17 participants and collected the data on how the players behave in the game.

Out of all logged features we restricted our attention to the three: the difficulty level, the score, and the health, as they seem to represent the most important aspects of the player's state. The log of each game instance k is in fact a time trace

$$T_k = ((t_1, f_1(t_1), \ldots, f_L(t_1)), \ldots, (t_N, f_1(t_N), \ldots, f_L(t_N))),$$

where $t_1 = 0$, $t_N \leq 100$, and $f_i(t_j)$ is the value of a corresponding feature (Figure 2). Therefore, to model the players we cluster provided by the testers time sequences.

5.1 Technical Considerations

Several complications arise from the characteristics of the collected data:

1. Irregularity of the time steps. To reduce the computational load the data is logged only when the game's or the player's state changes (in the case of a simple game used by us it may seem a trivial concern, but this is important to consider for the complex games). As a result for two different game instances k and $\hat{k}$ the time lines will be different:

$$t_{ik} \neq t_{i\hat{k}}.$$

2. Irregularity of the traces' durations. Since there are two criteria for the end of the game (health dropped to zero or the time limit of a hundred seconds is up), the durations of two different game instances k and $\hat{k}$ can be different:

$$t_{Nk} - t_{\hat{N}\hat{k}} \neq 0.$$

The second problem may appear irrelevant, but as described below it needs to be taken care of in order to create a nice, homogeneous set of data points to cluster.

To overcome the irregularity of the time steps we will construct a fit for each trace and then interpolate the data using the fit for every 0.1 of a second to produce the time sequences with identical time steps:

$$T_{k\,fitted} = ((t_1, f_1(t_1), \ldots, f_L(t_1)), \ldots, (t_N, f_1(t_N), \ldots, f_L(t_N))),$$

where $t_1 = 0$, $t_N \leq 100$, and for each $i \in [2, N]$ $t_i = t_{i-1} + 0.1$.

Now it becomes clear why we require the time traces to have equal durations. Since the longest game instances last for a hundred seconds, we need to be able to sample from all of the interpolated traces in the interval between zero and a hundred seconds to create a homogeneous data set. If the original trace was shorter than a hundred seconds, the resulting fitting function wouldn't necessarily provide us with the meaningful data outside of its duration region. Therefore, we augment original game traces in such a way that they all last for a hundred seconds, but the features retain their last achieved values (from the "game over" state):

$$T_k = ((t_1, f_1(t_1), \ldots, f_L(t_1)), \ldots, (t_N, f_1(t_N), \ldots, f_L(t_N)),$$
$$(t_{N+1}, f_1(t_N), \ldots, f_L(t_N)), \ldots, (100, f_1(t_N), \ldots, f_L(t_N))).$$

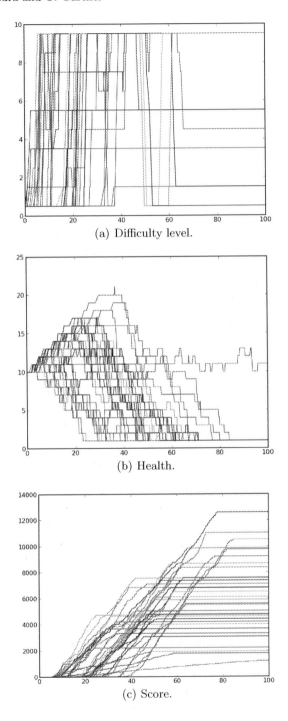

(a) Difficulty level.

(b) Health.

(c) Score.

Fig. 2. Game traces from one player. Different colours represent different game instances.

As mentioned in Section 4, after the augmenting step each time trace is split into two parts:

$$T_{kpre} = ((t_1, f_1(t_1), \ldots, f_L(t_1)), \ldots, (t_K, f_1(t_K), \ldots, f_L(t_K))),$$

$$T_{kpost} = ((t_{K+1}, f_1(t_{K+1}), \ldots, f_L(t_{K+1})), \ldots, (t_N, f_1(t_N), \ldots, f_L(t_N))),$$

where t_K is a predefined constant, in our experiments set to 30 seconds, that determines for how long the game observes the player before making a prediction. The *pre* parts of the traces are used for training and evaluating the predictor. The *post* parts of the traces are used for clustering.

6   Evaluation

To evaluate the performance of the SVM predictor we conduct a kind of "leave one out" cross-validation on the data. For each player presented we construct a following train/test split:

- training set consists of the game instances played by all players except this one;
- test set consists of all the game instances played by this player.

Constructing the train and test sets in this way models a real-life situation of adjusting the game to a previously unseen player. As a performance measure we use the mean absolute difference between the exhibited behaviour in the test instances and the behaviour described by the model of the predicted cluster. The mean is calculated over the test instances.

To provide the baselines for the performance evaluation, we construct for each test instance a sequence of "cheating" predictors: The first (best) one chooses a cluster that delivers a minimum possible absolute error (that is the difference between the predicted adjustment curve and the actual difficulty curve exhibited by this instance); the second best chooses the the cluster with the minimum possible absolute error from the remaining clusters, and so on. We call these predictors "cheating" because they have access to the test instances' data before they make the prediction. For each "cheating" predictor the error is averaged over all test instances and the error of the SVM predictor is compared to these values. As the result we can make some conclusion on which place in the ranking of the "cheating" predictors the SVM one takes.

Figure 3 illustrates the performance of the SVM predictor and the best and the worst baselines for a single player and 7 clusters. We can see from the plots that for each model the SVM predictor displays the performance close to the best cluster. Figure 4 shows that the performance of the SVM predictor averaged over all train/test splits demonstrates similar behaviour.

Statistical Tests

To verify our hypotheses, we performed proper statistical tests with the null hypothesis that the algorithms perform equally well. As suggested recently [6] we used the Wilcoxon signed ranks test.

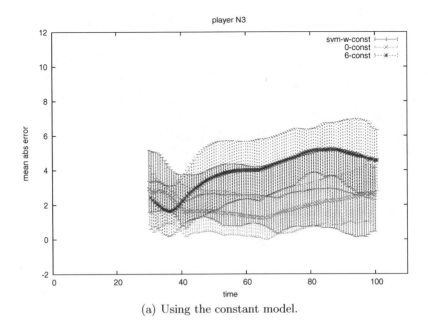

(a) Using the constant model.

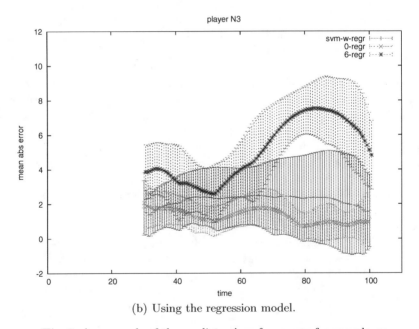

(b) Using the regression model.

Fig. 3. An example of the predictors' performances for one player

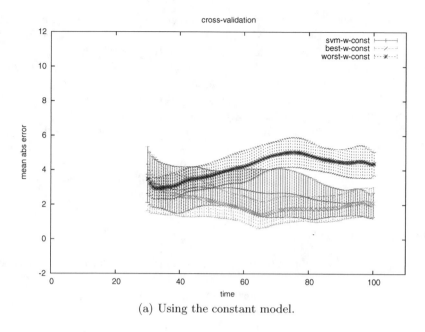

(a) Using the constant model.

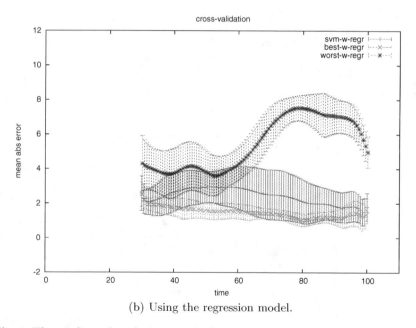

(b) Using the regression model.

Fig. 4. The predictors' performance averaged over all train/test splits for 7 clusters

The Wilcoxon signed ranks test is a nonparametric test to detect shifts in populations given a number of paired samples. The underlying idea is that under the null hypothesis the distribution of differences between the two populations is symmetric about 0. It proceeds as follows: (i) compute the differences between the pairs, (ii) determine the ranking of the absolute differences, and (iii) sum over all ranks with positive and negative difference to obtain W_+ and W_-, respectively. The null hypothesis can be rejected if W_+ (or $\min(W_+, W_-)$, respectively) is located in the tail of the null distribution which has sufficiently small probability.

For settings with a reasonably large number of measurements, the distribution of W_+ and W_- can be approximated sufficiently well by a normal distribution. Unless stated otherwise, we consider the 5% significance level ($t_0 = 1.78$).

Dynamic versus Static Difficulty

We first want to confirm the hypothesis that a dynamic difficulty function is more appropriate than a static one. To eliminate all other influences, we considered first and foremost only a single cluster. In this case, as expected, the dynamic adjustment significantly outperforms the static setting ($t = 2.67$).

We also wanted to compare the performance of dynamic and static difficulty adjustment for larger numbers of clusters. To again eliminate all other influences, we considered the best and the worst "cheating" predictor for either strategy. The t-values for these comparisons are displayed in Table 1.

While varying the amount of clusters from one to fifteen we found out that dynamic difficulty adjustment (the regression model) always significantly outperforms the static one (the constant model) for choosing the best cluster. The

Table 1. t-values for comparison of the constant model vs the regression model for the varying amount of clusters

c	best-const vs best-regr	worst-const vs worst-regr
1	8.46	8.46
2	6.12	9.77
3	5.39	12.64
4	5.26	11.37
5	4.90	12.62
6	4.77	11.05
7	4.80	10.38
8	4.62	6.83
9	4.61	7.20
10	4.63	4.36
11	4.55	0.71
12	4.68	-0.77
13	4.60	-9.16
14	4.50	-5.54
15	4.57	-13.26

same effect we can observe for the worst predictor, but only until the amount of clusters used is greater than ten. For more clusters the static model starts to outperform the dynamic one, probably due to there being insufficient amount of instances in some clusters to train a good regression model. Based on these results in the following we consider only the regression model and vary the amount of clusters from one to ten.

Right versus Wrong Choice of Cluster

As a sanity check, we next compared the performance of the best choice of a cluster versus the worst choice of cluster. To this end we found—very much unsurprisingly—that for any non-trivial number of clusters, the best always significantly outperforms the worst.

This means there is indeed room for a learning algorithm to fill. The best we can hope for is that in some settings the performance of the predicted cluster is close to, i.e., not significantly worse than, the best predictor while always being much, i.e., significantly, better than the worst predictor.

One versus Many Types of Players

The last parameter that we need to check before coming to the main part of the evaluation is the number of clusters. It can easily be understood that the quality of the best static model improves with the number of clusters while the quality of the worst degrades even further. Indeed, on our data, having more clusters was always significantly better than having just a single cluster for the best predictor using the regression model.

Under the assumption that we do not want to burden the players with choosing their difficulty, this implies that we do need a clever way to automatically choose the type of the player. Adjusting the game just to a single type is not sufficient.

Quality of Predicted Clusters

We are now ready to consider the main evaluation of how well the type of the player can be chosen automatically. As mentioned above the best we can hope for is that in some settings the performance of the predicted cluster is close to the best cluster while always being much better than the worst cluster. Another outcome that could be expected is that performance of the predicted cluster is far from that of the best cluster as well as from the worst cluster.

To illustrate the quality of the SVM predictor we look at its place in the ranking of the "cheating" predictors while varying the amount of clusters. The results of the comparison of the predictors' performance for the regression model are shown in Table 2. Each line in the table corresponds to the amount of clusters specified in the first column. The following columns contain values 'w', 's', and 'b', where 'w' means that the SVM predictor displayed the significantly worse performance than the corresponding "cheating" predictor, 'b' for the significantly

Table 2. Results of the significance tests for the comparison of performance of the SVM predictor and "cheating" predictors using the regression model

	1	2	3	4	5	6	7	8	9	10
1	s									
2	w	b								
3	w	s	b							
4	w	b	b	b						
5	w	s	b	b	b					
6	w	w	b	b	b	b				
7	w	s	b	b	b	b	b			
8	w	s	b	b	b	b	b	b		
9	w	w	s	b	b	b	b	b	b	
10	w	w	s	b	b	b	b	b	b	b

better performance, and 's' for the the cases where there was no significant difference. The columns are ordered according to the ranking of the "cheating" predictors, i.e. 1 stands for the best possible predictor, 2 for the second best, and so on.

We can observe a steady trend in the SVM predictor's performance: Even though it is always (apart from the trivial case of one cluster) significantly worse than that of the best possible predictor, it is also always significantly better than that of the most other predictors. In other words, regardless of the amount of clusters, the SVM predictor always chooses a reasonably good one.

This last investigation confirms our hypothesis that predicting the difficulty-type for each player based on short periods of gameplay is a viable approach to taking the burden of choosing the difficulty from the players.

7   Conclusion and Future Work

In this paper we investigated the use of supervised learning for dynamical difficulty adjustment. Our aim was to devise a difficulty adjustment algorithm that does not bother the actual players. Our approach to building a difficulty model consists of clustering different types of players, finding a good difficulty adjustment for each cluster, and predicting the cluster for short traces of gameplay. Our experimental results confirm that dynamic adjustment and cluster prediction together outperform the alternatives significantly.

One parameter left out in our investigation is the length of the prefix that is used for the prediction. We will investigate its influence on the predictors' performance in the future work. We also plan to collect and examine more players' data to see how transferable our algorithm is to the other games. Another direction for the future investigation is the comparison of our prediction model to the other algorithms employed for the time series predictions, such as neural networks or gaussian processes.

Acknowledgments

We would like to thank the anonymous reviewers for their helpful and insightful comments.

References

1. Biederman, I., Vessel, E.: Perceptual pleasure and the brain. American Scientist 94(3) (2006)
2. Charles, D., Black, M.: Dynamic player modeling: A framework for player-centered digital games. In: Proc. of the International Conference on Computer Games: Artificial Intelligence, Design and Education, pp. 29–35 (2004)
3. Cook, D.: The chemistry of game design. Gamasutra 07 (2007)
4. Cortes, C., Vapnik, V.: Support vector networks. Machine Learning 20, 273–297 (1995)
5. Danzi, G., Santana, A.H.P., Furtado, A.W.B., Gouveia, A.R., Leitão, A., Ramalho, G.L.: Online adaptation of computer games agents: A reinforcement learning approach. In: II Workshop de Jogos e Entretenimento Digital, pp. 105–112 (2003)
6. Demšar, J.: Statistical comparisons of classifiers over multiple data sets. Journal of Machine Learning Research 7(1) (2006)
7. Hartigan, J., Wong, M.: A k-means clustering algorithm. JR Stat. Soc., Ser. C 28, 100–108 (1979)
8. Herbrich, R., Minka, T., Graepel, T.: Trueskill^{tm}: A bayesian skill rating system. In: NIPS, pp. 569–576 (2006)
9. Hunicke, R., Chapman, V.: AI for dynamic difficulty adjustment in games. In: Proceedings of the Challenges in Game AI Workshop, Nineteenth National Conference on Artificial Intelligence (2004)
10. Rifkin, R.M.: Everything Old is new again: A fresh Look at Historical Approaches to Machine Learning. PhD thesis, MIT (2002)
11. Togelius, J., Nardi, R., Lucas, S.: Making racing fun through player modeling and track evolution. In: SAB 2006 Workshop on Adaptive Approaches for Optimizing Player Satisfaction in Computer and Physical Games, pp. 61–70 (2006)
12. Yannakakis, G.N., Maragoudakis, M.: Player modeling impact on player's entertainment in computer games. In: Ardissono, L., Brna, P., Mitrović, A. (eds.) UM 2005. LNCS, vol. 3538, p. 74. Springer, Heidelberg (2005)

Unsupervised Fuzzy Clustering for the Segmentation and Annotation of Upwelling Regions in Sea Surface Temperature Images

Susana Nascimento and Pedro Franco

Departamento de Informática and Centre for Artificial Intelligence (CENTRIA)
Faculdade de Ciências e Tecnologia, Universidade Nova de Lisboa
2829-516 Caparica, Portugal

Abstract. The Anomalous Pattern algorithm is explored as an initialization strategy to the Fuzzy K-Means (FCM), with the sequential extraction of clusters, that simultaneously allows the determination of the number of clusters. The composed algorithm, Anomalous Pattern Fuzzy Clustering (AP-FCM), is applied in the segmentation of Sea Surface Temperature (SST) images for the identification of Coastal Upwelling.

A set of features are constructed from the AP-FCM clustering segmentation taking into account domain knowledge and a threshold procedure is defined in order to identify the transition cluster whose frontline is automatically annotated on SST images to separate the upwelling regions from the background.

Two independent data samples in a total of 61 SST images covering large diversity of upwelling situations are analysed. Results show that by tuning the AP-FCM stop conditions it fits a good number of clusters providing an effective segmentation of the SST images whose spatial visualization of fuzzy membership closely reproduces the original images. Comparing the AP-FCM with the FCM using several validation indices shows the advantage of the AP-FCM avoiding under or over-segmented images. Quantitative assessment of the segmentations is accomplished through ROC analysis. Compared to FCM, the number of iterations of the AP-FCM is significantly decreased.

The automatic annotation of upwelling frontlines from the AP-FCM segmentation overcomes the subjective visual inspection made by the Oceanographers.

1 Introduction

In the coastal ocean of Portugal, during the summer, the upward movement of cool and nutrient rich waters toward the surface of the Ocean, due to the northern winds, leads to alterations in the distribution of the physical, chemical and biological properties. These alterations expressed in the movement of surface water masses along the horizontal direction perpendicular to the coast line, characterizes the coastal upwelling. Remote sensing is a widely applied technique in the detection of coastal upwelling [1]. Images of the Sea Surface Temperature

J. Gama et al. (Eds.): DS 2009, LNAI 5808, pp. 212–226, 2009.

(SST) are computed by the Oceanographers and a high resolution color scale is applied to each image. The color scale has to be manually tuned in order to get the better contrast definition for a good visualization of the signature of the phenomena, which is a very time-consuming process. Automatic detection tools are a demand due to the enormous amount of data daily collected and preprocessed by the Oceanographers, and due to the subjectivity inherent to visual inspection.

Remote sensing derived data is imprecise in nature due to the intricate interactions controlling its process. In particular, upwelling regions are often characterized by transition zones with smooth thermal boundaries, they correspond to multi-modal and very irregular histograms, their signatures express strong morphological variation and, due to the absence of a valid analytical model for the structures, one is confronted with a 'semantic-gap' between the implicit Oceanographer knowledge and a systematic working definition providing a 'gold-standard' set of images.

Image segmentation is considered one of the most critical steps in image processing and fuzzy clustering provides a mechanism to represent and manipulate uncertainty and ambiguity. With a fuzzy subset each pixel in an image has assigned a degree of membership to which it belongs to each region or each boundary [2]. The Fuzzy K-Means clustering (FCM) and its extensions are methods that have received much attention in image segmentation [3]-[8]. Other fuzzy segmentation algorithms include fuzzy histogram thresholding [9], fuzzy rule based approaches [10], and neuro-fuzzy systems including adaptive extensions with edge detection [11,12]. However, significant fine tunning of parameters in particular with membership functions, overfitting of cluster prototypes not representing any segmented region, and heavy computational complexity should be avoided in the current application due to the inherent characteristics of the problem.

Despite the FCM simplicity and its applicability as shown on upwelling detection from SST images [13], fuzzy clustering faces two important issues: (i) the definition of a strategy for choosing the initial cluster prototypes; and (ii) determination of a good number of clusters to be found in the grouping of data [5,14]. In this work, the Iterative Anomalous Pattern (IAP) algorithm [15] is explored as an initialization strategy to the FCM that simultaneously allows to determine the number of clusters. The composed algorithm, Anomalous Pattern Fuzzy Clustering (AP-FCM), is applied for an effective segmentation of the upwelling regions from SST images. Taking into account domain-knowledge about the upwelling phenomenon and the AP-FCM segmentation results (i.e. cluster's information), there have been defined features able to identify the transition cluster that separates the upwelling regions from the background. From that, it is introduced an iterative threshold procedure whose threshold values are established based on an information-gain attribute discretization criterion.

Section 2 introduces the SST data sets and the problem of upwelling visual identification. Section 3 points out some initialization and validation issues of the FCM. Section 4 describes the IAP algorithm, its initialization to the FCM, leading to the AP-FCM, and a scheme to visualize fuzzy segmented images. In Section 5 it is described the empirical feature definition process. Section 6 highlights the results of the AP-FCM algorithm applied to the segmentation of SST images, comparing with the FCM ones applying validation indices. Experimental analysis of the proposed features identifying the cluster of interest for the automatic annotation of upwelling frontlines are discussed. Concluding remarks are in Section 7.

2   SST Upwelling Images and Ground-Truth Maps

Two independent data sets with 30 and 31 SST images, representing two distinct upwelling seasons of the years 1998 and 1999, constitutes our benchmark. Each SST image, I, is represented by a 500×500 pixels map with a spatial resolution of 1.1Km $\times$ 1.1Km, with each sea pixel being a temperature in degrees Celsius. The contiguous white region on the right side of each SST image corresponds to land, whereas the white pixels in the Ocean region correspond to missing values during the satellite transmission, normally due to cloud cover.

The upwelling phenomenon ranges, in a direction perpendicular to the coast, from colder coastal upwelling waters, to warmer offshore upwelling waters, and the remaining even warmer temperature offshore waters. We are interested to identify the distinct upwelling regions with focus on the front region roughly characterized as a "relatively narrow" region with "relatively strong" horizontal thermal gradients, establishing a transition zone between colder coastal and warmer offshore surface waters.

Figure 1-a), c), e) shows three representative SST images selected from the benchmark data sample, illustrating the variability of upwelling situations. Specifically: i) SST images with a well characterized upwelling situation in terms of fairly sharp boundaries between cold and warm surface waters measured by relatively contrasting thermal gradients and continuity along the cost; ii) SST images showing distinct upwelling situations related to thermal transition zones offshore from the North toward the South and with smooth transition zones between upwelling regions; iii) noisy SST images with clouds, when information to define the upwelling front lacks.

The only domain knowledge provided by the Oceanographers, assigned to each SST image, is a color bar annotation (right sides of Figure 1-a),c), e)) corresponding to the color of a relatively strong thermal front that establishes the transition zone between upwelling regions and offshore non-upwelling waters. From this information a binary ground-truth map has been constructed for each SST image, with 1/0 pixel corresponding to an 'upwelling/non-upwelling' pixel, according to the Oceanographers' annotation. Even though the ground-truth map is binary constructed, an effective segmentation should comprise more than three clusters in order to identify the various upwelling regions. These maps support the assessment evaluation of results of experiments 2 and 3.

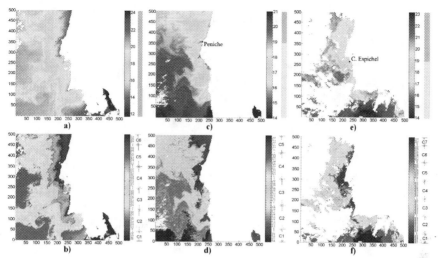

Fig. 1. SST images showing different upwelling situations: a), c), e); corresponding AP-FCM segmentations into 6, 5, 7 clusters, and fuzzy map visualization: b), d), f)

3   Fuzzy Clustering Initialization and Validation

We focus on two important issues in fuzzy K-means clustering: (i) the definition of a strategy for choosing the initial cluster prototypes; and (ii) determining a good number of clusters, K^*, to be found in the grouping of data. Even though there is no general agreement about a good initialization scheme to the FCM, it is common practice to choose a random selection because it guarantees the convergence of the iterative algorithm. To avoid the algorithm to be trapped into local minima, it is run several times starting from distinct initial seed points, and choose the partition that corresponds to the minimum value of FCM's clustering criterion. Concerning problem (ii), and in order to find the "best" number of clusters, designated as the *clustering validation problem*, the FCM algorithm is run from different number of clusters, ranging from $K = K_{min}, \cdots, K_{max}$, which leads to different partitions. Therefore, it is necessary to quantitatively evaluate the clustering results, by selecting a validity index, VI, and running a validation procedure as follows: compute VI for each K-partition, and choose the 'best" number of clusters, K^*, as the one that extremises the validity criterion VI. There is no general agreement on what value to use for K_{max}.

Many validity indices have been proposed in the literature [17,18]. We have selected two well recognized indices, the Fukuyama-Sugeno index, V_{FS}, and the Xie-Beni index, V_{XB}, both exploring the concepts of compactness and separation between clusters.

The process of repeatedly initializing the FCM dozens of times and *a posteriori* run it for different values of K, select an appropriate validation index (or compare the results of a collection of indices), is a very time-consuming and tedious process to arrive at a good (possibly sub-optimal) solution.

4   Anomalous Pattern Fuzzy K-Means for SST Image Segmentation

The task of image segmentation can be stated as the partition of an image into a number of potentially non-overlapping regions, each with distinct properties. In this work, the FCM algorithm and a proposed extension are applied as segmentation techniques to partition SST images, based on the temperature value of each pixel.

Each SST image is converted into a set $X = \{x_i\}_{i=1}^N$ of N feature vectors and, by pre-specifying the number of groups, K, the FCM organizes the N pixels into K homogeneous groups represented as fuzzy sets $\mathcal{F} = \{\mathcal{F}_1, \mathcal{F}_2, \cdots, \mathcal{F}_K\}$, which constitute a fuzzy K-partition of X. Starting from a set of K random initial prototypes $V^0 = \{v_k^0\}_{k=1}^K$ the goal of the FCM is to iteratively improve a sequence of sets $\mathcal{F}(1), \mathcal{F}(2), \cdots, \mathcal{F}(t)$ (with t the iteration step), of K fuzzy clusters, until FCM clustering criterion shows no further improvement (see e.g. Ref. [5] for a detailed description).

4.1   Iterative Anomalous Pattern for Initial Setting of FCM

The Iterative Anomalous Pattern (IAP) algorithm is explored as an initial configuration scheme to the FCM that simultaneously, provides an indicator of the number of clusters present in data.

The IAP algorithm sequentially extracts clusters from the data set as follows. Let Y denote the standardized data set, by shifting the origin of the original data X to the grand mean, $\overline{x}$. The feature vector $\overline{x}$ is taken as the reference point unvaried all over the sequential process, and take as seed point the data point that is farthest from the reference point. One crisp cluster, $\mathcal{C}_t$, is iteratively constructed, defined as the set of points that are closer to the seed point than to the reference point. After, the cluster seed is substituted by the cluster gravity center and the procedure is reiterated until it converges. The Anomalous Pattern procedure is reiterated over the residual data set taken as $Y^{t+1} = Y^t - \mathcal{C}_t$ until any of the following stop conditions is reached: C1) the residual data set, Y^{t+1}, is empty which means that all the entities had been clustered; C2) the contribution of the t-th cluster to the data scatter (equation (1)) is too small (i.e. less than a pre-specified threshold τ); C3) the number of clusters, t, has reached a pre-specified value K_{max}. The algorithm is described next.

Taking Y as the standardized data set, the total scatter of all data points (row-vectors in the $N \times p$ matrix Y) is defined as $T(Y) = \sum_{i=1}^N \sum_{h=1}^p y_{ih}^2$. In [15] it is derived how the total data scatter $T(Y)$ can be decomposed into an explained part due to the cluster structure retrieved from data Y and the unexplained part which corresponds to the K-means clustering criterion to be minimized. From that, it is defined the relative contribution of each individual cluster, $(\mathcal{C}_t, v_t)$, to the data scatter, such as:

$$W((\mathcal{C}_t, v_t)) = \frac{n_t \sum_{h=1}^p v_{th}^2}{T(Y)} = \frac{n_t \sum_{h=1}^p v_{th}^2}{\sum_{i=1}^N \sum_{h=1}^p y_{ih}^2}, \tag{1}$$

with n_t the cardinality of cluster $\mathcal{C}_t$.

Iterative Anomalous Pattern (IAP) Algorithm

```
 1  Given: τ; K_max;
 2  Set: t = 0, Y^t = standardize(X);
 3  Repeat
 4      t = t + 1;
 5      C_t = {};
 6      v_t = argmax_j d²(y_j, 0) for y_j ∈ Y^t;
 7      Repeat
 8        for (y_j ∈ Y^t)
 9          if (d²(y_j, v_t) < d²(y_j, 0) )
10            C_t = C_t ∪ {y_j};
11          endif
12        endfor
13        v'_t = v_t;
14        v_t = (Σ_{y_j ∈ C_t} y_j) / |C_t|;
15      until (v_t ≈ v'_t)
16      Y^{t+1} = Y^t - C_t;
17 until (Y^{t+1} == {} .or. W((C_t, v_t)) ≤ τ .or. t == K_max)
18 return {v_1, v_2, ⋯, v_t};
```

The IAP algorithm sets the initial prototypes of the FCM and simultaneously, by tuning the stop conditions C1 or C2, establishes the number of clusters. The FCM pre-specified number of clusters is fixed as $K = t$ and the set of centroids, $\{v_1, v_2, \cdots, v_t\}$ of the IAP, defines the seed prototypes to the FCM algorithm, i.e. $V^0 = \{v_k^0\}_{k=1}^K = \{v_k\}_{k=1}^t$. The composed algorithm is referred to as the Anomalous Pattern Fuzzy Clustering (AP-FCM).

4.2   Visualization of a Fuzzy Segmented Image

After running the AP-FCM/FCM algorithms, the corresponding fuzzy partition is defuzzified by assigning to each pixel its maximum grade of membership, mapping the pixel to the corresponding cluster. The defuzzified fuzzy partition is mapped onto the spatial grid of the image and visualized on a fuzzy color scale in accordance with the degrees of membership. This completes the segmentation process.

According to the thermal gradient spatial orientation of upwelling regions clusters are ordered according to their prototypes' value, and assign to each of them a crisp color label following that order. Then, each pixel is mapped onto the spacial grid of the image and colored according to its cluster color label combined with its membership value. Thus, each segmented region is assigned with a specific color and the corresponding pixels are assigned shade tones of the color according to their degree of membership in the cluster. Visualization of the AP-FCM fuzzy segmentation results are shown in Figure 1-b), d), f).

5   Feature Definition to Annotate Upwelling Fronts

The remaining problem concerns the separation between the objects of interest (i.e. the upwelling regions) and the background (i.e. non-upwelling offshore
waters), from the fuzzy segmented map.

Taking the domain-knowledge of the phenomenon and the segmentation results (i.e. clusters' information) we studied which features can be defined from
the clusters in order to identify the 'transition cluster' as the cluster that contains
the "relatively narrow" transition zone between upwelling and non-upwelling regions extending offshore no further than a certain limit.

Conform with the characterization of upwelling, its regions correspond to the
first k clusters ($k < K$), with the last, the 'transition cluster', $\mathcal{T}$. Two features
constructed from the fuzzy segmentation results of each SST image are as follows:

– Relative Difference between consecutive clusters prototypes, each one being
  the mean temperature of SST fuzzy homogeneous regions:

$$TDiff(\mathcal{T}) = \frac{v_{\mathcal{T}+1} - v_{\mathcal{T}}}{|\mathcal{C}_{\mathcal{T}}|}(1 \leq \mathcal{T} \leq K - 1),$$

  with $|\mathcal{C}_{\mathcal{T}}|$ the cardinality of defuzzified cluster $\mathcal{C}_{\mathcal{T}}$.
– Relative Cumulative Cardinality of the first $\mathcal{T}$ clusters:

$$CCard(\mathcal{T}) = \frac{\sum_{k=1}^{\mathcal{T}} |\mathcal{C}_k|}{|I|},$$

  with $|I|$ the number of pixels with an SST value (excluding NaN's) for an
  SST image I.

The threshold value of each feature, τ_d and τ_c, has been established using an
entropy-based attribute discretization procedure [16]. Specifically, let A denote
one of the features, and l one of its value. Given a data sample D (in our
study, the set of segmented maps of the 1998 data set), the threshold value τ
selected is the one that maximizes the information gain resulting from the split
binary partition in D_1 and D_2, corresponding to the samples of D satisfying the
conditions $A < \tau$ and $A \geq \tau$, respecting the class distribution of the tuples in the
partition. Notice that a binary class-label attribute (upwelling/non-upwelling)
is considered, taken from the ground-truth maps of the SST images. The split
value with the highest information gain is considered the most discriminating
value of the set, and consequently taken as the threshold value.

The extensive analysis of the segmented SST images showed that the tran-
sition cluster, $\mathcal{T}$, is one of the first three clusters ($\mathcal{T} \leq 3$). Given a segmented
SST image, S(I), its transition cluster is iteratively identified according to the
Transition-Cluster Threshold (TCT) procedure, as follows:

6   Experimental Study

The goal of this study is three-fold. First, the contribution of the deterministic IAP algorithm to improve the rate of convergence of the random-based

---

Transition-Cluster Threshold (TCT) Procedure

Step 1 *Set:* τ_d, τ_c, by information-gain based attribute discretization;
       Set: $T = 3$;
Step 2 if $(TDiff(T) \leq \tau_d)$ then $T = T - 1$ endif
Step 3 $CCard(T) = \sum_{k=1}^{T} |\mathcal{C}_k| / |I|$;
Step 4 while $(CCard(T) > \tau_c$ .and. $T > 1)$
Step 5 $CCard(T) = CCard(T) - |\mathcal{C}_T| / |I|$
Step 6 $T = max(T - 1, 1)$;
       endwhile
Step 7 return T;

---

initialization of the FCM is analysed. Second, the AP-FCM algorithm is explored as a good indicator of the number of clusters able to provide an effective segmentation of the images concerning the problem at hand. Third, to experimentally analyse the $TDiff(T)$ and $CCard(T)$ features constructed from the AP_{c2}-FCM segmentations, and apply the TCT procedure, leading to the recognition of the transition cluster and annotation of its frontline in the SST images, and accuracy analysis according to the Oceanographers' visual annotation of upwelling front.

In the studies, the 1999 data sample is taken for 'validation' of the results obtained with the 1998 data sample. Each data set X corresponds to the set of 500×500 pixels, with x_i a temperature value. For a fixed number of clusters, K, the FCM algorithm is run 10 times, starting from distinct initial configurations.

6.1   Experiment 1

In this set of experiments it is analysed the improvement in convergence of the FCM when initialized by the IAP algorithm. The range in the number of clusters has been set between $K_{min} = 2$ and $K_{max} = 8$. For that we compared the number of iterations for distinct experimental situations of the FCM/AP-FCM: i) number of iterations of the best partition of the FCM, R^*_{FCM}; ii) average of the number of iterations of the 10 runs of the FCM, $\overline{R}_{FCM}$; iii) number of iterations of the AP-FCM, $R_{AP_{c2}-FCM}$, $R_{AP_{c3}-FCM}$, for the case of the C2 and C3 stop conditions of the IAP algorithm.

The graphic in Figure 2-a) shows the mean values of those measures over the whole set of images. Analysis of these values shows that as the number of clusters increases, the number of iterations of the best run of the FCM, R^*_{FCM}, and the mean of runs, $\overline{R}_{FCM}$, is about twice as compared to the FCM runs of the AP-FCM. By comparing the AP_{c2}-FCM and AP_{c3}-FCM it can be said that the stop condition by the cluster's contribution to the explanation of the data scatter, provides a slight improvement in convergence when compared to the pre-specification of the number of clusters. Notice, that this is particularly relevant for the case where the AP_{c2}-FCM leads to $K = 6$ or $K = 7$ clusters, which covers 55/61 of the images. In order to get a global view of the effective number of iterations taken by each version of the algorithms, the graphic in

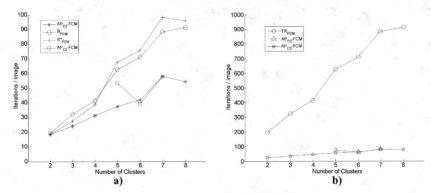

Fig. 2. a) Mean number of iterations, over the whole set of images, for distinct experimental situations of the FCM/AP-FCM. **b)** Total number of iterations for distinct experimental situations of the FCM/AP-FCM.

Figure 2-b) shows the total number of iterations taken by the 10 runs of the FCM, TR_{FCM}, versus the total number of iterations taken by the AP-FCM, $R_{AP_{c2}-FCM}$, $R_{AP_{c3}-FCM}$. One see that the effective number of runs of the FCM is 10 times higher than each version of the AP-FCM.

6.2 Experiment 2

A major aspect is the analysis of the effectiveness of the segmentation results provided by each of the AP-FCM/FCM algorithms. We focus on the AP-FCM with stop condition C2, in order to analyse its quality as an indicator of the number of clusters. The AP_{c2}-FCM scatter threshold τ parameter has been experimentally tuned, exclusively through the analysis of resulting segmentations for the year of 1998, at $\tau = 10^{-3}$.

Figure 1-b), d), f) shows the fuzzy segmentation and corresponding spatial visualization obtained by the AP_{c2}-FCM concerning the corresponding original SST images above. The results illustrate the kind of variability on original SST images and corresponding segmentations the AP-FCM is able to deal with, leading to partitions with $K = 6$, $K = 5$, and $K = 7$ clusters, respectively.

The result of running the AP_{c2}-FCM algorithm over the whole set of images lead to fuzzy K-partitions with $K = 5, 6, 7$. This strict interval corresponds to a good number of clusters for an effective segmentation of the SST images. These results are validated with the analysis of the AP_{c2}-FCM segmentation for the 31 images of 1999, for which threshold value τ was confirmed and result also in segmentations with $K = 5, 6, 7$ clusters.

When applying the FCM and corresponding validation procedure setting with $K_{min} = 2$ to $K_{max} = 10$ and applied the validation indices V_{FS}, V_{XB}, the obtained segmentations are typically under or over-segmented[1]. The histograms in

---

[1] Other validation indices, like the partition coefficient (PC), the partition entropy (PE) and Pakhira and co-authors (PBMF) indices [18] have been applied without success, since the corresponding selected partition always lead to under-segmented images.

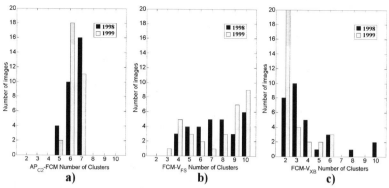

Fig. 3. Histograms with the absolute frequency of SST images segmented with a given number of clusters: a) AP_{c2}-FCM, b) FCM-V_{FS}, c) FCM-V_{XB} algorithms

Figure 3 show the absolute frequency of SST images with a given number of clusters resulting from their segmentation with the AP_{c2}-FCM, FCM-V_{FS}, FCM-V_{XB} algorithms, respectively, for the two years. The FCM-V_{FS}, FCM-V_{XB} segmentations for the images of 1999 have less quality than the segmentations of 1998, resulting in more over-segmented images (FCM-V_{FS}) or under-segmented images (FCM-V_{XB}), while with the AP_{c2}-FCM the differences between the data sets of 1998 and 1999 don't have any evident impact on the quality of the results.

In order to quantitatively evaluate the segmentation results of the AP_{c2}-FCM, FCM-V_{FS}, FCM-V_{XB}, over the two data sets, the corresponding ground-truth maps were taken. The matching between each fuzzy K-partition from the AP_{c2}-FCM or FCM, and the ground-truth 2-partition is established by making the correspondence of each fuzzy cluster prototype to the prototype of the binary partition. The AP-FCM/FCM clusters are then merged in order to obtain a 2-partition to compare against the ground-truth corresponding one.

Results in terms of sensitivity versus specificity are summarised in the ROC plots in Figure 4-a)-b) for the years of 1998 and 1999. The analysis shows that for the year of 1998 $33\% - 43\%$ of the classifications are very close to a perfect classification whereas for the segmented images of 1999, the corresponding percentage improves slightly ($42\% - 48\%$). Even though there is no significant difference between the results obtained by the AP_{C2}-FCM and the FCM-V_{FS}, FCM-V_{XB}, one can say that the AP_{C2}-FCM results are slightly more 'conservative' comparing with the FCM ones. That is, the AP_{C2}-FCM makes positive classifications (i.e union of upwelling regions), corresponding to strong evidence with few false positive errors. On the other hand, the FCM results tend to make more positive classifications with slightly weaker evidence. The more conservative tendency is preferable against the more liberal one, since $\approx 76\%$ of pixels in the total analysed images are negative (i.e. non-upwelling) and so, the performance in the top left-hand side of the ROC graphic is more interesting.

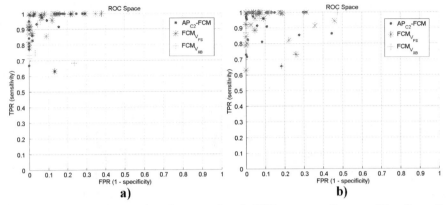

Fig. 4. ROC Plot of the classification of each SST segmentation resulting from the AP_{c2}-FCM, FCM-V_{FS}, FCM-V_{XB}, against ground-truth maps, for the two SST data sets: a) 1998; b) 1999

6.3   Experiment 3

Taking the 30 SST AP_{c2}-FCM segmentation results of the year of 1998, the threshold values of features $TDiff$ and $CCard$ had been experimentally fixed as $\tau_d = 4.5 \times 10^{-5}$ and $\tau_c = 0.52$, according to the information-gain based discretization procedure.

Starting by setting the 'transition cluster' at $\mathcal{T} = 3$, the TCT procedure is applied in order to find the transition cluster and the corresponding external frontier. Figure 5 a)-b) shows the scatter plots of $TDiff(3)$ feature values, for the 30 SST segmentations, separating between North (a) and South (b). The dash lines mark the threshold value τ_d. The upper legend bar indicates the index of the transition cluster containing the upwelling front, according to the Oceanographer pre-specified annotation, whereas the lower bar shows the corresponding indices after applying the $TDiff(\mathcal{T})$ rule (Step2 of the TCT procedure). Analysis of the

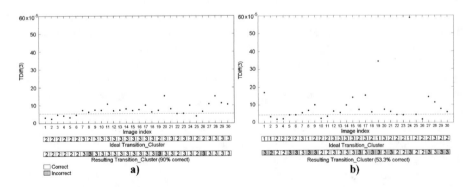

Fig. 5. Scatter plot of the $TDiff(3)$ feature for the 1998 SST images divided in: a) South and b) North

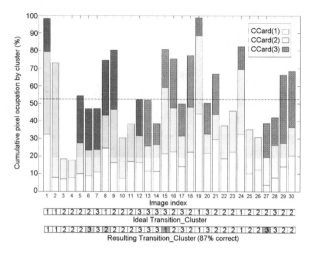

Fig. 6. Bar chart of the $CCard(\mathcal{T})$ constructed feature applied to the North region of the 1998 SST images

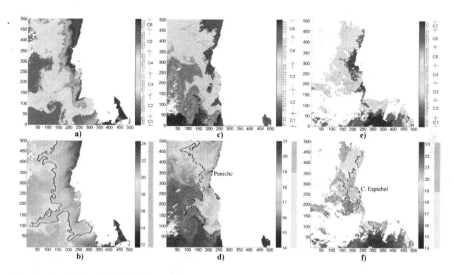

Fig. 7. a), c), e)- AP-FCM segmentations of the original SST images shown in Figure 1; b), d), f) corresponding upwelling frontline annotation over the original SST images after applying the transition cluster threshold procedure

results show that the $TDiff(\mathcal{T})$ rule leads to an accuracy of 90% of corrected transition clusters for the analysed segmentations in the case of South regions and 53% for the North regions.

In order to increase the accuracy of the results in the North, the $CCard(\mathcal{T})$ rule (Steps 3-6 of TCT procedure) is applied. The results are presented in the bar chart of Figure 6 whose vertical bars are divided according to the parcels of summation $CCard(\mathcal{T})$, and dash line marks the threshold $\tau_c = 0.52$. The horizontal lower legend bar shows the number of parcels corresponding to the index of the transition cluster calculated according to the $CCard(\mathcal{T})$ rule (i.e. number of fractions totally under τ_c). After this second stage one gets an increase to 87% accuracy of $\mathcal{T}$ indexes in the North segmented regions of the 1998 data set. Slightly worse results were obtained, on identifying the transition cluster $\mathcal{T}$ for the segmentations of the 1999 SST images, with accuracies of 84% (South) and 75% (North).

Finally, the frontline of the transition cluster $\mathcal{T}$ is defined by the set of pixels sharing a 4-neighborhood with its external adjacent cluster. This line, separating the upwelling regions from the background, is annotated over the SST images. Figure 7-b),d),f) illustrate the results for SST images with the upwelling frontlines determined and annotated according to the TCT procedure from the AP-FCM segmented results shown in Figure 7-a),c),e).

7   Conclusion and Future Work

The unsupervised fuzzy clustering proposed, AP-FCM, shows to be an effective approach for the segmentation and annotation of upwelling regions of the two independent SST image samples analysed in this study. Specifically:

1. The study of the stop condition to achieve a meaningless residual cluster contribution to the data scatter (AP$_{C2}$-FCM) fits a good number of clusters for an effective segmentation of the original SST images. The AP$_{C2}$-FCM prevents under or over-segmentation of the SST images, which is precisely the situation that occurs when applying the FCM with well-recognized validation indices, and provides a 'range of interest' of the number of clusters for FCM validation procedure.

2. Result of the unsupervised fuzzy segmentation, the fuzzy membership spatial visualization reproduces the original SST image very closely. Indeed, it prevents the subjective and labor intensive task of visual inspection by the Oceanographers to adjust a color scale to enhance the patterns to be automatically recognized.

3. To access the quality of segmented AP$_{C2}$-FCM, FCM-V$_{FS}$ and FCM-V$_{XB}$ images, they have been compared with corresponding binary ground-truth maps. ROC analysis shows that $\approx 41\%$ of the segmented images are very close to a perfect classification. The AP$_{C2}$-FCM makes positive classifications (i.e union of upwelling regions) corresponding to strong evidence, making few false positive errors. This is interesting since the majority of pixels in the analysed images are negative.

4. The features defined from the AP-FCM segmentation and the threshold procedure to identify the transition cluster containing the upwelling front provides a systematic way to annotate the upwelling fronts in the SST images and overcomes the subjective visual inspection made by the Oceanographers.

5. Compared to FCM the number of iterations of the AP-FCM decreases in the order of 10 times less.

The proposed method is part of an interactive system that allows the automatic recognition and annotation of upwelling regions in SST images. The interest of such a tool is to provide the Oceanographers a systematic method to construct good working definitions of upwelling, with visualization facilities of fuzzy clustering results, and to build a data-base of 'gold-standard' annotated images.

Acknowledgments. This work is a contribution of the LSTOP project (PTDC/EIA/68183/2006) funded by the Portuguese Foundation for Science & Technology. The colleagues of the Oceanographic Institute of Faculdade de Ciências da Universidade de Lisboa (IO-FC-UL) are acknowledge for providing the annotated images.

References

1. Ambar, I., Dias, J.: Remote Sensing of Coastal Upwelling in the North-Eastern Atlantic Ocean. In: Barale, V., Gade, M. (eds.) Remote Sensing of the European Seas, pp. 141–152. Springer, Netherlands (2008)
2. Pal, N.R., Pal, S.K.: A Review of Image Segmentation Techniques. Pattern Recognition 26(9), 1277–1294 (1993)
3. Wang, F.: Fuzzy supervised classification of remote sensing images. IEEE Trans. on Geoscience and Remote Sensing 28(2), 194–201 (1990)
4. Bezdek, J., Pal, S.K.: Fuzzy Models for Pattern Recognition. IEEE Press, Los Alamitos (1992)
5. Bezdek, J., Keller, J., Krishnapuram, R., Pal, T.: Fuzzy Models and Algorithms for Pattern Recognition and Image Processing. Kluwer Academic Publishers, Dordrecht (1999)
6. Chen, S., Zhang, D.: Robust image segmentation using FCM with spatial constraints based on new kernel-induced distance measure. IEEE Trans. on Systems, Man, and Cybernetics, Part B: Cybernetics 34(4), 1907–1916 (2004)
7. Caia, W., Chen, S., Zhanga, D.: Fast and robust fuzzy c-means clustering algorithms incorporating local information for image segmentation. Pattern Recognition 40(3), 825–838 (2007)
8. Schowengerdt, R.: Remote sensing: models and methods for image processing, 3rd edn. Academic Press, London (2007)
9. Tobias, O., Seara, R.: Image segmentation by histogram thresholding using fuzzy sets. IEEE Trans. on Image Processing 11(12), 1457–1465 (2002)
10. Karmakar, G., Dooley, L.: A generic fuzzy rule based image segmentation algorithm. Pattern Recognition Letters 23(10), 1215–1227 (2002)
11. Cinque, L., Foresti, G., Lombardi, L.: A clustering fuzzy approach for image segmentation. Pattern Recognition 37, 1797–1807 (2004)
12. Boskovitz, V., Guterman, H.: An adaptive neuro-fuzzy system for automatic image segmentation and edge detection. IEEE Trans. on Fuzzy Systems 10(2), 247–262 (2002)
13. Nascimento, S., Sousa, F., Casimiro, H., Boutov, D.: Applicability of Fuzzy Clustering for the Identification of Upwelling Areas on Sea Surface Temperature Images. In: Mirkin, B., Magoulas, G. (eds.) Procs. 2005 UK Workshop on Computational Intelligence, London, pp. 143–148 (2005)

14. Gath, I., Geva, A.: Unsupervised Optimal Fuzzy Clustering. IEEE Transactions on Pattern Analysis and Machine Intelligence 11(7), 773–780 (1989)
15. Mirkin, B.: Clustering for Data Mining: A Data Recovery Approach. Chapman & Hall/CRC Press, Boca Raton (2005)
16. Han, J., Kamber, M.: Data Mining: Concepts and Techniques, 2nd edn. The Morgan Kaufmann Series in Data Management Systems (2006)
17. Halkidi, M., Batistakis, Y., Vazirgiannis, M.: On Clustering Validation Techniques. Journal of Intelligent Information Systems 17(2-3), 107–145 (2001)
18. Wanga, W., Zhanga, Y.: On fuzzy cluster validity indices. Fuzzy Sets and Systems 158(19), 2095–2117 (2007)

Discovering the Structures of Open Source Programs from Their Developer Mailing Lists

Dinh Anh Nguyen, Koichiro Doi, and Akihiro Yamamoto

Graduate School of Informatics, Kyoto University
Yoshida-Honmachi, Sakyo-ku, Kyoto 606-8501 Japan
dinhanh_ng@iip.ist.i.kyoto-u.ac.jp,
{doi,akihiro}@i.kyoto-u.ac.jp

Abstract. This paper presents a method which discovers the structure of given open source programs from their developer mailing lists. Our goal is to help successive developers understand the structures and the components of open source programs even if documents about them are not provided sufficiently. Our method consists of two phases: (1) producing a mapping between the source files and the emails, and (2) constructing a lattice from the produced mapping and then reducing it with a novel algorithm, called PRUNIA (PRUNing Algorithm Based on Introduced Attributes), in order to obtain a more compact structure. We performed experiments with some open source projects which are originally from or popular in Japan such as Namazu and Ruby. The experimental results reveal that the extracted structures reflect very well important parts of the hidden structures of the programs.

Keywords: mailing lists, open source programs, extraction of structures, concept lattice.

1 Introduction

In open source software development, developer mailing lists, as well as tools for software configuration management, are essential. However, they are used independently and have no significant mutual complement. The reason lies in the fact that there is no helpful method to link them together in order to obtain a better use of the two. Besides, together with the weak obligation of documenting the development process, open source programs are often developed in an evolutionary approach, which does not strongly require the developers to produce pre-defined documents before coding. Therefore, the development process of open source programs is normally not documented. It leads to many difficulties for successive developers to understand the programs.

Until now, researches analyzing the processes of software development and software programs go mainly into two streams. The first stream contains researches whose objectives are to extract and to understand the software development process [14] by analyzing the documents created by developers. In the second stream are researches which attempt to reconstruct the structures of the

J. Gama et al. (Eds.): DS 2009, LNAI 5808, pp. 227–241, 2009.

programs [12,7,13], to visualize the evolution of the programs [10], or to extract other kind of information such as the reusable components [16] and dependency between components [2] by analyzing the source code archives. We argue that the former stream can extract topic-oriented development flows but fails to link the flows with the actual programs. The latter stream can extract syntax structures of programs but fails to give the structures semantic explanations.

Presuming that no document except the developer mailing lists is available for an open source program, we propose a method that uses the information from its developer mailing lists to extract the structure of the program and categorize the emails in the mailing lists into the components of the extracted structure. It is also an attempt to link the two research streams addressed above.

The key idea of the proposed method based on our observation that a lot of useful information about the programs such as the structures of the programs and the explanations about many parts of the programs hide in the contents of the emails created by developers during the development process. Moreover, we noticed that for a collection of source files, if there are many emails mentioning about them, those source files may constitute a sub-system. In this research, sub-systems are viewed as relatively independent program components that deliver some particular functions.

Figure 1(a) illustrates a project consisting of five source files $\{m_1, ..., m_5\}$ in its source code archive and five emails $\{d_1, ..., d_5\}$ in its developer mailing list. The arrow from d_i to m_j represents that d_i mentions about m_j. We assume that (1) a large number of co-relating emails to a set of modules indicates the relevance for that set to become a sub-system, and (2) a sub-system can be included in other larger sub-systems. As a result, as shown in Figure 1(b), the sets $\{m_2, m_3\}$, $\{m_1, m_2, m_3\}$, and $\{m_4, m_5\}$ are probably sub-systems because there are at least one email relating to each of them. In fact, such subsystems are called *formal concepts* in Formal Concept Analysis [17,3,4].

This paper is organized as follows. In the next section, some definitions and the method to extract the relation between emails and modules is explained. In Section 3, we give a brief introduction to Formal Concept Analysis and then

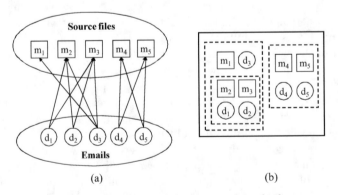

(a)   (b)

Fig. 1. Main idea of the proposed method

explain our method to extract the program structures. Experimental results are given in Section 4. Concluding remarks and future plans are presented in the last section.

2   Mapping Emails and Modules

In this research, an *email* is assumed to be a datum with two attributes: *subject* (the subject of the email) and *body* (the content in the body of the email). Both of them are string of characters. A developer mailing lists for a program is a set of emails and denoted by $\mathcal{D} = \{d_1, d_2, ..., d_D\}$.

A *module* refers to a source file in the latest available release. It is viewed as the smallest unit of the program. A module possesses two attributes: *name* (the name of the respective source file), and *body* (the content of the respective source file). Both of them are also strings of characters. The set of all modules is denoted by $\mathcal{M} = \{m_1, m_2, ..., m_M\}$.

If a part of an email mentions about a module, the email and the module are said to have relationship. The set of all email-module pairs that have relationship is denoted by $R^* \subseteq \mathcal{D} \times \mathcal{M}$.

A *program structure* is a digraph $G(\mathcal{C}, \mathcal{E})$ where $\mathcal{C} \subseteq 2^{\mathcal{M}} \times 2^{\mathcal{D}}$. Each element (s_1, s_2) of $\mathcal{C}$ is called a *subsystem* of the program. Here, s_1 represents the set of its constitual modules and s_2 represents the set of its explanatory emails. The paths in $\mathcal{E}$ represent the super-sub relationship between subsystems.

2.1   Evidences

The objective of this phase is to find the relation set $\mathcal{R}^*$. It begins with finding the set $\mathcal{R}_S$ containing all the pairs $(d_i, m_j) \in \mathcal{D} \times \mathcal{M}$ which satisfy one of the following conditions:

1. The name of m_j appears in the subject of d_i.
2. The name of m_j appears in the text area in the body of d_i.
3. The name of m_j appears in the code area in the body of d_i.
4. A discriminative substring of the body of m_j appears in the body of d_i.

A *text area* is defined as a continuous sequence of terms that are created and used by human for exchanging information. Meanwhile, a *code area* is defined as continuous sequence of terms which are created by or for computers. This different dealing is caused by our observation that text areas where module names appear are often the discussions between developers about the modules. Whereas, the appearances of module names in code areas could be unintentional includings when parts of source code are quoted.

If the pair (d_i, m_j) satisfies the k-th condition above, the pair is said to be *supported by Evidence k*. Figure 2 illustrates an example of the four kinds of evidences. The superscripts are added for explanation and they indicate the kinds of respective evidences. Code areas are bordered by the dashed lines.

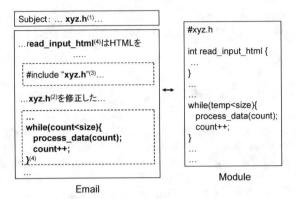

Fig. 2. An example of the four types of evidences

In order to obtain $\mathcal{R}^*$, the set $\mathcal{R}_S$ is futher filtered as

$$\mathcal{R}^* = \{(d_i, m_j) \in \mathcal{R}_S \mid conf(d_i, m_j) \geq 0\},$$

where $conf$ is the confidence degree which incorporates all the four evidences and will be explained later.

The set D_S and M_S are defined as the set of emails and modules which have at least one related module or related email according to R_S. Similarly, D^* and M^* are also defined. The following subsumptions hold:

$$\mathcal{R}^* \subseteq \mathcal{R}_S, \mathcal{D}^* \subseteq \mathcal{D}_S, \text{ and } \mathcal{M}^* \subseteq \mathcal{M}_S.$$

2.2   Finding Evidences

Evidence 1, 2 and 3. Depending on the naming convention, the root parts of the module names can be common keywords or very unique keywords. In order to avoid inaccurate extraction while not missing related emails, the requirement of module name with or without extension is decided manually for each project.

The identifying of the first kind of evidence is straight-forward. In Figure 2, xyz.h[1] is an example of Evidence 1. Relating to identifying Evidence 2, such as xyz.h[2], and Evidence 3, such as xyz.h[3], we need to separate text areas and code areas in an email. For this task, we adopt the following method:

- Tokenizing the body of emails into tokens with a part of speech assigned to each token.
- Sequences of *undefined* tokens with the lengths exceed a specified threshold are recognized as code areas. Other parts are recognized as text areas.

Evidence 4. In Figure 2, the two sequences marked with the superscript [4] are examples of Evidence 4. They are fragments of source code in the body of some module that appear in the body of the email. For identifying such fragments, we adopt the following method:

- Tokenizing the bodies of all modules by using a list of stop words together with using some matching rules for identifying the phrases that include stop words such as the comments and the pattern matching phrases.
- Sequences of code tokens $t_1 t_2 ... t_s$ which have the distinctive degree

$$\sum_{k=1}^{s} \log \frac{| \mathcal{M} |}{| \{ m_j | t_k \in Tks(m_j) \} |}$$

greater than $\log(| \mathcal{M} |)$ are selected. Here, $Tks(m_j)$ is the set of tokens appearing in the module m_j. This distinctive degree is a modification of the IDF measure for a sequence of terms, indicating the rarity of a sequence according to a set of modules. The threshold $\log(| \mathcal{M} |)$ implies that only the sequences expected to appear in the body of only one module should be selected.
- Selected sequence is matched with the sequence of token of each module by a normal exact substring matching algorithm. If there is at least one matching, the pair of the email and the module is supported by Evidence 4.

2.3   Filtering Relations

The set $\mathcal{R}^*$ is the set of email-module pairs in $\mathcal{R}_S$ which have the confidence degrees greater than 0. The confidence degree is defined by

$$conf(d_i, m_j) = b + \sum_{k=1}^{4} a_k \frac{e_k(d_i, m_j)}{\#n_{e_k}(d_i)}.$$

Here, $e_k(d_i, m_j) = 1$ if the k-th evidence found between d_i and m_j, otherwise, equals to 0. The number of modules supported by the email d_i by the k-th evidence is denoted by $\#n_{e_k}(d_i)$. This setting is based on our hypothesis that for an email, with a single type of evidence, the more modules they relate to, the less reliable the relationship is. This hypothesis will be tested in Section 4.

Besides, a_1, a_2, a_3 and a_4 are weight parameters and b is a bias satisfying $0 \leq a_1, a_2, a_3, a_4 \leq 1$ and $\sum_{i=1}^{4} a_i = 1$. These parameters are estimated from some real data.

3   Extracting Structure

We build a *concept lattice* from the extracted relation between emails and modules, then we prune it with PRUNIA (PRUNing Algorithm Based on Introduced Attributes) in order to receive a more compact and suitable structure.

3.1   Formal Concept Analysis

Formal Concept Analysis (FCA), a well-known technique for deriving ontology from a collection of objects and their attributes, was introduced by Wille [17] and then it was intensively researched and currently become an established research field[3,4]. In FCA, given information is structured into units which represent formal abstraction of concepts.

A *formal context* is a triple $(\mathcal{O}, \mathcal{A}, \mathcal{I})$ consisting of a set of *formal objects* $\mathcal{O}$, a set of *formal attributes* $\mathcal{A}$, and a binary relation $\mathcal{I} \subseteq \mathcal{O} \times \mathcal{A}$ which expresses the possessed attributes of each object in $\mathcal{O}$. A formal concept of $(\mathcal{O}, \mathcal{A}, \mathcal{I})$ is a pair (A, B) which satisfies

$$A \subseteq \mathcal{O}, \ B \subseteq \mathcal{A}, \ A' = B, \text{ and } A = B',$$

where A' is the set of attributes possessed by all the objects of A and B' is the set of objects possessing all the attributes of B.

The set A is called the *extent* of the formal concept (A, B), and B is called its *intent*. The order between two concepts of a given context is defined by

$$(A_1, B_1) \leq (A_2, B_2) \Leftrightarrow A_1 \subseteq A_2 (B_2 \subseteq B_1).$$

The set $\mathcal{B}(\mathcal{O}, \mathcal{A}, \mathcal{I})$ of all concepts of $(\mathcal{O}, \mathcal{A}, \mathcal{I})$ with this order is a complete lattice and called the *concept lattice* of $(\mathcal{O}, \mathcal{A}, \mathcal{I})$. In this paper, every element of B is called *possessed attribute* of (A, B), while every element of B but not being possessed by any super-concept of (A, B) is called *introduced attribute* of (A, B). The set of introduced attributes is denoted by $\hat{B}$.

An example of formal context is shown in Figure 4(a) with the objects are $m_1,...,m_4$ and the attributes are $d_1,...,d_{12}$. The concept lattice is shown in Figure 4(b). The nodes in solid and dashed border are all concepts. If the number i appears in a concept, the concept possesses the module m_i. The possessed attributes of concepts are shown on the right side, and among them, introduced attributes are underlined. The straight edges represent the intermediate super-sub concept relationship. The curve edges are used for other purpose and will be explained later.

3.2   Extracting Structure

The formal context is chosen as the triple $(\mathcal{M}^*, \mathcal{D}^*, \mathcal{R}^*)$ where $\mathcal{M}^*$, $\mathcal{D}^*$, and $\mathcal{R}^*$ are the sets obtained in the first phase. In this section, a concept is represented as the triple $(A, B, \hat{B})$ for our convenience of indicating the set of introduced attributes of each concept.

In general, the concept lattice is very complex because the number of its concepts is exponential to the number of its objects. In fact, only a small number of its concepts interest human developers. Therefore, we will remove those concepts that do not fit to become the sub-systems of a program by using the two following constraints:

1. The number of the possessed attributes is greater than or equals to a threshold σ.
2. The number of the introduced attributes is greater than or equals to a threshold τ.

	d_1	d_2	d_3	d_4	d_5	d_6	d_7
m_1	●	●	●	●	●	●	
m_2				●	●	●	●

Fig. 3. Example with two modules and seven emails

The first constraint reflects the observation addressed in the first chapter. It is a *monotone* constraint because for any concept satisfying this constraint, its sub-concepts also satisfy this constraint. This constraint has been intensively used in closed frequent itemset mining [9,19]. The second constraint is an enhancement for the first constraint because in many cases, even though the first criterion holds, actually most of those emails are not about the concept but its super-concepts. Figure 3 illustrates the context table of an example with two modules and seven emails where there are four emails relating m_2 but among them only d_7 does not relate to m_1. It implies that m_2 depends strongly in m_1 and must not be recognized as a subsytem. Different from the first constraint, this constraint is not *monotone*. In other words, there are cases that a concept does not satisfy the second constraint even though its super-concepts satisfy the second constraint. For this reason, an algorithm which simply removes non-satisfying concepts would obtain a structure with many divided components.

PRUNIA. Our proposed algorithm, shown in Algorithm 1, removes concepts which do not satisfy the two constraints explained above, and simultaneously, connects pairs of concepts that have super-sub concept relationship but being disconnected because some concepts in the middle have been removed. We introduce the following notations:

- pa: the number of possessed attributes.
- ia: the number of introduced attributes.
- $\mathcal{C}^i$: the set of concepts having exactly i objects.
- $c \succ_{\mathcal{E}} d \Leftrightarrow d$ can be reached from c by some paths in $\mathcal{E}$.

The inputs are the concept lattice and two parameters σ and τ. For shortening the term, members of $\mathcal{C}_{\sigma\tau}$ are called *good concepts* and the others are called *bad concepts*. Moreover, elements of $\mathcal{E}_{\sigma\tau}$ are called *good edges* while the others are called *bad edges*. Initially, both $\mathcal{C}_{\sigma\tau}$ and $\mathcal{E}_{\sigma\tau}$ contain no element. The algorithm proceeds upwardly, and begins with processing concepts at the first level. Cnd_c is the set of good nodes which can be reached from c by only bad edges. Max_c is the set of elements of Cnd_c that cannot be reached by good edges from c as well as from any other element of Cnd_c. The edges connecting c to all concepts in Max_c are added to a temporary set $\mathcal{E}_{add}$. At the end of each level, concepts in $\mathcal{C}_{add}$ are added to $\mathcal{C}_{\sigma\tau}$ and edges in $\mathcal{E}_{add}$ are added to $\mathcal{E}_{\sigma\tau}$.

Algorithm 1. PRUNIA

Input: σ, τ, $G(\mathcal{C}, \mathcal{E})$
Output: $G(\mathcal{C}_{\sigma\tau}, \mathcal{E}_{\sigma\tau})$
$\quad \mathcal{C}_{\sigma\tau} \leftarrow \emptyset,\ \mathcal{E}_{\sigma\tau} \leftarrow \emptyset$
$\quad h \leftarrow$ the number of objects of the top concept
$\quad \textbf{for } i = 1 \textbf{ to } h \textbf{ do}$
$\quad\quad \mathcal{C}_{add} \leftarrow \emptyset,\ \mathcal{E}_{add} \leftarrow \emptyset$
$\quad\quad \textbf{for all } c \in \mathcal{C}^i \text{ such that } pa(c) \geq \sigma \text{ and } ia(c) \geq \tau \textbf{ do}$
$\quad\quad\quad \mathcal{E}_{add} \leftarrow \mathcal{E}_{add} \cup \{(c, d) \in \mathcal{E} \mid d \in \mathcal{C}_{\sigma\tau}\}$
$\quad\quad\quad Cnd_c \leftarrow \{d \in \mathcal{C}_{\sigma\tau} \mid c \succ_{\mathcal{E} \setminus \mathcal{E}_{\sigma\tau}} d\}$
$\quad\quad\quad Max_c \leftarrow \{d \in Cnd_c \mid \neg \exists f \in (Cnd_c \cup \{c\}) \text{ such that } f \succ_{\mathcal{E}_{\sigma\tau}} d\}$
$\quad\quad\quad \mathcal{C}_{add} \leftarrow \mathcal{C}_{add} \cup \{c\}$
$\quad\quad\quad \mathcal{E}_{add} \leftarrow \mathcal{E}_{add} \cup \{(c, d) \mid d \in Max_c\}$
$\quad\quad \textbf{end for}$
$\quad\quad \mathcal{C}_{\sigma\tau} \leftarrow \mathcal{C}_{\sigma\tau} \cup \mathcal{C}_{add}$
$\quad\quad \mathcal{E}_{\sigma\tau} \leftarrow \mathcal{E}_{\sigma\tau} \cup \mathcal{E}_{add}$
$\quad \textbf{end for}$

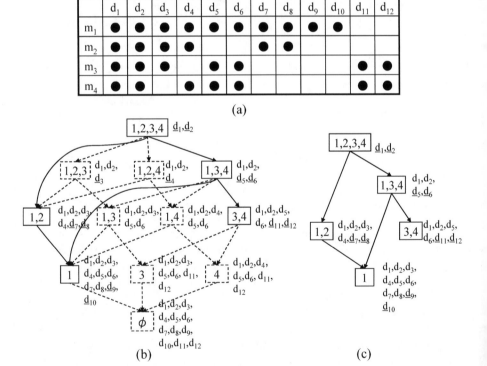

Fig. 4. Example of applying PRUNIA

Table 1. Summarization of processing at each level

Level	Concepts	Cnd	Max	C_{add}	$\mathcal{E}_{add}$
1	{1}	∅	∅	{1}	∅
2	{1,2}	∅	∅	{1,2}	({1,2},{1})
	{3,4}	∅	∅	{1,2},{3,4}	({1,2},{1})
3	{1,3,4}	{1}	{1}	{1,3,4}	({1,3,4},{3,4}),({1,3,4},{1})
4	{1,2,3,4}	{1},{1,2}	{1,2}	{1,2,3,4}	({1,2,3,4},{1,3,4}),({1,2,3,4},{1,2})

Example. Figure 4 shows an example of applying PRUNIA with the context table shown in 4(a) and the parameters set to $\sigma = 2$ and $\tau = 2$. In Figure 4(b), good concepts are nodes bordered by solid lines, good edges are those represented in straight solid lines and the newly added edges are those in curve solid lines. The processings at each level are summarized in Table 1. For example, at Level 4, concept {1} belongs to the set $Cnd_{\{1,2,3,4\}}$ but because it is reachable from {1,2} by good edges, it is excluded from $Max_{\{1,2,3,4\}}$. Figure 4(c) shows the graph obtained after pruning.

Lemma 1. *The super-sub concept order is preserved after pruning, that is, for all c, $d \in \mathcal{C}_{\sigma\tau}$, if $c \succ_{\mathcal{E}} d$ then $c \succ_{\mathcal{E}_{\sigma\tau}} d$.*

Proof. The proof is given in the mathematical induction form.

For all pairs of good concepts at Level 1 and Level 2, if they have super-sub concept relationship, they must be directly connected by an edge. On the other hand, all edges which connect pairs of good concepts are added to $\mathcal{E}_{\sigma\tau}$. Hence, the lemma holds at Level 1 and Level 2.

Assume that the lemma holds from Level 1 to Level k. Suppose that c is a good concept at Level $k + 1$ and d is one of its good sub-concepts. Consider all paths of both good and bad edges connecting c and d. If the concept f next to c is a good concept, there is already a path of good edges connecting f and d because f is at Level k and according to the assumption. Hence, c and d is connected by good edges. If the concept next to c is a bad concept, then the first good concept after c, let say f, in the path must be an element of Cnd_c. Moreover, according to the definition of Max_c, f must be the elements of Max_c or the sub-concept of some concepts in Max_c. Remembering that elements of Max_c are under or at Level k, hence all super-concepts of d in Max_c are already connected to d. After adding the edges which connect c to all elements of Max_c, c and d must be connected.

Lemma 2. *PRUNIA produces no new redundant path, that is, for all c, $d \in \mathcal{C}_{\sigma\tau}$ and $(c,d) \in \mathcal{E}_{\sigma\tau}$, there is no $f \in \mathcal{C}_{\sigma\tau}$ such that $(c,f) \in \mathcal{E}_{\sigma\tau}$ and $f \succ_{\mathcal{E}_{\sigma\tau}} d$.*

Proof. Because the algorithm performs upwardly, there will be no more edges which connect concepts under the level of c after it reaches to the level of c. On the other hand, by Lemma 1, f and d are finally connected by good edges so when the algorithm reaches to c, f and d must be already connected. Hence, d cannot be element of Max_c. Therefore no edge between c and d is added.

Table 2. Datasets used for experiments

Project	Language	Begin	Emails	Modules
HOS	C	2002	1515	186
Namazu	C,Perl	1997	9462	164
Ruby	C	1997	36399	255

Table 3. Number of extracted pairs

Project	Data $\frac{|\mathcal{D}|}{|\mathcal{M}|}$	Supported $\frac{|\mathcal{D}_S|}{|\mathcal{M}|}$	$\frac{|\mathcal{R}_S|}{|\mathcal{M}|}$	Filtered $\frac{|\mathcal{D}^*|}{|\mathcal{M}|}$	$\frac{|\mathcal{R}^*|}{|\mathcal{M}|}$
HOS	8.2	2.0	7.2	1.7	5.3
Namazu	57.8	15.05	93.3	9.5	33.1
Ruby	142.7	30.1	111.8	25.7	89.8

4   Experimental Results

4.1   Extracting Relation

The method explained in Section 2 is used for extracting the relation between emails and modules with three Japanese projects HOS [5], Namazu [8] and Ruby [11]. An outline of these projects is presented in Table 2. Each dataset consists of a collection of modules with the extension .h and .c (C source files), .pl and .in (Perl source files) from the latest release at the time and a collection of emails in Japanese extracted from their developer mailing lists. Relating to separation of text and code areas, we used Chasen [1] for tokenizing emails and assigning part-of-speechs. The minimum length for a sequence of undefined tokens to be recognized as a code area was set to 15, a value chosen from experience. Relating to identification of Evidence 4, a list of stop words including line-break, blank, ;, +, -, *, /, {, }, (,),... was used for tokenizing the bodies of modules.

We manually annotated 747 email-module pairs (from the supported relation obtained with Namazu dataset) to be *Yes* or *No* according to whether or not the modules are mentioned in the emails.

In order to evaluate the efficiency of the used classifying function, we performed an experiment using the linear classifier SimpleLogistic on the annotated data, and compare the obtained results with those of other three non-linear classifiers, NaiveBayes, J4.8, and MultilayerPerceptron. The tool Weka [18] was used for the experiment. Input data are provided in two forms: a binary form where only the binary values of four evidences are taken as attributes, and a weighted form where attributes are evidences being weighted with the inverse number of the related modules by different evidences. Our experimental results show that that all classifiers produce better results when the data are provided in weighted form. It proves that a classifier, which takes the number of modules related by emails into account, is more effective than the classifiers of the same learning model that do not. Moreover, when data are given in weighted form, SimpleLogistic outperforms other classifier (1.21% higher than the second). This result assures that the proposed classifying function is usable. In addition, the parameters of the linear classifying function, $conf$, were estimated as

$$a_1 = 0.141, a_2 = 0.459, a_3 = 0.373, a_4 = 0.027, \text{ and } b = -0.033$$

when SimpleLogistic builds its classifying model with the full annotated data. The most powerful evidence is the second evidence. This result agrees with our anticipation about the different meanings of module names in text areas and code areas. Surprisingly, the first evidence, the appearance of module names in the subject of email, is only at the third position. The reason is its recall ratio is quite small (only account for about 10% of the annotated pairs).

Table 3 shows the relative number of emails in the initial data and the relative size of email sets and relation sets. We can have some conclusions: (1) approximately, the proposed method can extract about 20% of the emails from the mailing lists, and (2) over the extracted pairs, on average, each email relates to about three modules.

4.2   Extracting Structure

For the richness of related emails, we choose Ruby to do the experiment with the method proposed above. We used Colibri [6], an open source tool developed by Lindig for constructing concept lattices and computing sets of introduced attributes. Colibri implements the NextClosure algorithm which was proposed by Ganter and appeared later in the textbook [4]. The extracted structures are evaluated with the following three criteria:

1. *Module recall ratio*: the ratio of the number of modules appearing in the extracted structure to the number of modules in the concept lattice before pruning.
2. *Email recall ratio*: the ratio of the number of emails remaining after pruning to the number of emails before pruning.
3. *Concept recall ratio*: the ratio of the number of concepts remaining after pruning to the number of concepts before pruning.

We performed experiments with many parameter combinations, but because there is not enough room, only the results obtained with the parameter combinations in the range $0 \leq \tau \leq \sigma \leq 10$ are provided in Table 4, 5 and 6. The results of the three recall ratios are shown respectively in the tables. The cell of the i-th row and the j-th column represents the ratio obtained when $\sigma = j$ and $\tau = i$.

As shown in Table 4, when σ and τ increase, the module recall ratio reduces quite fast. When both τ and σ are set to 10, only about one third of the modules remains. It implies that only a part of the modules have many emails relating to them. They are considered as the important modules of the structure.

As shown in Table 5, when σ and τ increase, the email recall ratio also reduces but the reducing speed is much slower than that of the module recall ratio.

Table 4. The module recall ratios of Ruby dataset (100%=188 modules)

$\tau\backslash\sigma$	0	1	2	3	4	5	6	7	8	9	10
0	100	100	87.77	81.91	76.06	70.74	66.49	63.83	61.70	58.51	56.38
1		100	85.11	77.13	70.74	64.89	61.70	60.64	57.98	54.79	53.19
2			73.40	68.09	67.02	62.77	60.64	59.57	57.45	54.26	52.66
3				61.17	61.17	57.45	56.38	54.79	53.19	51.06	49.47
4					52.13	50.00	50.00	49.47	48.94	47.34	46.81
5						46.81	46.81	46.28	45.74	45.21	44.15
6							42.55	42.02	42.02	42.02	40.96
7								39.36	39.36	39.36	39.36
8									37.77	37.77	37.77
9										37.23	37.23
10											37.23

Table 5. The email recall ratios of Ruby dataset (100%=6563 emails)

$\tau\backslash\sigma$	0	1	2	3	4	5	6	7	8	9	10
0	100	100	93.25	90.26	88.08	86.53	85.08	84.17	83.24	82.2	81.59
1		100	93.25	90.26	88.08	86.53	85.08	84.17	83.24	82.2	81.59
2			87.48	86.26	85.19	84.18	83.24	82.57	81.81	80.91	80.42
3				82.05	81.78	81.26	80.83	80.47	79.92	79.48	79.08
4					79.49	79.25	78.96	78.73	78.45	78.20	77.94
5						77.66	77.43	77.27	77.11	77.04	76.78
6							75.68	75.59	75.59	75.59	75.41
7								74.31	74.31	74.31	74.31
8									73.46	73.46	73.46
9										72.73	72.73
10											72.18

Moreover, when σ and τ become bigger, the reducing speed gets slower and the email recall ratio seem to converge at some value around 70%. It is because a lot of emails (about 70%) belong to some concepts whose the numbers of introduced attributes are larger than 10. For this reason, when σ and τ increase, those emails still remain in the structures.

The results in Table 6 prove that the number of the concepts were pruned efficiently. For example, at $\sigma = 10$ and $\tau = 10$, only 2.92% (81 concepts) of the concepts in the initial lattice remain. Moreover, we also can see that for parameter combinations with similar concept ratios, the one with higher τ value produces higher module recall ratio and higher email recall ratio. For example, even though $\sigma=3$, $\tau=3$ and $\sigma=9$, $\tau=2$ have the same concept ratio, that is 8.16%, the module recall ratio and the email recall ratio of the latter are higher than those of the former (61.17% to 54.26% and 82.05% to 80.91% respectively). This fact proves that by using both constraints, the modules and emails are preserved better in pruning.

Table 6. The concept recall ratios of Ruby dataset (100%=2771 concepts)

$\tau\backslash\sigma$	0	1	2	3	4	5	6	7	8	9	10
0	100	99.96	83.97	61.91	47.22	37.80	31.12	26.97	23.61	21.01	19.03
1		44.26	28.27	22.64	18.84	16.61	14.55	13.39	12.38	11.23	10.61
2			14.58	13.14	11.99	11.05	10.18	9.60	8.99	8.16	7.83
3				8.16	7.94	7.58	7.33	7.11	6.75	6.46	6.25
4					6.14	5.99	5.85	5.74	5.60	5.45	5.34
5						5.05	4.95	4.87	4.80	4.77	4.66
6							4.12	4.08	4.08	4.08	4.01
7								3.57	3.57	3.57	3.57
8									3.29	3.29	3.29
9										3.07	3.07
10											2.92

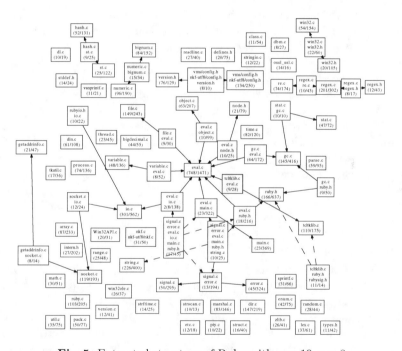

Fig. 5. Extracted structure of Ruby with $\sigma = 10, \tau = 8$

Figure 5 shows an extracted structure for Ruby with the parameters set as $\sigma = 10$ and $\tau = 8$. Each node represents a concept with its possessed modules, the number of introduced attributes and the number of possessed attributes. The arrows in dashed lines represent the edges added after pruning. The module recall ratio is 37.77% (71 modules), the email recall ratio is 73.46% (4821 emails), and the concept recall ratio is 3.29% (91 concepts). In fact, this structure consists of important modules of Ruby. The concept eval.c, which appears as a main hub, has as many as 747 emails relating to it, and actually, eval.c deals with evaluating expressions of Ruby. This module can be said as the most important component of Ruby. Furthermore, modules related closely are correctly grouped. For example, two concepts which have 6 modules, in fact, represent the kernel subsystems of Ruby. The concept with the four modules `io.c`, `socket.c`, `getaddrinfo.c`, and `rubyio.h` deals with the input-output and networking.

5 Concluding Remarks and Future Plans

Our method is an attempt to utilize both types of information: the mailing-lists and the source code archives in order to extract the structures of open source programs and categorize the emails in mailing-lists into the extracted structures. Our method discovers the structure of the program based on the view of the developers which can be obtained from the mailing-lists. This kind of structure is distinguished from the structures built from the code, such as the dependency graphs between source files. The advantage of our method is the usage of information about related emails of source files. We found that groups of source files which have large number of related emails are normally essential components. Therefore, by changing the pruning thresholds, we can extract the structure with some different degrees of summarization. As a result, it helps the user to understand the program more easily.

Our method for separating text area and code area is simple and shows good results when tested with Japanese emails. For emails in English, more sophisticated methods are required to do the separation. The reason is many computer-used terms are in English and it makes the problem more difficult. Tang et al. [15] proposed a different method which used pre-defined rules to identify the parts of source code of some particular programming languages in an email. However, this method is not suitable for our task because it can only indentify limited types of codes.

In the future, we would like to continue this research in several directions. The first direction is to give a stronger background theory to support the proposed PRUNIA algorithm and compare it with other existing similar algorithms if there are any. Another direction is to extend our method for extracting the evolution of software programs from their mailing lists. The time attributes, as well as all versions of source files will be taken into account in this research. Besides, automatic selection of parameters is also an interesting research problem.

Acknowledgments. This work was partially supported by Grant-in-Aid for Scientific Research (19300046) from JSPS.

References

1. Chasen, `http://chasen.naist.jp/hiki/ChaSen/`
2. Cimitile, A., Visaggio, G.: Software salvaging and the call dominance tree. Journal of Systems and Software 28(2), 117–127 (1995)

3. Ganter, B., Wille, R.: Applied lattice theory–Formal concept analysis. In: Gratzer, G. (ed.) General Lattice Theory. Birkhauser, Basel (1997)

4. Ganter, B., Wille, R.: Formal Concept Analysis–Mathematical Foundations. Springer, Heidelberg (1999)

5. HOS, http://sourceforge.jp/projects/hos/

6. Lindig, C.: Colibri–command line tool for concept analysis, http://www.st.cs.uni-saarland.de/~lindig/

7. Lindig, C., Snelting, G.: Assessing modular structure of legacy code based on mathematical concept analysis. In: Proceedings of the 19th International Conference on Software Engineering (ICSE 1997), pp. 349–359 (1997)

8. Namazu, http://www.namazu.org/

9. Nicolas, P., Yves, B., Rafik, T., Lotfi, L.: Efficient mining of association rules using closed itemset lattices. Information Systems 24, 25–46 (1999)

10. Rasinen, A., Hollmen, J., Mannila, H.: Analysis of Linux evolution using aligned source code segments. In: Todorovski, L., Lavrač, N., Jantke, K.P. (eds.) DS 2006. LNCS (LNAI), vol. 4265, pp. 209–218. Springer, Heidelberg (2006)

11. Ruby, http://www.ruby-lang.org/

12. Schwanke, R.W.: An intelligent tool for re-engineering software modularity. In: Proceedings of the 13th International Conference on Software Engineering (ICSE 1991), pp. 83–92. IEEE Computer Society Press, Los Alamitos (1991)

13. Snelting, G.: Concept analysis–A new framework for program understanding. In: Proceedings of the 1998 ACM SIGPLAN-SIGSOFT Workshop on Program Analysis for Software Tools and Engineering (PASTE 1998), pp. 1–10. ACM, New York (1998)

14. Tanaka, K., Akaishi, M., Takasu, A.: Topic change extraction and reorganization from problem-solving records. In: Proceedings of International Conference on Software Knowledge Information Management and Applications, pp. 153–158 (2006)

15. Tang, J., Li, H., Cao, Y., Tang, Z.: Email data cleaning. In: Proceedings of the 11th International Conference on Knowledge Discovery in Data Mining (KDD 2005), pp. 489–498 (2005)

16. Washizaki, H., Fukazawa, Y.: A technique for automatic component extraction from object-oriented programs by refactoring. Sci. Comput. Program. 56(1-2), 99–116 (2005)

17. Wille, R.: Restructuring lattice theory–An approach based on hierarchies of concepts. In: Rival, I. (ed.) Ordered Sets, pp. 445–470. Reidel, Dordrecht (1982)

18. Witten, I.H., Frank, E.: Data Mining: Practical Machine Learning Tools and Techniques, 2nd edn. Morgan Kaufmann Series in Data Management Systems. Morgan Kaufmann, San Francisco (2005)

19. Zaki, M.J.: Mining non-redundant association rules. Data Min. Knowl. Discov. 9(3), 223–248 (2004)

A Comparison of Community Detection Algorithms on Artificial Networks

Günce Keziban Orman[1,2] and Vincent Labatut[1]

[1] Galatasaray University, Computer Science Department, Ortaköy/İstanbul, Turkey
[2] TÜBİTAK, Gebze/Kocaeli, Turkey
keziban.orman@uekae.tubitak.gov.tr, vlabatut@gsu.edu.tr

Abstract. Community detection has become a very important part in complex networks analysis. Authors traditionally test their algorithms on a few real or artificial networks. Testing on real networks is necessary, but also limited: the considered real networks are usually small, the actual underlying communities are generally not defined objectively, and it is not possible to control their properties. Generating artificial networks makes it possible to overcome these limitations. Until recently though, most works used variations of the classic Erdős-Rényi random model and consequently suffered from the same flaws, generating networks not realistic enough. In this work, we use Lancichinetti *et al.* model, which is able to generate networks with controlled power-law degree and community distributions, to test some community detection algorithms. We analyze the properties of the generated networks and use the normalized mutual information measure to assess the quality of the results and compare the considered algorithms.

Keywords: Complex networks, Community detection, Algorithms comparison.

1 Introduction

Complex networks are now a popular tool to model a given system, by representing its components and their interactions with nodes and links, respectively. This model can then be analyzed or visualized thanks to some of the many tools designed for graph mining. Complex networks have been used in very different application domains, such as physics, biology, social science or computer networks [1].

Among the various approaches used to study complex networks properties, community detection has become one of the most popular ones. A community, or cluster, is generally defined as a subset of nodes densely interconnected relatively to the rest of the network [2]. Many different community detection algorithms have been defined to identify these subsets. They are generally based on classical clustering principles adapted to graphs, using hierarchical or optimization methods. Hierarchical approaches divide or merge communities by considering the distance or similarity between them, whereas optimization approaches partition the network according to a given criterion.

Authors traditionally test their community detection algorithms on a few real [3-11] or artificial [2-6, 10] networks. Limiting these tests to real networks can be

J. Gama et al. (Eds.): DS 2009, LNAI 5808, pp. 242–256, 2009.

considered as an issue for several reasons. First, building such networks is a costly and difficult task, and determining reference communities can only be done by experts. This leads to small networks, where actual communities are not always defined objectively, or even known. Second, a complex network is characterized by various statistics like its average degree, degree distribution, shortest average path, etc. By definition, it is not possible to control these features in a real network. This means the algorithm is tested on a very specific and limited set of features.

Artificial networks seem to overcome these limitations, because it is possible to randomly generate many of them, while controlling their properties. All that is needed is a generative model able to produce networks with features similar to those of real networks. Of course, artificial networks must not be seen as a substitute to real networks, but rather as a complement. In the context of testing community detection algorithms, the most popular generative model is the one defined by Newman and Girvan [4], which is used in all the works cited above. It is a variation of the classic Erdős-Rényi random model [12] (or Poisson random model), and it consequently suffers from the same limitation: the generated networks do not show a realistic topology [13, 14].

Some recent works tried to improve this by defining more realistic models, able to mimic some of the real networks features. In this work, we use the model proposed by Lancichinetti *et al.* [14], which is able to generate networks with controlled power-law degree and community distributions. Our purpose is to generate a set of artificial networks with various size and properties, and to use it to test the existing community detection algorithms. We use the normalized mutual information measure [15-17] to assess the quality of the results and compare the considered algorithms.

In section 2, we explain what the properties of a complex network are. It is of course an open question to decide how a complex network can be described by a few features, but we kept only the most widely used ones. In section 3, we focus on the community detection task. We first describe its general mechanisms, and the modularity measure, which is used as an optimization criterion in many algorithms. Then, we list the algorithms we chose to compare. In section 4, we explain how we generated our test set of artificial networks, and we give some explanation about the normalized mutual information measure we used to assess the algorithms performance. In section 5, we present and discuss our results, focusing first on the observed properties of the generated networks, and then on the comparison of the algorithms' performances.

2   Complex Networks Properties

Undirected real networks are known to share some common properties. In this section, we present the most prominent ones: small-worldness, transitivity, degree-related properties and community structure. Many other properties can be used to describe a network, either by analyzing some measure, like betweenness-centrality distribution [18] or network diameter [19], or by counting the number of occurrences of a given substructure like motifs in [20]. But their use is not really widespread, and we would consequently lack experimental values to exploit them in this work.

Small-World. A network is said to have the small-world property if, for a fixed average degree, the average distance (i.e. the length of the shortest path) between pairs of nodes increases logarithmically with the number of nodes n [1]. This property can be interpreted as propagation efficiency: spreading on the network remains relatively fast even if the network grows.

Transitivity. The transitivity property is measured by a transitivity coefficient, also called clustering coefficient [21]. Different versions of this coefficient exist, but they all try to assess the density of triangles in a network. The higher this coefficient, the more probable it is to observe a link between two nodes which are both connected to a third one. Independently of the considered coefficient version, a real network is supposed to have a higher transitivity than a Poisson random network (such as those generated by the Erdős–Rényi model [12]) with the same number of nodes and links, by a factor of order n [1].

Degree. Networks can also be described according to their degree distribution. In most real networks, this distribution follows either a power or an exponential law. In other terms, the probability for a node to have a degree k is either $p_k \sim k^{-\gamma}$ or $p_k \sim e^{-k/\kappa}$ [1]. Networks with a power-law degree distribution are the most common. They are called scale-free, because their degree distribution does not depend on their size (some other properties may, though). Experimental studies showed that the γ coefficient usually ranges from 2 to 3 [1, 19, 22]. It is known that for values of γ smaller than 3.48, there is a high probability the network contains one giant component and several small ones (a component is a separated subgraph), or even only one component (the network being completely connected) [22].

In a real network, the average and maximal degrees generally depend on the number of nodes it contains. For a scale-free network, it is estimated to be $\langle k \rangle \sim k_{max}^{-\gamma+2}$ [19, 22] and $k_{max} \sim n^{1/(\gamma-1)}$ [1], respectively.

The degree correlation of a network constitutes another interesting property. The question is to know how a node degree is related to its neighbors'. Real networks usually show a non-zero degree correlation. If it is positive, the network is said to have assortatively mixed degrees, whereas if it is negative, it is disassortatively mixed [1]. According to Newman, social networks tend to be assortatively mixed, while other kinds of networks are generally disassortatively mixed. Nodes with high degree are called hubs, because they have a more central position in the network.

Community. In this work, our focus is on detecting communities in networks. Of course, it is important to note that not all real networks have a community structure. According to Newman though, it is a common feature in biological and social networks [1]. When the community structure is present, the community size distribution seems to follow a power-law distribution [23] with a parameter β ranging from 1 to 2 [5, 24].

3 Community Detection

Complex networks have been used widely to model real-world systems in many application fields. When analyzing a complex network, the problem of identifying its communities is universal, and has consequently been raised in many domains, leading to different solutions. Many of them rely on Newman's modularity to assess the quality of their results, so we will first introduce this measure. Then, we will present the principles of community detection, and give a short description of the algorithms we chose to compare.

3.1 Modularity

The modularity measure has been presented by Newman and Girvan [2] to assess the quality of a network partition. They first define what could be called a community contingency matrix, whose elements p_{ij} represent the fraction of total links from a node in community i towards a node in community j. The fraction of links inside community i is therefore p_{ii}. Moreover, since we are considering undirected networks, we have $p_{ij} = p_{ji}$ and the matrix is symmetric.

Let p_{i+} and p_{+j} be the sums over row i and column j, respectively. If the network has no community structure, or if the considered communities are not defined accordingly to the network structure, then one can suppose the links are randomly distributed. Under this hypothesis, the expected fraction of links inside community i can be estimated as the probability for a link to start in community i, which is p_{i+}, multiplied by the probability to end in community i, which is p_{+i}. The matrix being symmetric, we have $p_{i+}p_{+i} = (p_{i+})^2$. The modularity measure is defined as the difference between the observed and expected fractions of links in each community, summed over all communities:

$$Q = \sum_i p_{ii} - \sum_i (p_{i+})^2 \tag{1}$$

When the communities are not better than a random partition, or when the network does not exhibit any community structure, Q is negative or zero. Its superior limit is 1, but it can be approached only if the network has a strong community structure and if the communities have been perfectly detected.

Interestingly enough, the modularity measure is similar to the numerator of chance-corrected measures used to assess the performance of classic classifiers, such as Cohen's κ coefficient [25]. The general formula for these measures is $(P_o - P_e)/(1 - P_e)$, where P_o is the observed agreement and P_e is the expected agreement between the classifier results and the classified data. But unlike modularity, chance corrected measures are normalized by the dividing term $(1 - P_e)$, which represents a perfect classifier result (reaching a 1 observed agreement). Of course, it is not possible to process the corresponding value in the case of modularity, because the superior limit for $\sum_i p_{ii}$ depends on the community structure of the network, and is usually less than one (whereas 1 is an absolute value for classic classifiers).

The modularity measure is known to have some flaws. For example, it is sensitive to community size [26] and it is possible to find partitions of Poisson random networks with relatively high modularity values [27] (although they have no community structure). However, many community detection algorithms use it as an optimization criterion, as we will see in the following section.

3.2 Algorithms for Community Detection

It is difficult to categorize the community detection algorithms, but one could group them in three different families: hierarchical, optimization, and others.

Early solutions are based on hierarchical approaches whose result is a tree of communities called dendrogram. Agglomerative approaches starts with as many communities as nodes, each node having its own community, and iteratively merge these communities until only one giant community remains. On the opposite, divisive approaches start with one community containing all nodes, and iteratively split the communities until each node constitute one community. The communities to be merged or split are chosen accordingly to some distance or similarity function which allows detecting which communities are similar (agglomerative approach) or heterogeneous (divisive approach). What distinguishes algorithms in this family is mainly the nature of the distance or similarity function. The result being a dendrogram, one still needs to find out where to perform a cut in order to get an actual partition. For instance, one can compute the modularity at each level, and use the partition with maximal modularity.

The optimization-based approaches use a measure to estimate the quality of a network partition. This measure is, most of the time, Newman's modularity [2]. The general algorithm consists in first processing several partitions of the network (randomly or by following a fitting function) and second keeping the best one according to the quality measure. This partition can then be refined in order to get a better quality. Modularity is a costly measure to process, hence the numerous algorithms defined for its optimization [3, 6, 24].

The last family contains all the remaining approaches. Some use different principles coming from classical clustering like density-based clustering [7]; some are agent-based [8]; some allows finding overlapping communities (one node can be a part of several communities at once) [28]; some use a latent space approach to process the probability for a node to belong to a community [9].

This work consists in comparing community detection algorithms on many generated networks, so we chose to focus first on the following algorithms, which are fast and simple.

Fast Greedy Algorithm. This algorithm was developed by Newman *et al.* [10, 24]. It is modularity-based and uses a hierarchical agglomerative approach. It is called fast greedy because thanks to a standard greedy method, it is significantly faster than older algorithms.

Walktrap Algorithm. This algorithm by Pons and Latapy [29] uses a hierarchical agglomerative method. Here, the distance between two nodes is defined in terms of random walk process. The basic idea is that if two nodes are in the same community, the probability to get to a third node k located in the same community through a random walk should not be very different for i and j. The distance is constructed by summing these differences over all nodes, with a correction for degree.

Eigenvector Algorithm. This algorithm by Newman [30] is modularity-based, and it uses an optimization method inspired by graph partitioning techniques. It relies on the eigenvectors of a so-called modularity matrix, instead of the graph Laplacian traditionally used in graph partitioning.

Label Propagation Algorithm. This algorithm by Raghavan *et al.* [11] uses the concept of node neighborhood and the diffusion of information in the network to identify communities. Initially, each node is labeled with a unique value. Then an iterative process takes place, where each node takes the label which is the most spread in its neighborhood. This process goes on until one of several conditions is met, for instance no label change. The resulting communities are defined by the last label values.

Spinglass Algorithm. This algorithm by Reichardt and Bornholdt [31] is an optimization method relying on an analogy between the statistical mechanics of complex networks and physical spin glass models.

4 Method

In order to compare the selected algorithms, we chose to generate a set of artificial networks. If we want the results to hold when the algorithms are applied on real networks, our artificial networks properties must be the most similar possible to those we previously described for real networks. Another important point is the assessment of the results quality, which must be reliable in order to compare efficiently the communities detected by the tested algorithms. In this section, we present the model we used to generate our test data and the measure we chose to assess the algorithms performance.

4.1 Network Generation

In many community detection works [3, 32, 33], artificial community-structured networks are generated with models similar to the one defined by Newman and Girvan [4, 10]. It relies on the principle of the Erdős-Rényi model [12]: each community corresponds to a Poisson random network, with a probability p_{in} to have a link between two of its nodes (an internal link). Another probability p_{out} is used to add links between nodes from different communities (external links). The probabilities are constrained so that $p_{in} > p_{out}$ and the average degree d of the resulting network tends towards a fixed value.

This model lacks some of the properties we described earlier: the degree distribution and the community size distribution do not follow a power-law, and we have no information about the other properties. For this reason, we chose to use a more recent model defined by Lancichinetti *et al.* [14] to generate our test set of artificial networks. It allows generating random networks with a community structure and a power-law degree distribution. Moreover, the size of the resulting communities also follows a power-law distribution.

This method needs the following compulsory parameters: the number of nodes n, the desired average $\langle k \rangle$ and maximum k_{max} degrees, the exponent γ for the degree distribution, the exponent β for the community size distribution, and a value μ called the mixing coefficient. The latter represents the average proportion of links between a node and nodes located outside its community, $1 - \mu$ being the proportion of links with nodes located in the same community. This leads to the concepts of internal and external degrees, corresponding to the number of links a node has inside and outside its community, respectively. For a node of degree k, we then have the values $(1 - \mu)k$ for the internal degree and μk for the external degree. Of course, these values hold in average, but can only be approximated when considering a given node. Two additional parameters, the minimum and maximum community sizes, can also be optionally precised. If this is not the case, they are automatically set to values smaller than the minimal degree and greater than the maximal degree, respectively. This way, every node can fit in a community, whatever its degree.

The generation is performed in three steps. First, the well-known configuration model [34] is used to generate a scale-free network corresponding to the specified γ parameter. Second, the community sizes are drawn in accordance with the β parameter, and each node is randomly affected to a compatible community. Compatible means here that the community size must be greater or equal to the node internal degree. Some specific mechanisms ensure the convergence of the processing, see [14] for more details. Third, some links are rewired in order to respect the mixing coefficient. For a given node, the total degree is not modified, but the ratio of internal and external links is changed so that the resulting proportion gets close to μ.

Our goal was to compare the performance of community detection algorithms, so we generated networks with parameters consistent with what is observed in real community-structured networks. We used the value 1000 for the number of nodes n. The β and γ exponents ranged from 1 to 2 and from 2 to 3, respectively. We used values of μ in $[0.05; 0.95]$ with a 0.05 step. For each set of parameters, we generated 25 networks in order to deal with possible discrepancies in the networks properties due to the random generation.

In rare occasions, we observed that some parameters can cause several components to appear in the same network. Some algorithms like Walktrap cannot be applied on such networks, so we decided to randomly connect these components in order to be able to apply all the algorithms.

4.2 Performance Assessment

As we stated before, the modularity measure is a standard for assessing the quality of a network partition. But it was designed to be an approximation of the partition quality, to guide community detection algorithms when the actual communities are unknown. The value computed for a given situation depends on both the quality of the detected communities and of the nature of the network community structure.

This dependence to the network structure prevents from using modularity to compare algorithm performances on different networks. Furthermore, we will use artificial networks, whose communities are known *a priori*. In this context, modularity is not an appropriate measure, because it does not make use of this important information. For interpretation purposes, we nevertheless processed the modularity for the various tests we performed (several tested algorithms use modularity during their processing).

Instead of modularity, we used the normalized mutual information measure (NMI). It was defined in the context of classical clustering to compare two different partitions of one data set [15, 16]. It was shown to be an efficient way to assess the quality of estimated network communities by Danon *et al.* [17].

The measure is derived from a confusion matrix whose element m_{ij} represents the number of nodes put in community i by the considered algorithm, whereas they actually belong to community j. This matrix is usually rectangular, because the algorithm does not necessary find the correct number of communities.

$$I = \frac{-2\sum_i \sum_j m_{ij} \log(nm_{ij}/m_{i+}m_{+j})}{\sum_i m_{i+}\log(m_{i+}/n) + \sum_j m_{+j}\log(m_{+j}/n)} \tag{2}$$

If the estimated communities correspond perfectly to reality, the measure takes the value 1, whereas it is 0 when the estimated communities are independent from the actual ones.

5 Results and Discussion

5.1 Generated Networks

The model from Lancichinetti *et al.* [14] allows controlling most of the network properties: number of nodes, degree distribution, maximal and average degrees and community size distribution. For these properties, we used realistic values in accordance with the literature (cf. the Complex Network Properties section). The question is to know whether the uncontrolled properties (average distance, transitivity, correlation degree), arising from the processing, are realistic too. Furthermore, we would like to know if and how changes in the controlled parameters affect the uncontrolled ones. This is an important matter, because such a change may influence the algorithms' performances, which could therefore be explained either by a direct or an indirect effect. By direct effect, we mean the observed performance modifications are related to the changed controlled properties. By indirect effect, we mean they are related to a change in some uncontrolled properties, caused itself by the change in controlled properties.

In the following, we will discuss separately the relation between each parameter and the uncontrolled properties. The numbers indicated in parenthesis correspond to the processed (Pearson's) correlation values between the considered parameter and uncontrolled property, for 1000 nodes networks.

The variations in the average and maximal degrees have little or no effect on the degree correlation and transitivity coefficient (-0.14 and 0.09, respectively), but there is a direct relation with the average distance (-0.66). Unsurprisingly, it decreases dramatically when the average degree increases, certainly due to the rise in the number of links.

The β parameter has little or no effect on the average path length (0.01) and the transitivity coefficient (0.05), but it relatively affects the degree correlation (0.37). The β parameter controls the homogeneity of the community sizes: when it increases, the communities tend to be more uniform in terms of size [14]. Our interpretation is that with a small beta, we have many small communities with no hubs, much less medium communities with a few hubs, and a few big communities with more hubs.

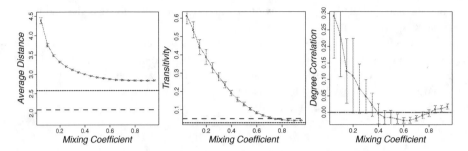

Fig. 1. Influence of the mixing coefficient μ on the properties of the generated networks. The controlled parameters are $n = 1000$, $\langle k \rangle = 30$, $k_{max} = 90$, $\beta = 2$ and $\gamma = 3$. Each point corresponds to an average over **25** generated networks. The dotted and dashed horizontal lines represent the expected values for the same properties in networks generated with the configuration model [34] and Poisson model [12], respectively, using similar parameters.

Medium community hubs have less chance to get linked with other hubs, because there are only a few hubs in their community, and links between communities are rarer, which prevents them to get linked with hubs in other communities. When beta increases, this chance also increases because the number of hubs in the same community gets larger.

The γ parameter has little or no effect on average distance (0.07) and transitivity (−0.06), but it relatively affects the degree correlation (−0.26). When γ increases, the network degree distribution becomes more homogeneous, so this is consistent with the fact that degree correlation is close to zero in Poisson random networks.

The most influent parameter is the mixing coefficient μ, as shown in Fig. 1. The computed correlations are not necessarily high, but the plots show a non-linear relationship between μ and all three uncontrolled properties. As shown on the plot, the average distance decreases when μ increases. However, we performed additional measurements on networks with sizes between 100 and 100000 nodes, and observed a clear logarithmic relationship between the size and the average distance, which is consistent with real networks features. The transitivity is very high for low μ values, but gets down to the level of Poisson random networks when μ reaches 0.7. In the same way, for low μ values, the degree correlation is relatively high, but quickly decreases until μ reaches 0.4 or 0.5, and then stays close to zero. Interestingly, $\mu = 0.5$ corresponds to a limit above which the proportion of external links is higher than the proportion of internal links. In other terms, when μ goes above this limit, the communities are not well defined anymore, and we have a scale-free network with no community structure. Here, we must recall Lancichinetti *et al.* method consists in using the configuration model to generate a scale-free network, which is then partially rewired to create a community structure. For a given node, there are usually many more nodes outside than inside its community. Therefore, the higher μ and the lesser the original network is modified. Put differently: when μ grows, the generated networks get more similar to scale-free networks generated by the configuration model. The configuration model is known to produce networks with no degree correlation [35]. Furthermore, Newman [1] showed that when it is used to generate scale-free networks, for $\gamma > 7/3$ the transitivity tends toward zero as the number of nodes is increasing. Our measures show close to zero degree correlation and transitivity when μ gets close to 1, which is consistent with the previous remarks. The average distance is also close to what is expected from a configuration model-produced network [36]. Using smaller μ values, i.e. defining more distinct communities, makes all three properties grow. The effect on degree correlation could be due to the apparition of hub-to-hub links between communities. The definition of community used here relies on stronger inner density, and is therefore related to the concept of transitivity, which may explain its increase. The disappearance of shortcut links between the communities could explain the observed decrease in average distance.

To conclude these observations, we can state the generated networks show some reasonably realistic properties when μ is relatively small. However, increasing this parameter not only causes communities to become less distinct, but also makes the whole network becoming less realistic, its average distance, transitivity and degree correlation decreasing rapidly.

5.2 Algorithms Performance

The results from the five algorithms are presented in Fig. 2. We can distinguish three kinds of results: Spinglass and Walktrap perform generally very well; Label Propagation also performs well, but is more sensitive to decreases in μ; Eigenvector and Fast greedy are clearly below the others, especially for networks with high degrees. More generally, all the algorithms are sensitive to changes in the average and maximal degree, and have better performances when it increases, as Lancichinetti *et al.* previously noticed on different algorithms [14]. But this sensibility is not the same for all of them, as we can observe different decreases in performance when μ is increasing. This general sensitivity to μ is not surprising, since an increase in μ means the communities are vanishing. Spinglass and Walktrap are the most robust, with NMI results remaining at 1 until they suddenly drop between $\mu = 0.6$ and 0.8 for the two higher values of $\langle k \rangle$ and k_{max} (last two rows). For the lower degrees values (first row), the decrease is more regular and starts from $\mu = 0.05$.

For Eigenvector and Fast greedy, the performance drop takes place sooner, and is almost linear starting from $\mu = 0.05$, with all three tested degree values. Label Propagation behavior is apart: it performs almost as well as Spinglass and Walktrap, but its performance drop happens sooner, between $\mu = 0.5$ and 0.6, and is more sudden.

When observing the joint effect of the mixing coefficient and the average and maximum degrees on the performance, it is interesting to observe the reversal taking place around $\mu = 0.75$, as illustrated by Fig. 3 for Walktrap and Spinglass. Below this limit, the higher the degree and the better the performance. Above this limit, the lower the degree and the better the performance. This means high density helps discovering community structure when it is strong, whereas it hides it when it is weak. But as we stated in the previous section, the generated networks become less realistic when μ increases, so the observed change in performance could actually be caused not directly by the degree variations, but by consequent decreases in the transitivity or degree correlation.

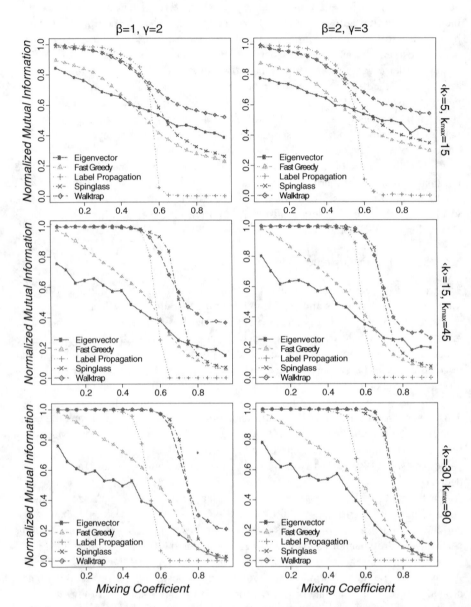

Fig. 2. Comparison of the five algorithms results for $n = 1000$. On the left column: $\beta = 1$ and $\gamma = 2$, on the right column $\beta = 2$ and $\gamma = 3$. On the first row $\langle k \rangle = 5$ and $k_{max} = 15$, on the second one $\langle k \rangle = 15$ and $k_{max} = 45$, and on the third one $\langle k \rangle = 30$ and $k_{max} = 90$. Each point corresponds to an average over **25** generated networks.

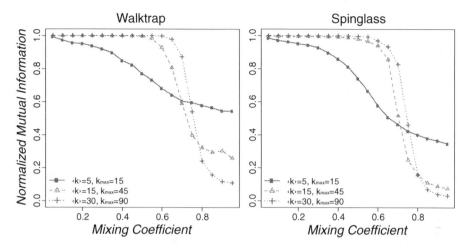

Fig. 3. Joint influence of the mixing coefficient μ and the average degree $\langle k \rangle$ on the performance of two algorithms: Walktrap on the left and Spinglass on the right. Each point corresponds to an average over **25** generated networks, with $\beta = 2$ and $\gamma = 3$.

The β and γ parameters do not seem to affect any of the algorithms (correlation smaller than 0.06 for all five), except for Walktrap, for which it looks like γ has an effect similar to the degree effect observed before. In other terms, the scale-free property makes it easier for Walktrap to discover communities when the community structure is strong, but makes it more difficult when it is weak.

6 Conclusion

In this paper, we compared five different community detection algorithms. We used a set of artificial networks generated with the model defined by Lancichinetti *et al.* [14], which allows randomly producing networks with a community structure and power-law degree and community size distributions. To our knowledge, this type of comparative study was never conducted on such realistic networks before. We used the normalized mutual information measure [15-17] to assess the performance of the algorithms. Our results show that among the Fast Greedy [10, 24], Walktrap [29], Eigenvector [30], Label Propagation [11] and Spinglass [31] algorithms, Walktrap and Spinglass get generally the best results. They succeed in identifying the communities even for high mixing coefficient values. Label Propagation has also excellent results, but its performance drop happens before Spinglass and Walktrap. Fast greedy and Eigenvector are clearly outclassed by all three other algorithms.

After having analyzed the data, we concluded the mixing coefficient and average and maximum degrees have a strong joint effect on the algorithms results. A higher density tends to improve community finding when the communities are distinct, but makes it harder to find them when the community structure is weak. In these algorithms and in the modularity measure, a community is defined as a subset of nodes densely interconnected relatively to the rest of the network. This definition does not hold anymore when $\mu > 0.5$, which means above this limit, the network structure does not reflect the community structure. In other terms, the information conveyed by the network links is not pertinent anymore, and this can explain the observed joint effect. Moreover, increases in the mixing coefficient also make the networks becoming less realistic, which could as well be a cause for the observed drop in performance.

Our work can be seen as a first attempt at comparing community detection algorithms, and can be extended in several ways. The generative model we used is more realistic than earlier ones, but we observed that increasing the mixing coefficient μ makes the produced networks less realistic (strong decrease in the average distance, degree correlation and transitivity). We suppose this is due to the use of the configuration model [34] by Lancichinetti et al. [14] to produce an initial network, which is then modified to create the community structure. Maybe this could be corrected by using another model instead, such as preferential attachment [37] (or one of its variations), able to generate networks with more realistic properties. Of course there is no certainty about whether or not these properties would resist the modifications performed on the initial network.

We only considered a few properties to analyze the artificially generated networks, and some additional properties, maybe more community-oriented (see [19]) could be used to have a more precise idea of their realism. Moreover, real networks properties are usually described commonly, but there may be strong differences between the various types of real networks such as social networks, biological networks, information networks, etc. [1]. In that case, a proper test should compare algorithms on different types of corresponding artificial networks.

We compared the algorithms on networks containing only 1000 nodes. Real networks are generally much bigger, in the order of tens of thousands or millions of nodes. For more significance, the algorithms should be tested on this type of networks, but this raises two problems: 1) processing community detection on such huge networks is significantly more time expensive, and 2) determining realistic average and maximal degrees is difficult because of the heterogeneity observed in real networks for these properties. The second point is important, since we observed the performance of a given algorithm could vary strongly in function of these sole properties. We also limited this comparison to the fastest algorithms, again for computability and time reasons. A proper exhaustive test should consider more expensive algorithms (see [17, 19]).

Finally, we sometimes observed extreme disagreements between the final modularity measure processed by the various algorithms and the information measure corresponding to their performance. It should be interesting to process the optimal modularity over all the tested algorithms and to study how it evolves relatively to the networks properties and the measured performances.

Acknowledgments. The authors would like to thank the TÜBİTAK (Scientific and Technological Research Council of Turkey) for its support, and the anonymous referees for their valuable comments.

References

1. Newman, M.E.J.: The structure and function of complex networks. SIAM Review 45, 167–256 (2003)
2. Newman, M.E.J., Girvan, M.: Finding and evaluating community structure in networks. Physical Review E 69, 026113 (2004)
3. Duch, J., Arenas, A.: Community detection in complex networks using extremal optimization. Physical Review E 72, 027104 (2005)
4. Girvan, M., Newman, M.E.J.: Community structure in social and biological networks. Proceedings of the National Academy of Sciences of the United States of America 99, 7821–7826 (2002)
5. Palla, G., Derenyi, I., Farkas, I., Vicsek, T.: Uncovering the overlapping community structure of complex networks in nature and society. Nature 435, 814–818 (2005)
6. Reichardt, J., Bornholdt, S.: Detecting Fuzzy Community Structures in Complex Networks with a Potts Model. Physical Review Letters 93, 218701 (2004)
7. Falkowski, T., Barth, A., Spiliopoulou, M.: DENGRAPH: A Density-based Community Detection Algorithm. In: IEEE/WIC/ACM International Conference on Web Intelligence, pp. 112–115 (2007)
8. Liu, Y., Wang, Q., Wang, Q., Yao, Q., Liu, Y.: Email Community Detection Using Artificial Ant Colony Clustering. In: Chang, K.C.-C., Wang, W., Chen, L., Ellis, C.A., Hsu, C.-H., Tsoi, A.C., Wang, H. (eds.) APWeb/WAIM 2007. LNCS, vol. 4537, pp. 287–298. Springer, Heidelberg (2007)
9. Hoff, P., Raftery, A., Handcock, M.: Latent space approaches to social network analysis. Journal of the American Statistical Association 97, 1090–1098 (2002)
10. Newman, M.E.J.: Fast algorithm for detecting community structure in networks. Physical Review E 69, 066133 (2004)
11. Raghavan, U.N., Albert, R., Kumara, S.: Near linear time algorithm to detect community structures in large-scale networks. Phys. Rev. E 76, 036106 (2007)
12. Erdos, P., Renyi, A.: On random graphs. Publicationes Mathematicae 6, 290–297 (1959)
13. Danon, L., Diaz-Guilera, A., Arenas, A.: The effect of size heterogeneity on community identification in complex networks. Journal of Statistical Mechanics-Theory and Experiment 11010 (2006)
14. Lancichinetti, A., Fortunato, S., Radicchi, F.: Benchmark graphs for testing community detection algorithms. Phys. Rev. E Stat. Nonlin Soft. Matter Phys. 78, 46110 (2008)
15. Kuncheva, L.I., Hadjitodorov, S.T.: Using diversity in cluster ensembles. In: IEEE International Conference on Systems, Man and Cybernetics, vol. 2, pp. 1214–1219 (2004)
16. Fred, A.L.N., Jain, A.K.: Robust Data Clustering. In: IEEE Computer Society Conference on Computer Vision and Pattern Recognition, vol. 2, pp. 128–136. IEEE Computer Society, Los Alamitos (2003)
17. Danon, L., Díaz-Guilera, A., Duch, J., Arenas, A.: Comparing community structure identification. J. Stat. Mech P09008 (2005)
18. Holme, P., Kim, B.J., Yoon, C.N., Han, S.K.: Attack vulnerability of complex networks. Physical Review E 65, 026139 (2002)
19. Boccaletti, S., Latora, V., Moreno, Y., Chavez, M., Hwang, D.: Complex networks: structure and dynamics. Physics Reports 424, 175–308 (2006)
20. Milo, R., Shen-Orr, S., Itzkovitz, S., Kashtan, N., Chklovskii, D., Alon, U.: Network Motifs: Simple Building Blocks of Complex Networks. Science 298, 824–827 (2002)
21. Watts, D., Strogatz, S.H.: Collective dynamics of 'small-world' networks. Nature 393, 409–410 (1998)

22. Barabasi, A., Albert, R.: Statistical mechanics of complex networks. Reviews of Modern physics 74, 47–96 (2002)
23. Guimerà, R., Danon, L., Díaz-Guilera, A., Giralt, F., Arenas, A.: Self-similar community structure in a network of human interactions. Physical Review E 68, 065103 (2003)
24. Clauset, A., Newman, M.E.J., Moore, C.: Finding community structure in very large networks. Physical Review E 70, 066111 (2004)
25. Cohen, J.: A Coefficient of Agreement for Nominal Scales. Educational Psychology Measurement 20, 37–46 (1960)
26. Fortunato, S., Barthelemy, M.: Resolution limit in community detection. Proceedings of the National Academy of Science of the USA 104, 36–41 (2007)
27. Guimerà, R., Sales-Pardo, M., Amaral, L.A.N.: Modularity from fluctuations in random graphs and complex networks. Physical Review E 70, 025101 (2004)
28. Derenyi, I., Palla, G., Vicsek, T.: Clique percolation in random networks. Physical Review Letters 94 (2005)
29. Pons, P., Latapy, M.: Computing communities in large networks using random walks. In: Yolum, p., Güngör, T., Gürgen, F., Özturan, C. (eds.) ISCIS 2005. LNCS, vol. 3733, pp. 284–293. Springer, Heidelberg (2005)
30. Newman, M.E.J.: Finding community structure in networks using the eigenvectors of matrices. Phys. Rev. E 74, 036104 (2006)
31. Reichardt, J., Bornholdt, S.: Statistical mechanics of community detection. Phys. Rev. E 74, 016110 (2006)
32. Radicchi, F., Castellano, C., Cecconi, F., Loreto, V., Parisi, D.: Defining and identifying communities in networks. Proceedings of the National Academy of Sciences of the United States of America 101, 2658–2663 (2004)
33. Donetti, L., Munoz, M.A.: Detecting network communities: a new systematic and efficient algorithm. Journal of Statistical Mechanics: Theory and Experiment P10012 (2004)
34. Molloy, M., Reed, B.: A critical point for random graphs with a given degree sequence. Random Structures and Algorithms 6, 161–179 (1995)
35. Serrano, M., Boguñá, M.: Weighted Configuration Model. In: AIP Conference Proceedings, vol. 776, p. 101 (2005)
36. Chung, F., Lu, L.: The average distances in random graphs with given expected degrees. PNAS 99, 15879–15882 (2002)
37. Barabási, A.-L., Albert, R.: Emergence of Scaling in Random Networks. Science 286, 509–512 (1999)

Towards an Ontology of Data Mining
Investigations

Pan" Panov[1], Larisa N. Soldatova[2], and Sašo Džeroski[1]

[1] Jožef Stefan Institute, Jamova cesta 39, SI-1000 Ljubljana, Slovenia
{Pance.Panov,Saso.Dzeroski}@ijs.si
[2] Aberystwyth University, Penglais, Aberystwyth, SY23 3DB, Wales, UK
lss@aber.ac.uk

Abstract. Motivated by the need for unification of the domain of data
mining and the demand for formalized representation of outcomes of
data mining investigations, we address the task of constructing an ontology of data mining. In this paper we present an updated version of the
OntoDM ontology, that is based on a recent proposal of a general framework for data mining and it is aligned with the ontology of biomedical
investigations (OBI) . The ontology aims at describing and formalizing
entities from the domain of data mining and knowledge discovery. It includes definitions of basic data mining entities (e.g., datatype, dataset,
data mining task, data mining algorithm etc.) and allows extensions with
more complex data mining entities (e.g. constraints, data mining scenarios and data mining experiments). Unlike most existing approaches to
constructing ontologies of data mining, OntoDM is compliant to best
practices in engineering ontologies that describe scientific investigations
(e.g., OBI) and is a step towards an ontology of data mining investigations. OntoDM is available at:http://kt.ijs.si/panovp/OntoDM/.

1 Introduction

Traditionally, ontology has been defined as the philosophical study of what exists: the study of kinds of entities in reality, and the relationships that these
entities bear to one another [21]. In recent years use of term ontology has become prominent in the area of computer science research and the application
of computer science methods in management of scientific and other kinds of information. In this sense the term ontology has the meaning of a standardized
terminological framework in terms of which the information is organized.

The ontological problem is adopting a set of basic categories of objects, determining what kinds of entities fall within each of these categories of objects, and
determining what relationships hold within and among different categories in
the ontology. The ontological problem for computer science is identical to many
of the problems in philosophical ontology, and the success of constructing such
an ontology is achievable by applying methods, insights and theories of philosophical ontology. When one sets out to construct an ontology then, what one
is doing is designing a representational artifact that is intended to represent the

J. Gama et al. (Eds.): DS 2009, LNAI 5808, pp. 257–271, 2009.

universals and relations amongst universals that exist, either in a given domain of reality (e.g data mining domain) or across such domains.

The engineering of ontologies is still a relatively new research field and some of the steps in ontology design remain manual and can be considered as an art by itself. Recently there was a significant progress in automatic ontology learning [14], application of text mining [17], and ontology mapping [13]. However the construction of a good quality ontology with the use of automatic and even semi-automatic techniques still requires manual definition of the key upper level entities of the domain of interest. Good practices in ontology development are: following an upper level ontology as a template, the use of formally defined relations between the entities and not allowing multiple inheritances [25].

In the domain of data mining and knowledge discovery, researchers have tried to construct ontologies describing data mining entities that were targeted to solve specific problems. Most of the developments are with the aim of automatic planning of data mining workflows [1,30,11,8]. Some of the developments are concerned with description of data mining services on the GRID [6,5].

Current proposals for ontology of data mining are not based on upper level categories nor have used a predefined set of relations based on a upper level ontology. Most of the semantic representations for data mining proposed so far are based on so called light-weight ontologies [15]. Light-weight ontologies are often shallow, without rigid relations between the defined entities, but they are relatively easy to develop by semi/automatic methods and they still greatly facilitate computer applications. The reason why these type of ontologies are more frequently developed then heavy-weight ontologies is that process of development is more difficult and time consuming. In contrast to many other domains, data mining requires elaborate inference over its entities, and hence requires rigid heavy-weight ontologies with the aim of improving the KDD (Knowledge Discovery in Databases) process and providing support for development of new data mining approaches and techniques.

While KDD and data mining have enjoyed great popularity and success in recent years, there is a distinct lack of a generally accepted framework that would cover and unify the data mining domain. The present lack of such a framework is perceived as an obstacle to the further development of the field. In [29], Yang and Wu collected the opinions of a number of outstanding data mining researchers about the most challenging problems in data mining research. Among the ten topics considered most important and worthy of further research, the development of an unifying framework for data mining is listed first. One step towards developing a general framework for data mining is constructing an ontology of data mining.

In this paper we propose an extended and updated version of the ontology of data mining named OntoDM. Our ontology design takes into consideration the best practices in ontology engineering. We use an upper level ontology BFO (Basic Formal Ontology)[1] to define the upper level classes, the OBO Relational

---

[1] BFO: http://www.ifomis.org/bfo

Ontology (RO)[2] to define the semantics of the relationships between the data mining entities, and provide is-a completeness and single is-a inheritance for all DM entities. We also developed our ontology in the most general fashion in order to be able to represent the complex entities in data mining that are becoming more and more popular research areas such as mining structured data and constraint-based mining.

In previous work [16] we presented an initial version of OntoDM sufficient for the representation of data mining tasks and complex data types. The ontology is based on the proposal for a general framework for data mining presented in [9]. The initial version of OntoDM was using the philosophy of Ontology of Scientific Experiments (EXPO) [26] and ontology of biomedical investigations (OBI)[3] for identification and organization of entities in a *is-a* class hierarchy.

The version described in the current paper has been sufficiently updated in several ways. First, the structure of the ontology was aligned with the top level structure of the OBI ontology. This procedure requested revising the representation of some data mining entities and also introduced new entities in the ontology (e.g., the entity data mining algorithm was split into three entities each capturing different dimension of a description; algorithm specification, algorithm implementation and algorithm description). Second, we extended the set of relations used in the initial version with relations defined in the OBI ontology in order to express the relations between informational entities, entities that are realized in a process and processes. Finally, we extended the OBI classes with data mining specific classes for describing complex entities (e.g., data mining scenarios, queries).

The rest of the paper is structured as follows: Section 2 provides the background for this work. Section 3 presents the ontology design principles and we provide a detailed description of the alignment with OBI ontology and description of upper level classes and relations. Section 4 presents an example of representation of a data mining algorithm in OntoDM based on the alignment with OBI ontology and Section 5 discusses the representation of complex data mining entities. In Section 6 we give a roadway for future research and development of the ontology.

2   Background

2.1   Motivation

The motivation for developing an ontology of data mining is multi-fold. Firstly, as it was mentioned in the introduction, the area of data mining is developing rapidly and one of the most challenging problems deals with developing a general framework for data mining. By developing an ontology of data mining we are taking one step towards solving this problem. The ontology would define and formalize what are the basic entities (e.g., dataset, data mining algorithm) in

---

[2] RO: http://www.obofoundry.org/ro/
[3] OBI: http://obi-ontology.org/

data mining and define the relations between the entities. After the basic entities are identified and defined, we can build upon them and define more complex entities (e.g. data mining query, data mining scenario and experiment). All the defined data mining entities organized in the form of an ontology would be a backbone of the systems for automated data mining.

Secondly, there exist several proposals for ontologies of data mining but all of them are light-weight, aimed at covering a particular use-case in data mining, are of a limited scope and highly use-case dependent. Data mining is a domain that needs a heavy-weight ontology with a broader scope, where much attention is paid to the rigorous meaning of each entity, semantically rigorous relations between entities and compliance to an upper level ontology and the domains of application (e.g., biology, environmental sciences).

Finally, an ontology of data mining should define what is the minimum information required for the description of a data mining investigation. Biology is leading the way in developing standards for recording and representation of scientific data and biological investigations (e.g., already more than 50 journals require compliance of papers reporting microarray experiments to the Minimum Information About a Microarray Experiment - MIAME standard). The researchers in the domain of data mining should follow this good practice and the ontology of data mining would support development of standards for performing and recording of data mining investigations.

2.2   State-of-the-Art

Formalizing scientific investigations. In recent years, there is an increased need for formalized representations of the domain of data mining and formal representation of outcomes of research in general. There exist several formalisms for describing scientific investigations and outcomes of research. In this part we will focus on two proposals that are relevant for describing data mining investigations: Ontology for Biomedical Investigations (OBI) and Ontology of Scientific Experiments (EXPO).

Ontology of biomedical investigations - OBI. The OBI(`http://obi-ontology.org/`) ontology aims to provide a standard for the representation of biological and biomedical investigations. OBI is developed through collaboration of 19 biomedical communities (transcriptomics, proteomics, metabolomics, etc.). They are developing a set of universal terms that are applicable across various biological and technological domains and domain specific terms relevant only to a given domain. The ontology supports consistent annotation of biomedical investigations regardless of particular field of the study. It aims to represent design of an investigation, the protocols and used instrumentation, used materials, generated data and type of analysis performed on it.

The OBI ontology employs rigid logic and semantics as it uses an upper level ontology BFO and the RO relations to define the top classes and a set of relations. OBI defines occurrences (processes) and continuants (materials, instruments, qualities, roles, functions) relevant to biomedical domains. OBI is fully compliant with the existing formalisms in biomedical domains. OBI is a part of OBO Foundry [22] which requires all member ontologies follow the same design principles, the same set of relations, the same upper ontology, and to define a single class only once within OBO to facilitate integration and automatic reasoning.

The Data Transformation Branch is an OBI branch with the scope of identifying and ontologising entities and relations to describe processes which produce output data given some input data, and the work done by this branch is related to the proposal presented in this paper.

Ontology of experiments EXPO and LABORS. The formal definition of experiments for analysis, annotation and sharing of results is a fundamental part of science practice. A generic ontology of experiments EXPO [26] tries to define the principal entities for representation of scientific investigations. The EXPO ontology is of a general value in describing experiments from various areas of research. This was demonstrated with the use of the ontology for the description of high-energy physics and phylogenetics investigations [26].The ontology uses a subset of SUMO[4] suitable for scientific representations as an upper level ontology and a minimized set of relations in order to provide compliance with the existing formalisms. An ontology LABORS is an extension of EXPO for the description of automated investigations (the Robot Scientist Project [5]).

LABORS defines such research units as investigation, study, test, trial, replicate which are required for the description of complex multilayered investigations carried out by a robot. For example an investigation resulted in a fully automatic discovery of new gene functions consists of >10,000 such research units [12]. LABORSs logical definions of the research units properties, hypotheses, results, conclusions and data base of the experimental observations and results are translated into datalog for the reasoning over all data and metadata.

Ontology of experiment actions - EXACT. An ontology of experiment actions (EXACT) [24] aims to provide a structured vocabulary of concepts for the description of protocols in biomedical domains. The main contribution of this ontology is the formalizing biological laboratory protocols in order to enable repeatability and reuse of already published experiment protocols. This work is related with the descriptions of data mining scenarios and workflows.

Describing data mining entities. Main developments in description of data mining entities in a form of an ontology are in the area of semi automatic data mining workflow construction and description of data mining services and resources on the GRID. Other research includes description of machine learning experiments in context of experiment databases and identification of entities using collection of data mining literature. We will briefly describe all the mentioned approaches.

---

[4] SUMO: http://www.ontologyportal.org/

[5] http://www.aber.ac.uk/compsci/Research/bio/robotsci/

Describing data mining workflows. In [1] the authors propose a prototype of an Intelligent Discovery Assistant (IDA) which provides users with systematic enumerations of valid data mining processes (sequences of data mining operators) and effective rankings of the processes by different criteria, in order to facilitate the choice of data mining processes to execute to solve a concrete data mining task. This automated system takes the advantage of an explicit ontology of data mining operators (algorithms). The ontology that is designed is a lightweight ontology that contains only a hierarchy of data mining operators divided into three main classes: preprocessing operators, induction algorithms and post processing operators. The leaves of the hierarchy are the actual operators. The ontology does not contain information about the internal structure of the operators and the taxonomy is produced only according to the role that the operator has in the knowledge discovery process.

In [11] the authors build upon the work presented in [1] and propose an intelligent data mining assistant that combines planning and meta-learning for automatic design of data mining workflows. A knowledge driven planner relies on a knowledge discovery ontology [1], to determine the valid set of operators for each step in the workflow. The probabilistic meta-learner is proposed for selecting the most appropriate operators by using relational similarity measures and kernel functions based on past data mining experiments.

The work in [30] also addresses the problem of semiautomatic design of workflows for complex knowledge discovery tasks. The idea is to automatically propose workflows for the given type of inputs and required outputs of the discovery process. This is done by formalizing the notions of a knowledge type and data mining algorithm in the form of an ontology. The planning algorithm accepts task descriptions expressed using the vocabulary of the ontology.

Describing data mining services and resources. In [5] the authors introduce an ontology-based framework for automated construction of complex interactive data mining workflows as a means of improving productivity of GRID-enabled data systems. For this purpose they develop a data mining ontology which is based on concepts from industry standards like: predictive model mark-up language (PMML)[6], WEKA [28] and Java data mining API.

In the context of GRID programming in [6] the authors propose a design and implementation of an ontology of data mining. The motivation for building the ontology comes from the context of the author's work in Knowledge GRID [7]. The main goals of the ontology are to allow the semantic search of data mining software and other data mining resources and to assist the user by suggesting the software to use on the basis of the user's requirements and needs. The proposed DAMON (DAta Mining ONtology) ontology is built through a characterization of available data mining software.

In [8] the authors introduce a semantic based, service oriented framework for tools sharing and reuse, in order to give support for the semantic enrichment through semantic annotation of KDD tools and deployment of tools as web

---

[6] http://www.dmg.org/

services. For describing the domain the authors propose an ontology named KD-DONTO which is developed having in mind the central role of a KDD algorithm and their composition similar to work in [1,30].

Experiment databases. As data mining and machine learning are experimental sciences, lot of insight of the performance of a particular algorithm is obtained by implementing it and studying how it behaves on different datasets. In [2,3] the authors propose an experimental methodology based on experiment database in order to allow repeatability of experiments and generalizability of experimental results in machine learning. In [27] the authors propose an XML based language for describing classification and regression experiments. In this process the authors identified the main entities for describing a machine learning experiment, which is the first step towards including the experimental entities in an ontology.

Identification of data mining entities using collections of DM literature. In [18] the authors survey a large collection of data mining and knowledge discovery literature in order to identify and classify the data mining entities into high-level categories using grounded theory approach and validating the classification using document clustering. As a result of the research study the authors have identified eight main areas of data mining and knowledge discovery: data mining tasks, learning methods and tasks, mining complex data, foundations of data mining, data mining software and systems, high-performance and distributed data mining, data mining applications and data mining process and project.

3   OntoDM Design and Description

Our ontology of data mining (OntoDM) aims to provide a structured vocabulary of entities sufficient for the description of data mining scenarios and workflows. OntoDM aims to follow the OBO Foundry principles[7] in ontology engineering that are widely accepted in the biomedical domains. The main OBO Foundry principles state that "the ontology is open and available to be used by all", "is in a common formal language", "includes textual definition of all terms", "uses relations which are unambiguously defined", "is orthogonal to OBO ontologies" and "follows a naming convention" [20]. In this way, OntoDM will be built on a sound theoretical foundation, will be compliant with other (e.g., biological) domains and can be widely re-usable. Our ontology intends to be compatible with other formalisms, to share and reuse already formalized knowledge. OntoDM is available at: http://kt.ijs.si/panovp/OntoDM/.

OntoDM is expressed in OWL-DL and is being developed using the Protege ontology editor[8]. It consists of three main components: classes, a hierarchical structure (*is-a* relations) of classes and relations (other than *is-a* relations) between instances. All three major components are described in the following subsections.

---

[7] OBO Foundry: http://ontoworld.org/wiki/OBO_foundry
[8] Protege: http://protege.stanford.edu

3.1   Identifying Basic Data Mining Entities

OntoDM is based on the proposal of a general framework for data mining by Džeroski [9]. From the framework proposal we identified a set of basic entities of data mining. The basic entities identified are the following (please consult [9] for a detailed description of the entities):

- dataset, which consists of data items;
- datatype, which can be primitive (nominal, boolean, numeric), or structured (set, sequence, tree, graph);
- data mining task, which includes predictive modeling, pattern discovery, clustering and probability distribution estimation;
- generalization, the output of a data mining algorithm, which can be: predictive model, pattern, clustering, probability distribution;
- data mining algorithm, which solves a data mining task and produces generalizations from a dataset and includes components of algorithms such as: distance function, kernel function, refinement operator;
- function, which can be: an aggregation function, prototype function, evaluation function, cost function etc;
- constraint, which include evaluation and language constraint (hard constraint, soft constraint, optimization constraint) and
- data mining scenarios, related to queries and inductive queries.

The entities listed above are used to describe different dimensions of data mining. These are all orthogonal dimensions and different combinations among these should be facilitated. Through combination of these basic entities, one should be able to describe most of the diversity present in data mining approaches today.

3.2   Upper Level Concepts

In the initial version of the ontology [16] the structure was grounded by the following upper level classes: <*informational entity*>, <*agregate*>, <*procedure*>, <*process*>, <*quality*>, <*representation*> and <*role*>.

In this version of the ontology we mapped the entities more closely to the structure of the OBI ontology. We use BFO upper level classes to represent entities which exist in the real world (i.e., processes, informational entities created in human brain), and in addition we use extensions of EXPO <*abstract entity*> to represent mathematical entities. Recently, due to the limitations of BFO in dealing with information, an Information Artifact Ontology (IAO) has been proposed as a spin-off of the OBI project[9]. Currently IAO is available only in a draft version, but we have included the most stable and relevant classes into OntoDM.

Figure 1 shows the part of the OntoDM class hierarchy. The OntoDM ontology contains 292 classes (including imported upper level classes), and all of the OntoDM classes are extensions of the upper level classes from BFO, OBI, IAO, and EXPO.

---

[9] IAO:http://code.google.com/p/information-artifact-ontology/

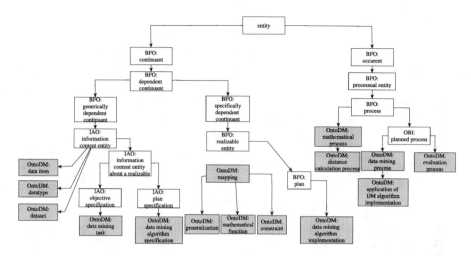

Fig. 1. Part of the OntoDM class hirearchy (*is-a* hierarchy): OntoDM classes are extensions of BFO, OBI, IAO and EXPO top level classes

3.3   Alignment of OntoDM with OBI

Information content entity. The class <*information content entity*> was recently introduced into OBI and denotes all entities that are generically dependent on some artifact and stand in relation of aboutness to some entity. In the domain of data mining we have identified and extended the <*information content entity*> class with the following sub-classes: <*datatype*>, <*data item*> and others. The class <*dataset*> is an information content entity that is an aggregate of data items.

Realizable entity and information entity about a realizable. Realizable entities include all entities that can be executed (manifested, actualized, realized) in concrete occurrences (e.g processes). Realizable entities are entities of a type whose instances are typically such that in the course of their existence they contain periods of actualization, when they are manifested through processes in which their bearers participate.

We have identified and extended the class <*realizable entity*> and its sub-classes <*plan*>, <*role*>, <*function*> with data mining specific entities. Basic realizable data mining classes are: <*generalization*>, <*data mining algorithm implementation*>, <*constraint*>, <*mathematical function*>, <*query*>, <*data mining scenario*>. Here we just briefly describe <*generalization*> and <*data mining algorithm implementation*>.

The class <*generalization*> represent entities that are products of a data mining process (e.g., the application of a data mining algorithm implementation on a concrete dataset with concrete parameter settings) and includes entities:<*predictive model*>, <*pattern*>, <*clustering*> and <*probability distribution*>. These entities are realized in the <*generalization interpretation process*> where an input to a process is a <*data item*> and the output is a result of applying of the generalization to the data item (e.g.,the prediction of a predictive models).

The class <*data mining algorithm implementation*> is a subclass of the class <*plan*>. It describes a concrete implementation of a <*data mining algorithm specification*>, subclass of <*plan specification*> and is realized though a data mining process <*application of data mining algorithm*>.

Information entities that concern a realizable entity include: objective specification, plan specification, action specification, etc. A plan specification includes parts such as: objective specification, action specifications and conditional specifications. When concretized, it is executed in a process in which the bearer tries to achieve the objectives, in part by taking the actions specified. An objective specification describes an intended process endpoint.

We have identified and extended the <*information entity about a realizable*> and its subclasses, <*objective specification*> and <*plan specification*>, with data mining specific entities. Basic information entities about a realizable are: <*data mining task*>, subclass of <*objective specification*>, and <*data mining algorithm specification*>, which is a subclass of <*plan specification*>.

Process. Process entities represent occurrences that have a specified beginning and end. A planned process is the realization of a plan borne by an agent that initiates this process in order to bring about the objective(s) specified as part of the plan specification. Process entities have as participants continuants and can be also performed by an agent. In the case of data mining, processes have inputs and outputs that can be informational entities and realizable entities. We have identified and extended the <*process*> and <*planned process*> classes with data mining specific classes. Basic data mining process entities described in our ontology include: <*application of a data mining algorithm implementation*>, <*evaluation process*>, <*distance function calculation*> etc.

3.4   Ontological Relations

The consistent use of rigorous definitions to characterize formal relations is a major step towards enabling the achievement of interoperability among ontologies in support of automated reasoning across data derived from multiple domains. For, if a fruitful exchange of information to be possible between such ontologies and the data annotated with their terms, each of the system involved must treat the relations in the same way. A relational expression must always stand for one and the same relation, even if it is used in multiple ontologies.

The OntoDM ontology includes and different types of formaly defined ontological relations in order to achieve the desired level of expressiveness. The initial version of the ontology [16] included: fundamental relations (*is-a, part-of*), relations from RO [23] *has-participant, has-agent*, relations from EXPO/LABORS [26] (*has-representation*), relations from EXACT[24](*has-information*) and relations from OBI (*has-role, has-quality, has-specified-input,has-specified-output*).

The fundamental relations *is-a* and *has-part* are used to express subsumption and part-whole relationships between entities. The relations *has-participant* and *has-agent* express the relationship between a process and participants in a process, that can be passive or active. Other relations, *has-specified-input* and *has-specified-output*, are specific for relating data mining processes with special types of participants that are inputs and outputs of the data mining process. These two relations have been recently introduced in the OBI ontology.

The relation between an entity and a dependent continuant is expressed via the relation *bearer-of* (defined in the OBI ontology) and this relation is more general and replaces the relations *has-role* and *has-quality* used in the inital version of the ontology.

For expression of informational properties of entities we are using the relation *has-information* and for expression of a representational properties of entities we use the relation *has-representation*, both defined in the EXAT and EXPO/LABORS ontologies.

In this version of the ontology we include relations for expressing relationships between: a process and realizable entity (*realizes*), a planned process and objective specification (*achieves-planned-objective*) and informational entity about a realizable and a realizable entity (*is-concretized-as*). These relations are defined in the OBI ontology.

4   The Example Representation of a Data Mining Algorithm

In this section we give an example of the representation of a concrete algorithm using the OntoDM ontology terms (see Figure 2). We describe how to represent the well known C4.5 algorithm [19] for learning decision tree predictive models and its concrete implementation in the WEKA data mining system [28].

When describing a data mining algorithm, one has to have in mind three different aspects. First aspect is the data mining algorithm specification, e.g. *<c45 algorithm specification>*, which is a subclass of the *<information entity>* class about a realizable entity that describes declarative aspects of an algorithm, e.g. has as a part *<predictive modeling>* information about a data mining task in hand. The second aspect is the concrete implementation of an algorithm, e.g. *<wekaJ48 algorithm implementation>*, which is a realizable entity. The third aspect is the process aspect where we describe an application of a concrete data mining algorithm (e.g *<application of wekaJ48>*) on a dataset under concrete algorithm parameter settings. It is necessary to have all three aspects represented separately in the ontology as they have distinctly different nature and this will facilitate different usage of the ontology. The process aspect can be used for constructing data mining workflows and definition of participants of workflows and its parts; the specification aspect can be used to reason about components of data mining algorithms; the implementation aspect can be used for search over implementations of data mining algorithms and to compare various implementations.

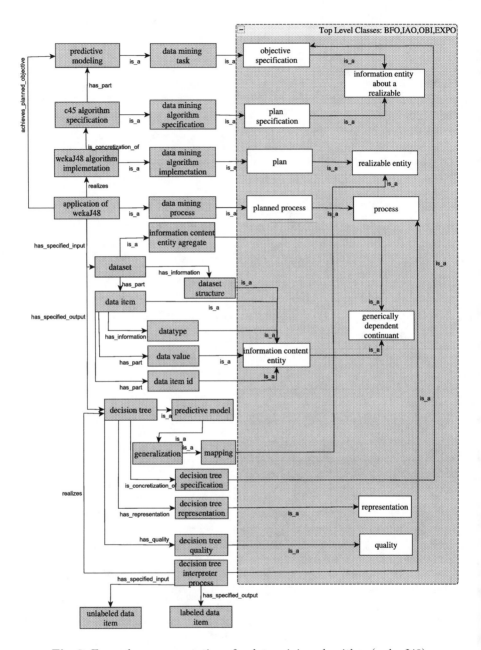

Fig. 2. Example - representation of a data mining algorithm (weka.J48)

The relations between the classes representing different aspects of data mining algorithm are as follows:

<wekaJ48 alg. implementation> is-concretization-of <c45 alg. specification>
<wekaJ48 application > realizes <wekaJ48 alg. implementation>
<wekaJ48 application > achieves planned objective <predictive modeling>

Figure 2 presents the process aspect of a data mining algorithm in more detail. Each process has defined input and output entities which are linked to the process via *has-specified-input* and *has-specified-output* relations correspondingly. An input to an application of data mining algorithm is a dataset and parameter values and as output we get a generalization (e.g., <decision tree>). A dataset has as parts data items that are characterized with a datatype (e.g., <tuple of primitives>). In the case of propositional learning, the datatype of data items is a tuple of primitive data types (nominal values, numeric values, boolean values). A generalization entity has also two aspects. One is connected with looking at it as a data structure and in that case we have a generalization specification (e.g. <decision tree specification>) and generalization representation (e.g. <decision tree representation>). Another aspect is the functional aspect, when we apply a concrete generalization to a new data item (e.g., prediction using a decision tree). In this case a generalization is realized through a generalization interpreter process (e.g. <decision tree interpreter process>) where the input to the process is an unlabeled data item and the output is a labeled data item.

5   Complex Data Mining Entities in OntoDM

Our proposal for an ontology of data mining includes descriptions of basic data mining entities. These basic entities are to define more complex entities e.g., entities from the area of inductive databases. The concept of an inductive database [10] employs a database perspective on knowledge discovery, where the knowledge discovery process is composed of query sessions. In this case ordinary queries can be used to access and manipulate the data, while inductive queries (data mining queries) can be used to generate (mine), manipulate and apply generalizations.

Real life applications of data mining typically require interactive sessions and involve formulation of a complex sequence of inter-related inductive queries, which we call a KDD scenario [4]. KDD scenarios can be described at different level of detail and precision and can serve multiple purposes. At the most detailed level of description, KDD scenarios can serve to document the exact sequence of data mining operations undertaken by a human analyst on a specific task. At higher level of abstraction, the scenarios enable the re-use of already performed analyses,e.g., on a new dataset of the same type. The explicit storage and manipulation of scenarios would greatly facilitate the KDD process in whole. Our proposed ontology can be used for formalizing and describing KDD scenarios at various levels of abstraction.

6   Conclusion and Further Work

In this paper we present updated and modified version of the OntoDM ontology, which is based on a recent proposal of a general framework for data mining, and includes definitions of basic data mining entities and it also allows for the definition of more complex entities, e.g., constraints in constraint-based data mining, sets of such constraints (inductive queries) and data mining scenarios (sequences of inductive queries).

OntoDM is general-purpose and has been designed with as broad as possible use in mind and can be used to support a number of relevant activities, such as describing data mining services and resources, data mining experiments/investigations, as well as data mining scenarios/workflows.

The ontology OntoDM as presented here is in its early stages of development and hence much work remains to be done. We first need to populate the proposed classes of data mining entities with individuals, identify shortcomings of our ontology in the process and refine the structure of OntoDM as needed in order to describe different aspects of data mining.

Formalizing the knowledge about the domain of data mining and building of a heavy weight ontology of data mining is a time and resource consuming task and should be a community effort. Our goal is to have a mature ontology of data mining that is sufficient and expressive enough to describe the current trends in data mining. This would be also be a helpful step in developing standards for data mining and would lead towards an ontology of data mining investigations.

References

1. Bernstein, A., Provost, F., Hill, S.: Toward intelligent assistance for a data mining process: An ontology-based approach for cost-sensitive classification. IEEE Trans. on Knowl. and Data Eng. 17(4), 503–518 (2005)
2. Blockeel, H.: Experiment databases: A novel methodology for experimental research. In: Bonchi, F., Boulicaut, J.-F. (eds.) KDID 2005. LNCS, vol. 3933, pp. 72–85. Springer, Heidelberg (2006)
3. Blockeel, H., Vanschoren, J.: Experiment databases: Towards an improved experimental methodology in machine learning. In: Kok, J.N., Koronacki, J., Lopez de Mantaras, R., Matwin, S., Mladenič, D., Skowron, A. (eds.) PKDD 2007. LNCS (LNAI), vol. 4702, pp. 6–17. Springer, Heidelberg (2007)
4. Boulicaut, J.-F., Klemettinen, M., Mannila, H.: Modeling KDD processes within the inductive database framework. In: Data Warehousing and Knowledge Discovery, pp. 293–302 (1999)
5. Brezany, P., Janciak, I., Tjoa, A.: Ontology-Based Construction of Grid Data Mining Workflows. In: Data Mining with Ontologies: Implementations, Findings and Frameworks. IGI Global (2007)
6. Cannataro, M., Comito, C.: A data mining ontology for grid programming. In: Proceedings of (SemPGrid2003), pp. 113–134 (2003)
7. Cannataro, M., Talia, D.: The knowledge GRID. Commun. ACM 46(1), 89–93 (2003)

8. Diamantini, C., Potena, D.: Semantic annotation and services for KDD tools sharing and reuse. In: ICDMW 2008, Washington, DC, USA, 2008, pp. 761–770. IEEE Computer Society Press, Los Alamitos (2008)
9. Džeroski, S.: Towards a general framework for data mining. In: Džeroski, S., Struyf, J. (eds.) KDID 2006. LNCS, vol. 4747, pp. 259–300. Springer, Heidelberg (2006)
10. Imielinski, T., Mannila, H.: A database perspective on knowledge discovery. Comm. Of The ACM 39, 58–64 (1996)
11. Kalousis, A., Bernstein, A., Hilario, M.: Meta-learning with kernels and similarity functions for planning of data mining workflows. In: Proceedings of the Second PlanLearn Workshop 2008, pp. 23–28 (2008)
12. King, R.D., et al.: The Automation of Science. Science 324(5923), 85–89 (2009)
13. Lister, A., Lord, Ph., Pocock, M., Wipat, A.: Annotation of SMBL models through rule-based semantic integration. In: Proc. of Bio-ontologies SIG/ ISMB 2009 (2009)
14. Malaia, E.: Engineering ontology: domain acquisition methodology and practice. VDM Saarbrucken (2009)
15. Mizoguchi, R.: Tutorial on ontological engineering - part 3: Advanced course of ontological engineering. New Generation Comput 22(2) (2004)
16. Panov, P., Džeroski, S., Soldatova, L.: OntoDM: An ontology of data mining. In: ICDMW 2008, pp. 752–760 (2008)
17. Cimiano, P., Buitelaar, P. (eds.): Ontology learning and population: bridging the gap between text and knowledge. IOS Press, Netherlands (2008)
18. Peng, Y., Kou, G., Shi, Y., Chen, Z.: A descriptive framework for the field of data mining and knowledge discovery. International Journal of Information Technology & Decision Making (IJITDM) 7(04), 639–682 (2008)
19. Quinlan, R.: C4.5: programs for machine learning. Morgan Kaufmann, San Francisco (1993)
20. Schober, D., Kusnierczyk, W., Lewis, S.E., Lomax, J.: Towards naming conventions for use in controlled vocabulary and ontology engineering. In: Proceedings of BioOntologies SIG, ISMB 2007, pp. 29–32 (2007)
21. Smith, B.: Ontology. In: Blackwell Guide to the Philosophy of Computing and Information, pp. 155–166. Oxford Blackwell, Malden (2003)
22. Smith, B., et al.: The OBO foundry: coordinated evolution of ontologies to support biomedical data integration. Nature Biotechnology 25(11), 1251–1255 (2007)
23. Smith, B., et al.: Relations in biomedical ontologies. Genome Biology 6(5) , (2005)
24. Soldatova, L., Aubrey, W., King, R.D., Clare, A.: The exact description of biomedical protocols. Bioinformatics, 24(13) (2008)
25. Soldatova, L., King, R.D.: Are the current ontologies in biology good ontologies? Nature Biotechnology 23(9), 1095–1098
26. Soldatova, L., King, R.D.: An ontology of scientific experiments. Journal of the Royal Society Interface 3(11), 795–803 (2006)
27. Vanschoren, J., Blockeel, H., Pfahringer, B., Holmes, G.: Experiment databases: Creating a new platform for meta-learning research. In: Proceedings of the Second PlanLearn Workshop 2008, pp. 10–15 (2008)
28. Witten, I., Frank, E.: Data Mining: Practical Machine Learning Tools and Techniques, 2nd edn. (June 2005)
29. Yang, Q., Wu, X.: 10 challenging problems in data mining research. International Journal of Information Technology and Decision Making 5(4), 597–604 (2006)
30. Zakova, M., Kremen, P., Zelezny, F., Lavrač, N.: Planning to learn with a knowledge discovery ontology. In: Proceedings of the Second Planning to Learn Workshop, pp. 29–34 (2008)

OMFP: An Approach for Online Mass Flow Prediction in CFB Boilers

Indrė Žliobaitė[1], Jorn Bakker[1], and Mykola Pechenizkiy[1,2]

[1] Department of Computer Science, TU Eindhoven
P.O. Box 513, NL-5600 MB, Eindhoven, The Netherlands
{i.zliobaite,j.bakker,m.pechenizkiy}@tue.nl
[2] Dept. of MIT, U. Jyväskylä, P.O. Box 35, FIN-40014 Finland

Abstract. Fuel feeding and inhomogeneity of fuel typically cause process fluctuations in the circulating fluidized bed (CFB) boilers. If control systems fail to compensate the fluctuations, the whole plant will suffer from fluctuations that are reinforced by the closed-loop controls. Accurate estimates of fuel consumption among other factors are needed for control systems operation. In this paper we address a problem of online mass flow prediction. Particularly, we consider the problems of (1) constructing *the ground truth*, (2) handling noise and abrupt concept drift, and (3) learning an accurate predictor. Last but not least we emphasize the importance of having the domain knowledge concerning the considered case. We demonstrate the performance of OMPF using real data sets collected from the experimental CFB boiler.

1 Introduction

Online estimation of fuel consumption in mechanical devices is a challenging task due to noise, presence of outliers and non-stationarity of the signal. Mechanical devices typically are comprised of moving parts. The movements cause interference in the observed sensor data. The challenge is to filter out the *true* signal from the measured noise. In this study we develop a generic approach for online prediction of the true signal values from the sensor measurements under concept drift assumption. In particular, we address online mass flow estimation problem for a circulating fluidized bed (CFB) boiler.

Different amounts of fuel can be added to the boiler at irregular time intervals resulting in sudden drifts in a signal. Since the fuel is added mechanically (feeding), the start and the end time of this process is not necessarily (as in our case) available from the sensors as a direct measurement. Hence, in order to estimate accurately the amount of fuel in the container at each moment in time the algorithms should be able explicitly or implicitly handle these changes.

There is a lot of work on change detection and outlier detection, see e.g. a recent review [3]. However, the boiler problem exposes specific combination of change points and outliers at which existing change detection methods may fail. Statistical change detection methods, which are based on comparing pieces of raw data (e.g. [2]) do not take signal trends into account, which contain significant

J. Gama et al. (Eds.): DS 2009, LNAI 5808, pp. 272–286, 2009.

part of discriminatory information in the boiler problem. The noise and outliers are not normally distributed making it hard to use statistical methods that assume a particular distribution of the data [1]. Learner based change detection (e.g. [5]) is not directly suitable for this problem due to the nature of the signal: noise, trends and specific outliers. Burning and feeding stages, which are observed in the fuel mass signal, are very different in nature and timing.

We design an online signal prediction method, which takes into account the properties of mass flow signal (noise, trends, specific outliers, switch between operational stages). The method is equipped with a tailored change detection, which is needed to drop out the old signal from the training sample of the predictor. In this study we take a data mining approach, we use no additional input data from the boiler except the noisy signal itself.

For evaluation of the performance of signal estimators labeled data is needed. There is no hard evaluation method for the actual amount of fuel present. It could be generated by the domain experts. It is difficult to extract the actual signal, since the data includes the effects of external influencers. In our approach we use an offline best fit method as internal validation for the estimators.

The rest of the paper is organized as follows. In Section 2 we overview the problem of a mass flow prediction in CFB boiler. In Section 3 we present our solution for online mass flow prediction. In Section 4 the experimental evaluation is presented and the results are discussed. We conclude and point out open problems in Section 5.

2   Problem Description and Related Work

To better understand and control the operation of CFB boiler it is important to know how much fuel mass is in the furnaces. Direct measurement is hardly possible in practice from the technological perspective. Therefore, this is done by estimating mass flow in the system that is equivalent to predicting the amount of fuel in the fuel feeding system at each point in time.

We start by briefly explaining how the input signal is generated, discuss the properties of the data and available solutions.

2.1   The Input Signal

The automatically available mass signal is a noisy estimate of fuel mass at each operation time point. The mass of the fuel inside the container is measured by a scale, sampled with a sample rate of 1 Hz.

The boiler is fed with fuel from the fuel container ('bunker') as depicted in Figure 1. The fuel inside the container is mixed using a mixing screw. There is a feeding screw at the outlet of the container, which transfers the fuel from the container to the boiler. During the burning stage the mass of fuel inside the container decreases (reflected by a decreasing amount of fuel in the data signal, as pointed by arrow (1)). As new fuel is added to the container (the burning process continues), the fuel feeding stage starts that is reflected by a rapid mass increase (arrow (2)).

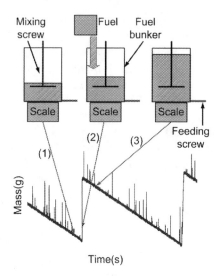

Fig. 1. The origin of the input signal

There are three main sources of changes in the signal.

First, fuel feeding is a manual and non standardized process, which is not necessarily smooth, it can have short interruptions (see Figure 2). Each operator can have different habits. Besides, the feeding speed depends on the type of fuel.

Second, the feeding screw rotation adds noise to the measured signal. Besides, fuel particle jamming often happens, slowing down the screw for some seconds and distorting the signal estimate. Therefore, the reported mass inside the container is not accurate, the signal contains extreme upward outliers in the original signal, that can be seen in Figure 3.

Third, there is a low amplitude rather periodic noise, which is caused by the mechanical rotation of the system parts. These amplitudes may become higher depending on the burning setup.

2.2 Data Properties

Due to the processes described above, the fuel mass signal has the following characteristics:

1. There are two types of change points: an abrupt change from burning to feeding and slower but still abrupt change from feeding to burning.
2. There are asymmetric outliers, oriented upwards. In online settings the outliers can be easily mixed with the changes from burning to feeding.
3. There is a symmetric high frequency signal noise.

Non stationarity of a signal can be regarded as a form of *concept drift* [15,8]. We focus on analyzing abrupt changes of a signal, which are caused by interchange of the boiler operation stages (burning and feeding).

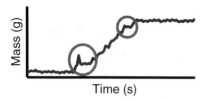

Fig. 2. An example of a short burning within the feeding stage

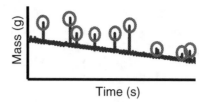

Fig. 3. Upward outliers due to jamming of the screw

Algorithmic change detection is not trivial as it might seem from the visual inspection of the signal. The signal would be elevated if approximated directly due to the asymmetric nature of the outliers (no opposite negative outliers).

Besides, there are short burning periods within the feeding stages, due to possible pauses in a feed, which depend on human operator behavior. These interruptions can vary from 5 to 20 seconds are difficult to identify.

We assume that the mass flow signal has a nonzero second derivative. It implies that the speed of the mass change depends on the amount of fuel in the container. The more fuel is in the container, the higher is the acceleration, thus the more fuel gets into the screw. The weight of the fuel at higher levels of the tank compresses the fuel in the lower levels and in the screw, the fuel density is increased. Besides, compression and thus the burning speed depends on the type and quality of the fuel.

Our ultimate goal is to learn an accurate online prediction of the signal given that we can (1) catch and handle the changing behavior due to a process change, and (2) ignore the noisy patterns generated by anomalous behavior or the influence of moving parts.

2.3   Connection to the Related Work

Data mining approaches can be used to develop better understanding of the underlying processes in CFB boilers, or learning a model to optimize its efficiency [13]. Fundamental studies develop mathematical models for boiler operation [12,10,7,14], incorporating operational parameters in the models.

In this study we take a data driven approach for modelling the signal for online operation using only the historical data. A straightforward approach to non stationary time series prediction would be SARIMA model [4]. Seasonal periodicity is expected there, but in case of boiler mass flow prediction fuel feeding periods are not regular. The patterns might differ in every feeding round as well as the periods between two feedings. Our preliminary experiments confirmed that SARIMA indeed did not give satisfactory results.

In our previous work [1], performance of several change detection methods was compared in terms of detection accuracy and lag. In this study we develop a tailor made online method for the signal prediction and do a thorough quantitative evaluation.

Our present study differs from the previous works in the following way.

1. A change detection method tailored for trendy noisy drifting time series is proposed.
2. Experiments with two different fuel types are carried out.
3. Quantitative comparison of the alternative prediction methods is performed.

3   Online Mass Flow Prediction

In this section we present our solution to online mass flow prediction. We start with setting up a general framework, followed by depicting the base model, change and outlier detection mechanisms.

3.1   General Framework

Let's have the original signal $\mathbf{X} = (x_1, x_2, \ldots, x_t, \ldots, x_n)$. Having $\mathbf{X}$ as input we want to obtain the actual mass flow signal $\mathbf{Y}$ that can be achieved by learning a functional mapping of noisy sensor measurements to the true actual signal, so that $\mathbf{Y} = \mathcal{F}(\mathbf{X})$.

This problem has connection to the problem of *concept drift* [8,15] that refers to unforeseen changes over time in the phenomenon of interest. Once a change in the system stage happens (reasons are described in Section 2) the functional mapping $\mathcal{F}$ might become outdated. The learners capable of handling concept drift can be classified into proactive (explicitly detecting the change and dropping out the old training sample) or reactive (using forgetting heuristics at each time step to have the best adapted learner) [9]. The boiler data exhibits abrupt changes, thus we employ a proactive approach.

The intuition behind the model is the following: at each point in time t we fit a model $\mathcal{F}(x)$, using all or a subset of the historical data $\mathbf{X}$. If a change is detected, the old portion of the historical data is dropped out. A simplified estimation procedure is presented in Figure 4, the steps are explained in more detail in the following subsections.

3.2   Elimination of Outliers

The outliers are asymmetric, they do not have zero mean with respect to the signal. If not eliminated before fitting the model (step 1 in Figure 4), they can lead to significant distortion of the prediction, which as a result, will be elevated.

We know that the outliers are oriented upwards. For online detection of the outliers we check if the difference between the given point and moving average of the signal exceeds a threshold Tr_{out}. We replace the detected outliers with an average of the two nearest neighbors.

Note that in an online setting the nearest neighbors for calculating moving average are available only from the past, but not from the future, thus the detection accuracy is expected to be lower than it would be offline. It is obvious at the start of a feeding stage, when the distinction between the change and the outlier can be noted only after some time lag.

Online Mass Flow Prediction (OMFP) for a time point $t + 1$

input: historical signal $\mathbf{X} = \{x_1, \ldots, x_t\}$.

1. Eliminate the outliers from $\mathbf{X}$ to obtain $\mathbf{X}^* = \{x_1, \ldots, x^*_{oi}, \ldots, x^*_{oj}, \ldots, x_t\}$, where $x^*_{oi} = \frac{x_{i-1}+x_{i+1}}{2}$ is a replacement for an outlier.
2. Find the last change point c.
3. Learn the model $\mathcal{F}(x)$, using $\{x_c, \ldots, x_t\}$ as a training sample.
4. Cast the prediction $\hat{y}_{t+1} = \mathcal{F}(x_{t+1})$.

output: $\hat{y}_{t+1}$.

Fig. 4. Online Mass Flow Prediction

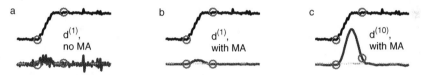

a
d(1),
no MA
b
d(1),
with MA
c
d(10),
with MA

Fig. 5. Change detection using L^{th} order signal differences $d^{(L)}$ and moving averages (MA). The upper (black) line represents the original signal and the lower (blue) is the differentiated signal. Dashed line (green) is the threshold for a change. Circles indicate *the ground truth* change.

3.3 Change Detection

The data exhibits trends, therefore change detection based on comparison of raw data subsets fails, when applied directly. We are interested in detecting the feeding stages, which are characterized by a steep increase in the signal value. An intuitive solution would be to take the first order differences of the signal $d_t^{(1)} = x_t - x_{t-1}$ and threshold these values. If $d_t^{(1)} > 0$ the system is at feeding stage, if $d_t^{(1)} < 0$ the system is at burning stage.

Unfortunately, due to signal noise, the stages are undistinguishable directly (see Figure 5a). We can try replacing the original signal with the moving average, before taking the first order differences, this already gives apparent feed regions, but that still is noisy (see Figure 5b).

We propose using L^{th} order differences $d_t^{(L)} = x_t - x_{t-L}$, applied on moving averaged signal for detection of stage changes. The more noisy the signal is, the larger lag is needed. In this case study we use $L = 10$ (see Figure 5c). Then we use a threshold Tr_{ch} to discriminate between feeding and burning stages. We use a high threshold $Tr_{ch} = 100$ to avoid false positives. The values were chosen based on preliminary experiments with the training set.

Tr_{ch} for changes is not to be mixed with Tr_{out} for outliers. The first is applied to a differentiated signal, while the second is applied to raw data.

Change detection might be equipped with a prior probability of switching the stages, based on the total amount of mass present in the container.

3.4 The Predictor

The functional mapping $\mathcal{F}$ (step 3 in Figure 4) is designed as follows.

In Section 2 we assumed that the mass flow signal has a nonzero second derivative. The true signal *in a single stage* can be modeled using the following equation:

$$y_t = \frac{a \cdot t^2}{2} + v_0 \cdot t + m_0 + A \cdot \sin(\omega_{feed} \cdot t + \alpha_{feed}) + B \cdot \sin(\omega_{mix} \cdot t + \alpha_{mix}) + e(t), \quad (1)$$

where y_t denotes the output of the scales at time t, a is acceleration of the mass change, v_0 stands for the speed of the mass change at time t_0, m_0 is the initial mass at time t_0; A and B, ω_{feed} and ω_{mix}, α_{feed} and α_{mix} are amplitude, frequency and phase of the fluctuations caused by feeding and mixing screws, respectively; $e(t)$ denotes the random peaked high amplitude noise caused by the jamming of the fuel particle at time t. We assume t_0 was the time of switch in the feeding/burining stages (change point X_c).

Since we are not interested in estimating the signal generated by the oscillations of the screw and the noise signal, we make a simplifying assumption that these parts can be treated as a signal noise. Thus we choose the following model:

$$\hat{y}_t = \frac{a \cdot t^2}{2} + v_0 \cdot t + m_0 + E(t), \quad (2)$$

where $E(t)$ is the aggregated noise component and the other terms are as in (1).

In our estimator we use a linear regression approach with respect to the second order polynomial given by (2). The model is inspired by the domain knowledge of the underlying process in the boiler, therefore seem more reasonable choice than alternative autoregressive models.

3.5 Learning the Predictor

To learn a regressor, the Vandermonde matrix [6] $\mathbf{V}$, which elements $v_{i,j}$ are the powers of independent variable x, can be used. In our case the independent variable is time $x_i = t_{i-1} - t_0$, $i = 1, \ldots, T$, where T denotes the number of the time steps. If the linear regression is applied for a polynomial of order n $(p^n(x) = p_n x^n + p_{n-1} x^{n-1} + \ldots + p_1 x + p_0)$, $\mathbf{V}$ is computed from the observed time series of the independent variable as follows:

$$v_{i,j} = x_i^{n-j+1}, \quad i = 1, \ldots, T, \ j = 1, \ldots, n+1, \quad (3)$$

where i and j run over all time samples and powers, respectively. Provided with $\mathbf{V}$ the problem of polynomial interpolation is solved by solving the system of linear equations $\mathbf{V}\mathbf{p} \cong \mathbf{y}$ with respect to $\mathbf{p}$ in the least square sense:

$$\hat{\mathbf{p}} = \text{argmin}_{\mathbf{p}} \sum_{i=1}^{T} (\sum_{j=1}^{n+1} V_{i,j} p_{n-j+1} - y_i)^2 \quad (4)$$

Here, $\mathbf{p} = [p_n \; p_{n-1} \; \ldots \; p_1 \; p_0]^T$ denotes the vector of the coefficients of the polynomial, and $\mathbf{y} = [y(x_1) \; y(x_2) \; \ldots y(x_T)]^T = [y_1 \; y_2 \; \ldots y_T]^T$ is the time series of the dependent variable that is indication of the scales. Provided that the $n+1$ columns of the matrix $\mathbf{V}$ are linearly independent, this minimization problem has a unique solution given by solving the normal equation [11]:

$$(\mathbf{V}^T\mathbf{V})\widehat{\mathbf{p}} = \mathbf{V}^T\mathbf{y}. \tag{5}$$

This procedure is used to estimate the mass flow signal between change points. If the process switches from fuel feeding to fuel burning or the other way around, a new model is learnt on the new data.

3.6   Constructing *the Ground Truth*

The mass flow prediction is an unsupervised learning task in a way that the need for prediction arises from the fact that there is no method to measure *the ground truth*. However, to verify the validity of the model we still need a benchmark.

To obtain an approximation to *the ground truth* we use all the data set at once offline. We employ similar procedure as presented in Section 3.2. We identify the outliers by comparing the difference between the signal and the moving average against a threshold Tr_{out}. Then we take a moving average of the modified signal to obtain an approximation to *the ground truth*, which we associate as $\mathbf{Y}$.

Next we identify the change points from burning to feeding stage and vice versa (C_{feed} and C_{burn}). We employ different approach than in the online change detection. We use ADWIN method [2], which showed to be robust to false positives in semi-online settings [1]. We do not use it in online settings, because the lag needed to detect the change after it happened is too large.

Given a sequence of signals, ADWIN checks whether there are statistically significant differences between the means of each possible split of the sequence. If statistically significant difference is found, the oldest portion of the data backwards from the detected point is dropped and the splitting procedure is repeated recursively until there are no significant differences in any possible split of the sequence. More formally, suppose m_1 and m_2 are the means of the two subsequences as a result of a split. Then the criterion for a change detection is $|m_1 - m_2| > \epsilon_{cut}$, where

$$\epsilon_{cut} = \sqrt{\frac{1}{2m} \log \frac{4n}{\delta}}, \tag{6}$$

here m is the harmonic mean of the windows $m = \frac{1}{\frac{1}{n_1} + \frac{1}{n_2}}$, n is total size of the sequence, while n_1 and n_2 are sizes of the subsequences respectively. Note that $n = n_1 + n_2$. $\delta \in (0, 1)$ is a hyper-parameter of the model. In our experiments we used $\delta = 0.3$, $n = 200$ which were set during the preliminary experiments using the training data.

ADWIN identifies C_{feed} approximately. To get the exact change points we search for a maximum and minimum of the moving average in the neighborhood of the points identified by ADWIN. We validate *the estimated ground truth* by visual inspection of a domain expert.

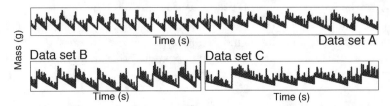

Fig. 6. The three complete data sets A, B, and C used in the experiments

4   Experimental Evaluation

4.1   Data sets

In this study we use three mass signal data sets A,B and C, which are plotted in Figure 6. The total length of A is different from B and C. A summary of the data sets is provided in Figure 7. Number of feeds means the number of feeding stages in the data set.

Data set A is used for training the model and selecting the model parameters. Data sets B and C are used as testing sets, the model trained on A with the same set of parameters is applied. Note that the level of noise and outliers in the data sets are different. B and C represent two fuel tanks, operating in parallel, therefore there are nearly twice as much of noise sources as in A.

Using training data set A we construct a representation of an average feeding stage pattern, which is depicted in Figure 8. This pattern is obtained by partitioning *the approximated ground truth* data into separate feeding sections. Then the partitions are matched by the change points from burning to feeding and averaged across.

Name	Size	Number of feeds	Fuel
A	50 977	24	bio
B	25 197	9	bio
C	25 197	6	coal

Fig. 7. Data sets used

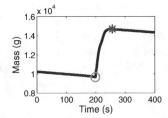

Fig. 8. An average feeding stage pattern

4.2   Experimental Setup

We conduct numerical experiments to test for prediction accuracy and for change detection accuracy. We chose moving average prediction as a 'naive' method to compare the performance.

In addition to next step $(t + 1)$ prediction experiments, we conduct a set of experiments allowing a delay D in predictions. For example, having $D = 5$ we would predict (filter) the signal x_t, but will have the historical data available up to time x_{t+4} inclusive. This gives a smoother moving average (nearest neighbors from both sides are available) as well as it allows more firm verification of outlier

Table 1. Mean average prediction accuracies. The best accuracies for each delay are bold; the best overall accuracy over a single experiment is underlined.

Delay	t+1	now	t-2	t-4	t-9	t+1	now	t-2	t-4	t-9	t+1	now	t-2	t-4	t-9
Data	A					B					C				
Overall performance															
OMFP	**34.1**	**29.4**	**27.8**	**27.6**	**29.0**	23.8	20.9	16.6	**16.3**	31.4	12.9	13.0	10.3	**10.1**	16.3
MA3	64.0	64.0	66.4			48.5	47.2	46.9			36.3	35.6	35.2		
MA5	63.1	51.9		39.9		49.7	45.3		41.7		35.9	33.9		32.5	
MA10	59.1	54.8			33.2	58.1	53.7			34.9	39.3	37.2			28.5
win50	53.0	45.0	44.4	44.4	44.4	40.3	34.3	32.0	32.0	32.0	19.7	16.7	15.2	15.2	**15.2**
all	1271	1269	1267	1265	1261	1313	1310	1308	1306	1301	1022	1021	1019	1019	1016
known	34.8	32.0	30.6	31.3	41.8	50.7	47.9	45.1	44.6	65.5	18.0	16.5	15.7	16.3	22.0
Feeding stages															
OMFP	463	325	229	231	321	1531	952	601	682	968	519	334	182	180	325
MA3	**308**	**181**	**115**			**733**	**510**	**434**			**260**	**163**	**119**		
MA5	438	294		**77**		1118	713		**359**		374	236		**105**	
MA10	781	640			**60**	2081	1714			**171**	714	578			**61**
win50	751	646	577	577	577	1867	1602	1248	1248	1248	860	735	645	645	645
all	1757	1753	1748	1744	1731	3259	3253	3248	3242	3225	2264	2259	2255	2250	2237
known	441	315	236	244	290	1493	924	561	594	752	464	306	249	269	296
Burning stages															
OMFP	30.0	28.7	28.5	**28.2**	**28.1**	22.1	19.7	**16.9**	17.2	34.9	10.7	11.4	**10.5**	11.3	19.1
MA3	60.8	62.4	65.8			48.1	46.9	46.8			35.6	35.0	34.8		
MA5	57.7	48.4		39.4		48.3	44.4		41.5		34.5	32.8		32.1	
MA10	48.3	45.9			32.8	54.6	50.6			34.9	36.0	34.3			28.3
win50	42.8	37.0	37.1	37.7	39.1	39.4	33.1	32.4	33.3	35.5	15.6	12.9	12.8	13.8	**16.3**
all	1264	1262	1261	1260	1257	1320	1317	1315	1314	1311	1015	1013	1013	1013	1013
known	**29.1**	**28.0**	**28.1**	29.3	40.7	50.6	47.7	45.7	46.0	69.1	16.2	15.0	15.5	16.9	25.1

and change detection. D is not to be mixed with L, which is a lag used by change detection method itself (Section 3.3).

We do the following verification: the stage (feeding or burning) is defined to be consistent if it lasts for not less than D time steps. Say at time t the system is at burning stage and at time $t + 1$ we detect the feeding stage. Having a delay $D = 5$ we are able to see the next four examples before casting the signal prediction for time $t + 1$. Thus we check if the feeding stage sustains at time $t + 2, \dots, t + 5$. If positive, we fix the change point, if negative, we cancel the detected change and treat this as an outlier.

The domain experts suggested that maximum possible delay (D) in prediction could be 10 sec.

Once a change is detected, old portion of the data is dropped out of the training sample. We do not start using the 2nd order polynomial model until we pass 10 samples after the change. For the first 2 samples we use simple moving average rule: $x_{t+1} = x_t + s$, where s is a linear intercept term obtained using an average feeding stage pattern of the training data (A), which is presented in Figure 8. For burning stage $s_c = -2$ is used, for feeding stage $s_f = 81$. If from 2 to 10 historical data points are available after the change, we fit the 1st order polynomial model, since the 2nd order approximation is too noisy with this few amount of points.

4.3 Prediction Accuracy

The mean absolute errors (MAE) with respect to our approximation to *the ground truth* (described in Section 3.6) are listed in Table 1. We present MAE for the whole data sets and then present MAE's for feeding and burning stages separately. Delay $t + 1$ means prediction of change one second ahead, 'now' means real time signal estimation and $t - 2$, $t - 4$ and $t - 9$ means estimation with respective delay of 2, 4 or 9 seconds.

For online prediction we set the following parameters. For outlier detection a moving average with a lag of 9 and a threshold $Tr_{out} = 400$ is used. For change detection a moving average with a lag of 8 and a threshold $Tr_{ch} = 100$ was used. the parameters were obtained from the preliminary experiments with the training data set A.

'MA3', 'MA5' and 'MA10' stand for simple prediction by moving averages, the number indicates how many instances are averaged. 'win50' uses the 2nd order prediction model presented in Section 3, but instead of change detection a simple

Table 2. Confusion matrixes of detecting changes to feeding (φ) and burning (κ) stages and outlier detection (o). P - positive, N - negative, T - true, F - false.

Training data set A

		φ P N	κ P N	o P N
t+1	T	24 50946	12 50934	659 49784
	F	26 0	38 12	543 10
now	T	24 50946	12 50934	659 49784
	F	26 0	38 12	543 10
t-2	T	24 50967	10 50953	659 49783
	F	5 0	19 14	544 10
t-4	T	24 50969	10 50955	658 49783
	F	3 0	17 14	544 11
t-9	T	24 50972	8 50956	660 49782
	F	0 0	16 16	545 9

Testing data set B / Testing data set C

		φ P N	κ P N	o P N	φ P N	κ P N	o P N
t+1	T	6 25162	2 25158	475 24597	6 25176	2 25172	362 24750
	F	26 3	30 7	104 21	15 0	19 4	75 10
now	T	6 25162	2 25158	475 24597	6 25176	2 25172	362 24750
	F	26 3	30 7	104 21	15 0	19 4	75 10
t-2	T	6 25165	2 25161	477 24597	6 25177	1 25172	364 24750
	F	23 3	27 7	104 19	14 0	19 5	75 8
t-4	T	6 25165	2 25161	477 24597	6 25177	1 25172	364 24750
	F	23 3	27 7	104 19	14 0	19 5	75 8
t-9	T	6 25183	2 25179	489 24594	6 25191	1 25186	372 24746
	F	5 3	9 7	107 7	0 0	5 5	79 0

moving window of the 50 last instances is used for the model training at each time step. 'all' uses the 2nd order prediction model with no change detection at all, it retrains the model at every time step. Finally we include a benchmark of the 2nd order model assuming known change points ('known'). We assume with this method that the change detection is 100% accurate.

MAE in 'overall performance' is rather close to MAE in 'burning stages' and very different from 'feeding stages'. This is because of uneven distribution of the stages in the data. 'Burning stages' comprise less than 2% of the data.

4.4   Change Detection Accuracy

We report the performance of the change detection in online settings in Table 2. For each method we present confusion matrixes of detecting sudden changes in feeding (φ) and burning (κ) stages and detecting of outliers (o). For φ and κ we allow 10 sec deviation. If a change is detected within the allowed region it is considered as identified correctly. We require the outlier detection to be precise.

We visualize change and outlier detection in Figure 9. The solid (blue) lines represent the true positives (TP) divided by the actual number of changes, the dashed (red) lines represent the number of false positives (FP) divided by the actual number of changes. The dotted black lines show the level of true changes (i. e. 24 change points for data set A, 9 for B, 6 for C).

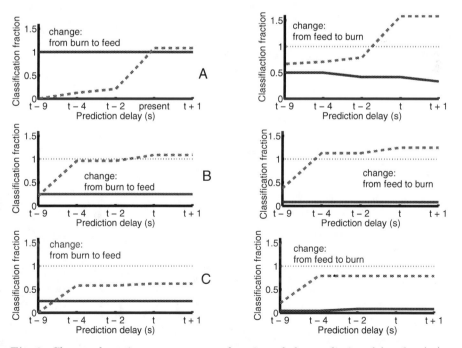

Fig. 9. Change detection accuracy as a function of the prediction delay for A (top row), B (middle row) and C (bottom row) data sets. Solid lines (blue) represent true positives, dashed lines (red) represent true negatives.

The number of FP is decreasing along with the increase in allowed prediction delay. A delay allows to inspect the following signal values after the detected change and if necessary cancel the alarm within the delay period.

The number of false negatives (FN) is relatively large. However, this does not mean that the changes from feed to burn were not detected at all. In this setting it means that they were not detected in time (within 10 sec interval).

4.5   Discussion

OMFP outperforms the competitive methods in terms of overall accuracy. However, for the feeding stage, simple moving average is the most accurate. Note that the approximation to *the ground truth* was constructed using moving averages, thus it could be expected that moving average performs well in this test setup.

OMFP method performance gets worse having a large delay in predictions. This is likely due to a fixed number of the nearest neighbors for moving average calculations, as we are using the same parameter settings for all the experiments.

Degradation of OMFP performance along with the increase in prediction delay also suggests, that there might be more accurate cutting points than just the change points themselves. Note that having a delay we allow canceling the detected changes.

In Figure 10 extraction from the prediction outputs is provided. In $t + 1$ prediction (a) the prediction follows previous points almost as a straight line. It is reasonable to expect, since in the fitted function 2nd order coefficient is mostly 0. Prediction $t - 2$ (b) is more curvy than $t - 9$ (c) likely due to more change points identified and therefore more cuts in history.

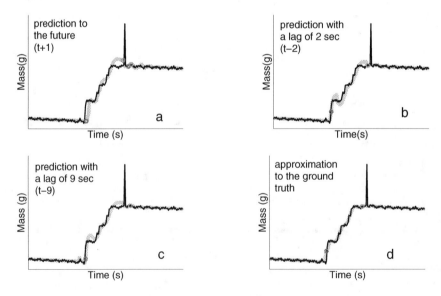

Fig. 10. The original signal (black) and OMFP estimation (grey)

Overall accuracy indicates that OMFP method is more favorable than moving average or alternative methods in the burning stage (see Table 1), while in the feeding stage optimal model selection is yet to be seeked. Separate handling of prediction in feeding and burning stages might be advantageous.

5 Conclusion

We developed and experimentally evaluated an online method for mass flow prediction during the boiler operation. We evaluated the performance of the method on three real data sets, including two distinct fuel types and two distinct operating stages (single vs multiple fuel).

One of the challenges in this task is coming up with approximation for constructing *the ground truth* for the signal, which we handle by a combination of moving average and responding for change and outlier points. We use this approximation to evaluate the performance of the online predictors.

Change detection is sufficiently accurate in transition from the burning to the feeding stage, where the incline in signal is rather sharp. However, the reverse detection still has room for improvement.

OMFP method clearly outperforms the competitive methods in terms of overall accuracy, while at the feeding stage simple moving average is a more accurate approach. The results suggest that separate handling of prediction in feeding and burning stages is needed.

The next steps of the research would be to employ the presented method in operational settings to see, what is the generalization on unseen cases. In addition, the effects of the rotation screw on the signal will be explored. Further, it would be interesting to come up with different models for different fuel types.

Acknowledgements. This research is partly supported by NWO HaCDAIS and TEKES DYNERGIA projects. The authors are thankful to Andriy Ivannikov (U. Jyväskylä, Finland) and the domain experts Timo Leino and Mikko Jegoroff (VTT, Finland) for their contribution to this work.

References

1. Bakker, J., Pechenizkiy, M., Zliobaite, I., Ivannikov, A., Karkkainen, T.: Handling outliers and concept drift in online mass flow prediction in cfb boilers. In: Proceedings of the 3rd Int. Workshop on Knowledge Discovery from Sensor Data (SensorKDD 2009), pp. 13–22 (2009)
2. Bifet, A., Gavalda, R.: Learning from time-changing data with adaptive windowing. In: Proc. of the 7th SIAM International Conference on Data Mining (SDM 2007), pp. 443–448 (2007)
3. Chandola, V., Banerjee, A., Kumar, V.: Anomaly detection - a survey. ACM Computing Surveys 41 (to appear, 2009)
4. Chatfield, C., Prothero, D.L.: Box-jenkins seasonal forecasting: Problems in a case-study. Journal of the Royal Statistical Society. Series A (General) 136(3), 295–336 (1973)

5. Gama, J., Castillo, G.: Learning with local drift detection. In: Li, X., Zaïane, O.R., Li, Z.-h. (eds.) ADMA 2006. LNCS (LNAI), vol. 4093, pp. 42–55. Springer, Heidelberg (2006)
6. Horn, R.A., Johnson, C.R.: Topics in matrix analysis. Cambridge University Press, Cambridge (1991)
7. Huilin, L., Guangboa, Z., Rushana, B., Yongjina, C., Gidaspowb, D.: A coal combustion model for circulating fluidized bed boilers. Fuel 79, 165–172 (2000)
8. Kubat, M., Widmer, G.: Adapting to drift in continuous domains. In: Proc. of the 8th European Conference on Machine Learning, pp. 307–310 (1995)
9. Kuncheva, L.I.: Classifier ensembles for detecting concept change in streaming data: Overview and perspectives. In: Proc. 2nd Workshop on Supervised and Unsupervised Ensemble Methods and their Applications (SUAMA ECAI 2008), pp. 5–10 (2008)
10. Kusiak, A., Burns, A.: Mining temporal data: A coal-fired boiler case study. In: Khosla, R., Howlett, R.J., Jain, L.C. (eds.) KES 2005. Part III LNCS (LNAI), vol. 3683, pp. 953–958. Springer, Heidelberg (2005)
11. Lawson, Ch.L., Hanson, R.J.: Solving Least Squares Problems. Prentice-Hall, Englewood Cliffs (1974)
12. Park, C.K., Bas, P.: A model for prediction of transient response to the change of fuel feed rate to a circulating fluidized bed boiler furnace. Chemical Engineering Science 52, 3499–3509 (1997)
13. Pechenizkiy, M., Tourunen, A., Karkkainen, T., Ivannikov, A., Nevalainen, H.: Towards better understanding of circulating fluidized bed boilers: a data mining approach. In: Proc. ECML/PKDD Workshop on Practical Data Mining, pp. 80–83 (2006)
14. Saastamoinen, J.: Modelling of dynamics of combustion of biomass in fluidized beds. Thermal Science 8, 107 (2004)
15. Tsymbal, A.: The problem of concept drift: definitions and related work. Technical Report TCD-CS-2004-15, Department of Computer Science, Trinity College Dublin, Ireland (2004)

C-DenStream: Using Domain Knowledge on a Data Stream

Carlos Ruiz[1], Ernestina Menasalvas[1], and Myra Spiliopoulou[2]

[1] Facultad de Informtica, Universidad Politcnica de Madrid, Spain
cruiz@cettico.fi.upm.es, emenasalvas@fi.upm.es
[2] Faculty of Computer Science, Magdeburg University, Germany
myra@iti.cs.uni-magdeburg.de

Abstract. Stream clustering algorithms are traditionally designed to process streams efficiently and to adapt to the evolution of the underlying population. This is done without assuming any prior knowledge about the data. However, in many cases, a certain amount of domain or background knowledge is available, and instead of simply using it for the external validation of the clustering results, this knowledge can be used to guide the clustering process. In non-stream data, domain knowledge is exploited in the context of *semi-supervised clustering*.

In this paper, we extend the static semi-supervised learning paradigm for streams. We present C-DenStream, a density-based clustering algorithm for data streams that includes domain information in the form of constraints. We also propose a novel method for the use of background knowledge in data streams. The performance study over a number of real and synthetic data sets demonstrates the effectiveness and efficiency of our method. To our knowledge, this is the first approach to include domain knowledge in clustering for data streams.

1 Introduction

The rapid growth and complexity of information and communication systems in all areas of society has led to on-line and real-time applications where huge amounts of evolving data is constantly generated at high speed over time. Data of this nature, known as data streams [3], pose new challenges for data mining. There are multiple applications that generate data streams: among others, financial applications, Web applications, sensor networks, security and performance control in networks, monitoring environmental sensors, detecting gamma rays in astrophysics [1,9]. More specific examples are the analysis of 20 million sales transactions per day of Walmart, the 70 million searches on Google or the 275 million calls at AT&T [15].

In data streams, underlying data distribution does not remain stationary, but can evolve over time, forcing discovered patterns to be updated. As in the stationary case, the validity of the patterns is not only determined by statistical features of the model but by the expert's perception and expectations regarding the domain knowledge. Domain information is used to establish the desired criteria for validity, usefulness and success of the models obtained and therefore,

J. Gama et al. (Eds.): DS 2009, LNAI 5808, pp. 287–301, 2009.

effective methods to include domain knowledge during the process of analysis and construction of the models are required.

Specifically, clustering with constraints or semi-supervised clustering seeks to alleviate the problem of interpretation and evaluation including domain knowledge within clustering methods allowing practitioners to move from unsupervised solutions toward semi-supervised solutions [20].

Semi-supervised clustering methods exploit background knowledge to guide the clustering process. Instance-level constraints are a specific and popular form of background knowledge: They refer to instances that must belong to the same cluster (Must-Link constraints) and those that must be assigned to different clusters (Cannot-Link constraints) [29]. Research into semi-supervised clustering with instance-level constraints has shown that the selection of the input constraints has a substantial influence on the quality of the output clustering model [12].

In this paper we propose the use of semi-supervised approach including domain information in order to exploit expert knowledge in the area of density-based clustering algorithms for data streams. Constraints are a very appropriate instrument to guide the cluster adaptation and reconstruction process across a data stream. However, current models of instance-level constraints cannot support this process, because they mostly refer to specific data records. In this paper, we extend the notion of instance-level constraints from static data to data streams, focusing on density-based clustering and cluster evolution in streams. For this reason, we apply this semi-supervised approach to extend the DenStream algorithm [8]

The rest of the paper is organized as follows: Section 2 presents the main approaches to semi-supervised clustering with emphasis on those methods proposed for data streams. In Section 3 the DenStream algorithm is presented as it is the basis of the C-DenStream approach that will be further presented in Section 4. Results of the experimentation showing the improvements introduced by the algorithm presented are shown in Section 5. To end with Section 6 discuses the presented approach and outlines future work.

2   Related Work

Approaches on clustering with instance-level constraints designed for static data include partitioning algorithms [29], hierarchical algorithms [11], and density-based algorithms [27] either can be classified as *constraint-based* or *distance-based* methods.

In the so-called "Constraint-based" approaches, the original objective function is modified into one that satisfies as many of the constraints as possible [29,5,11]. On the other hand, in "distance-based" approaches, the algorithm is trained on the data involved in the constraints; thus a new metric is learned and used for clustering. In the new function, instances associated with a Must-Link constraint are "pushed closer" to each other, while instances associated with a Cannot-Link constraint are "pulled away" from each other. This approach is used in [22,30].

Both approaches deal with constraints in different ways. When constraints are embedded into the distance function, their violation can be penalized but not prohibited per se. On the other hand, algorithms that embed constraints in the objective function may fail to deliver a solution, if they cannot satisfy all constraints. Hybrid methods have emerged such as [6,7,21] to overcome these problems.

However, all these methods have been designed for static data. Stream clustering algorithms have concentrated thus far rather on efficient data processing and model adaptation by: i) processing data in a single pass, ii) deriving models incrementally, iii) detecting model changes over time, iv) keeping the use of memory and computing time low, and v) automating the evaluation process.

Stream clustering algorithms may strive for a model that satisfies an overall optimization criterion and must thus be re-learned as new data arrive. Variations of K-means or K-medians for streams, as e.g. [19], belong to this category. Other algorithms build incremental clusters by means of summaries or synopsis and assigning data points to the locally optimal cluster, as they encounter them: CluStream [2] which initially defines the use of micro-clusters, and DenStream [8] belong to this category. TECNO-STREAMS [24], TRAC-STREAMS [23], and DUCStream [18] are other incremental single-pass algorithms within this category.

For the exploitation of domain knowledge during stream clustering we have opted for a single-pass density-based algorithm: in [27] we have shown that a density-based clustering algorithm for static data lends itself almost naturally to the incorporation of instance-level constraints. In [27] we have extended DB-SCAN with constraints. Our new C-DENSTREAM constraint-based stream clustering algorithm is based on the stream clustering variation of DBSCAN, namely DenStream [8].

3 DenStream - Density-Based Clustering on an Evolving Data Stream with Noise

DenStream [8] is a density-based clustering solution for data streams, which extends the micro-cluster concept [2]. Instead of using the number of points that are in the neighborhood as density concept like DBSCAN [16], micro-cluster density is based on weighting areas of points in the neighborhood as a result of an exponential decay function over time. This allows core-micro-clusters as (w, c, r) to be defined at the time t for a group of close points $p_{i_1}, \ldots, p_{i_n}$ with time stamps $T_{i_1}, \ldots, T_{i_n}$ and μ and ϵ as DBSCAN parameters:

- $w = \sum_{j=1}^{n} f(t - T_{i_j})$ is the core micro-cluster weight with $w \geq \mu$.
- $c = \frac{\sum_{j=1}^{n} f(t - T_{i_j}) \cdot p_{i_j}}{w}$ is the core-micro-cluster center.
- $r = \frac{\sum_{j=1}^{n} f(t - T_{i_j}) \cdot dist(p_{i_j}, c)}{w}$ is the core-micro-cluster radius, with $r < epsilon$ and $dist(p_{i_j}, c)$ is the euclidean distance between the point p_{i_j} and the center c.

Given the evolving nature of data streams, new clusters can appear or old instances can become clusters, so DenStream introduces a structure for outliers, known as potential core-micro-clusters. A potential core-micro-cluster is defined as $(w, \overline{CF^1}, \overline{CF^2})$ at time point t for a set of close points $p_{i_1}, \ldots, p_{i_n}$ with time stamps $T_{i_1}, \ldots, T_{i_n}$ where:

- $w = \sum_{j=1}^{n} f(t - T_{i_j})$ is the potential core-micro-cluster weight with $w > \beta\mu$, where $0 < \beta < 1$ is the parameter to determine the threshold to consider a micro-cluster as a potential core-micro-cluster or an outlier micro-cluster.
- $\overline{CF^1} = \sum_{j=1}^{n} f(t-T_{i_j}) \cdot p_{i_j}$ is the linear sum of the weighting of the instances.
- $\overline{CF^2} = \sum_{j=1}^{n} f(t-T_{i_j}) \cdot p_{i_j}^2$ is the sum of the square weighting of the instances.
- In addition, the center of a potential core-micro-cluster is: $c = \frac{\overline{CF^1}}{w}$, while its radius is: $r = \sqrt{\frac{|\overline{CF^2}|}{w} - \left(\frac{|\overline{CF^1}|}{w}\right)^2}$.

On the other hand, an outlier micro-cluster is defined as $\overline{CF^1}, \overline{CF^2}, w, t_0$ at t time for a set of close points $p_{i_1}, \ldots, p_{i_n}$ with time stamps $T_{i_1}, \ldots, T_{i_n}$ where: w and $\overline{CF^1}$ y $\overline{CF^2}$ have the same definition as given in potential core-micro-cluster, while $t_0 = T_{i_1}$ denotes the time of creation used to determine its existence point of time.

DenStream divides the process of clustering into two parts: first, an online maintenance step of core-micro-clusters is carried out followed by an offline step generates the final clusters using the clustering algorithm DBSCAN [8].

4   C-DenStream - An Approach to Include Domain Knowledge in Data Stream Clustering

We propose a clustering algorithm for data streams based on the extension of DenStream [8], allowing the inclusion of domain information, at the same time as requirements for data streams algorithms are satisfied. For this purpose, we extend the notion of the instance-level constraint for data streams. To do this, we use the concept of constraint between micro-clusters [2], based on the fact that elements involved in a constraint are generally representative of their local neighborhood [22].

The ability of DenStream to discover clusters of arbitrary size and shape together with the fact that it satisfies the requirements of data stream clustering, made us choose it as the candidate to include constraints. Our approach is based on the modification of the algorithm in 2 issues: i) adapting the offline step of DenStream based on DBSCAN to use C-DBSCAN [27] to include the constraints ii) include constraints in the DenStream phase in which micro-clusters are created, removed and maintained.

4.1   From Instance-Level Constraints to Micro-cluster Constraints

The main challenges of using instance-level constraints in the data stream domain are related to the use of memory: On the one hand how to generate

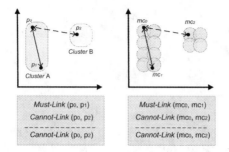

Fig. 1. C-DenStream: Transformation step for micro-cluster-level constraints

instance-level constraints that refer to items arriving at each time stamp, on the other, how to deal with a constraint stream that can potentially be of the same magnitude in size as that the one of instances.

Consequently we propose to use a transformation step in which instance-level constraints are translated into micro-cluster-level constraints using the micro-cluster membership of each instance. Based on the fact that the local neighborhood is represented by a micro-cluster, we proceed as follows: for each pair of instances involved in a Must-Link or Cannot-link constraint we produce a constraint between the micro-clusters to which they belong. The constraint will have a weighting depending on the arrival time and the number of instances between the two micro-clusters. An example of the approach is set out in Figure 1. We call these constraints "micro-cluster-level constraint" or "micro constraints". The main aim behind this transformation step is to minimize the use of memory by constraints as well as minimizing its complexity by providing an approximate solution. To that end, a matrix is used to store constraints between micro-clusters, as well as their weighting. As micro constraints are also symmetrical (e.g. if there is a constraint that links the micro-cluster (A) to the micro-cluster (B), it also means that the micro-cluster (B) links to the micro-cluster (A)), only a triangular matrix is required.

Definition 1. *Micro-cluster constraint (also referred to as micro constraint). A micro-cluster constraint regarding mc' and mc'', $RNM(mc',''mc)$, at the time stamp t for a set of instance-level constraint with the same semantic as $\{constraint_{i0}, \ldots, constraint_{in}\}$, and with time stamps $T_{i1}, \ldots, T_{in}$, is defined as the tuple $\{type, weight\}$, where:*

- *type, is the level of instances that relates both micro-clusters, namely whether it is a Cannot-Link or Must-Link constraints.*
- *weight, is the weight of the relationship between the two micro-clusters depending on the amount of points that belong to this relationship and based on a decay function, such that: $w = \sum_{j=1}^{n} f(t - T_{i_j})$.*

As in [8], a damper window model is considered, where a decay function is used to define the age of an instance. In this case, the weight of each item of the data stream decreases exponentially with time t using the decay function $f(t) = 2^{-\lambda t}$, con $\lambda > 0$. The higher the value of λ, the lesser weight of historical data compared to recent data. In this way, the total weight of a data stream is constant so that:

$$W = \upsilon \left(\sum_{t=0}^{t=t_c} 2^{-\lambda i} \right) = \frac{\upsilon}{1 - 2^{\lambda}}$$

where t_c is the current time stamp, and υ shows the arrival speed.

4.2   C-DenStream Algorithm

We take into account the original algorithm DenStream, the fade function for instance-level constraints, and the micro-cluster-level constraint given in Definition 1, to design an algorithm C-DenStream that includes domain information, for data stream clustering. The algorithm has been divided into three main steps (cf. Algorithm 1):

Step 1. Initialization of micro-clusters and micro-cluster-level constraints. This process corresponds to the Procedure 2 and it creates and initializes the main structures for micro-clusters and micro-cluster-level constraints. When running the algorithm, if there are instance-level constraints, they are used to generate an initial clustering with C-DBSCAN. These clusters are then transformed into potential-micro-cluster, and initial instance-level constraints to micro constraints.

Step 2. Online maintenance of micro-clusters and micro constraints. This second step tracks the evolution of the micro-clusters and micro constraints in the data stream, capturing the density of the data as they arrive. As in DenStream, and based on the observation that the new items typically belong to the potential micro-clusters, the algorithm keeps a complete list of micro-clusters called outlier micro-cluster to cover the possibility that any of them can be absorbed by the potential micro-cluster or application of some constraint. To the same effect, it always stores all micro-constraints for potential micro-cluster and outlier micro-cluster. As new items and constraints arrive, the weights of the micro-clusters and micro constraints are updated using their fade function. This updating does not need to be done constantly according to [8], and it is enough to do it at T_p time in which a potential micro-cluster can become outlier micro-cluster. This same interval of time will be used to check the status of the micro constraints, which will be updated according to changes in the micro-clusters. There are two choices in updating of micro constraints between micro-clusters (reflected in Procedure 4), either updating micro-clusters first and then micro constraints or the other way round updating first the micro-constraints. In our case, we firstly update the micro-clusters, so that, micro constraints can inherit the fade mechanisms.

Step 3. Generation of final clusters through C-DBSCAN and micro-cluster-level constrains. As a final step, and at the request of the user, it is necessary to generate the final clusters. Although the technique of keeping micro-cluster structures during the on-line phase allows the notion of local density to be suitably collected and the changes that occur during the data arrival in the stream, it is necessary to transfer these data stream structures to a more appropriate approach, identifying the nature and number of clusters in a more overall fashion. We use C-DBSCAN in order to generate these final clusters using the potential micro-cluster and the micro constraints.

Algorithm 1. C-DENSTREAM - A Constrained Density-Based Clustering Algorithm Over Evolving Data Stream.

Data:

DS, a data stream.
ϵ, neighborhood radius.
μ, minimum number of points in neighborhood.
β, outlier radius.
λ, decay function.
CO, a stream of instance-level constraints (Must-Link and Cannot-Link)

begin

> *p-buffer*, the set of p-micro-clusters.
> *o-buffer* the set of o-micro-clusters.
> *CO-MC*, constraint matrix between micro-clusters.
>
> **1. Setting up micro-clusters and micro constraints**.
> Using *InitPoints* and constraints CO from DS:
> *INICIALIATE(InitPoints, CO)*.
>
> **2. Online step: Maintaining micro-clusters and micro constraints.**
> $T_p = \left\lceil \frac{1}{\lambda} log(\frac{\beta \, \mu}{\beta \, \mu \, - \, 1}) \right\rceil$.
>
> **Adding new points:**
> $\forall p \in DS$: *MERGE(p, p-buffer, o-buffer)*.
> **Adding new constraints:**
> $\forall co \in DS$: *MERGE-CONSTRAINTS(co, CO)*.
>
> **Updating micro-clusters and micro constraints:**
> **if** *(t mod T_p = 0)* **then**
> > *UPDATING MICRO-CLUSTERS (p-buffer, o-buffer)*.
> > *UPDATING MICRO CONSTRAINTS (CO-MC)*.
>
> **end**
>
> **3. Offline step: generating final clusters.**
> **if** *user request* **then**
> > C-DBSCAN *(p-buffer, CO-MC)*.
>
> **end**

end

4.3 Strategies for Constraint Arrival in Data Streams

Given the described scenario where data and instance-level constraints arrive over time, we have identified five cases of constraint arrival with five different policies and their corresponding actions to manage the micro-cluster-level constraints:

Procedure. INITIALIZE(InitPoints, CO).

Data:

 InitPoints, initial set of points.

 CO-MC, initial set of constraints.

begin

 C-DBSCAN (*InitPoints, CO*) → *clustering* $\{P\}$.

 $\forall p \in \{P\}$:

 if *(weight$_\epsilon$(p) > $\beta\mu$)* **then**

  New micro-cluster C_p in *p-micro-cluster* with p points in ϵ
  neighborhood.

  Add C_p to *p-buffer*.

 else

  New micro-cluster C_o in *o-micro-cluster* with p points in ϵ neighborhood.

  Add C_o to *o-buffer*.

 $\forall$ *constraint(x,y)* $\in CO$: If $x \in cmc_x$ e $y \in cmc_y$

 CO-MC[mc$_x$, mc$_y$].type = constraint.

 CO-MC[mc$_x$, mc$_y$].weight = t$_c$.

end

Procedure. MERGE MICRO CONSTRAINTS(*co, CO-MC*).

Data:

 co, a instance-level constraint.

 CO, a set of micro constraints.

begin

 A constraint *co(x,y)*: $x \in cmc_x$ e $y \in cmc_y$.

 if *(no exists CO-MC[cmc$_x$, cmc$_y$]* **then**

  CO-MC[cmc$_x$, cmc$_y$].type = co(x,y).type.

  CO-MC[cmc$_x$, cmc$_y$].weight = t$_c$.

 else

  if *(CO-MC[cmc$_x$, cmc$_y$].type = co(x,y).type)* **then**

   CO-MC[cmc$_x$, cmc$_y$].weight++.

  else

   CO-MC[cmc$_x$, cmc$_y$].type = co(x,y).type.

   CO-MC[cmc$_x$, cmc$_y$].weight = t$_c$.

end

Procedure. UPDATING MICRO CONSTRAINTS(*CO-MC*).

Data:

 CO-MC, a set of micro constraints.

begin

 foreach p-micro-cluster C_p *delete* $\in$ p-buffer **do**

  $\forall C_q \in$ *p-micro-cluster*: Remove *CO-MC(C$_p$, C$_q$)*.

 foreach o-micro-cluster C_o *delete* $\in$ o-buffer **do**

  $\forall C_q \in$ *o-micro-cluster*: Remove *CO-MC(C$_p$, C$_q$)*.

end

Fig. 2. C-DenStream: Exception for Cannot-Link constraint

1. Some instance-level constraints arrive, after the transformation step in micro constraints, these are not reflected in the constraint matrix. In this case, the corresponding entry is initialized with those micro constraints with the value indicated by the weight function.
2. Some instance-level constraints arrive, after the corresponding transformation step in micro constraints, these are reflected in the constraint matrix, and their semantic is satisfied: if the constraint matrix shows a Must-Link micro constraint, the new constraints are also Must-Link constraints; if the constraint matrix is a Cannot-Link constraint, the new constraints are also Cannot-Link constraints. As the semantic relationship in the instance-level constraints and micro constraints are the same (Must-Link and Must-Link constraints, or, Cannot-Link and Cannot-Link constraints), the semantic is maintained, and their weights are updated.
3. Some instance-level constraints arrive, after the transformation step into micro constraints, these are reflected in the constraint matrix, but their semantic is not satisfied: if the constraint matrix shows a Must-Link micro constraint, the new constraints are Cannot-Link constraints; whether the constraint matrix is a Cannot-Link constraint, the new constraints are Must-Link constraints. Since the semantic is contradictory (Must-Link and Cannot-Link constraints, or, Cannot-Link and Must-Link constraints), and in this case, the domain information is provided by a domain expert, the constraint matrix is refreshed with the new arrived constraints.
4. There is no new Must-Link or Cannot-Link constraints on arrival. In this case, the corresponding fade function is applied to the constraint matrix.
5. A Cannot-Link constraint arrives, and the instances involved belong both to the same micro-cluster. In this case, the micro-cluster has a constraint that goes against its structure, so we propose to split the micro-cluster and create two new micro-clusters where the elements are located. An example of this case is shown in Figure 2 where there is a restriction Cannot-Link in a micro-cluster.

5   Experimentation and Results

5.1   Datasets

To evaluate the impact of domain information on data streams and how the proposed algorithm deal with it, some interesting synthetic and real datasets are used.

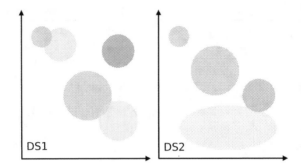

Fig. 3. C-DenStream: Synthetic datasets used for experimentation

On the one hand, the two synthetic datasets, DS1 and DS2, are depicted in Figure 3. Each of them contains 10,000 points. The objectives pursued with each dataset are the following: DS1 has four spherical overlapping clusters. For this case, DenStream performance should deteriorate over time, while the use of domain information should enable C-DenStream to obtain better results; DS2 is similar to DS1, but with non-spherical clusters.

On the other hand, we will also use two well-known data streams sets used in [2,8,17,18], both part of the KDD Cup - International Knowledge Discovery and Data Mining Tools Competition[1]:

- KDD Cup'98 Charitable Donation. This dataset (PVA) with 481 attributes in total was used to predict users that were more likely to donate to charity. The dataset describes those who have donated or not, over the years to the PVA.
- KDD Cup'99 Network Intrusion. This data set [28] had the goal to build a network intrusion detector, capable to distinguish attack, intrusions, and the rest type of the connections.

In both cases, as in [17,18], we use their continuous attributes for our experiments.

5.2   Evaluation Measurements

We use the Rand Index [26] to compare results against the initial labeled class within the synthetic and real datasets. The Rand Index computes the similarity of partitions, having the highest value 1 when the clusters are exactly the same. Partial results are given as new item arrives in the data stream. Taking into account that each point is forgotten over time, the results are computed only taking into account those items that have arrived within a predefined numbers of windows. Other approaches have shown that the results are insensitive to the number of windows [8].

---

[1] http://www.kdd.org/kddcup/

5.3   Constraint Generation

There are no instance-level constraints related to the datasets described, so we use available labels for the constraint generation. This approach is typically used in the static case [21]. There are two possible strategies to feed the clustering algorithm with constraints, which matches two feasible situations in real systems:

- Feeding the algorithm with constraints each time the final clustering in C-DenStream is obtained. This strategy simulates an oracle delivering constraints at the same time the final clustering is requested. This situation reflects a specific supervision and not a constant monitoring by the expert.
- Feeding the algorithm with constraints each time new points arrive to the data stream. This strategy simulates the existence of an intelligent system that generates the constraints over time. In this case, constraints are generated from a 5% to 10% of the size of the window used, following an approach similar to that used in the classical approach.

5.4   Experimentation

Synthetic datasets. For each of the synthetic datasets DS1, and DS2, with 10,000 points, we simulate data arrival in a data stream using a window $Size_{window} = 100$ points, which implies a number of windows so that $Num_{Window} = 100$ per window, and where final clustering will be done at every $Clustering_{final} = 10$ windows. We use the second strategy presented to generate constraints, which uses 5% of the instances of each window. The implementation of DenStream used, is in accordance to the specifications of [8], and this is the one that has been used as a basis for the deployment of C-DenStream. The parameters for DenStream and C-DenStream, λ (decay factor) and β (outlier threshold), are 0.2 and 0.25, respectively. The parameters for DBSCAN and C-DBSCAN, ϵ and μ , are 0.6 and 5, respectively.

Results for DS1 and DS2 datasets are depicted in Figure 4(a) and in Figure 4(b), respectively. The horizontal axis indicates the arrival of the points in the data stream, showing the final clustering for DenStream and C-DenStream using the corresponding Rand Index.

Figure 4(a) shows the results for the case of DS1, where there is a cluster overlap, and where the use of data stream clustering with constraints by C-DenStream, improves the partial results in each of the milestones. A similar case can be observed for the DS2 dataset in Figure 4(b). A note on the empirical comparison: we have included DBSCAN, a non-stream clustering algorithm, as a performance baseline to ilustrate a default behaviour to compare stream methods with. For this case, DBSCAN takes the same parameters used for DBSCAN within DenStream.

The above results show how C-DenStream partially removes the drawbacks found in DenStream because of the lack of domain information. It is worth

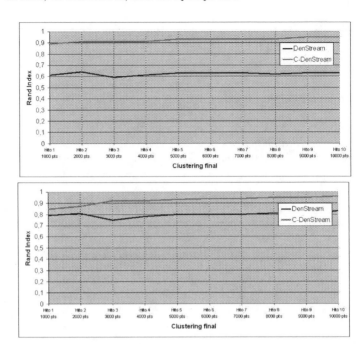

Fig. 4. C-DenStream: Results for (a) DS1 and (b) DS2

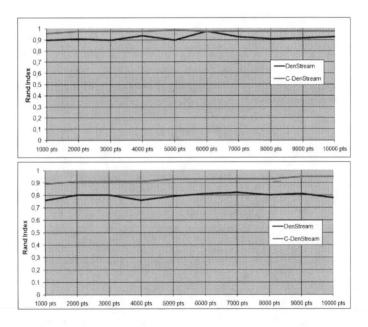

Fig. 5. C-DenStream: Results for KDD CUP'99 Network Intrusion and KDD CUP'98 Charitable Donation

noting the fact that a small amount of domain information in the form of constraints, transformed to a micro-level constraints, is enough to get a better performance.

Experiments with datasets from data stream repositories. For each dataset from the KDD Cup'98 and KDD Cup'99 competition, 65,536 instances have been used, as in the synthetic datasets, we simulate the data arrival as a data stream with a window $Size_{Window} = 100$ points, representing a number of windows so that $Num_{Window} = 655$ per window and a final clustering at $Clustering_{final} = 65$ windows. The parameters for DenStream and C-DenStream are chosen to be the same as those adopted in [8]. The second strategy of generating restrictions has again been used, which uses 5 % of the instances of each window for the constraint generation for C-DenStream. Figure 5 sets out the results for the first 10,000 points for each dataset showing again the improvememts as a result of the domain information use.

Figure 4 and Figure 5 show the performance results of C-DenStream and DenStream in a fixed small horizon for synthetic and real datasets, and small amount of constraints. For both, C-DenStream and DenStream, we have used the same input parameters, λ and β, for C-DBSCAN and DBSCAN. It can be seen that C-DenStream has a very good clustering quality and outperforms the results obtained with DenStream, the approach with no background. A few cases show that the result for C-DenStream and DenStream are similar, this is because there are sometimes no interpretation problems.

6   Conclusion

Data streams pose new and interesting challenges for data mining systems [4]. Consequently solutions to deal with computational and storage challenges have appeared. On the one hand, some algorithms address specific computational requirements, for example, with a single pass in the analysis in order to minimise the use of memory [13,14,25]; on the other hand some solutions are based on structures that summarise the properties of different clusters such as micro-clusters or wavelets [1,8].

The use of domain information as an effective mechanism to deliver results with the vision and expectations of the experts, has attracted a lot of attention in data mining. In particular, semi-supervised clustering or clustering with constraints [29] uses domain information in the form of instance-level constraints to improve the clustering results and performance [10].

In this paper we have proposed C-DenStream, a clustering algorithm based on DenStream for data streams which makes use of domain information. In this paper we have shown the challenges of using contraints on data streams: an infinite number of constraints cannot be stored and an effective management way is needed. Thus we have shown how to transform the succesful approach of instance-level constraints into micro-cluster-level constraints. Experiments presented show that results are highly satisfactory.

References

1. Aggarwal, C.C. (ed.): Data Streams: Models and Algorithms, Advances in Database Systems. Springer, Heidelberg (2007)
2. Aggarwal, C.C., Han, J., Wang, J., Yu, P.S.: A Framework for Clustering Evolving Data Streams. In: VLDB 2003: Proc. of the 29th in Very Large Data Bases Conf (2003)
3. Aguilar-Ruiz, J.S., Gama, J.: Data Streams. Journal of Universal Computer Science 11(8), 1349–1352 (2005)
4. Barbará, D.: Requirements for clustering data streams. SIGKDD Explor. Newsl. 3(2), 23–27 (2002)
5. Basu, S., Banerjee, A., Mooney, R.J.: Semi-supervised Clustering by Seeding. In: ICML 2002: Proc. Int. Conf. on Machine Learning, pp. 19–26 (2002)
6. Basu, S., Bilenko, M., Mooney, R.J.: A Probabilistic Framework for Semi-Supervised Clustering. In: KDD 2004: Proc. of 10th Int. Conf. on Knowledge Discovery in Databases and Data Mining, pp. 59–68 (2004)
7. Bilenko, M., Basu, S., Mooney, R.J.: Integrating Constraints and Metric Learning in Semisupervised Clustering. In: ICML 2004: Proc. of the 21th Int. Conf. on Machine Learning, pp. 11–19 (2004)
8. Cao, F., Ester, M., Qian, W., Zhou, A.: Density-based clustering over an evolving data stream with noise. In: SIAM 2006: SIAM Int. Conf. on Data Mining (2006)
9. Chaudhry, N., Shaw, K., Abdelguerfi, M. (eds.): Stream Data Management, Advances in Database Systems. Springer, Heidelberg (2005)
10. Davidson, I., Basu, S.: Clustering with Constraints: Theory and Practice. In: KDD 2006: Tutorial at The Int. Conf. on Knowledge Discovery in Databases and Data Mining (2006)
11. Davidson, I., Ravi, S.S.: Clustering with Constraints: Feasibility Issues and the k-Means Algorithm. In: SIAM 2005: Proceeding of the SIAM Int. Conf. on Data Mining Int. Conf. in Data Mining (2005)
12. Davidson, I., Wagstaff, K.L., Basu, S.: Measuring Constraint-Set Utility for Partitional Clustering Algorithms. In: Fürnkranz, J., Scheffer, T., Spiliopoulou, M. (eds.) PKDD 2006. LNCS (LNAI), vol. 4213, pp. 115–126. Springer, Heidelberg (2006)
13. Domingos, P., Hulten, G.: Mining High-Speed Data Streams. In: KDD 2000: Proc. of the ACM 6th Int. Conf. on Knowledge Discovery and Data Mining, pp. 71–80 (2000)
14. Domingos, P., Hulten, G.: A General Method for Scaling Up Machine Learning Algorithms and its Application to Clustering. In: ICML 2001: Proc. of the 18th Int. Conf. on Machine Learning, San Francisco, CA, USA, pp. 106–113. Morgan Kaufmann Publishers Inc., San Francisco (2001)
15. Domingos, P., Hulten, G.: Catching Up with the Data: Research Issues in Mining Data Streams. In: DMKD 2001: Workshop on Research Issues in Data Mining and Knowledge Discovery (2001)
16. Ester, M., Kriegel, H.-P., Sander, J., Xu, X.: A Density-Based Algortihm for Discovering Clusters in Large Spatial Database with Noise. In: KDD 1996: Proc. of 2nd Int. Conf. on Knowledge Discovery in Databases and Data Mining (1996)
17. Gaber, M., Krishnaswamy, S., Zaslavsky, A.: Ubiquitous Data Stream Mining. In: Proc. of the Current Research and Future Directions Workshop in PAKDD 2004, pp. 37 – 46 (2004)

18. Gao, J., Li, J., Zhang, Z., Tan, P.-N.: An Incremental Data Stream Clustering Algorithm Based on Dense Units Detection. In: Proc. of the Current Research and Future Directions Workshop held in PAKDD 2005, pp. 420–425 (2005)
19. Guha, S., Meyerson, A., Mishra, N., Motwani, R., O'Callaghan, L.: Clustering Data Streams: Theory and Practice. IEEE Transactions on Knowledge and Data Engineering 15(3), 515–528 (2003)
20. Gunopulos, D., Vazirgiannis, M., Halkidi, M.: Novel Aspects in Unsupervised Learning: Semi-Supervised and Distributed Algorithms. In: ECML/PKDD 2006: Tutorial at 17th European Conf. on Machine Learning and the 10th European Conf. on Principles and Practice of Knowledge Discovery in Databases (2006)
21. Halkidi, M., Gunopulos, D., Kumar, N., Vazirgiannis, M., Domeniconi, C.: A Framework for Semi-Supervised Learning Based on Subjective and Objective Clustering Criteria. In: ICDM 2005: Proc. of the 5th IEEE Int. Conf. on Data Mining, pp. 637–640 (2005)
22. Klein, D., Kamvar, S.D., Manning, C.: From instance-level constraints to space-level constraints: making the most of prior knowledge in data clustering. In: ICML 2002: Proc. of the 19th Int. Conf. on Machine Learning, pp. 307–314 (2002)
23. Nasraoui, O., Rojas, C.: Robust Clustering for Tracking Noisy Evolving Data Streams. In: SDM 2006: Proc. of the 6th SIAM Int. Conf. on Data Mining, pp. 80–89 (2006)
24. Nasraoui, O., Uribe, C.C., Coronel, C.R., Gonzalez, F.: TECNO-STREAMS: Tracking Evolving Clusters in Noisy Data Streams with a Scalable Immune System Learning Model. In: ICDM 2003: Proc. of the 3rd IEEE Int. Conf. on Data Mining, Washington, DC, USA, p. 235. IEEE Computer Society Press, Los Alamitos (2003)
25. Ordonez, C.: Clustering binary data streams with K-means. In: DMKD 2003: Proc. of the 8th ACM SIGMOD Workshop on Research Issues in Data Mining and Knowledge Discovery, pp. 12–19. ACM Press, New York (2003)
26. Rand, W.M.: Objective Criteria for the Evalluation of Clustering Methods. Journal of the American Statistical Association 66, 846–850 (1971)
27. Ruiz, C., Spiliopoulou, M., Menasalvas, E.: C-DBSCAN: Density-Based Clustering with Constraints. In: RSFDGrC 2007: Proc. of the Int. Conf. on Rough Sets, Fuzzy Sets, Data Mining and Granular Computing (2007)
28. Stolfo, S.J., Fan, W., Lee, W., Prodromidis, A., Chan, P.K.: Cost-based Modeling and Evaluation for Data Mining With Application to Fraud and Intrusion Detection: Results from the JAM Project. Technical report, U. of Columbia (1998)
29. Wagstaff, K., Cardie, C., Rogers, S., Schroedl, S.: Constrained K-means Clustering with Background Knowledge. In: ICML 2001: Proc. of 18th Int. Conf. on Machine Learning, pp. 577–584 (2001)
30. Xing, E.P., Ng, A.Y., Jordan, M.I., Russell, S.: Distance Metric Learning, with Application to Clustering with Side-Information. Advances in Neural Information Processing Systems 15, 505–512 (2003)

Discovering Influential Nodes for SIS Models in Social Networks

Kazumi Saito[1], Masahiro Kimura[2], and Hiroshi Motoda[3]

[1] School of Administration and Informatics, University of Shizuoka
52-1 Yada, Suruga-ku, Shizuoka 422-8526, Japan
k-saito@u-shizuoka-ken.ac.jp
[2] Department of Electronics and Informatics, Ryukoku University
Otsu, Shiga 520-2194, Japan
kimura@rins.ryukoku.ac.jp
[3] Institute of Scientific and Industrial Research, Osaka University
8-1 Mihogaoka, Ibaraki, Osaka 567-0047, Japan
motoda@ar.sanken.osaka-u.ac.jp

Abstract. We address the problem of efficiently discovering the influential nodes in a social network under the *susceptible/infected/susceptible (SIS) model*, a diffusion model where nodes are allowed to be activated multiple times. The computational complexity drastically increases because of this multiple activation property. We solve this problem by constructing a layered graph from the original social network with each layer added on top as the time proceeds, and applying the bond percolation with pruning and burnout strategies. We experimentally demonstrate that the proposed method gives much better solutions than the conventional methods that are solely based on the notion of centrality for social network analysis using two large-scale real-world networks (a blog network and a wikipedia network). We further show that the computational complexity of the proposed method is much smaller than the conventional naive probabilistic simulation method by a theoretical analysis and confirm this by experimentation. The properties of the influential nodes discovered are substantially different from those identified by the centrality-based heuristic methods.

1 Introduction

Social networks mediate the spread of various information including topics, ideas and even (computer) viruses. The proliferation of emails, blogs and social networking services (SNS) in the World Wide Web accelerates the creation of large social networks. Therefore, substantial attention has recently been directed to investigating information diffusion phenomena in social networks [1,2,3].

Overall, finding influential nodes is one of the most central problems in social network analysis. Thus, developing methods to do this on the basis of information diffusion is an important research issue. Widely-used fundamental probabilistic models of information diffusion are the *independent cascade (IC) model* and the *linear threshold (LT) model* [4,5]. Researchers investigated the problem of finding a limited number of influential nodes that are effective for the spread of information under the above models [4,6]. This combinatorial optimization problem is called the *influence maximization*

J. Gama et al. (Eds.): DS 2009, LNAI 5808, pp. 302–316, 2009.

problem. Kempe et al. [4] experimentally showed on large collaboration networks that the greedy algorithm can give a good approximate solution to this problem, and mathematically proved a performance guarantee of the greedy solution (i.e., the solution obtained by the greedy algorithm). Recently, methods based on bond percolation [6] and submodularity [7] were proposed for efficiently estimating the greedy solution. The influence maximization problem has applications in sociology and "viral marketing" [3], and was also investigated in a different setting (a descriptive probabilistic model of interaction) [8,9]. The problem has recently been extended to influence control problems such as a contamination minimization problem [10].

The IC model can be identified with the so-called *susceptible/infected/recovered (SIR) model* for the spread of a disease [11,5]. In the SIR model, only infected individuals can infect susceptible individuals, while recovered individuals can neither infect nor be infected. This implies that an individual is never infected with the disease multiple times. This property holds true for the LT model as well. However, there exist phenomena for which the property does not hold. For example, consider the following propagation phenomenon of a topic in the blogosphere: A blogger who has not yet posted a message about the topic is interested in the topic by reading the blog of a friend, and posts a message about it (i.e., becoming infected). Next, the same blogger reads a new message about the topic posted by some other friend, and may post a message (i.e., becoming infected) again. Most simply, this phenomenon can be modeled by an *susceptible/infected/susceptible (SIS) model* from the epidemiology. Like this example, there are many examples of information diffusion phenomena for which the SIS model is more appropriate, including the growth of hyper-link posts among bloggers [2], the spread of computer viruses without permanent virus-checking programs, and epidemic disease such as tuberculosis and gonorrhea [11].

We focus on an information diffusion process in a social network $G = (V, E)$ over a given time span T on the basis of an SIS model. Here, the SIS model is a stochastic process model, and the *influence* of a set of nodes H at time-step t, $\sigma(H, t)$, is defined as the expected number of infected nodes at time-step t when all the nodes in H are initially infected at time-step $t = 0$. We refer to σ as the *influence function* for the SIS model. Developing an effective method for estimating $\sigma(\{v\}, t)$, $(v \in V, t = 1, \ldots, T)$ is vital for various applications. Clearly, in order to extract influential nodes, we must estimate the value of $\sigma(\{v\}, t)$ for every node v and time-step t. Thus, we proposed a novel method based on the bond percolation with an effective pruning strategy to efficiently estimate $\{\sigma(\{v\}, t); v \in V, t = 1, \ldots, T\}$ for the SIS model in our previous work [12].

In this paper, we consider solving the influence maximization problems on a network $G = (V, E)$ under the SIS model. Here, unlike the cases of the IC and the LT models, we define two influence maximization problems, the *final-time maximization problem* and the *accumulated-time maximization problem*, for the SIS model. We introduce the greedy algorithm for solving the problems according to the work of Kempe et al. [4] for the IC and the LT models. Now, let us consider the problem of influence maximization at the final time step T (i.e., final-time maximization problem) as an example. We then note that for solving this problem by the greedy algorithm, we need a method for not only evaluating $\{\sigma(\{v\}, T); v \in V\}$, but also evaluating the *marginal influence gains* $\{\sigma(H \cup \{v\}, T) - \sigma(H, T); v \in V \setminus H\}$ for any non-empty subset H of V. Needless to say, we can naively estimate the marginal influence gains for any non-empty subset H of V by simulating the SIS model[1]. However, this naive simulation method is overly inefficient and not practical at all. In this paper, by incorporating the new techniques (the pruning and the burnout methods) into the bond percolation method, we propose a method to efficiently estimate the marginal influence gains for any non-empty subset H of V, and apply it to approximately solve the two influence maximization problems for the SIS model by the greedy alogrithm. We show that the proposed method is expected to achieve a large reduction in computational cost by theoretically comparing computational complexity with other more naive methods. Further, using two large real networks, we experimentally demonstrate that the proposed method is much more efficient than the naive greedy method based on the bond percolation method. We also show that the discovered nodes by the proposed method are substantially different from and can result in considerable increase in the influence over the conventional methods that are based on the notion of various centrality measures.

2   Information Diffusion Model

Let $G = (V, E)$ be a directed network, where V and E ($\subset V \times V$) stand for the sets of all the nodes and (directed) links, respectively. For any $v \in V$, let $\Gamma(v; G)$ denote the set of the child nodes (directed neighbors) of v, that is,

$$\Gamma(v; G) = \{w \in V; (v, w) \in E\}.$$

2.1   SIS Model

An SIS model for the spread of a disease is based on the cycle of disease in a host. A person is first *susceptible* to the disease, and becomes *infected* with some probability when the person encounters an infected person. The infected person becomes susceptible to the disease soon without moving to the immune state. We consider a discrete-time SIS model for information diffusion on a network. In this context, infected nodes mean that they have just adopted the information, and we call these infected nodes *active* nodes.

We define the SIS model for information diffusion on G. In the model, the diffusion process unfolds in discrete time-steps $t \geq 0$, and it is assumed that the state of a node is either active or inactive. For every link $(u, v) \in E$, we specify a real value $p_{u,v}$ with $0 < p_{u,v} < 1$ in advance. Here, $p_{u,v}$ is referred to as the *propagation probability* through link (u, v). Given an initial set of active nodes X and a time span T, the diffusion process proceeds in the following way. Suppose that node u becomes active at time-step t ($< T$). Then, node u attempts to activate every $v \in \Gamma(u; G)$, and succeeds with probability $p_{u,v}$. If node u succeeds, then node v will become active at time-step $t + 1$. If multiple active nodes attempt to activate node v in time-step t, then their activation attempts are sequenced in an arbitrary order. On the other hand, node u will become or remain inactive at time-step $t + 1$ unless it is activated from an active node in time-step t. The process terminates if the current time-step reaches the time limit T.

---

[1] Note that the method we proposed in [12] does not perform simulation.

2.2 Influence Function

For the SIS model on G, we consider a diffusion sample from an initially activated node set $H \subset V$ over time span T. Let $S(H,t)$ denote the set of active nodes at time-step t. Note that $S(H,t)$ is a random subset of V and $S(H,0) = H$. Let $\sigma(H,t)$ denote the expected number of $|S(H,t)|$, where $|X|$ stands for the number of elements in a set X. We call $\sigma(H,t)$ the *influence* of node set H at time-step t. Note that σ is a function defined on $2^{|V|} \times \{0, 1, \cdots, T\}$. We call the function σ the *influence function* for the SIS model over time span T on network G. In view of more complex social influence, we need to incorporate a number of social factors with social networks such as rank, prestige and power. In our approach, we can encode such factors as diffusion probabilities of each node.

It is important to estimate the influence function σ efficiently. In theory we can simply estimate σ by the simulations based on the SIS model in the following way. First, a sufficiently large positive integer M is specified. For each $H \subset V$, the diffusion process of the SIS model is simulated from the initially activated node set H, and the number of active nodes at time-step t, $|S(H,t)|$, is calculated for every $t \in \{0, 1, \cdots, T\}$. Then, $\sigma(H,t)$ is estimated as the empirical mean of $|S(H,t)|$'s that are obtained from M such simulations. However, this is extremely inefficient, and cannot be practical.

3 Influence Maximization Problem

We mathematically define the influence maximization problems on a network $G = (V, E)$ under the SIS model. Let K be a positive integer with $K < |V|$. First, we define the *final-time maximization problem*: Find a set H_K^* of K nodes to target for initial activation such that $\sigma(H_K^*; T) \geq \sigma(H; T)$ for any set H of k nodes, that is, find

$$H_K^* = \arg \max_{\{H \subset V; |H|=K\}} \sigma(H; T). \tag{1}$$

Second, we define the *accumulated-time maximization problem*: Find a set H_K^* of K nodes to target for initial activation such that $\sigma(H_K^*; 1) + \cdots + \sigma(H_K^*; T) \geq \sigma(H; 1) + \cdots + \sigma(H; T)$ for any set H of k nodes, that is, find

$$H_K^* = \arg \max_{\{H \subset V; |H|=K\}} \sum_{t=1}^{T} \sigma(H; t). \tag{2}$$

The first problem cares only how many nodes are influenced at the time of interest. For example, in an election campaign it is only those people who are convinced to vote the candidate at the time of voting that really matter and not those who were convinced during the campaign but changed their mind at the very end. Maximizing the number of people who actually vote falls in this category. The second problem cares how many nodes have been influenced throughout the period of interest. For example, maximizing the amount of product purchase during a sales campaign falls in this category.

4   Proposed Method

Kempe et al. [4] showed the effectiveness of the greedy algorithm for the influence maximization problem under the IC and LT models. In this section, we introduce the greedy algorithm for the SIS model, and describe some techniques (the bond percolation method, the pruning method, and the burnout method) for efficiently solving the influence maximization problem under the greedy algorithm, together with some arguments for evaluating the computational complexity for these methods.

4.1   Greedy Algorithm

We approximately solve the influence maximization problem by the greedy algorithm. Below we describe this algorithm for the final-time maximization problem:

Greedy algorithm for the final-time maximization problem:
$\mathcal{A}$1. Set $H \leftarrow \emptyset$.
$\mathcal{A}$2. For $k = 1$ to K do the following steps:
$\mathcal{A}$2-1. Choose a node $v_k \in V \setminus H$ maximizing $\sigma(H \cup \{v\}, T)$.
$\mathcal{A}$2-2. Set $H \leftarrow H \cup \{v_k\}$.
$\mathcal{A}$3. Output H.

Here we can easily modify this algorithm for the accumulated-time maximization problem by replacing step $\mathcal{A}$2-1 as follows:

Greedy algorithm for the accumulated-time maximization problem:
$\mathcal{A}$1. Set $H \leftarrow \emptyset$.
$\mathcal{A}$2. For $k = 1$ to K do the following steps:
$\mathcal{A}$2-1'. Choose a node $v_k \in V \setminus H$ maximizing $\sum_{t=1}^{T} \sigma(H \cup \{v\}, t)$.
$\mathcal{A}$2-2. Set $H \leftarrow H \cup \{v_k\}$.
$\mathcal{A}$3. Output H.

Let H_K denote the set of K nodes obtained by this algorithm. We refer to H_K as the *greedy solution* of size K. Then, it is known that

$$\sigma(H_K, t) \geq \left(1 - \frac{1}{e}\right) \sigma(H_K^*, t),$$

that is, the quality guarantee of H_k is assured [4]. Here, H_k^* is the exact solution defined by Equation (1) or (2).

To implement the greedy algorithm, we need a method for estimating all the marginal influence degrees $\{\sigma(H \cup \{v\}, t); v \in V \setminus H\}$ of H in step $\mathcal{A}$2-1 or $\mathcal{A}$2-1' of the algorithm. In the subsequent subsections, we propose a method for efficiently estimating the influence function σ over time span T for the SIS model on network G.

4.2   Layered Graph

We build a layered graph $G^T = (V^T, E^T)$ from G in the following way (see Figure 1). First, for each node $v \in V$ and each time-step $t \in \{0, 1, \cdots, T\}$, we generate a copy v_t

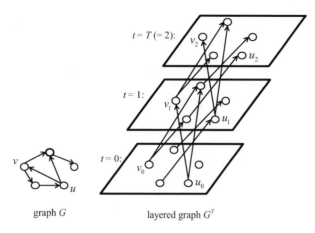

Fig. 1. An example of a layered graph

of v at time-step t. Let V_t denote the set of copies of all $v \in V$ at time-step t. We define V^T by $V^T = V_0 \cup V_1 \cup \cdots \cup V_T$. In particular, we identify V with V_0. Next, for each link $(u, v) \in E$, we generate T links (u_{t-1}, v_t), $(t \in \{1, \cdots, T\})$, in the set of nodes V^T. We set $E_t = \{(u_{t-1}, v_t); (u, v) \in E\}$, and define E^T by $E^T = E_1 \cup \cdots \cup E_T$. Moreover, for any link (u_{t-1}, v_t) of the layered graph G^T, we define the occupation probability q_{u_{t-1}, v_t} by $q_{u_{t-1}, v_t} = p_{u,v}$.

Then, we can easily prove that the SIS model with propagation probabilities $\{p_e; e \in E\}$ on G over time span T is equivalent to the *bond percolation process (BP) with occupation probabilities* $\{q_e; e \in E^T\}$ on G^T.[2] Here, the BP process with occupation probabilities $\{q_e; e \in E^T\}$ on G^T is the random process in which each link $e \in E^T$ is independently declared "occupied" with probability q_e. We perform the BP process on G^T, and generate a graph constructed by occupied links, $\tilde{G}^T = (V^T, \tilde{E}^T)$. Then, in terms of information diffusion by the SIS model on G, an occupied link $(u_{t-1}, v_t) \in E_t$ represents a link $(u, v) \in E$ through which the information propagates at time-step t, and an unoccupied link $(u_{t-1}, v_t) \in E_t$ represents a link $(u, v) \in E$ through which the information does not propagate at time-step t. For any $v \in V \setminus H$, let $F(H \cup \{v\}; \tilde{G}^T)$ be the set of all nodes that can be reached from $H \cup \{v\} \in V_0$ through a path on the graph $\tilde{G}^T$. When we consider a diffusion sample from an initial active node $v \in V$ for the SIS model on G, $F(H \cup \{v\}; \tilde{G}^T) \cap V_t$ represents the set of active nodes at time-step t, $S(H \cup \{v\}, t)$.

4.3 Bond Percolation Method

Using the equivalent BP process, we present a method for efficiently estimating influence function σ. We refer to this method as the *BP method*. Unlike the naive method, the BP method simultaneously estimates $\sigma(H \cup \{v\}, t)$ for all $v \in V \setminus H$. Moreover, the BP method does not fully perform the BP process, but performs it partially. Note first

---

[2] The SIS model over time span T on G can be exactly mapped onto the IC model on G^T [4]. Thus, the result follows from the equivalence of the BP process and the IC model [11,4,6].

that all the paths from nodes $H \cup \{v\}$ ($v \in V \setminus H$) on the graph $\tilde{G}^T$ represent a diffusion sample from the initial active nodes $H \cup \{v\}$ for the SIS model on G. Let L' be the set of the links in G^T that is not in the diffusion sample. For calculating $|S(H \cup \{v\}, t)|$, it is unnecessary to determine whether the links in L' are occupied or not. Therefore, the BP method performs the BP process for only an appropriate set of links in G^T. The BP method estimates σ by the following algorithm:

BP method:
$\mathcal{B}$1. Set $\sigma(H \cup \{v\}, t) \leftarrow 0$ for each $v \in V \setminus H$ and $t \in \{1, \cdots, T\}$.
$\mathcal{B}$2. Repeat the following procedure M times:
$\mathcal{B}$2-1. Initialize $S(H \cup \{v\}, 0) = H \cup \{v\}$ for each $v \in V \setminus H$, and set $A(0) \leftarrow V \setminus H$, $A(1) \leftarrow \emptyset, \cdots, A(T) \leftarrow \emptyset$.
$\mathcal{B}$2-2. For $t = 1$ to T do the following steps:
$\mathcal{B}$2-2a. Compute $B(t-1) = \bigcup_{v \in A(t-1)} S(H \cup \{v\}, t-1)$.
$\mathcal{B}$2-2b. Perform the BP process for the links from $B(t-1)$ in G^T, and generate the graph $\tilde{G}_t$ constructed by the occupied links.
$\mathcal{B}$2-2c. For each $v \in A(t-1)$, compute $S(H \cup \{v\}, t) = \bigcup_{w \in S(H \cup \{v\}, t-1)} \Gamma(w; \tilde{G}_t)$, and set $\sigma(H \cup \{v\}, t) \leftarrow \sigma(H \cup \{v\}, t) + |S(H \cup \{v\}, t)|$ and $A(t) \leftarrow A(t) \cup \{v\}$ if $S(H \cup \{v\}, t) \neq \emptyset$.
$\mathcal{B}$3. For each $v \in V \setminus H$ and $t \in \{1, \cdots, T\}$, set $\sigma(H \cup \{v\}, t) \leftarrow \sigma(H \cup \{v\}, t)/M$, and output $\sigma(H \cup \{v\}, t)$.

Note that $A(t)$ finally becomes the set of information source nodes that have at least an active node at time-step t, that is, $A(t) = \{v \in V \setminus H; S(H \cup \{v\}, t) \neq \emptyset\}$. Note also that $B(t-1)$ is the set of nodes that are activated at time-step $t-1$ by some source nodes, that is, $B(t-1) = \bigcup_{v \in V} S(H \cup \{v\}, t-1)$.

Now we estimate the computational complexity of the BP method in terms of the number of the nodes, N_a, that are identified in step $\mathcal{B}$2-2a, the number of the coin-flips, N_b, for the BP process in step $\mathcal{B}$2-2b, and the number of the links, N_c, that are followed in step $\mathcal{B}$2-2c. Let $d(v)$ be the number of out-links from node v (i.e., out-degree of v) and $d'(v)$ the average number of occupied out-links from node v after the BP process. Here we can estimate $d'(v)$ by $\sum_{w \in \Gamma(v;G)} p_{v,w}$. Then, for each time-step $t \in \{1, \cdots, T\}$, we have

$$N_a = \sum_{v \in A(t-1)} |S(H \cup \{v\}, t-1)|, \quad N_b = \sum_{w \in B(t-1)} d(w), \quad N_c = \sum_{v \in A(t-1)} \sum_{w \in S(H \cup \{v\}, t-1)} d'(w) \quad (3)$$

on average.

In order to compare the computational complexity of the BP method to that of the naive method, we consider mapping the naive method onto the BP framework, that is, separating the coin-flip process and the link-following process. We can easily verify that the following algorithm in the BP framework is equivalent to the naive method:

A method that is equivalent to the naive method:
$\mathcal{B}$1. Set $\sigma(H \cup \{v\}, t) \leftarrow 0$ for each $v \in V \setminus H$ and $t \in \{1, \cdots, T\}$.
$\mathcal{B}$2. Repeat the following procedure M times:
$\mathcal{B}$2-1. Initialize $S(H \cup \{v\}, 0) = H \cup \{v\}$ for each $v \in V \setminus H$, and set $A(0) \leftarrow V \setminus H$, $A(1) \leftarrow \emptyset, \cdots, A(T) \leftarrow \emptyset$.

$\mathcal{B}$2-2. For $t = 1$ to T do the following steps:

$\mathcal{B}$2-2b'. For each $v \in A(t-1)$, perform the BP process for the links from $S(H \cup \{v\}, t-1)$ in G^T, and generate the graph $\tilde{G}_t(v)$ constructed by the occupied links.

$\mathcal{B}$2-2c'. For each $v \in A(t-1)$, compute $S(H \cup \{v\}; t) = \bigcup_{w \in S(H \cup \{v\}, t-1)} \Gamma(w; \tilde{G}_t(v))$, and set $\sigma(H \cup \{v\}, t) \leftarrow \sigma(H \cup \{v\}, t) + |S(H \cup \{v\}, t)|$ and $A(t) \leftarrow A(t) \cup \{v\}$ if $S(H \cup \{v\}, t) \neq \emptyset$.

$\mathcal{B}$3. For each $v \in V \setminus H$ and $t \in \{1, \cdots, T\}$, set $\sigma(H \cup \{v\}, t) \leftarrow \sigma(H \cup \{v\}, t)/M$, and output $\sigma(H \cup \{v\}, t)$.

Then, for each $t \in \{1, \cdots, T\}$, the number of coin-flips, $N_{b'}$, in step $\mathcal{B}$2-2b' is

$$N_{b'} = \sum_{v \in A(t-1)} \sum_{w \in S(H \cup \{v\}, t-1)} d(w), \qquad (4)$$

and the number of the links, $N_{c'}$, followed in step $\mathcal{B}$2-2c' is equal to N_c in the BP method on average. From equations (3) and (4), we can see that $N_{b'}$ is much larger than $N_{c'} = N_c$, especially for the case where the diffusion probabilities are small. We can also see that $N_{b'}$ is generally much larger than each of N_a and N_b in the BP method for a real social network. In fact, since such a network generally includes large clique-like subgraphs, there are many nodes $w \in V$ such that $d(w) \gg 1$, and we can expect that $\sum_{v \in A(t-1)} |S(H \cup \{v\}, t-1)| \gg |\bigcup_{v \in A(t-1)} S(H \cup \{v\}, t-1)|$ $(= |B(t-1)|)$. Therefore, the BP method is expected to achieve a large reduction in computational cost.

4.4   Pruning Method

In order to further improve the computational efficiency of the BP method, we introduce a pruning technique and propose a method referred to as the *BP with pruning method*. The key idea of the pruning technique is to utilize the following property: Once we have $S(H \cup \{u\}, t_0) = S(H \cup \{v\}, t_0)$ at some time-step t_0 on the course of the BP process for a pair of information source nodes, u and v, then we have $S(H \cup \{u\}, t) = S(H \cup \{v\}, t)$ for all $t > t_0$. The BP with pruning method estimates σ by the following algorithm:

BP with pruning method:

$\mathcal{B}$1. Set $\sigma(H \cup \{v\}, t) \leftarrow 0$ for each $v \in V \setminus H$ and $t \in \{1, \cdots, T\}$.

$\mathcal{B}$2. Repeat the following procedure M times:

$\mathcal{B}$2-1''. Initialize $S(H \cup \{v\}; 0) = H \cup \{v\}$ for each $v \in V \setminus H$, and set $A(0) \leftarrow V \setminus H$, $A(1) \leftarrow \emptyset, \cdots, A(T) \leftarrow \emptyset$, and $C(v) \leftarrow \{v\}$ for each $v \in V \setminus H$.

$\mathcal{B}$2-2. For $t = 1$ to T do the following steps:

$\mathcal{B}$2-2a. Compute $B(t-1) = \bigcup_{v \in A(t-1)} S(H \cup \{v\}, t-1)$.

$\mathcal{B}$2-2b. Perform the BP process for the links from $B(t-1)$ in G^T, and generate the graph $\tilde{G}_t$ constructed by the occupied links.

$\mathcal{B}$2-2c''. For each $v \in A(t-1)$, compute $S(H \cup \{v\}, t) = \bigcup_{w \in S(H \cup \{v\}, t-1)} \Gamma(w; \tilde{G}_t)$, set $A(t) \leftarrow A(t) \cup \{v\}$ if $S(H \cup \{v\}, t) \neq \emptyset$, and set $\sigma(H \cup \{u\}, t) \leftarrow \sigma(H \cup \{u\}, t) + |S(H \cup \{v\}, t)|$ for each $u \in C(v)$.

$\mathcal{B}$2-2d. Check whether $S(H \cup \{u\}, t) = S(H \cup \{v\}, t)$ for $u, v \in A(t)$, and set $C(v) \leftarrow C(v) \cup C(u)$ and $A(t) \leftarrow A(t) \setminus \{u\}$ if $S(H \cup \{u\}, t) = S(H \cup \{v\}, t)$.

$\mathcal{B}$3. For each $v \in V \setminus H$ and $t \in \{1, \cdots, T\}$, set $\sigma(H \cup \{v\}, t) \leftarrow \sigma(H \cup \{v\}, t)/M$, and output $\sigma(H \cup \{v\}, t)$.

Basically, by introducing step $\mathcal{B}$2-2d and reducing the size of $A(t)$, the proposed method attempts to improve the computational efficiency in comparison to the original BP method. For the proposed method, it is important to implement efficiently the equivalence check process in step $\mathcal{B}$2-2d. In our implementation, we first classify each $v \in A(t)$ according to the value of $n = |S(H \cup \{v\}, t)|$, and then perform the equivalence check process only for those nodes with the same n value.

4.5   Burnout Method

In order to further improve the computational efficiency of the BP with pruning method, we additionally introduce a burnout technique and propose a method referred to as the *BP with pruning and burnout method*. More specifically, we focus on the fact that maximizing the marginal influence degree $\sigma(H \cup \{v\}, t)$ with respect to $v \in V \setminus H$ is equivalent to maximizing the marginal influence gain $\phi_H(v, t) = \sigma(H \cup \{v\}, t) - \sigma(H, t)$. Here on the course of the BP process for a newly added information source node v, maximizing $\phi_H(v, t)$ reduces to maximizing $|S(H \cup \{v\}, t) \setminus S(H, t)|$ on average. The BP with pruning and burnout method estimates ϕ_H by the following algorithm:

BP with pruning and burnout methods:
C1. Set $\phi_H(v, t) \leftarrow 0$ for each $v \in V \setminus H$ and $t \in \{1, \cdots, T\}$.
C2. Repeat the following procedure M times:
C2-1. Initialize $S(H; 0) = H$, and $S(\{v\}; 0) = \{v\}$ for each $v \in V \setminus H$, and set $A(0) \leftarrow V \setminus H$, $A(1) \leftarrow \emptyset, \cdots, A(T) \leftarrow \emptyset$, and $C(v) \leftarrow \{v\}$ for each $v \in V \setminus H$.
C2-2. For $t = 1$ to T do the following steps:
C2-2a. Compute $B(t-1) = \bigcup_{v \in A(t-1)} S(\{v\}, t-1) \cup S(H, t-1)$.
C2-2b. Perform the BP process for the links from $B(t-1)$ in G^T, and generate the graph $\tilde{G}_t$ constructed by the occupied links.
C2-2c. Compute $S(H, t) = \bigcup_{w \in S(H, t-1)} \Gamma(w; \tilde{G}_t)$, and for each $v \in A(t-1)$, compute $S(\{v\}, t) = \bigcup_{w \in S(\{v\}, t-1)} \Gamma(w; \tilde{G}_t) \setminus S(H, t)$, set $A(t) \leftarrow A(t) \cup \{v\}$ if $S(\{v\}, t) \neq \emptyset$, and set $\phi_H(\{u\}, t) \leftarrow \phi_H(\{u\}, t) + |S(\{v\}, t)|$ for each $u \in C(v)$.
C2-2d. Check whether $S(\{u\}, t) = S(\{v\}, t)$ for $u, v \in A(t)$, and set $C(v) \leftarrow C(v) \cup C(u)$ and $A(t) \leftarrow A(t) \setminus \{u\}$ if $S(\{u\}, t) = S(\{v\}, t)$.
C3. For each $v \in V \setminus H$ and $t \in \{1, \cdots, T\}$, set $\phi_H(\{v\}, t) \leftarrow \phi_H(\{v\}, t)/M$, and output $\phi_H(\{v\}, t)$.

Intuitively, compared with the BP with pruning method, by using the burnout technique, we can substantially reduce the size of the active node set from $S(H \cup \{v\}, t)$ to $S(\{v\}, t)$ for each $v \in V \setminus H$ and $t \in \{1, \cdots, T\}$. Namely, in terms of computational costs described by Equation (3), we can expect to obtain smaller numbers for $\mathcal{N}_a$ and $\mathcal{N}_c$ when $H \neq \emptyset$. However, how effectively the proposed method works will depend on several conditions such as network structure, time span, values of diffusion probabilities, and so on. We will do a simple analysis later and experimentally show that it is indeed effective.

5   Experimental Evaluation

In the experiments, we report our evaluation results on the final-time maximization problem due to the space limitation.

5.1   Network Data and Settings

In our experiments, we employed two datasets of large real networks used in [10], which exhibit many of the key features of social networks.

The first one is a trackback network of Japanese blogs. The network data was collected by tracing the trackbacks from one blog in the site "goo (http://blog.goo.ne.jp/)" in May, 2005. We refer to the network data as the blog network. The blog network was a strongly-connected bidirectional network, where a link created by a trackback was regarded as a bidirectional link since blog authors establish mutual communications by putting trackbacks on each other's blogs. The blog network had 12,047 nodes and 79,920 directed links.

The second one is a network of people that was derived from the "list of people" within Japanese Wikipedia. We refer to the network data as the Wikipedia network. The Wikipedia network was also a strongly-connected bidirectional network, and had 9,481 nodes and 245,044 directed links.

For the SIS model, we assigned a uniform probability p to the propagation probability $p_{u,v}$ for any link $(u, v) \in E$, that is, $p_{u,v} = p$. According to [4,2], we set the value of p relatively small. In particular, we set the value of p to a value smaller than $1/\bar{d}$, where $\bar{d}$ is the mean out-degree of a network. Since the values of $\bar{d}$ were about 6.63 and 25.85 for the blog and the Wikipedia networks, respectively, the corresponding values of $1/\bar{d}$ were about 0.15 and 0.03. We decided to set $p = 0.1$ for the blog network and $p = 0.01$ for the Wikipedia network. Also, for the time span T, we set $T = 30$.

For the bond percolation method, we need to specify the number M of performing the bond percolation process. According to [12], we set $M = 10,000$ for estimating influence degrees for the blog and Wikipedia networks.

All our experimentation was undertaken on a single PC with an Intel Dual Core Xeon X5272 3.4GHz processor, with 32GB of memory, running under Linux.

5.2   Comparison Methods

First, we compared the proposed method with three heuristics from social network analysis with respect to the solution quality. They are based on the notions of "degree centrality", "closeness centrality", and "betweenness centrality" that are commonly used as influence measure in sociology [13]. Here, the betweenness of node v is defined as the total number of shortest paths between pairs of nodes that pass through v, the closeness of node v is defined as the reciprocal of the average distance between v and other nodes in the network, and the degree of node v is defined as the number of links attached to v. Namely, we employed the methods of choosing nodes in decreasing order of these centralities. We refer to these methods as the *betweenness method*, the *closeness method*, and the *degree method*, respectively.

Next, to evaluate the effectiveness of the pruning and the burnout strategies, we compared the proposed method with the naive greedy method based on the BP method with respect to the processing time. Hereafter, we refer to the naive greedy method based on the BP method as the BP method for short.

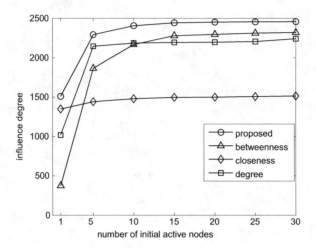

Fig. 2. Comparison of solution quality for the blog network

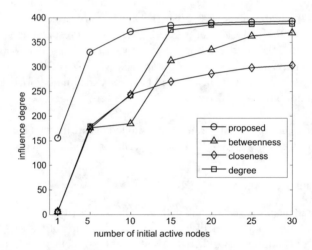

Fig. 3. Comparison of solution quality for the Wikipedia network

5.3 Solution Quality Comparison

We first compared the quality of the solution H_K of the proposed method with that of the betweenness, the closeness, and the degree methods for solving the problem of the influence maximization at the final time step T. Clearly, the quality of H_K can be evaluated by the influence degree $\sigma(H_K, T)$. We estimated the value of $\sigma(H_K, T)$ by using the bond percolation method with $M = 10,000$ according to [12].

Figures 2 and 3 show the influence degree $\sigma(H_K, T)$ as a function of the number of initial active nodes K for the blog and the Wikipedia networks, respectively. In the figures, the circles, triangles, diamonds, and squares indicate the results for the proposed, the betweenness, the closeness, and the degree methods, respectively. The proposed method performs the best for both networks, while the betweenness method follows for the blog dataset and the degree method follows for the Wikipedeia dataset. Note that how each of the conventional heuristics performs depends on the characteristics of the network structure. These results imply that the proposed method works effectively, and outperforms the conventional heuristics from social network analysis.

It is interesting to note that the k nodes ($k = 1, 2, ..., K$) that are discovered to be most influential by the proposed method are substantially different from those that are found by the conventional centrality-based heuristic methods. For example, the best node ($k = 1$) chosen by the proposed method for the blog dataset is ranked 118 for the betweenness method, 659 for the closeness method and 6 for the degree method, and the 15th node ($k = 15$) by the proposed method is ranked 1373, 8848 and 507 for the corresponding conventional methods, respectively. The best node ($k = 1$) chosen by the proposed method for the Wikipedia dataset is ranked 580 for the betweenness method, 2766 for the closeness method and 15 for the degree method, and the 15th node ($k = 15$) by the proposed method is ranked 265, 2041, and 21 for the corresponding conventional methods, respectively. It is hard to find a correlation between these rankings, but for the smaller k, it appears that degree centrality measure is better than the other centrality measures, which can be inferred from Figures 2 and 3.

5.4   Processing Time Comparison

Next, we compared the processing time of the proposed method (BP with pruning and burnout method) with that of the BP method. Let $\tau(K, T)$ denote the processing time of a method for solving the problem of the influece maximization at the final time step T, where K is the number of initial active nodes. Figures 4 and 5 show the processing time difference $\Delta\tau(K, T) = \tau(K, T) - \tau(K - 1, T)$ as a function of the number of initial active nodes K for the blog and the Wikipedia networks, respectively. In these figures, the circles, and crosses indicate the results for the proposed and the BP methods, respectively.

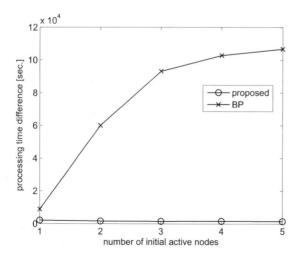

Fig. 4. Comparison of processing time for the blog network

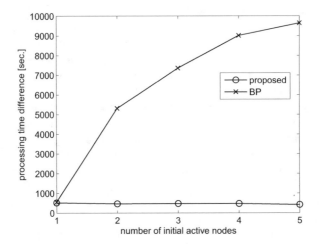

Fig. 5. Comparison of processing time for the Wikipedia network

Note that $\Delta\tau(K,T)$ decreases as K increases for the proposed method, whereas $\Delta\tau(K,T)$ increases for the BP method. This means that the difference in the total processing time becomes increasingly larger as K increases. In case of the blog dataset, the total processing time for $K = 5$ is about 2 hours for the proposed method and 100 hours for the BP methods. Namely, the proposed method is about 50 times faster than the BP method for $K = 5$. The same is true for the Wikipedia dataset. The total processing time for $K = 5$ is about 0.5 hours for the proposed method and 9 hours the BP methods, and the proposed method is about 18 times faster than the BP method for $K = 5$. These results confirm that the proposed method is much more efficient than the BP method, and can be practical.

6   Discussion

The influence function $\sigma(\cdot, T)$ is submodular [4]. For solving a combinatorial optimization problem of a submodular function f on V by the greedy algorithm, Leskovec et al. [7] have recently presented a lazy evaluation method that leads to far fewer (expensive) evaluations of the marginal increments $f(H \cup \{v\}) - f(H)$, $(v \in V \setminus H)$ in the greedy algorithm for $H \neq \emptyset$, and achieved an improvement in speed. Note here that their method requires evaluating $f(v)$ for all $v \in V$ at least. Thus, we can apply their method to the influence maximization problem for the SIS model, where the influence function $\sigma(\cdot, T)$ is evaluated by simulating the corresponding random process. It is clear that 1) this method is more efficient than the naive greedy method that does not employ the BP method and instead evaluates the influence degrees by simulating the diffusion phenomena, and 2) further the both methods become the same for $K = 1$ and empirically estimate the influence function $\sigma(\cdot, T)$ by probabilistic simulations. These methods also require M to be specified in advance as a parameter, where M is the number of simulations. Note that the BP and the simulation methods can estimate influence degree $\sigma(v, t)$ with the same accuracy by using the same value of M (see [12]). Moreover, as shown in [12], estimating influence function $\sigma(\cdot, 30)$ by $10,000$ simulations needed more than 35.8 hours for the blog dataset and 7.6 hours for the Wikipedia dataset, respectively. However, the proposed method for $K = 30$ needed less than 7.0 hours for the blog dataset and 3.2 hours for the Wikipedia dataset, respectively. Therefore, it is clear that the proposed method can be faster than the method by Leskovec [7] for the influence maximization problem for the SIS model.

7 Conclusion

Finding influential nodes is one of the most central problems in the field of social network analysis. There are several models that simulate how various things, e.g., news, rumors, diseases, innovation, ideas, etc. diffuse across the network. One such realistic model is the *susceptible/infected/susceptible (SIS) model*, an information diffusion model where nodes are allowed to be activated multiple times. The computational complexity drastically increases because of this multiple activation property, e.g., compared with the *susceptible/infected/recovered (SIR) model* where once activated nodes can never be deactivated/reactivated. We addressed the problem of efficiently discovering the influential nodes under the SIS model, i.e., estimating the expected number of activated nodes at time-step t for $t = 1, \cdots, T$ starting from an initially activated node set $H \in V$ at time-step $t = 0$. We solved this problem by constructing a layered graph from the original social network by adding each layer on top of the existing layers as the time proceeds, and applying the bond percolation with a pruning strategy. We showed that the computational complexity of the proposed method is much smaller than the conventional naive probabilistic simulation method by a theoretical analysis. We applied the proposed method to two different types of influence maximization problem, i.e. discovering the K most influential nodes that together maximize the expected influence degree at the time of interest or the expected influence degree over the time span of interest. Both problems are solved by the greedy algorithm taking advantage of the submodularity of the objective function. We confirmed by applying to two real world networks taken from blog and Wikipedia data that the proposed method can achieve considerable reduction of computation time without degrading the accuracy compared with the naive simulation method, and discover nodes that are more influential than the nodes identified by the conventional methods based on the various centrality measures. Just as a key task on biology is to find some important groups of genes or proteins by performing biologically plausible simulations over regulatory networks or metabolic pathways, our proposed method can be a core technique for the discovery of influential persons over real social networks, which can contribute to a progress on social science.

Acknowledgments

This work was partly supported by Asian Office of Aerospace Research and Development, Air Force Office of Scientific Research, U.S. Air Force Research Laboratory under Grant No. AOARD-08-4027, and JSPS Grant-in-Aid for Scientific Research (C) (No. 20500147).

References

1. Adar, E., Adamic, L.A.: Tracking information epidemics in blogspace. In: Proceedings of the 2005 IEEE/WIC/ACM International Conference on Web Intelligence, WI 2005, pp. 207–214 (2005)
2. Leskovec, J., McGlohon, M., Faloutsos, C., Glance, N., Hurst, M.: Patterns of cascading behavior in large blog graphs. In: Proceedings of the 2007 SIAM International Conference on Data Mining (SDM 2007), pp. 551–556 (2007)
3. Agarwal, N., Liu, H.: Blogosphere: Research issues, tools, and applications. SIGKDD Explorations 10(1), 18–31 (2008)
4. Kempe, D., Kleinberg, J., Tardos, E.: Maximizing the spread of influence through a social network. In: Proceedings of the 9th ACM SIGKDD International Conference on Knowledge Discovery and Data Mining (KDD 2003), pp. 137–146 (2003)
5. Gruhl, D., Guha, R., Liben-Nowell, D., Tomkins, A.: Information diffusion through blogspace. In: Proceedings of the 13th International World Wide Web Conference (WWW 2004), pp. 107–117 (2004)
6. Kimura, M., Saito, K., Nakano, R.: Extracting influential nodes for information diffusion on a social network. In: Proceedings of the 22nd AAAI Conference on Artificial Intelligence (AAAI 2007), pp. 1371–1376 (2007)
7. Leskovec, J., Krause, A., Guestrin, C., Faloutsos, C., VanBriesen, J., Glance, N.: Cost-effective outbreak detection in networks. In: Proceedings of the 13th ACM SIGKDD International Conference on Knowledge Discovery and Data Mining (KDD 2007), pp. 420–429 (2007)
8. Domingos, P., Richardson, M.: Mining the network value of customers. In: Proceedings of the 7th ACM SIGKDD International Conference on Knowledge Discovery and Data Mining (KDD 2001), pp. 57–66 (2001)
9. Richardson, M., Domingos, P.: Mining knowledge-sharing sites for viral marketing. In: Proceedings of the 8th ACM SIGKDD International Conference on Knowledge Discovery and Data Mining (KDD 2002), pp. 61–70 (2002)
10. Kimura, M., Saito, K., Motoda, H.: Blocking links to minimize contamination spread in a social network. ACM Transactions on Knowledge Discovery from Data 3(2), 9:1–9:23 (2009)
11. Newman, M.E.J.: The structure and function of complex networks. SIAM Review 45, 167–256 (2003)
12. Kimura, M., Saito, K., Motoda, H.: Efficient estimation of influence functions fot SIS model on social networks. In: Proceedings of the 21st International Joint Conference on Artificial Intelligence (IJCAI 2009) (to appear, 2009)
13. Wasserman, S., Faust, K.: Social network analysis. Cambridge University Press, Cambridge (1994)

An Empirical Comparison of Probability Estimation Techniques for Probabilistic Rules

Jan-Nikolas Sulzmann and Johannes Fürnkranz

Department of Computer Science, TU Darmstadt
Hochschulstr. 10, D-64289 Darmstadt, Germany
{sulzmann,juffi}@ke.informatik.tu-darmstadt.de

Abstract. Rule learning is known for its descriptive and therefore comprehensible classification models which also yield good class predictions. However, in some application areas, we also need good class probability estimates. For different classification models, such as decision trees, a variety of techniques for obtaining good probability estimates have been proposed and evaluated. However, so far, there has been no systematic empirical study of how these techniques can be adapted to probabilistic rules and how these methods affect the probability-based rankings. In this paper we apply several basic methods for the estimation of class membership probabilities to classification rules. We also study the effect of a shrinkage technique for merging the probability estimates of rules with those of their generalizations.

1 Introduction

The main focus of symbolic learning algorithms such as decision tree and rule learners is to produce a comprehensible explanation for a class variable. Thus, they learn concepts in the form of crisp IF-THEN rules. On the other hand, many practical applications require a finer distinction between examples than is provided by their predicted class labels. For example, one may want to be able to provide a confidence score that estimates the certainty of a prediction, to rank the predictions according to their probability of belonging to a given class, to make a cost-sensitive prediction, or to combine multiple predictions.

All these problems can be solved straight-forwardly if we can predict a probability distribution over all classes instead of a single class value. A straightforward approach to estimate probability distributions for classification rules is to compute the fractions of the covered examples for each class. However, this naïve approach has obvious disadvantages, such as that rules that cover only a few examples may lead to extreme probability estimates. Thus, the probability estimates need to be smoothed.

There has been quite some previous work on probability estimation from decision trees (so-called *probability-estimation trees (PETS)*). A very simple, but quite powerful technique for improving class probability estimates is the use of *m*-estimates, or their special case, the Laplace-estimates (Cestnik, 1990). Provost and Domingos (2003) showed that unpruned decision trees with

J. Gama et al. (Eds.): DS 2009, LNAI 5808, pp. 317–331, 2009.
© Springer-Verlag Berlin Heidelberg 2009

Laplace-corrected probability estimates at the leaves produce quite reliable decision tree estimates. Ferri et al. (2003) proposed a recursive computation of the m-estimate, which uses the probability disctribution at level l as the prior probabilities for level $l + 1$. Wang and Zhang (2006) used a general shrinkage approach, which interpolates the estimated class distribution at the leaf nodes with the estimates in interior nodes on the path from the root to the leaf.

An interesting observation is that, contrary to classification, class probability estimation for decision trees typically works better on unpruned trees than on pruned trees. The explanation for this is simply that, as all examples in a leaf receive the same probability estimate, pruned trees provide a much coarser ranking than unpruned trees. Hüllermeier and Vanderlooy (2009) have provided a simple but elegant analysis of this phenomenon, which shows that replacing a leaf with a subtree can only lead to an increase in the area under the ROC curve (AUC), a commonly used measure for the ranking capabilities of an algorithm. Of course, this only holds for the AUC estimate on the training data, but it still may provide a strong indication why unpruned PETs typically also outperform pruned PETs on the test set.

Despite the amount of work on probability estimation for decision trees, there has been hardly any systematic work on probability estimation for rule learning. Despite their obvious similarility, we nevertheless argue that a separate study of probability estimates for rule learning is necessary.

A key difference is that in the case of decision tree learning, probability estimates will not change the prediction for an example, because the predicted class only depends on the probabilities of a single leaf of the tree, and such local probability estimates are typically monotone in the sense that they all maintain the majority class as the class with the maximum probability. In the case of rule learning, on the other hand, each example may be classified by multiple rules, which may possibly predict different classes. As many tie breaking strategies depend on the class probabilities, a local change in the class probability of a single rule may change the global prediction of the rule-based classifier.

Because of these non-local effects, it is not evident that the same methods that work well for decision tree learning will also work well for rule learning. Indeed, as we will see in this paper, our conclusions differ from those that have been drawn from similar experiments in decision tree learning. For example, the above-mentioned argument that unpruned trees will lead to a better (training-set) AUC than pruned trees, does not straight-forwardly carry over to rule learning, because the replacement of a leaf with a subtree is a local operation that only affects the examples that are covered by this leaf. In rule learning, on the other hand, each example may be covered by multiple rules, so that the effect of replacing one rule with multiple, more specific rules is less predictable. Moreover, each example will be covered by some leaf in a decision tree, whereas each rule learner needs to induce a separate default rule that covers examples that are covered by no other rule.

The rest of the paper is organized as follows: In section 2 we briefly describe the basics of probabilistic rule learning and recapitulate the estimation techniques used for rule probabilities. In section 3 we explain our two approaches for the generation of a probabilistic rule set and describe how it is used for classification. Our experimental setup and results are analyzed in section 4. In the end we summarize our conclusions in section 5.

2   Rule Learning and Probability Estimation

This section is divided into two parts. The first one describes briefly the properties of conjunctive classification rules and of its extension to a probabilistic rule. In the second part we introduce the probability estimation techniques used in this paper. These techniques can be divided into basic methods, which can be used stand-alone for probability estimation, and the meta technique shrinkage, which can be combined with any of the techniques for probability estimation.

2.1   Probabilistic Rule Learning

In classification rule mining one searches for a set of rules that describes the data as accurately as possible. As there are many different generation approaches and types of generated classification rules, we do not go into detail and restrict ourselves to conjunctive rules. The *premise* of these rules consists of a conjunction of number of conditions, and in our case, the *conclusion* of the rule is a single class value. So a conjunctive classification rule r has basically the following form:

$$condition_1 \wedge \cdots \wedge condition_{|r|} \implies class \tag{1}$$

The size of a rule $|r|$ is the number of its conditions. Each of these conditions consists of an attribute, an attribute value belonging to its domain and a comparison determined by the attribute type. For our purpose, we consider only nominal and numerical attributes. For nominal attributes, this comparison is a test of equality, whereas in the case of numerical attributes, the test is either less (or equal) or greater (or equal). If all conditions are met by an instance, the instance is covered by the rule ($r \supseteq x$) and the class value of the rule is predicted for the instance. Consequently, the rule is called a *covering rule* for this instance.

   This in mind, we can define some statistical values of a data set which are needed for later definitions. A data set consists of $|C|$ classes and n instances from which n^c belong to the class c respectively ($n = \sum_{c=1}^{|C|} n^c$). A rule r covers n_r instances which are distributed over the classes, so that n_r^c instances belong to class c ($n_r = \sum_{c=1}^{|C|} n_r^c$).

   A probabilistic rule is an extension of a classification rule, which does not only predict a single class value, but a set of *class probabilities*, which form a probability distribution over the classes. This probability distribution estimates all probabilities that a covered instance belongs to any of the class in the data set, so we get one class probability per class. The example is then classified with the most probable class. The probability that an instance x covered by rule r belongs to c can be viewed as a conditional probability $\Pr(c|r \supseteq x)$.

   In the next section, we discuss some approaches for estimating these class probabilities.

2.2   Basic Probability Estimation

In this subsection we will review three basic methods for probability estimation. Subsequently, in section 2.3, we will describe a technique known as shrinkage, which is known from various application areas, and show how this technique can be adapted to probabilistic rule learning.

All of the three basic methods we employed, calculate the relation between the number of instances covered by the rule n_r and the number of instances covered by the rule but also belong to a specific class n_r^c. The differences between the methods are the minor modifications of the calculation of this relation.

The simplest approach to rule probability estimation directly estimates a class probability distribution of a rule with the fraction of examples that belong to each class.

$$\Pr_{\text{naïve}} (c|r \supseteq x) = \frac{n_r^c}{n_r} \qquad (2)$$

This naïve approach has several well-known disadvantages, most notably that rules with a low coverage may be lead to extreme probability values. For this reason, Cestnik (1990) suggested the use of the Laplace- and m-estimates.

The Laplace estimate modifies the above-mentioned relation by adding one additional instance to the counts n_r^c for each class c. Hence the number of covered instances n_r is increased by the number of classes $|C|$.

$$\Pr_{\text{Laplace}} (c|r \supseteq x) = \frac{n_r^c + 1}{n_r + |C|} \qquad (3)$$

It may be viewed as a trade-off between $\Pr_{\text{naïve}}(c|r \supseteq x)$ and an *a priori* probability of $\Pr(c) = 1/|C|$ for each class. Thus, it implicitly assumes a uniform class distribution.

The m-estimate generalizes this idea by making the dependency on the prior class distribution explicit, and introducing a parameter m, which allows to trade off the influence of the *a priori* probability and $\Pr_{\text{naïve}}$.

$$\Pr_{m}(c|r \supseteq x) = \frac{n_r^c + m \cdot \Pr(c)}{n_r + m} \qquad (4)$$

The m-parameter may be interpreted as a number of examples that are distributed according to the prior probability, which are added to the class frequencies n_r^c. The prior probability is typically estimated from the data using $\Pr(c) = n^c/n$ (but one could, e.g., also use the above-mentioned Laplace-correction if the class distribution is very skewed). Obviously, the Laplace-estimate is a special case of the m-estimate with $m = |C|$ and $\Pr(c) = 1/|C|$.

2.3   Shrinkage

Shrinkage is a general framework for smoothing probabilities, which has been successfully applied in various research areas.[1] Its key idea is to "shrink" probability

---

[1] Shrinkage is, e.g., regularly used in statistical language processing (Chen and Goodman, 1998; Manning and Schütze, 1999).

estimates towards the estimates of its generalized rules r_k, which cover more examples. This is quite similar to the idea of the Laplace- and m-estimates, with two main differences: First, the shrinkage happens not only with respect to the prior probability (which would correspond to a rule covering all examples) but interpolates between several different generalizations, and second the weights for the trade-off are not specified *a priori* (as with the m-parameter in the m-estimate) but estimated from the data.

In general, shrinkage estimates the probability $\Pr(c|r \supseteq x)$ as follows:

$$\Pr_{Shrink}(c|r \supseteq x) = \sum_{k=0}^{|r|} w_c^k \Pr(c|r_k) \tag{5}$$

where w_c^k are weights that interpolate between the probability estimates of the generalized rules r_k. In our implementation, we use only generalizations of a rule that can be obtained by deleting a final sequence of conditions. Thus, for a rule with length $|r|$, we obtain $|r| + 1$ generalizations r_k, where r_0 is the rule covering all examples, and $r_{|r|} = r$.

The weights w_c^k can be estimated in various ways. We employ a shrinkage method proposed by Wang and Zhang (2006) which is intended for decision tree learning but can be straight-forwardly adapted to rule learning. The authors propose to estimate the weights w_c^k with an iterative procedure which averages the probabilities obtained by removing training examples covered by this rule. In effect, we obtain two probabilities per rule generalization and class: the removal of an example of class c leads to a decreased probability $\Pr_-(c|r_k \supseteq x)$, whereas the removal of an example of a different class results in an increased probability $\Pr_+(c|r_k \supseteq x)$. Weighting these probabilities with the relative occurrence of training examples belonging to this class we obtain a smoothed probability

$$\Pr_{Smoothed}(c|r_k \supseteq x) = \frac{n_r^c}{n_r} \cdot \Pr_-(c|r_k \supseteq x) + \frac{n_r - n_r^c}{n_r} \cdot \Pr_+(c|r_k \supseteq x) \tag{6}$$

Using these smoothed probabilities, this shrinkage method computes the weights of these nodes in linear time (linear in the number of covered instances) by normalizing the smoothed probabilities separately for each class.

$$w_c^k = \frac{\Pr_{Smoothed}(c|r_k \supseteq x)}{\sum_{i=0}^{|r|} \Pr_{Smoothed}(c|r_i \supseteq x)} \tag{7}$$

Multiplying the weights with their corresponding probability we obtain "shrinked" class probabilities for the instance.

Note that all instances which are classified by the same rule receive the same probability distribution. Therefore the probability distribution of each rule can be calculated in advance.

3 Rule Learning Algorithm

For the rule generation we employed the rule learner Ripper (Cohen, 1995), arguably one of the most accurate rule learning algorithms today. We used Ripper both in ordered and in unordered mode:

Ordered Mode: In ordered mode, Ripper learns rules for each class, where the classes are ordered according to ascending class frequencies. For learning the rules of class c_i, examples of all classes c_j with $j > i$ are used as negative examples. No rules are learned for the last and most frequent class, but a rule that implies this class is added as the default rule. At classification time, these rules are meant to be used as a decision list, i.e., the first rule that fires is used for prediction.

Unordered Mode: In unordered mode, Ripper uses a one-against-all strategy for learning a rule set, i.e., one set of rules is learned for each class c_i, using all examples of classes $c_j, j \neq i$ as negative examples. At prediction time, all rules that cover an example are considered and the rule with the maximum probability estimate is used for classifying the example. If no rule covers the example, it classified by the default rule predicting the majority class.

We used JRip, the Weka (Witten and Frank, 2005) implementation of Ripper. Contrary to William Cohen's original implementation, this re-implementation does not support the unordered mode, so we had to add a re-implementation of that mode.[2] We also added a few other minor modifications which were needed for the probability estimation, e.g. the collection of statistical counts of the sub rules.

In addition, Ripper (and JRip) can turn the incremental reduced error pruning technique (Fürnkranz and Widmer, 1994; Fürnkranz, 1997) on and off. Note, however, that with turned off pruning, Ripper still performs pre-pruning using a minimum description length heuristic (Cohen, 1995). We use Ripper with and without pruning and in ordered and unordered mode to generate four set of rules. For each rule set, we employ several different class probability estimation techniques.

In the test phase, all covering rules are selected for a given test instance. Using this reduced rule set we determine the most probable rule. For this purpose we select the most probable class of each rule and use this class value as the prediction for the given test instance and the class probability for comparison. Ties are solved by predicting the least represented class. If no covering rules exist the class probability distribution of the default rule is used.

4   Experimental Setup

We performed our experiments within the WEKA framework (Witten and Frank, 2005). We tried each of the four configuration of Ripper (unordered/ordered and pruning/no pruning) with 5 different probability estimation techniques, Naïve (labeled as Precision), Laplace, and m-estimate with $m \in \{2, 5, 10\}$, both used as a stand-alone probability estimate (abbreviated with B) or in combination with

---

[2] Weka supports a general one-against-all procedure that can also be combined with JRip, but we could not use this because it did not allow us to directly access the rule probabilities.

shrinkage (abbreviated with S). As a baseline, we also included the performance of pruned or unpruned standard JRip accordingly. Our unordered implementation of JRip using Laplace stand-alone for the probability estimation is comparable to the unordered version of Ripper (Cohen, 1995), which is not implemented in JRip.

We evaluated these methods on 33 data sets of the UCI repository (Asuncion and Newman, 2007) which differ in the number of attributes (and their categories), classes and training instances. As a performance measure, we used the weighted area under the ROC curve (AUC), as used for probabilistic decision trees by Provost and Domingos (2003). Its key idea is to extend the binary AUC to the multi-class case by computing a weighted average the AUCs of the one-against-all problems N_c, where each class c is paired with all other classes:

$$AUC(N) = \sum_{c \in C} \frac{n_c}{|N|} AUC(N_c) \tag{8}$$

For the evaluation of the results we used the Friedman test with a post-hoc Nemenyi test as proposed in (Demsar, 2006). The significance level was set to 5% for both tests. We only discuss summarized results here, detailed results can be found in the appendix.

4.1 Ordered Rulesets

In the first two test series, we investigated the ordered approach using the standard JRip approach for the rule generation, both with and without pruning. The basic probability methods were used standalone (B) or in combination with shrinkage (S).

The Friedman test showed that in both test series, the employed combinations of probability estimation techniques showed significant differences. Considering the CD chart of the first test series (Figure 1), one can identify three groups of equivalent techniques. Notable is that the two best techniques, the m-Estimate used stand-alone with $m = 2$ and $m = 5$ respectively, belong only to the best group. These two are the only methods that are significantly better than the two worst methods, Precision used stand-alone and Laplace combined with shrinkage.

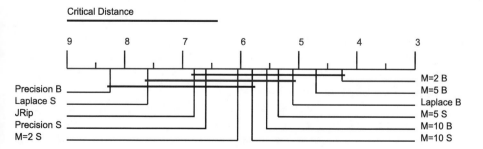

Fig. 1. CD chart for ordered rule sets without pruning

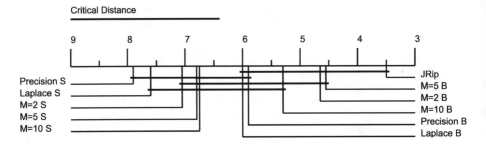

Fig. 2. CD chart for ordered rule sets with pruning

On the other hand, the naïve approach seems to be a bad choice as both techniques employing it rank in the lower half. However our benchmark JRip is positioned in the lower third, which means that the probability estimation techniques clearly improve over the default decision list approach implemented in JRip.

Comparing the stand-alone techniques with those employing shrinkage one can see that shrinkage is outperformed by their stand-alone counterparts. Only precision is an exception as shrinkage yields increased performance in this case. In the end shrinkage is not a good choice for this scenario.

The CD-chart for ordered rule sets with pruning (Figure 2) features four groups of equivalent techniques. Notable are the best and the worst group which overlap only in two techniques, Laplace and Precision used stand-alone. The first group consists of all stand-alone methods and JRip which dominates the group strongly covering no shrinkage method. The last group consists of all shrinkage methods and the overlapping methods Laplace and Precision used stand-alone. As all stand-alone methods rank before the shrinkage methods, one can conclude that they outperform the shrinkage methods in this scenario as well. Ripper performs best in this scenario, but the difference to the stand-alone methods is not significant.

4.2   Unordered Rule Sets

Test series three and four used the unordered approach employing the modified JRip which generates rules for each class. Analogous to the previous test series the basic methods are used as stand-alone methods or in combination with shrinkage (left and right column respectively). Test series three used no pruning while test series four did so. The results of the Friedman test showed that the techniques of test series three and test series four differ significantly.

Regarding the CD chart of test series three (Figure 3), we can identify four groups of equivalent methods. The first group consists of all stand-alone techniques, except for Precision, and the m-estimates techniques combined with shrinkage and $m = 5$ and $m = 10$, respectively. Whereas the stand-alone methods dominate this group, $m = 2$ being the best representative. Apparently these methods are the best choices for this scenario. The second and third consist

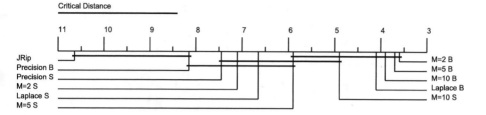

Fig. 3. CD chart for unordered rule sets without pruning

mostly of techniques employing shrinkage and overlap with the worst group in only one technique. However our benchmark JRip belongs to the worst group being the worst choice of this scenario. Additionally the shrinkage methods are outperformed by their stand-alone counterparts.

The CD chart of test series four (Figure 4) shows similar results. Again four groups of equivalent techniques groups can be identified. The first group consists of all stand-alone methods and the m-estimates using shrinkage and $m = 5$ and $m = 10$ respectively. This group is dominated by the m-estimates used stand-alone with $m = 2$, $m = 5$ or $m = 10$. The shrinkage methods are distributed over the other groups, again occupying the lower half of the ranking. Our benchmark JRip is the worst method of this scenario.

4.3   Unpruned vs. Pruned Rule Sets

Rule pruning had mixed results, which are briefly summarized in Table 1. On the one hand, it improved the results of the unordered approach, on the other hand it worsened the results of the ordered approach. In any case, in our experiments, contrary to previous results on PETs, rule pruning was not always a bad choice. The explanation for this result is that in rule learning, contrary to decision tree learning, new examples are not necessarily covered by one of the learned rules. The more specific rules become, the higher is the chance that new examples are not covered by any of the rules and have to be classified with a default rule. As these examples will all get the same default probability, this is a bad strategy

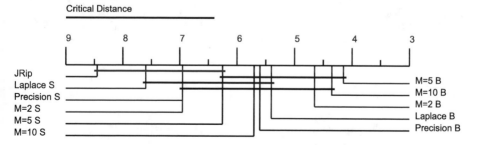

Fig. 4. CD chart for unordered rule sets with pruning

Table 1. Unpruned vs. pruned rule sets: Win/Loss for ordered (top) and unordered (bottom) rule sets

	Jrip	Precision	Laplace		M 2		M 5		M 10	
Win	26	23	19	20	19	18	20	19	20	19	20
Loss	7	10	14	13	14	15	13	14	13	14	13
Win	26	21	9	8	8	8	8	8	8	8	6
Loss	7	12	24	25	25	25	25	25	25	25	27

for probability estimation. Note, however, that JRip without pruning, as used in our experiments, still performs an MDL-based form of pre-pruning. We have not yet tested a rule learner that performs no pruning at all, but, because of the above deliberations, we do not expect that this would change the results with respect to pruning.

5   Conclusions

The most important result of our study is that probability estimation is clearly an important part of a good rule learning algorithm. The probabilities of rules induced by JRip can be improved considerably by simple estimation techniques. In unordered mode, where one rule is generated for each class, JRip is outperformed in every scenario. On the other hand, in the ordered setting, which essentially learns decision lists by learning subsequent rules in the context of previous rules, the results were less convincing, giving a clear indication that the unordered rule induction mode should be preferred when a probabilistic classfication is desirable.

Amongst the tested probability estimation techniques, the m-estimate typically outperformed the other methods. Among the tested values, $m = 5$ seemed to yield the best overall results, but the superiority of the m-estimate was not sensitive to the choice of this parameter. The employed shrinkage method did in general not improve the simple estimation techniques. It remains to be seen whether alternative ways of setting the weights could yield superior results. Rule pruning did not produce the bad results that are known from ranking with pruned decision trees, presumably because unpruned, overly specific rules will increase the number of uncovered examples, which in turn leads to bad ranking of these examples.

Acknowledgements

This research was supported by the *German Science Foundation (DFG)* under grant FU 580/2.

References

Asuncion, A., Newman, D.J.: UCI machine learning repository (2007)

Cestnik, B.: Estimating probabilities: A crucial task in Machine Learning. In: Aiello, L. (ed.) Proceedings of the 9th European Conference on Artificial Intelligence (ECAI 1990), Stockholm, Sweden, pp. 147–150. Pitman (1990)

Chen, S.F., Goodman, J.T.: An empirical study of smoothing techniques for language modeling. Technical Report TR-10-98, Computer Science Group, Harvard University, Cambridge, MA (1998)

Cohen, W.W.: Fast effective rule induction. In: Prieditis, A., Russell, S. (eds.) Proceedings of the 12th International Conference on Machine Learning (ML 1995), Lake Tahoe, CA, pp. 115–123. Morgan Kaufmann, San Francisco (1995)

Demsar, J.: Statistical comparisons of classifiers over multiple data sets. Journal of Machine Learning Research 7, 1–30 (2006)

Ferri, C., Flach, P.A., Hernández-Orallo, J.: Improving the AUC of probabilistic estimation trees. In: Proceedings of the 14th European Conference on Machine Learning, Cavtat-Dubrovnik, Croatia, pp. 121–132 (2003)

Fürnkranz, J.: Pruning algorithms for rule learning. Machine Learning 27(2), 139–171 (1997)

Fürnkranz, J., Flach, P.A.: Roc 'n' rule learning-towards a better understanding of covering algorithms. Machine Learning 58(1), 39–77 (2005)

Fürnkranz, J., Widmer, G.: Incremental Reduced Error Pruning. In: Cohen, W., Hirsh, H. (eds.) Proceedings of the 11th International Conference on Machine Learning (ML 1994), New Brunswick, NJ, pp. 70–77. Morgan Kaufmann, San Francisco (1994)

Hand, D.J., Till, R.J.: A simple generalisation of the area under the roc curve for multiple class classification problems. Machine Learning 45(2), 171–186 (2001)

Hüllermeier, E., Vanderlooy, S.: Why fuzzy decision trees are good rankers. IEEE Transactions on Fuzzy Systems (to appear, 2009)

Manning, C.D., Schütze, H.: Foundations of Statistical Natural Language Processing. The MIT Press, Cambridge (1999)

Provost, F.J., Domingos, P.: Tree induction for probability-based ranking. Machine Learning 52(3), 199–215 (2003)

Wang, B., Zhang, H.: Improving the ranking performance of decision trees. In: Fürnkranz, J., Scheffer, T., Spiliopoulou, M. (eds.) ECML 2006. LNCS (LNAI), vol. 4212, pp. 461–472. Springer, Heidelberg (2006)

Witten, I.H., Frank, E.: Data Mining: Practical machine learning tools and techniques, 2nd edn. Morgan Kaufmann, San Francisco (2005)

A   Detailed Experimental Results (Tables)

Table 2. Weighted AUC results with rules from ordered, unpruned JRip

Name	Jrip	Precision		Laplace		M 2		M 5		M 10	
		B	S	B	S	B	S	B	S	B	S
Anneal	.983	.970	.971	.970	.970	.970	.970	.971	.971	.970	.970
Anneal.orig	.921	.917	.920	.920	.919	.920	.919	.920	.920	.919	.919
Audiology	.863	.845	.843	.832	.836	.840	.844	.839	.841	.831	.832
Autos	.904	.907	.901	.900	.891	.907	.902	.904	.902	.903	.898
Balance-scale	.823	.801	.812	.821	.812	.820	.811	.821	.812	.821	.815
Breast-cancer	.591	.577	.581	.578	.580	.578	.580	.578	.579	.577	.579
Breast-w	.928	.930	.929	.935	.930	.935	.931	.935	.932	.935	.933
Colic	.736	.739	.741	.747	.746	.748	.746	.748	.745	.748	.746
Credit-a	.842	.849	.857	.861	.859	.861	.859	.861	.862	.861	.864
Credit-g	.585	.587	.587	.587	.587	.587	.587	.587	.587	.587	.587
Diabetes	.642	.654	.656	.655	.656	.655	.656	.655	.656	.655	.655
Glass	.806	.803	.795	.790	.787	.794	.797	.793	.799	.792	.795
Heart-c	.762	.765	.775	.796	.777	.796	.777	.796	.780	.796	.789
Heart-h	.728	.737	.755	.758	.757	.758	.755	.758	.757	.758	.757
Heart-statlog	.763	.759	.781	.806	.782	.806	.783	.806	.790	.806	.791
Hepatitis	.679	.661	.661	.660	.663	.660	.665	.660	.663	.660	.663
Hypothyroid	.971	.973	.974	.974	.974	.974	.974	.974	.974	.973	.974
Ionosphere	.884	.885	.897	.903	.900	.903	.899	.903	.900	.903	.902
Iris	.957	.889	.876	.889	.878	.889	.878	.889	.878	.889	.878
Kr-vs-kp	.993	.994	.994	.995	.994	.995	.994	.995	.994	.995	.994
Labor	.812	.800	.810	.794	.810	.793	.806	.793	.795	.793	.783
Lymph	.750	.739	.748	.748	.745	.746	.748	.744	.746	.749	.746
Primary-tumor	.649	.636	.652	.615	.638	.645	.656	.641	.653	.642	.662
Segment	.983	.964	.944	.967	.943	.966	.944	.967	.943	.966	.943
Sick	.922	.928	.929	.929	.929	.929	.929	.929	.929	.929	.929
Sonar	.774	.771	.779	.784	.778	.783	.778	.783	.779	.783	.781
Soybean	.962	.971	.972	.966	.971	.973	.972	.967	.973	.967	.971
Splice	.938	.934	.938	.943	.938	.943	.938	.943	.938	.943	.939
Vehicle	.772	.799	.811	.811	.816	.812	.813	.811	.816	.812	.819
Vote	.952	.954	.950	.955	.949	.955	.949	.955	.952	.953	.956
Vowel	.884	.906	.909	.909	.906	.909	.910	.911	.910	.910	.907
Waveform	.847	.850	.853	.872	.854	.872	.854	.873	.855	.873	.858
Zoo	.916	.899	.916	.902	.897	.908	.900	.907	.895	.899	.890
Average	.834	.830	.834	.836	.832	.837	.834	.837	.834	.836	.834
Average Rank	6.79	8.24	6.62	5.11	7.62	4.26	6.03	4.68	5.33	5.53	5.79

Table 3. Weighted AUC results with rules from ordered, pruned JRip

Name	Jrip	Precision		Laplace		M 2		M 5		M 10	
		B	S	B	S	B	S	B	S	B	S
Anneal	.984	.981	.980	.981	.981	.981	.980	.981	.980	.980	.980
Anneal.orig	.942	.938	.937	.936	.936	.937	.936	.936	.937	.935	.936
Audiology	.907	.865	.854	.810	.776	.852	.840	.839	.826	.834	.801
Autos	.850	.833	.836	.821	.829	.829	.830	.823	.830	.821	.819
Balance-scale	.852	.812	.810	.815	.810	.815	.810	.816	.811	.816	.811
Breast-cancer	.598	.596	.597	.596	.597	.596	.597	.598	.599	.598	.602
Breast-w	.973	.965	.956	.965	.956	.964	.956	.964	.957	.961	.957
Colic	.823	.801	.808	.804	.815	.809	.815	.813	.815	.816	.816
Credit-a	.874	.872	.874	.873	.874	.874	.874	.874	.873	.875	.874
Credit-g	.593	.613	.612	.613	.612	.613	.612	.613	.612	.613	.612
Diabetes	.739	.734	.736	.734	.736	.734	.736	.734	.736	.734	.736
Glass	.803	.814	.810	.822	.825	.820	.818	.820	.817	.820	.812
Heart-c	.831	.837	.818	.843	.818	.842	.818	.845	.823	.847	.825
Heart-h	.758	.739	.742	.740	.740	.740	.742	.741	.742	.742	.741
Heart-statlog	.781	.792	.776	.790	.776	.790	.776	.791	.775	.790	.773
Hepatitis	.664	.600	.596	.600	.596	.599	.596	.599	.595	.597	.586
Hypothyroid	.988	.990	.990	.990	.990	.990	.990	.990	.990	.990	.990
Ionosphere	.900	.904	.909	.907	.909	.908	.909	.910	.910	.910	.909
Iris	.974	.888	.889	.890	.891	.890	.891	.890	.891	.890	.891
Kr-vs-kp	.995	.994	.993	.994	.993	.994	.993	.994	.994	.994	.994
Labor	.779	.782	.755	.782	.761	.781	.764	.768	.759	.746	.745
Lymph	.795	.795	.767	.788	.772	.790	.773	.779	.773	.777	.774
Primary-tumor	.642	.626	.624	.622	.627	.630	.622	.627	.622	.629	.628
Segment	.988	.953	.932	.953	.933	.954	.932	.953	.932	.953	.933
Sick	.948	.949	.949	.950	.949	.950	.949	.950	.950	.950	.950
Sonar	.759	.740	.734	.742	.737	.743	.737	.746	.740	.744	.744
Soybean	.981	.980	.970	.968	.965	.978	.970	.971	.967	.969	.966
Splice	.967	.956	.953	.957	.953	.957	.953	.957	.954	.957	.954
Vehicle	.855	.843	.839	.844	.843	.844	.842	.843	.843	.842	.844
Vote	.942	.949	.947	.949	.947	.949	.947	.949	.947	.949	.947
Vowel	.910	.900	.891	.898	.891	.904	.892	.905	.893	.898	.892
Waveform	.887	.880	.862	.880	.863	.880	.862	.881	.863	.881	.863
Zoo	.925	.889	.909	.887	.895	.895	.902	.895	.901	.889	.893
Average	.855	.843	.838	.841	.836	.843	.838	.842	.838	.841	.836
Average Rank	3.52	5.88	7.92	5.98	7.62	4.65	7.06	4.55	6.79	5.29	6.74

Table 4. Weighted AUC results with rules from unordered, unpruned JRip

Name	Jrip	Precision B	Precision S	Laplace B	Laplace S	M 2 B	M 2 S	M 5 B	M 5 S	M 10 B	M 10 S
Anneal	.983	.992	.989	.992	.991	.994	.989	.994	.989	.994	.989
Anneal.orig	.921	.987	.984	.990	.983	.993	.984	.993	.984	.993	.984
Audiology	.863	.910	.887	.877	.874	.909	.895	.903	.894	.892	.889
Autos	.904	.916	.915	.926	.914	.927	.914	.929	.918	.930	.926
Balance-scale	.823	.874	.865	.908	.873	.908	.866	.909	.871	.908	.882
Breast-cancer	.591	.608	.587	.633	.605	.633	.589	.632	.606	.632	.617
Breast-w	.928	.959	.966	.953	.966	.953	.967	.953	.969	.953	.969
Colic	.736	.835	.840	.855	.851	.855	.849	.855	.849	.859	.849
Credit-a	.842	.890	.909	.913	.911	.913	.911	.913	.914	.913	.917
Credit-g	.585	.695	.717	.716	.716	.716	.716	.716	.716	.716	.718
Diabetes	.642	.760	.778	.783	.780	.783	.779	.783	.781	.783	.783
Glass	.806	.810	.826	.808	.833	.808	.825	.808	.827	.809	.830
Heart-c	.762	.790	.813	.861	.827	.861	.823	.861	.831	.861	.844
Heart-h	.728	.789	.803	.851	.839	.853	.819	.849	.835	.852	.837
Heart-statlog	.763	.788	.811	.845	.805	.841	.805	.841	.820	.841	.829
Hepatitis	.679	.774	.817	.799	.819	.802	.821	.802	.817	.802	.816
Hypothyroid	.971	.991	.994	.994	.993	.994	.994	.994	.993	.994	.993
Ionosphere	.884	.918	.932	.938	.931	.938	.931	.938	.931	.939	.935
Iris	.957	.968	.973	.978	.980	.978	.976	.978	.980	.978	.980
Kr-vs-kp	.993	.998	.997	.999	.997	.999	.997	.999	.997	.999	.997
Labor	.812	.818	.806	.777	.803	.778	.803	.778	.790	.778	.775
Lymph	.750	.843	.852	.891	.857	.887	.848	.881	.852	.884	.878
Primary-tumor	.649	.682	.707	.671	.690	.693	.712	.694	.711	.691	.711
Segment	.983	.991	.989	.997	.990	.997	.989	.997	.990	.997	.990
Sick	.922	.958	.979	.981	.984	.982	.979	.982	.980	.982	.980
Sonar	.774	.823	.826	.841	.826	.841	.826	.841	.828	.841	.836
Soybean	.962	.979	.981	.982	.979	.985	.981	.984	.981	.985	.981
Splice	.938	.964	.968	.974	.968	.974	.968	.974	.969	.974	.970
Vehicle	.772	.851	.879	.888	.881	.888	.879	.888	.881	.888	.884
Vote	.952	.973	.967	.982	.968	.983	.968	.983	.975	.983	.978
Vowel	.884	.917	.919	.922	.920	.922	.921	.922	.920	.922	.920
Waveform	.847	.872	.890	.902	.890	.902	.890	.902	.890	.902	.893
Zoo	.916	.964	.965	.965	.970	.984	.982	.984	.982	.987	.988
Average	.834	.875	.883	.891	.885	.893	.885	.893	.887	.893	.890
Average Rank	10.67	8.15	7.45	4.08	6.65	3.58	7.08	3.68	5.88	3.88	4.91

Table 5. Weighted AUC results with rules from unordered, pruned JRip

Name	Jrip	Precision		Laplace		M 2		M 5		M 10	
		B	S	B	S	B	S	B	S	B	S
Anneal	.984	.987	.988	.984	.986	.987	.985	.986	.986	.986	.986
Anneal.orig	.942	.990	.983	.985	.980	.989	.983	.988	.982	.984	.982
Audiology	.907	.912	.889	.891	.878	.895	.893	.889	.885	.883	.881
Autos	.850	.889	.882	.891	.889	.894	.888	.892	.889	.891	.889
Balance-scale	.852	.888	.861	.899	.864	.895	.860	.900	.861	.901	.864
Breast-cancer	.598	.562	.555	.557	.555	.557	.555	.557	.555	.560	.558
Breast-w	.973	.962	.972	.963	.973	.963	.973	.963	.973	.961	.974
Colic	.823	.782	.831	.799	.830	.793	.836	.801	.837	.812	.837
Credit-a	.874	.876	.878	.877	.877	.877	.878	.879	.879	.881	.879
Credit-g	.593	.702	.711	.703	.711	.703	.711	.703	.711	.705	.711
Diabetes	.739	.740	.729	.742	.729	.742	.729	.741	.730	.739	.731
Glass	.803	.819	.821	.821	.826	.819	.821	.824	.824	.828	.825
Heart-c	.831	.827	.816	.827	.804	.829	.816	.828	.810	.830	.807
Heart-h	.758	.739	.740	.735	.736	.737	.738	.736	.737	.735	.736
Heart-statlog	.781	.806	.815	.816	.813	.816	.812	.823	.819	.824	.827
Hepatitis	.664	.766	.790	.769	.793	.771	.790	.764	.795	.768	.789
Hypothyroid	.988	.984	.993	.992	.993	.987	.994	.992	.993	.992	.993
Ionosphere	.900	.918	.915	.921	.917	.922	.918	.926	.923	.926	.923
Iris	.974	.975	.969	.975	.969	.975	.969	.975	.970	.975	.973
Kr-vs-kp	.995	.999	.995	.999	.995	.999	.995	.999	.996	.998	.997
Labor	.779	.837	.820	.815	.811	.812	.818	.812	.812	.809	.803
Lymph	.795	.858	.832	.849	.833	.853	.836	.851	.842	.851	.856
Primary-tumor	.642	.703	.701	.679	.694	.709	.704	.710	.706	.708	.707
Segment	.988	.991	.989	.995	.990	.995	.990	.995	.990	.995	.990
Sick	.948	.949	.934	.948	.938	.948	.935	.948	.937	.948	.937
Sonar	.759	.827	.815	.827	.814	.827	.815	.824	.813	.824	.818
Soybean	.981	.989	.981	.988	.981	.990	.981	.989	.981	.989	.981
Splice	.967	.973	.967	.974	.967	.974	.967	.974	.968	.974	.968
Vehicle	.855	.892	.891	.893	.890	.893	.890	.893	.890	.893	.890
Vote	.942	.947	.956	.961	.957	.952	.957	.960	.956	.961	.958
Vowel	.910	.921	.915	.924	.915	.925	.915	.925	.916	.924	.915
Waveform	.887	.897	.877	.899	.878	.898	.877	.899	.878	.900	.880
Zoo	.925	.973	.989	.960	.969	.987	.989	.987	.989	.987	.989
Average	.855	.875	.873	.874	.871	.876	.873	.877	.874	.877	.874
Average Rank	8.45	5.61	6.95	5.38	7.59	4.67	6.95	4.14	6.23	4.33	5.7

Precision and Recall for Regression

Luis Torgo[1] and Rita Ribeiro[2]

[1] FC / LIAAD-Inesc Porto LA, University of Porto, R. de Ceuta, 118, 6., 4050-190
Porto, Portugal
ltorgo@liaad.up.pt
[2] LIAAD-Inesc Porto LA, University of Porto, R. de Ceuta, 118, 6., 4050-190 Porto,
Portugal
rpribeiro@liaad.up.pt

Abstract. Cost sensitive prediction is a key task in many real world
applications. Most existing research in this area deals with classification
problems. This paper addresses a related regression problem: the prediction of rare extreme values of a continuous variable. These values are
often regarded as outliers and removed from posterior analysis. However, for many applications (e.g. in finance, meteorology, biology, etc.)
these are the key values that we want to accurately predict. Any learning method obtains models by optimizing some preference criteria. In
this paper we propose new evaluation criteria that are more adequate
for these applications. We describe a generalization for regression of the
concepts of precision and recall often used in classification. Using these
new evaluation metrics we are able to focus the evaluation of predictive
models on the cases that really matter for these applications. Our experiments indicate the advantages of the use of these new measures when
comparing predictive models in the context of our target applications.

1 Introduction

Several important predictive data mining applications involve handling non-
uniform costs and benefits of the predictions. This is almost always the case
in event-based applications like prediction of ecological or meteorological catas-
trophes, fraud detection, network intrusions, financial forecasting, etc.. Many of
these tasks are particular cases of regression problems where the continuous target variable values have differentiated importance. Often these prediction tasks
are related to the anticipation of a critical phenomenon that is inherently continuous and for which an alarm may be triggered by a specific range of values of
a continuous target variable. This type of applications requires techniques that
are able to cope with differentiated costs and benefits of predictions.

In this paper we have as main goal to address a particular and highly relevant
sub-class of non-uniform cost/benefit prediction tasks. These applications associate higher cost or benefit with rarity. For these applications the most (and often
solely) important cases are the ones associated with unusual values of the target
variable. We are thus facing a task of predicting outlier values of a continuous
target variable.

J. Gama et al. (Eds.): DS 2009, LNAI 5808, pp. 332–346, 2009.

Handling applications with differentiated costs and benefits of predictions is not new and many cost-sensitive techniques have been proposed in the literature (e.g. [6,7]). Still, most of these works focus on predictive classification tasks. For regression the most common setup considers that the cost of predictions is uniform across the domain of the target variable and solely dependent on the magnitude of the prediction errors themselves.

Addressing cost-sensitive applications involves two major issues: i) defining proper evaluation metrics to correctly assert the merits of alternative models given the application preference biases; and ii) defining learning strategies to better tune the models towards these biases. These two issues have been throughly addressed within classification problems. However, they have been essentially ignored in research on regression. The goal of this paper is to address the first of these issues: the selection of proper evaluation metrics. The main contributions of the paper are: i) increasing the awareness of the research community for these important tasks and in general to cost-sensitive regression; ii) exposing the risks of using standard regression evaluation metrics on cost sensitive applications; iii) proposing a new evaluation framework for the prediction of rare extreme values of a continuous variable.

2   Problem Statement

In predictive data mining the goal is to learn a model of an unknown function that maps a set of predictor variables into a target variable. This model is to be obtained using a training set containing examples of this mapping. The training data is used to obtain the model parameters that minimise some preference criterion. The preference criteria that are commonly used in regression are the mean squared error, $MSE = \frac{1}{n} \sum_{i=1}^{n} (y_i - \hat{y}_i)^2$, and the mean absolute deviation, $MAD = \frac{1}{n} \sum_{i=1}^{n} |y_i - \hat{y}_i|$. These are average estimators of the true mean squared and absolute error of the model, respectively.

In this paper we are interested in a particular sub-class of regression problems. The main particularity of this sub-class of problems lies on their focus on the predictive performance at rare extreme values of the continuous target variable, i.e. extreme low and/or high values. Performance on the other more frequent values is basically irrelevant for the end user of these applications.

We claim that standard error measures, such as MSE and MAD, are not suitable for these tasks. They take all the prediction errors equally across the domain of the target variable, assuming that the magnitude of the error is the decisive factor for the "cost" of a prediction. We argue that while this magnitude is important it should be weighed by the "relevance" of the values involved in the prediction.

Let us illustrate our claim by a small example. In Table 1 we present the predictions of two artificial models (M_1 and M_2) for a set of 10 hypothetical returns of some financial asset given in percentage daily variation. For this prediction problem it is very clear that we want to be particularly accurate at predicting the large variations (positive or negative) as these are the ones on which we can

Table 1. The predictions of two artificial models

True	-5.29	-2.65	-2.43	-0.20	-0.03	0.03	0.51	1.46	2.53	2.94
M_1	-4.40	-2.06	-2.20	0.10	-0.23	-0.27	0.97	2.00	1.86	2.15
M_2	-5.09	-2.95	-2.89	0.69	-0.82	0.70	-0.08	0.92	2.83	3.17

earn some money if they are correct. Smaller variations, even if correctly predicted are most of the times not tradable given the transaction costs. From the observation of this table, we can say that M_1 has more accurate predictions at smaller returns (in absolute terms), while M_2 achieves more accurate predictions at the larger variations. However, if we calculate the values of both MAD and MSE of these two models we observe that they are exactly the same, 0.497 and 0.29893, respectively, meaning that these two metrics tag these two models as having the same performance. The reason for this is that both models obtain the same total error magnitude value and thus both have same average error. This is a clearly misleading "conclusion" for this type of applications, as model M_2 is obviously more useful. This small example provides a simple illustration of the problem of assuming that the error amplitudes cost the same across all the domain of the target variable (as it is the case of all standard error metrics). For our target applications this is clearly not the case and, therefore, it is necessary to have an error metric that is sensitive to where the errors occur within the range of the target variable, i.e. that copes with differentiated relevance across the domain of this variable.

Another further problem with standard error metrics, not illustrated in the above example, is the fact that even though some model may have a clear advantage on extreme values, given their rarity, this advantage may well be diluted by its poorer performance on the "irrelevant" (but very frequent) normal values.

3   Existing Approaches to the Problem

3.1   Case Weights

Within the regression learning setup described in Section 2, there are a few alternatives to the standard error measures that could be considered more adequated to our applications. One such alternative is to use case weights. Some learning algorithms allow the user to attach a weight to each observation of the training sample. Model parameters can then be obtained by minimizing a criterion that takes into account these weights. Training cases with a target variable value that is more "relevant" should have higher weights. In the case of rare extreme values prediction this would mean to give more weight to the extreme values.

Assuming we can easily obtain the values of these weights this would apparently lead to a proper evaluation of the models' performance. However, the main drawback of this approach is that it only sees one side of the problem, the true values. In effect, this method does not try to avoid (or penalize) the situation where a "relevant" value is predicted by the model, but the true value is "normal". This is a kind of false alarm and would correspond, for instance, to predict a high return for some stock that then turns out to have a really irrelevant (very small) return. This drawback stems from the fact that the weights are dependent solely on the true value of the cases, y_i, instead of being dependent on both y_i and $\hat{y}_i$. Because of this, the above example would have a low penalization (as the true value is irrelevant), which is contradictory to the application objectives where we clearly want to avoid these costly mistakes.

3.2   Special-Purpose Loss Functions

Some authors (e.g. [4]) have addressed the issue of differentiated prediction costs by the use of so-called asymmetric loss functions. Their main goal was to be able to distinguish two types of errors, and assign costs accordingly, namely, the cost of under-predictions ($\hat{y} < y$) and the cost of over-predictions ($\hat{y} > y$). That is the case of the $LINLIN$ loss function, presented in Equation 1.

$$LINLIN = \begin{cases} c_o|y - \hat{y}|, & \text{if } \hat{y} > y; \\ 0, & \text{if } \hat{y} = y; \\ c_u|y - \hat{y}|, & \text{if } \hat{y} < y. \end{cases} \qquad (1)$$

where c_o and c_u are constants for penalizing over- and under-predictions.

   In spite of its use for some type of applications, the $LINLIN$ loss function is far from being a general cost-sensitive approach for any regression task as it only distinguishes between two types of differentiated costs: under- and over-predictions. Moreover, even on these situations it considers all under-(over-) predictions as equally serious, only looking at the error amplitude as "standard" error metrics. For instance, in stock market forecasting, predicting a future price change of -1% for a true value of 1%, has the same error amplitude as predicting 6% for a true value of 8%, and both are under-predictions. Nonetheless, they may lead to very different trading actions, and thus different costs/benefits.

4   Precision and Recall for Regression

Our target applications are driven by rare events - the occurrence of rare extreme values of a continuous variable. Within research on classification, this type of event-driven prediction tasks are usually evaluated using the notions of *precision* and *recall*, which are preferred over other alternatives when in presence of large skew in the class distribution [5]. The main advantage of these statistics is that they are focused on the performance of the models on the events, completely ignoring their accurate predictions for the non-event classes. Informally, precision measures the proportion of events signalled by the model that are real events. Recall measures the proportion of events occurring in the domain that are "captured" by the models. There is usually a trade-off between these two statistics (always outputting an event signal will get you 100% recall but with a very poor precision as most signals will be wrong), and often the two are put together in a single weighted score like for instance the F-measure [11]. Conceptually, our proposal in this paper is to provide the equivalents of these two statistics for regression problems in order to properly evaluate the performance of the models on the values that really matter.

4.1   Our Proposal

The standard setup for event-driven classification is to have a so-called "positive" class that represents the target events while the "negative" class represents all non-events. Confusion matrices provide a good characterization of the performance of a model. The numbers in this matrix can be used to calculate several statistics among which are precision and recall [8]. Table 2 shows a general confusion matrix for these type of applications. Recall is defined as the ratio TP/POS, while precision as the ratio $TP/PPOS$.

Table 2. The 2-classes confusion matrix

	Predicted Pos	Predicted Neg	
Pos	TP	FN	POS
Neg	FP	TN	NEG
	PPOS	PNEG	

In these classification problems, relevance (importance) is established by declaring the "target" class. This enumeration strategy is not possible in regression given the infinite domain of the target variable. We propose the use of a relevance function, $\phi()$, that maps the original domain of the target variable into a continuous scale of relevance[1],

$$\phi(Y) : \quad]-\infty, \infty[\rightarrow [0,1] \tag{2}$$

This function allows the specification of different degrees of relevance with the obvious advantages in terms of sensibility of the method with respect to the different values of the target variable.

We can also describe the strategy followed in classification using this notion of relevance. In effect, from this perspective it corresponds to specifying the following relevance function,

$$\phi(Y) = I(Y = C_E) \tag{3}$$

where $I()$ is the indicator function given 1 if its argument is true and 0 otherwise, and C_E is the label of the class describing the events (i.e. the positive class).

The information on the relevance function is obviously domain-dependent. In classification this information consists of choosing the positive class. In regression, given the infinite nature of the domain of the target variable, a real valued function makes more sense. Specifying such function in an analytical way may not be

---

[1] We use the value of zero for completely irrelevant values, and one for maximally relevant values.

always easy for a user. Still, for some applications we can come up with a reasonable automatically generated relevance function. That is the case of our target applications. In effect, in these domains relevance is associated with rarity and extremeness of the values. In this context we may say that the relevance function is the complement of the probability distribution function (*pdf*) of the target variable. Box plots provide key information on this *pdf* in particular regards extreme values. In effect, they are at the basis of a parametric test for outliers, the box-plot rule. This test assumes a Gaussian distribution of the variable and tags as outliers all values above the high adjacent value given by $adj_H = Q_3 + 1.5 \cdot IQR$, where Q_3 is the third quartile and $IQR = Q_3 - Q_1$. Equivalently, all values below the low adjacent value, $adj_L = Q_1 - 1.5 \cdot IQR$, are also tagged as outliers. These values correspond to rare high (low) extreme values. For our target applications we may have both types of outliers or only high (low) outliers. Our proposal consists of using a sigmoid-like relevance function whose shape is a function of these adjacent values for each of these two "sides" of extremeness. Let us see how we can derive this function from the training sample we have available for each application.

The relevance function is based in the following sigmoid,

$$f(Y) = \frac{1}{1 + \exp^{-s \cdot (Y - c)}} \tag{4}$$

where c is the center of the sigmoid and s is the shape of the sigmoid. The values of these parameters are also dependent on the type of extremes the variable has (low, high or both types of extremes). For applications with only low or high extremes the relevance function is defined by a single sigmoid, while for applications with both types of extremes (like stock market prediction tasks) we will have two of these sigmoids defining $\phi(Y)$.

The parameter c, the center of the sigmoid, represents the value where $\phi(Y) = 0.5$. The meaning of c is that of a threshold above which the values of target variable start to be more relevant. We set the c values of the sigmoids to the values of the respective adjacent values, i.e. $c_L = adj_L$ and $c_H = adj_H$.

With respect to the parameter s we want to set it in such a way that for the high extreme values $\phi(c - c \cdot k) \simeq 0$, and for low extremes $\phi(c + c \cdot k) \simeq 0$, where k is a kind of decay factor that determines how fast the sigmoid decays to 0. By selecting a certain precision value Δ (e.g. $1e - 04$) and solving the equation in order to s we get,

$$s = \pm \frac{\ln(\Delta^{-1} - 1)}{|c \cdot k|} \tag{5}$$

where the $+$ signal is used for high extremes, while the $-$ signal is for low extremes.

In the case of applications with both extremes, each sigmoid is obtained using the parameter values described above. Figure 1 shows two relevance functions generated using this method for two types of applications: only with high extremes ; and with both types of extremes. We provide R code[2] that implements

---

[2] Available in http://www.liaad.up.pt/~ltorgo/DS09 .

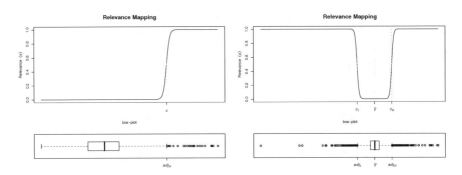

Fig. 1. Two examples of relevance functions generated from box plots

this type of relevance functions that are adequate for applications of predicting rare extreme values.

Please note that our proposal in no way depends on this illustrative and heuristic definition of a relevance function. This definition is only of use in cases where the user does not have a precise notion of relevance/importance of his application, simply having the "intuition" that relevance is associated with extreme and rare values. In these cases, this heuristic function we have described may help in defining a relevance function that is required for the application of our evaluation framework that we now describe.

Recall is informally defined as the proportion of relevant events that are retrieved by a model. Having defined events as a function of relevance we can say that relevant events are those for which $\phi(Y) \geq t_E$, where t_E is a domain-dependent threshold on relevance. In classification, as relevance is usually a 0/1 function (c.f. Equation 3), this threshold is 1. For regression, this will most probably be a value near 1, depending on the values we want to consider as the targets of our prediction task.

We now need to clarify the notion of "events that are retrieved by a model" in order to fully define recall. In classification this consists of achieving a correct prediction that is asserted by the usual 0/1 loss function, i.e. having $L_{0/1}(\hat{y}_i, y_i) = 0 \Leftrightarrow \hat{y}_i = y_i$. In regression loss functions are usually metric with domain $[0, \infty[$. Imposing that $\hat{y}_i = y_i$ will be, in most cases, too strict. Generally, we can say that a prediction is "correct" in regression if $L(\hat{y}_i, y_i) \leq t_L$, where t_L is a threshold on the range of the loss function. We may generalize even further this notion by allowing different degrees of "accuracy" within the interval of "admissible" errors, i.e. errors that are less than t_L. The value of t_L is again domain-dependent.

Having defined the two general concepts involved in the notion of recall we can now propose a general definition for this statistic that can cope with both classification and regression tasks,

$$Recall = \frac{\sum_{\phi(y_i) \geq t_E} \alpha(\hat{y}_i, y_i) \cdot \phi(y_i)}{\sum_{\phi(y_i) \geq t_E} \phi(y_i)} \tag{6}$$

where $\alpha()$ is a function that defines the accuracy of a prediction.

In classification the $\alpha()$ function is defined as follows,

$$\alpha(\hat{y}_i, y_i) = I(L_{0/1}(\hat{y}_i, y_i) = 0) \tag{7}$$

where $L_{0/1}()$ is a standard 0/1 loss function.

Given the definition of relevance for classification problems we have described above, and this definition of what is an accurate prediction in classification, it is easy to see that our proposed definition of Recall reduces to the standard proportion TP/POS.

For regression we may define $\alpha()$ using a similar indicator function,

$$\alpha(\hat{y}_i, y_i) = I(L(\hat{y}_i, y_i) \leq t_L) \tag{8}$$

where t_L is the above mentioned threshold defining an admissible error within the domain a metric loss function $L()$ (e.g. the absolute deviation).

Alternatively, we may use a smoother notion of accuracy by using a continuous function in the interval $[0, 1]$, instead of the 0/1 function of Equation 8. This allows a more accurate assessment of the quality of the signals of a regression model. There are many ways of mapping the loss function values in the interval $[0, t_L]$ into a $[1, 0]$ scale. Examples include variations of linear interpolation or the ramping function. Another alternative is to use a variant of the complementary error function [1], that has a Gaussian-type shape that we think is more adequate for our goals,

$$\alpha(\hat{y}_i, y_i) = I(L(\hat{y}_i, y_i) \leq t_L) \cdot \left(1 - \exp^{-k \cdot \frac{(L(\hat{y}_i, y_i) - t_L)^2}{t_L^2}}\right) \tag{9}$$

where k is a positive integer that determines the shape of the function. Larger values lead to steeper decreases.

Precision is the proportion of the events retrieved by a model that are effective events. We have already seen what is an event in both classification and regression. The only difference here is that we are talking about "retrieved" events and not the "real" events (i.e. predictions and not true values). Some of these correspond to "real" events but others not, and the goal of precision is to assert this proportion. In classification a retrieved event is a prediction of the "positive" class. In regression this is a prediction of a value whose relevance is greater than the user-defined relevance threshold t_E. As we have seen, both can be described by the same condition using the relevance function. In this context, we propose the following generalized definition of precision,

$$Precision = \frac{\sum_{\phi(\hat{y}_i) \geq t_E} \alpha(\hat{y}_i, y_i) \cdot \phi(\hat{y}_i)}{\sum_{\phi(\hat{y}_i) \geq t_E} \phi(\hat{y}_i)} \tag{10}$$

You may have noticed that the numerators of definitions of Precision and Recall we are proposing are different (c.f. Equations 6 and 10), which is not in aggrement with the standard definitions of recall and precision that have in the numerator the number of true positives (TP). However, for the settings used in classification the numerators of these equations we propose are in effect equal. The $\alpha()$ function used for classification is a 0/1 function that is 1 if the classification is accurate, which implies that $\hat{y}_i = y_i$. This in turn implies that $\phi(\hat{y}_i) = \phi(y_i)$ and thus the numerators are equal. However, we should remark that this may not be the case for regression setups where an accurate prediction may not mean that $\hat{y}_i = y_i$, namely if $t_L > 0$.

Precision and recall may be aggregated into composite measures, like for instance the F-measure [11],

$$F = \frac{(\beta^2 + 1) \cdot Precision \cdot Recall}{\beta^2 \cdot Precision + Recall} \tag{11}$$

where $0 \le \beta \le 1$, controls the relative importance of recall to precision.

These composite measures have the advantage of facilitating comparisons among models as they provide a single score.

5   Experimental Analysis

5.1   Artificial Data

On Table 1 we have presented an artificial example on stock returns prediction with the predictions of two models that, in spite of their clearly different approach to rare extreme values, had exactly the same score in terms of standard error metrics like MSE and MAD. Let us examine this example with our new proposed measures of recall and precision. Let us suppose that we use as threshold for events (t_E) a value of relevance greater than 0.75 ,i.e. $\phi(Y) \ge 0.75$. We will use an automatically generated relevance function for extremes (c.f. Equation 4). The generated function uses a larger sample of values than those shown on Table 1. Using this sample we estimate $adj_L = -1.5$ and $adj_H = 1.5$. These values setup the value of the c parameter of the function and together with a value of $k = 0.5$ we define our relevance function (c.f. Equation 4). We will also use the smooth $\alpha()$ function defined in Equation 9 with a threshold for accurate predictions of half percent return, i.e. $t_L = 0.5$. In this context, we come up with the results show in Table 3.

These values correspond to a recall of 0.178 for model M_1 and of 0.670 for M_2. Precision is of 0.292 for model M_1 and of 0.668 for M_2. These scores provide a completely different (and more correct with respect to the preference bias of this application) perspective on the performance of the models, which according to both MSE and MAD are equal.

5.2   Predicting Stock Market Returns

In this section we illustrate the use of the proposed precision and recall statistics in the context of the prediction of rare extreme returns of a set of stocks. The purpose of this study is to illustrate both the "danger" of using standard

Table 3. Evaluating the two artificial models with the new metrics

True	-5.29	-2.65	-2.43	-0.20	-0.03	0.03	0.51	1.46	2.53	2.94
$\phi(Y)$	1.00	1.00	0.98	0.00	0.00	0.00	0.00	0.01	0.99	1.00
M_1	-4.40	-2.06	-2.20	0.10	-0.23	-0.27	0.97	2.00	1.86	2.15
$\phi(\hat{Y}_1)$	1.00	0.63	0.86	0.00	0.00	0.00	0.00	0.50	0.22	0.80
$L(\hat{Y}_1, Y)$	0.89	0.59	0.23	0.30	0.20	0.30	0.46	0.54	0.67	0.79
$\alpha(\hat{Y}_1, Y)$	0.00	0.00	0.90	0.72	0.94	0.72	0.05	0.00	0.00	0.00
M_2	-5.09	-2.95	-2.89	0.69	-0.82	0.70	-0.08	0.92	2.83	3.17
$\phi(\hat{Y}_2)$	1.00	1.00	1.00	0.00	0.00	0.00	0.00	0.00	1.00	1.00
$L(\hat{Y}_2, Y)$	0.20	0.30	0.46	0.89	0.79	0.67	0.59	0.54	0.30	0.23
$\alpha(\hat{Y}_2, Y)$	0.94	0.72	0.05	0.00	0.00	0.00	0.00	0.00	0.72	0.90

regression evaluation statistics in this type of problems, as well as presenting and measuring the advantages of our proposals.

The Data. The base data we will use in our study are the standard daily quotes of four companies: International Business Machines (IBM), Coca-Cola (KO), Boeing (BA) and General Motors (GM). This daily data was obtained from Yahoo finance[3] and it contains the usual quotes and volume information.

Most applications of this type based on daily data focus on predicting the Adjusted Close prices of the stocks. Namely, a common procedure consists predicting the h-days returns defined as,

$$R_h(t) = \frac{Close(t) - Close(t-h)}{Close(t-h)} \tag{12}$$

Using this time series of returns we have defined a prediction task consisting of trying to predicted the future value of these returns, $R_h(t+h)$, using a set of p previous values of the time series (usually known as an embed of the time series). In our experiments we used an embed of 24 days back of the $R_h(t)$ variable. This modelling task was selected without any particular concern on whether this was the best setup for predicting future returns. That is not our main goal here. Our objective is to compare alternative modelling techniques on the same stock market prediction problems and check the model rankings we obtain when using both the standard evaluation metrics and our new proposals. Our hypothesis is that the model rankings obtained with our metrics are "better" from the perspective of the application objectives, which are being accurate at the rare extreme returns where profitable trading can take place.

Using this approach we have obtained datasets for the 1-, 3- and 5-days returns of the four companies used in our study, i.e. 12 regression tasks.

The Experimental Methodology. The used quotes data covers the period from 1970-01-02 till 2008-07-11, in a total of 9725 daily sessions.

---

[3] http://finance.yahoo.com.

In order to provide an accurate estimate of the statistics that we will use to compare our alternative models we have divided the period mentioned above in two main consecutive time windows. The first spans from the first date till 1990-01-01. The second time window goes from this latter date till 2008-07-11. The first time window (first 20 years) will be used for obtaining the prediction models, while the second window (around 18 and a half years) will be used to evaluate and compare the models.

The Modelling Tools. All tools we have used are available in the (free) R statistical environment[4], which allows easy replication of our results. We have considered 4 different regression techniques, each with several parameter variants, in a total of 57 different models being compared for each data set.

Artificial Neural Networks. We have used the neural networks provided by the nnet package of R. This package has a function to obtain feed-forward neural networks with one hidden layer using the back-propagation learning algorithm.

Regarding model tuning we have considered 15 alternatives varying the number of inner nodes (parameter Size) of the hidden layer between 5, 10, 15, 20 and 30, and also the learning rate (parameter Decay) between 0.01, 0.05 and 0.1.

Multivariate adaptive regression splines. The package mda of R has a re-implementation of MARS [9] done by Trevor Hastie and Robert Tibshirani. We have used this system in our experiments.

Regarding model tuning we have considered 16 variants formed by different combinations of the parameter setting the penalty for extra degrees of freedom (parameter Pen which was used with values 1,2,3,4), and of the parameter specifying the forward stepwise stopping threshold (parameter Thr that was tried with values 0.01, 0.005, 0.001 and 0.0005).

Support Vector Machines. Package e1071 of R includes a function implementing SVMs [10]. This implementation provides an interface to the award-winning libsvm library by Chang and Lin [3].

We have considered 16 variants of SVMs during our model tuning experiments. These variants were chosen according to the suggestions given in [10]. They include different values for the parameter Cost (tried values 400, 500, 600 and 700) and Gamma (tried values 0.01, 0.005, 0.001 and 0.0005). The former is a constraints violation parameter, while the latter is the radial basis function kernel parameter.

Random Forests. Package randomForest of R includes a function that implements random forests [2] based on original Fortran code by L. Breiman and A. Cutler.

We have considered 10 variants of these models by setting the parameter ntree, which controls the number of trees in the ensembles, to values from 50 to 500 in steps of 50.

---

[4] http://www.R-project.org

The Results. We have obtained the 57 model variants using the experimental methodology described before on the returns data sets. The main hypothesis that we are trying to check is that the model rankings obtained by using our proposed metrics are significantly different from the rankings obtained with standard regression statistics. Moreover, that these rankings obtained with our metrics are clearly advantageous in terms of the application preference bias, that in this case is related to having good "signals" of rare and extreme movements of the markets.

In terms of our evaluation framework we have used the following settings. We have assumed that, giving the transaction costs, users of these applications are not willing to trade on returns smaller than 2% (-2%) for buying (selling) actions. In this context, we have setup the notion of rare extreme values around these two thresholds. Namely, with respect to the relevance function we have used as centers of the two sigmoids the values $c_L = -0.02$ and $c_H = 0.02$, while for the shapes of the sigmoids we have calculated them using Equation 5 with $k = 0.5$ and $\Delta = 1e - 04$. In the context of precision and recall we have used $t_E = 0.5$ (thus any return above 2% or below -2% will be considered an event, given the definition of the $\phi()$ function), and $t_L = 0.005$ (i.e. errors above 0.5% are not considered, c.f. Equation 9).

The first results we show are designed to test the hypothesis concerning the different rankings. We have used the MAD statistic as a representative of the "standard" approaches, and the composite F-measure (with $\beta = 0.5$ that gives twice importance to precision compared to recall, as inaccurate trading signals may be costly) as representing our proposals. For all 12 experimental setups (4 companies and 3 forecasting scenarios), we have obtained the two model rankings according to these two statistics. Due to lack of space we can not present all graphs illustrating these 12 experimental setups[5]. All setups follow a similar results trend. We have selected 1 setup that is shown in Figure 2. The figure has two graphs. The graph on the left shows the scores of the best five models according to the two statistics. We should remark that for MAD, lower values are better, contrary to what happens with the F measure. On the X-axis we have the identifiers (a number from 1 to 57) of the top 5 models according to each statistic (the 5 on the left according to MAD and the other 5 according to F). Ideally these two sets of numbers should be different indicating that the best 5 models according to the two statistics are also different. On the graph we plot the actual values of these 10 models for the two statistics: circles and left Y-scale for MAD; and triangles and right Y-scale for the F measure. The second graph presented on the figures shows a global perspective (on all 57 models) of the two rankings produced by the statistics. On both axis we have the possible ranking positions (from 1 to 57). The coordinates of each of the 57 dots shown on the graphs are obtained using the rank position assigned by MAD (X coordinate), and the corresponding rank position assigned by the F measure (Y coordinate). If for any of the 57 models both statistics give it the same ranking position, the respective dot should lie in the dashed diagonal line. The vertical and horizontal

---

[5] All graphs may be obtained at `http://www.liaad.up.pt/$\sim$ltorgo/DS09`

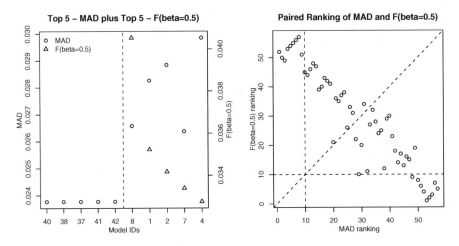

Fig. 2. The results for 3-days returns of Boeing (BA)

dashed lines highlight the results for the top 10 rank positions (left of the vertical line, and below the horizontal line) according to the two statistics.

Analysing the results in Figure 2, namely the left graph, we observe that the top 5 models according to the two statistics are completely different. Moreover, we see that their concrete scores on the statistics are also very different. For instance the best models according to MAD achieve a much worse score in terms of F[6], when compared to the best 5 models according to this later measure. In terms of overall ranking we also observe a general tendency for all ranking positions to be different as most points in the right graph are far from the diagonal. In particular the top 10 models according to MAD are all below the 45th position in the F ranking. These results clearly indicate that the two metrics are evaluating very different aspects of the performance of the models.

We have also carried out a formal statistical test of the differences between the model rankings. On all 12 data sets we have observed some evidence of disagreement between the rankings, with only 4 lacking proper statistical significance. In summary, our experiments have confirmed the hypothesis that the two considered metrics (MAD and F-measure based on our proposed Recall and Precision statistics), often obtain significantly different model rankings on this type of applications. Moreover, we should remark that this experimental setup is not particularly favorable to our proposals. In effect, we are comparing 57 models that optimize some variant of the squared error. This means that these models are not particularly focused on predicting rare extreme values. Even on these conditions we have observed that our proposals are able to detect models that have some ability at predicting rare extreme values. We can expect that the differences would be even more marked if among the 57 models we had some

---

[6] Actually, no score at all because they do not produce any event signal, i.e. predictions with relevance higher than t_E, and thus they have no precision score.

that were particularly competent at predicting rare extremes (e.g. if they were optimizing our F measure instead of squared errors).

What are the advantages of comparing a set of alternative models using our proposed metrics in alternative to standard statistics? Or in other words, what are the costs a user can expect if he uses a measure like MAD to select the model to apply for trading on stock markets? The results we have shown previously provided evidence that our metrics rank the 57 models we have considered, differently. However, do these ranking differences lead to better trading performance? In other words if a user uses the best model according to our F-measure, instead of the best model according to MAD, what does he have to gain or loose? For each of the 12 experimental setups we have selected the two best models according to MAD and F, respectively. We have used their respective predictions of the future returns for the 18 years and have calculated a set of trading-related statistics. We have assumed that we are going to trade with futures (thus allowing both short and long positions, i.e. trading when we predict the market goes down or up, respectively). Moreover, considering trading costs we only "trade" when a model predicts a future return above (below) 2% (-2%), i.e. we are going to take these situations as indicators for buying (selling). Under these conditions each model outputs a set of trading signals (predictions above 0.02 or below -0.02). The predicted signals were then compared to the "true" signals, i.e. did the prices go up (down) as predicted?

These experiments have confirmed the advantages of our metrics. In effect, the models "selected" by MAD almost never issue a single signal during the 18 testing years! On the contrary, the models selected using our metrics issue several trading signals during this period. Still, the accuracy of these signals is far from ideal as expected. This is expectable because: i) the candidate models are optimizing squared errors; ii) the information used to obtain the models (embed of 24 days) is clearly sub-optimal; and iii) predicting stock returns is a very difficult task!

6   Conclusions

This paper has presented a study on the prediction of rare extreme values of a continuous target variable that can be regarded as outliers. Our study is focused on the development of proper evaluation metrics for these tasks, which is a key step in addressing these problems.

We have described a generalization of the notions of precision and recall for regression tasks. These intuitive concepts are ideal for addressing our target problems as they focus the evaluation solely on the important events (the rare extreme values). Our proposals incorporate the standard definitions used in classification as particular cases.

We have illustrated the use of these metrics in the context of stock market forecasting applications. Namely, we have used our metrics to compare a large set of models in several experimental setups. Our experiments have confirmed that our evaluation metrics provide a significantly different perspective of the performance of the models, when compared to standard evaluation statistics. Moreover, this perspective is more adjusted to the preference biases of this type of applications. Our experimental results have also shown the danger of using standard evaluation metrics in this class of problems.

Acknowledgements

This work was partially supported by FCT projects oRANKI (PTDC/EIA/ 68322/2006) and MORWAQ (PTDC/EIA/68489/2006) , by a sabbatical scholarship of the Portuguese government (FCT/BSAB/388/2003) to L. Torgo and by a PhD scholarship of the Portuguese government (SFRH/BD/1711/2004) to R. Ribeiro.

References

1. Abramowitz, M., Stegun, I.A.: Handbook of Mathematical Functions. Dover, New York (1972)
2. Breiman, L.: Random forests. Machine Learning 1(45), 5–32 (2001)
3. Chang, C., Lin, C.: Libsvm: a library for support vector machines (2001), http://www.csie.ntu.edu.tw/cjlin/libsvm, detailed documentation, http://www.csie.ntu.edu.tw/cjlin/libsvm.ps.gz
4. Christoffersen, P., Diebold, F.: Further results on forecasting and model selection under asymmetric loss. Journal of Applied Econometrics 11, 561–571 (1996)
5. Davis, J., Goadrich, M.: The relationship between precision-recall and roc curves. In: Proceedings of 23rd International Conference on Machine Learning (2006)
6. Domingos, P.: Metacost: A general method for making classifiers cost-sensitive. In: Proceedings of the 5th International Conference on Knowledge Discovery and Data Mining (KDD 1999), pp. 155–164. ACM Press, New York (1999)
7. Elkan, C.: The foundations of cost-sensitive learning. In: Proc. of 7th International Joint Conference of Artificial Intelligence (IJCAI 2001), pp. 973–978 (2001)
8. Flach, P.: The geometry of roc space: understanding machine learning metrics through roc isometrics. In: Proceedings of the 20th International Conference on Machine Learning (2003)
9. Friedman, J.: Multivariate adaptive regression splines. The Annals of Statistics 19(1), 1–141 (1991)
10. Meyer, D.: Support Vector Machines, the interface to libsvm in package e1071. Technische Universitat Wien, Austria (2002)
11. Van Rijsbergen, C.: Information Retrieval, 2nd edn., Dept. of Computer Science, University of Glasgow (1979)

Mining Local Correlation Patterns in Sets of Sequences

Antti Ukkonen

Helsinki University of Technology & HIIT
antti.ukkonen@hiit.fi

Abstract. Given a set of (possibly infinite) sequences, we consider the problem of detecting events where a subset of the sequences is correlated for a short period. In other words, we want to find cases where a number of the sequences output exactly the same substring at the same time. Such substrings, together with the sequences in which they are contained, form a *local correlation pattern*. In practice we only want to find patterns that are longer than γ and appear in at least σ sequences.

Our main contribution is an algorithm for mining such patterns in an online case, where the sequences are read in parallel one symbol at a time (no random access) and the patterns must be reported as soon as they occur.

We conduct experiments on both artificial and real data. The results show that the proposed algorithm scales well as the number of sequences increases. We also conduct a case study using a public EEG dataset. We show that the local correlation patterns capture essential features that can be used to automatically distinguish subjects diagnosed with a genetic predisposition to alcoholism from a control group.

1   Introduction

Multidimensional time series and streams arise in a number of applications, such as finance (prices of securities at a stock exchange), medicine (multichannel EEG measurements) or telecommunications systems. Mining patterns in such data is a well studied topic, see for example [2,8,4,14,7].

In this paper we consider a case where the input consists of a set of sequences over some finite alphabet that are each read one symbol at a time. We propose a novel pattern class that represents local correlations among a subset of such sequences. More specifically, given the sequences, we consider the problem of finding subsets of sequences that are correlated for short periods of time by containing the same substring starting at the same position. We call such (subset, substring) pairs *local correlation patterns*.

For example, consider six time series that show the daily stock price of six different companies $C_1, \ldots, C_6$ over a number of days. We can create a modified set of time series where we mark for each day only whether the price of the stock went **up**, **down**, or stayed the **same** when compared to the previous quote. This gives us six sequences over the alphabet $\{u, d, s\}$. Below are the values of these sequences over a period of seven days:

J. Gama et al. (Eds.): DS 2009, LNAI 5808, pp. 347–361, 2009.

$$\begin{array}{llllllll} & -6 & -5 & -4 & -3 & -2 & -1 & 0 \\ C_1: & \text{u} & \text{u} & \textbf{d} & \textbf{u} & \textbf{s} & \text{s} & \text{d} \\ C_2: & \text{d} & \textbf{u} & \textbf{d} & \textbf{u} & \textbf{s} & \text{u} & \text{s} \\ C_3: & \text{s} & \text{s} & \text{u} & \text{d} & \text{s} & \text{d} & \text{s} \\ C_4: & \text{u} & \text{u} & \text{d} & \text{d} & \text{u} & \text{u} & \text{d} \\ C_5: & \text{s} & \textbf{u} & \textbf{d} & \textbf{u} & \textbf{s} & \text{s} & \text{u} \\ C_6: & \text{u} & \text{u} & \text{d} & \text{s} & \text{s} & \text{d} & \text{d} \end{array}$$

The last column, labeled with a 0, indicates the current day, while the column labeled with a -6 contains values from six days ago. The elements indicated in bold form a local correlation pattern starting at -5 with companies C_1, C_2 and C_5 and the substring $\langle \textbf{u}, \textbf{d}, \textbf{u}, \textbf{s} \rangle$. Another example is the substring $\langle \textbf{u}, \textbf{u}, \textbf{d} \rangle$ that starts at -6 and concerns companies C_1, C_4 and C_6.

In practice we have to be more specific when defining what counts as a local correlation pattern. Obviously it is possible that we observe temporary correlations in the sequences merely due to random chance, especially if there are many (e.g. hundreds of) of them. The first criteria we use is the *length* of the common substring. The second one is the *support*, i.e., the set of sequences that all contain the substring at the given position. That is, we do not expect to see a large number of sequences behaving in exactly the same way for several time steps simply by coincidence. More precisely: *We say that a string and its support form a local correlation pattern if the string is longer, and the support is larger than specified threshold values.*

The task is to efficiently find all local correlation patterns given that we obtain one symbol of each sequence at a time in an "online" fashion. Note that we could also consider an "offline" version of the problem, where all sequences support random access. However, this variant is not so interesting as it can be solved efficiently by existing algorithms. The main contribution of this paper is an efficient algorithm for the online setting.

Also, the data does not necessarily have to consist of multiple parallel sequences for our approach to be of interest. We can construct an input of the format discussed above from a single (long) string s by letting the suffixes of s be the individual sequences. That is, the suffix of s starting at position i is the ith sequence of the input. With this construction we can use our algorithm to find substrings that occur frequently inside a window of predetermined size in the string s.

In the experiments we give an example where local correlation patterns are used to classify EEG measurements. It turns out that a simple nearest mean classifier using features computed from sets of local correlation patterns can accurately distinguish subjects diagnosed with a genetic predisposition to alcoholism from the control group (see Section 4.3). This is quite interesting as we make no domain specific assumptions about the structure of the streams.

The rest of this paper is structured as follows. We give formal definitions for the problems of finding local correlation patterns in Section 2. The proposed algorithm for mining local correlation patterns is discussed in Section 3. Empirical experiments and their results are described in Section 4. Related work is covered in Section 5, and Section 6 is a short conclusion.

2   Problem Definition

Let Σ be an alphabet of size $|\Sigma|$, and let s be a sequence of symbols from Σ. Symbols of Σ are denoted with letters a, b, c, ... Denote by $s(i)$ the i:th symbol of s, and by $|s|$ the length of s. Sequences are indexed starting from 1. Let the pair (i, p) denote a *pattern* where i is a positive integer and p a string over the alphabet Σ. We say the sequence s *supports* the pattern (i, p) if p appears as a substring in s starting at position i. That is, if we have

$$s(i + j - 1) = p(j) \text{ for all } j \in \{1, \ldots, |p|\}.$$

Let $D = \{s_1, \ldots, s_n\}$ be a set of sequences over Σ that are all of the same length. Denote by $\theta(i, p)$ the set of sequences in D that support the pattern (i, p). The pattern (i, p) is a *local correlation pattern* in D given the parameters γ and σ, if and only if $|p| \geq \gamma$ and $|\theta(i, p)| \geq \sigma$.

We first briefly consider the problem of finding all local correlation patterns in a set D that supports random access to the sequences.

Problem 1. LCP-OFFLINE: Given D, γ and σ, find all local correlation patterns in D.

This can be solved using existing string-indexing techniques. We transform the strings in D by replacing the symbol $s(i)$ with $(i, s(i))$, that is, we create an extended alphabet where the positions are encoded in the symbols. Denote the new set of strings by D'. To solve LCP-OFFLINE for D, we simply find all frequent substrings in D' that are at least of length γ. This can be done efficiently by constructing either a suffix tree [13,10,12] or a suffix array [6] over D'.

A more interesting variant of the problem concerns an online setting where we can only read one symbol from each sequence at every time step. That is, at time step t, we read the symbol $s(t)$ from each $s \in D$. Moreover, the length of the sequences may be unbounded.

Problem 2. LCP-ONLINE: Given γ, σ, and n sequences that each output one symbol from Σ at each time step, find all local correlation patterns and output them as soon as they appear.

In practice this definition is somewhat inconvenient, because it is possible that a local correlation pattern found at step t is only a prefix of a pattern found at step $t + 1$. Consider the stock price example in the introduction. If we have set $\gamma = 3$ and $\sigma = 3$, we would first output the pattern $(2, \langle u\ d\ u \rangle)$ at step 4, and the pattern $(2, \langle u\ d\ u\ s \rangle)$ at step 5. This behavior is clearly undesirable.

To overcome this issue we propose to find only the *maximal local correlation patterns*. Let ap and pa denote the sequence p with the symbol a appended to its beginning and end, respectively. The pattern (i, p) is maximal if there is no

$a \in \Sigma$, such that $|\theta(i, pa)| \geq \sigma$ or $|\theta(i - 1, ap)| \geq \sigma$. In other words, the pattern (i, p) is maximal if it is not the prefix or suffix of another local correlation pattern.

Problem 3. MAXIMAL-LCP-ONLINE: Given γ, σ, and n sequences that each output one symbol from Σ at each time step, find all maximal local correlation patterns and output them as soon as they appear.

3   An Algorithm for MAXIMAL-LCP-ONLINE

In this section we describe an algorithm for the MAXIMAL-LCP-ONLINE problem. First we give an overview of the algorithm, and subsequently add details that address some issues with the general approach.

3.1   A General Approach

The algorithm we propose maintains a set of candidate patterns that are prefixes of strings that may later result in a local correlation pattern with respect to some of the input sequences. The candidates must all have a support at least of size σ, but they are in general shorter than γ. Some of the candidates may be longer than γ, because we want to find maximal local correlation patterns. The candidates may thus qualify as local correlation patterns themselves, but before returning them as new patterns, we have to make sure that they can not be extended without reducing the size of their support below σ. It is easy to see that the number of candidates is trivially upper bounded by the number of sequences and the support threshold σ. Given n sequences we can have at most n/σ candidates at any given time.

Consider the following situation at step $t+1$. Suppose that $p = p(1)p(2) \ldots p(k)$ is a candidate. That is, there are at least σ sequences that all behave as specified by p, starting from $t - k + 1$ and ending at t. The length of p may or may not exceed γ. At time $t + 1$ we must check for all $a \in \Sigma$ what happens with the support of the extended string pa.

Obviously there are two alternatives. If the support of pa remains above σ it becomes a candidate itself and we are done. However, if the support of pa drops below σ we have to form a new candidate, and possibly output p as a maximal local correlation pattern if $|p| \geq \gamma$. To find the new candidate, note that some suffix of pa may be the prefix of some other maximal local correlation pattern. The new candidate is *the longest suffix of pa with a support larger than σ*. We may obtain several candidates based on p depending how pa behaves for different $a \in \Sigma$.

A high-level description of this idea is given in Algorithm 1. At every step $t > 1$ we call LCP0 with the set of candidates $\mathcal{C}_{t-1}$ obtained in the previous step. At step 1 we set $\mathcal{C}_1 = \{a \in \Sigma : |supp(1, a)| \geq \sigma\}$, that is, every symbol of the alphabet with a large enough support forms a candidate by itself in the beginning. The algorithm returns an updated set of candidates $\mathcal{C}_t$ and outputs maximal local correlation patterns if any are found. Note that Algorithm 1 is

---

Algorithm 1. LCP0: A high-level algorithm for solving MAXIMAL-LCP-ONLINE.

1: LCP0($\mathcal{C}_{t-1}, \sigma, \gamma, t$)
2: $\mathcal{C}_t \leftarrow \emptyset$
3: **for** $p \in \mathcal{C}_{t-1}$ **do**
4:     extensionFound $\leftarrow$ **false**
5:     **for** $a \in \Sigma$ **do**
6:         **if** $|\theta(t - |pa| + 1, pa)| \geq \sigma$ **then**
7:             $\mathcal{C}_t \leftarrow \mathcal{C}_t \cup \{pa\}$
8:             extensionFound $\leftarrow$ **true**
9:         **else**
10:            $p' \leftarrow$ longest suffix $\hat{p}$ of pa, st. $|\theta(t - |\hat{p}| + 1, \hat{p})| \geq \sigma$
11:            $\mathcal{C}_t \leftarrow \mathcal{C}_t \cup \{p'\}$
12:         **end if**
13:     **end for**
14:     **if** extensionFound = **false** and $|p| \geq \gamma$ **then**
15:         output pattern $(t - |p| + 1, p)$
16:     **end if**
17: **end for**
18: **return** $\mathcal{C}_t$

---

only meant to illustrate the general approach. It does not specify any details on how to compute the supports and what parts of the sequences to store for processing.

3.2 A Detailed Algorithm

Now we address some details that are needed to develop an efficient implementation of LCP0. Most importantly, we must define what the algorithm has to keep in memory in order to process the sequences. Of course a trivial implementation of LCP0 could simply store everything it reads and perform the support computations on this stored data.

Problems with LCP0. The first problem we address is related to representing the candidate set. First we observe that Algorithm 1 can do some unnecessary work on line 11 by adding the same string p' multiple times as a new candidate pattern. To see this, note that some candidates will have the same suffix. This suffix, appended with some $a \in \Sigma$, can be added multiple times to $\mathcal{C}_t$.

For example, let $\Sigma = \{+, -\}$, and consider the four sequences given on the left side of Figure 1. Let $\sigma = 2$. When we are at step 3 the set of candidates from the previous step is $\mathcal{C}_2 = \{+-, --\}$. First LCP0 processes the candidate $+-$. It checks the supports of both $+ - +$ and $+ - -$, and finds both to be of size less than σ. The longest suffixes of $+ - +$ and $+ - -$ with enough support are $-+$ and $--$, respectively. These are both added to $\mathcal{C}_3$. After this the algorithm processes the second candidate in $\mathcal{C}_2$, namely $--$. Again it finds that neither $- - +$ nor $- - -$ have a support large enough, but the suffixes $-+$ and $--$, both of which were already added to $\mathcal{C}_3$ when processing $+-$, are again found to have enough support and are added to $\mathcal{C}_3$ for a second time.

Also, it can happen that redundant candidates are added to C_t. These are strings that are *suffixes of some other candidates* that are added to C_t as well. Such a candidate is redundant, because they will be added to C at some later step anyway. Returning to the example, suppose that in Figure 1 the symbol appearing in sequence $s2$ at step 3 is a $-$ instead of a $+$. The set of candidates is still $C_2 = \{+-, --\}$. This means that when LCP0 extends the 1st candidate with a $-$, it finds that $+--$ has enough support and adds it as a new candidate to C_3. However, when appending a third $-$ to the 2nd candidate, it turns out that the support of $---$ is no longer at least σ. It's suffix $--$ has a support of size 3, and will be added as a new candidate, and we end up with $C_3 = \{+--, --\}$. At step 4 the candidate $+--$ can not be extended with either $+$ or $-$ without decreasing the support below σ, but it's suffix $--$, appended with a $+$ has a large enough support, and $--+$ will be added to C_4. But this will happen twice, as $--$ is the other candidate in C_3.

The candidate trie. To avoid finding duplicate or redundant candidates, we represent C with a trie having symbols of Σ appear as labels of its edges. Let T_t be the trie that corresponds to the set of candidates C_t. We will denote an internal node of the trie by N, and a leaf by L. To each node is associated a symbol $a \in \Sigma$, denoted $\mathrm{sym}(N)$, that is the label of the edge leading to the node. Let $C(N)$ denote the set of child nodes of the node N.

We construct T_t so that every path starting from its root and ending at a leaf corresponds to *the reversal* of one candidate string in C_t. That is, for every $p \in C_t$ of length k, we have in T_t a path starting from the root and ending at a leaf L, so that the edges on the path are labeled with the symbols $p(k), p(k-1), \ldots, p(1)$. Moreover, a path in T_t starting from the root and ending at an arbitrary node N corresponds to the suffix of a candidate, or the common suffix of a number of candidates.

We also associate to every leaf L of T_t the set of sequences that support the string defined by the path from the root of T_t to L. Denote this by $\theta(L)$. For example, if L is the node of T_t that is reachable from the root by first following the edge labeled with a $-$ and then the edge labeled with a $+$, the set $\theta(L)$ contains the identifiers of all sequences that have the symbol $+$ at position $t-1$ and the symbol $-$ at position t. The trie T_2 that corresponds to $C_2 = \{+-, --\}$ of the previous example is depicted on the right side of Figure 1.

In practice we also need the supports of the suffixes of each candidate. Of course a suffix of candidate p may be supported by a number of sequences that do not belong to $\theta(p)$. This is illustrated by an example in Figure 2. The trie on the left shows all possible strings of length 2, together with their supports in some imaginary data that is not shown. Since $++$ and $--$ are only supported by one sequence each, we do not consider them frequent with $\sigma = 2$. According to the definition of T_t given above, the leafs corresponding to $++$ and $--$ are not stored at all. Still it is clear that $+$, which is a suffix of candidate $-+$, is supported by sequences $s1$, $s2$ and $s5$. To represent this in T_t, we include $s1$ to the node that corresponds to $+$, as shown in the final candidate trie on the right in Fig. 2. We call this the *local support* of N, denoted $\theta_l(N)$. Given a trie defined

Fig. 1. Left: An example data of four sequences ($s1$, $s2$, $s3$ and $s4$) of length 4 over $\Sigma = \{+, -\}$. Right: A trie representation of the candidate set $\mathcal{C}_2 = \{+-, --\}$ of the data on the left.

Fig. 2. Left: Trie showing four strings together with their supports. Two of these, $-+$ and $+-$, are frequent at $\sigma = 2$ and are hence considered as candidates. Right: The pruned trie showing only the candidates. Additional identifiers of sequences that support candidate suffixes ($+$ and $-$ in this case) are stored in the internal nodes.

in this way we can read the support of a string represented by the internal node N simply by computing the the union of its local support and the support of its child nodes. That is, we have

$$\theta(N) = \{\theta_l(N) \cup \bigcup_{N' \in C(N)} \theta(N')\}.$$

Before discussing the algorithm, we make some remarks about the size of $\mathcal{T}_t$. We already stated that the size of $\mathcal{C}_t$ is upper bounded by n/σ. This means that $\mathcal{T}_t$ can have at most n/σ leaf nodes. However, the length of the paths leading to the leaf nodes from the root of $\mathcal{T}_t$ can in theory be unbounded. In practice we can set an upper bound $h_{\max} > \gamma$ on the height of $\mathcal{T}_t$. Thus, $\mathcal{T}_t$ requires $O(h_{\max}n/\sigma) = O(n)$ space. Another simple but important observation is that any sequence can support at most one candidate at a time, and thus appears in at most one support list associated with a leaf node of $\mathcal{T}_t$. If a sequence does not support a candidate, it supports a suffix, and therefore has to appear in a local support list associated with some internal node of $\mathcal{T}_t$. As a consequence the support lists also need $O(n)$ space in total.

Algorithm LCP1. Now, instead of processing the candidates in $\mathcal{C}_{t-1}$ individually, we traverse $\mathcal{T}_{t-1}$ to produce the updated set of candidates $\mathcal{C}_t$, represented by the trie $\mathcal{T}_t$. This algorithm is given in Algorithm 2. In short, LCP1 updates the set of candidate patterns by traversing $\mathcal{T}_{t-1}$ once for each $a \in \Sigma$ starting from its root node. Each of these traversals, implemented by the PROCESS_TRIE function shown in Algorithm 3, corresponds to appending the symbol $a \in \Sigma$ to the end of the candidates. PROCESS_TRIE returns the root node of an updated

Algorithm 2

1: LCP1(T_{t-1})
2: $T_t \leftarrow$ new trie
3: $N \leftarrow$ root of T_{t-1}
4: **for** $a \in \Sigma$ **do**
5:     Let $S(a,t)$ be the set of sequences with symbol a at position t.
6:     $N' \leftarrow$ PROCESS_TRIE($N, S(a,t)$)
7:     Add N' as a new subtree to root of T_t.
8: **end for**
9: For all leaves L of T_{t-1} that were *not* extended by PROCESS_TRIE above, output the corresponding local correlation pattern if the path to L is at least of length γ.
10: **return** T_t

Algorithm 3

1: PROCESS_TRIE(N, S)
2: $\tilde{N} \leftarrow$ new trie node
3: **for** $N_c \in C(N)$ **do**
4:     $N'_c \leftarrow$ PROCESS_TRIE(N_c, S)
5:     **if** $|\theta(N'_c)| \geq \sigma$ **then**
6:         $C(\tilde{N}) \leftarrow C(\tilde{N}) \cup N'_c$
7:     **else**
8:         $\theta_l(\tilde{N}) \leftarrow \theta_l(\tilde{N}) \cup \theta(N'_c)$
9:     **end if**
10: **end for**
11: $\theta_l(\tilde{N}) \leftarrow \theta_l(\tilde{N}) \cup \{\theta_l(N) \cap S\}$
12: **if** N is a leaf and $|\theta(\tilde{N})| \geq \sigma$ **then**
13:     mark N as *extended*
14: **end if**
15: **return** $\tilde{N}$

trie that will be added to T_t. Once all symbols in Σ have been considered, we traverse T_{t-1} one more time and output the maximal local correlation patterns. These can be found at those leaf nodes of T_{t-1} that become infrequent for every $a \in \Sigma$. More precisely, every leaf L of T_{t-1} for which $|\theta(L) \cap S(a,t)| < \sigma$ for all $a \in \Sigma$ represents a maximal local correlation pattern.

The actual work of updating the candidate set is carried out in the PRO-CESS_TRIE function shown in Algorithm 3. It will return a new trie rooted at the node $\tilde{N}$ that initially has no local support or child nodes. First the algorithm recursively processes the children of the node N. For each child N_c we obtain the new trie rooted at N'_c (line 4). This will be added as a child of $\tilde{N}$ (line 6) if it has a large enough support. Otherwise we only add it's support to the local support of $\tilde{N}$ (line 8). On line 11 we update the support of the candidate that corresponds to node N. Lines 12–14 are needed to find those leaf nodes of T_{t-1} that can be returned as local correlation patterns.

Above we argued that the size of a candidate trie T is of order $O(n)$, where n is the number of sequences. In particular, the number of leafs is upper bounded by n and σ. Thus, the trie can be traversed in time $O(n)$ if we consider $h_{\max}$ a constant. On a first look it would seem that the overall complexity of PRO-CESS_TRIE is higher than this, since on line 11 we compute the intersection of the local support $\theta_l(N)$ and S. But since each sequence identifier appears only at one node N (most of them appear at the leafs), the total cost of line 11 over all recursive calls of PROCESS_TRIE is $O(n)$. Also, implementing lines 5 and 8 in time $O(1)$ requires some additional bookkeeping that is not shown in the pseudo-code of Algorithm 3. Essentially when PROCESS_TRIE returns it must in addition to $\tilde{N}$ also return a list of sequence identifiers that can be found in the subtrie below $\tilde{N}$. These lists must be constructed in such a way, that computing the union on line 8 is simply a matter of concatenation. Here we also use the property that a sequence identifier can appear in the trie only once.

Hence, the overall complexity of LCP1 is $O(|\Sigma|n)$, since we must traverse $\mathcal{T}$ once for each $a \in \Sigma$. Note that simply reading the next symbol from each sequence is an $O(n)$ operation.

4   Experiments

4.1   Performance of LCP1

In this section we study the behavior of LCP1 with different parameters of the input. We are interested in how the size (number of nodes) of the candidate trie $\mathcal{T}$ behaves and how the number of found patterns varies for different values of σ and γ. The implementation used is written in Java, and can be obtained from the web site of the author[1]. The experiments are run on a 2.2GHz Intel CPU.

Artificial data. Artificial data is generated using a model with n sequences that each output a uniformly at random chosen symbol of Σ independent of each other at every step. We do not plant any patterns into the input, and hence the test indicates only how the algorithm responses to noise. We let $n \in \{50, 200, 500, 1000\}$, $\sigma \in \{2, 4, 8, 16, 32\}$, and $|\Sigma| \in \{2, 4, 8, 16, 32\}$, and run LCP1 for 10000 steps with every combination of n, σ, and $|\Sigma|$. In each case we measure the running time and average size of the candidate trie.

Results for both are shown in Table 1. On the left of Table 1 we show the average number of time steps that LCP1 processes in one second for various parameter combinations. Clearly the algorithm becomes faster when σ is increased, since the size of $\mathcal{T}_i$ dramatically decreases due to the n/σ upper bound on the number of candidates. Another observation is that larger alphabets are slower to process despite the fact that the average size of $\mathcal{T}_i$ decreases when $|\Sigma|$ increases. This is also obvious, as the complexity of the algorithm is $O(|\Sigma|n)$, because on ever step we traverse the trie once for every $a \in \Sigma$. Finally we note that these numbers represent idealized conditions, since the input is being generated on the fly, and thus no data was read from any device, which is bound to be the bottleneck in many real applications.

[1] http://www.cis.hut.fi/aaukkonen

Table 1. *Left:* Average number of steps processed per second by LCP1 for different combinations of parameter values with random inputs. *Right:* Average size of the candidate trie with different combinations of parameters for randomly generated inputs.

n	$|\Sigma|$	σ 2	4	8	16	32	n	$|\Sigma|$	σ 2	4	8	16	32
50	2	13661	22471	27777	30959	31347	50	2	35.6	11.6	4.6	1.8	1.5
	4	14925	19920	26385	26455	26385		4	17.9	5.8	2.6	2.5	2.5
	8	13280	18867	19157	19193	19120		8	11.7	4.7	4.5	4.5	4.5
	16	9442	10893	10905	10869	10881		16	10.4	8.2	8.2	8.2	8.2
	32	5599	5837	5743	5685	5698		32	13.8	13.2	13.3	13.2	13.2
200	2	3367	6906	9469	10952	11918	200	2	143.5	47.5	20.0	8.9	3.8
	4	4012	6925	8650	9765	10111		4	72.1	23.7	10.4	4.0	2.5
	8	3376	5414	6910	6973	7032		8	48.6	16.9	5.0	4.5	4.5
	16	2461	3996	4196	4206	4171		16	34.7	9.5	8.5	8.5	8.5
	32	1674	2079	2076	2071	2067		32	25.3	16.5	16.5	16.5	16.5
500	2	874	2464	3910	4580	5055	500	2	359.7	119.6	51.0	23.6	11.3
	4	1164	2731	3442	3987	4130		4	180.4	58.9	27.3	10.7	6.3
	8	1033	2122	2525	3003	3012		8	118.0	39.5	21.2	4.7	4.5
	16	751	1417	1800	1803	1804		16	98.0	25.7	8.6	8.5	8.5
	32	569	918	936	933	934		32	62.8	17.3	16.5	16.5	16.5
1000	2	377	903	1542	1975	2228	1000	2	720.4	240.1	102.5	47.8	23.0
	4	515	1018	1468	1670	1960		4	360.7	121.9	48.0	26.4	10.5
	8	476	886	1110	1214	1419		8	245.8	71.1	36.3	20.4	4.5
	16	373	554	809	845	849		16	179.8	78.9	14.3	8.5	8.5
	32	249	439	469	469	468		32	155.0	25.5	16.5	16.5	16.5

Table 2. Average size of T_i for different σ in the Dow Jones data ($n = 30$, $|\Sigma| = 2$)

σ	4	6	8	10	12
avg. size of T_i	15.69	9.12	6.39	4.97	4.14

Real data. For this experiment we consider a stock market data[2] that contains the daily opening and closing prices of the Dow Jones 30 index between years 1985 and 2003. The index consists of 30 selected companies. We modify the data so that each day is labeled with a + if the price of the stock went up or with a − if the price went down. We recognize that this approach to discretization is not without problems as the magnitude of the variation is hidden, but for the purposes of demonstration we consider it sufficiently accurate. In real applications one might think of using a more sophisticated approach to discretizing the data.

First we investigate how the size of the candidate trie behaves in time. We run our algorithm on the Dow Jones data set again using different values for σ and γ, and record the size of T_t at every t. The average size of the trie is independent of γ, and is shown in Table 2 for different values of σ. We can also study the number of local correlation patterns found. This is shown for the Dow Jones

---

[2] http://lib.stat.cmu.edu/datasets/DJ30-1985-2003.zip (May 15. 2009)

Table 3. Number of maximal local correlation patterns found in the Dow Jones data with different combinations of σ and γ

	$\sigma = 4$	$\sigma = 6$	$\sigma = 8$	$\sigma = 10$	$\sigma = 12$
$\gamma = 4$	8510	3712	1638	650	286
$\gamma = 6$	2970	738	201	56	11
$\gamma = 8$	766	119	19	4	0
$\gamma = 10$	159	17	0	0	0
$\gamma = 12$	35	0	0	0	0

data in Table 3 for different combinations of σ and γ. Obviously the number of patterns found for small parameter values is orders of magnitude larger than for larger ones. Increasing γ has a stronger effect.

4.2   Using Local Correlation Patterns to Compare the Sequences

In this section we give an example on how to use local correlation patterns for comparing the sequences in the input. Consider the Dow Jones data used in the previous experiment. For each company X, we can look at the set of patterns that contain the sequence corresponding to X in their support. Denote this set by $P(X)$. Given a set of patterns, we can describe it by using a feature vector, where the features are some simple characteristics of the patterns. These could be the average length of patterns, the average number of alternations of the symbols in the patterns, the average size of the support, etc. Given all patterns that are found from the DJ data with $\sigma = 4$ and $\gamma = 4$, we compute the aforementioned features based on $P(X)$ for every company X. We obtain a small data set with one row for each company. Figure 3 shows a scatterplot with the

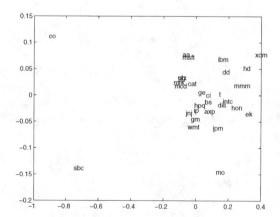

Fig. 3. A principal components plot showing ticker symbols of companies in the DJ data. The data is based on the local correlation patterns. We observe that SBC and CO are different from the other companies given the set of local correlation patterns they belong to.

1st and 2nd principal component of this data on the X and Y-axis, respectively. Clearly companies with the ticker symbols CO and SBC differ somehow from the rest. When looking at the actual values of the features, it turns out that these companies tend to belong to supports of patterns that are longer and contain a larger number of alternations between the symbols.

4.3   Using Local Correlation Patterns for Classification

In this section we study how local correlation patterns can be used to classify time series. We use data of EEG measurements made available by Henri Begleiter at the Neurodynamics Laboratory at the State University of New York Health Center at Brooklyn. The data can be downloaded from the UCI KDD repository[3]. Originally the data was used in [15].

The data consists of EEG measurements conducted with a number of subjects performing object recognition tasks. Part of the subjects had been diagnosed with a genetic predisposition to alcoholism (group $\mathcal{A}$), while the remaining ones belong to a control group (group $\bar{\mathcal{A}}$). Our aim in this experiment is to study if the local correlation patterns differ between the two groups, and if the patterns can be used to build a classifier for predicting the condition of unseen subjects. We point out that classification of this particular EEG data has been studied successfully in existing literature (see e.g. [5]), and we do not claim that the approach discussed here is superior to existing techniques. The purpose of this experiment is to study the applicability of local correlation patterns for classification in general.

There are 123 subjects in total, each of whom has completed 120 measurements (up to a small number of exceptions). Number of subjects in groups $\mathcal{A}$ and $\bar{\mathcal{A}}$ is 77 and 45, respectively. Each measurement contains 256 samples on 64 channels that represent one second of activity in the brain while the subject was exposed to a visual stimulus. The EEG output consists of floating point values which we discretize to obtain sequences over the alphabet $\{-, +\}$, where a $-$ ($+$) means that the measurement value decreased (increased) from the previous step.

First we compare the sets of patterns found, and see if there is any difference between the two groups of subjects. To this end we compute the set P_{ij} of local correlation patterns with different values of σ and γ for every measurement j of every subject i. Using the patterns in P_{ij} we compute the feature vector f_{ij}. The features we use are the number of patterns in P_{ij} denoted ϕ_n, average length of the patterns in P_{ij} denoted ϕ_l, average size of the support of the patterns in P_{ij} denoted ϕ_s, and average number of alternations between a $+$ and $-$ in the patterns denoted ϕ_a. All features are normalized to zero mean and unit variance.

We compare the within group means of each feature in Table 4. There are a number of things worth noting. First, the average number of patterns (ϕ_n) is larger in group $\bar{\mathcal{A}}$ with larger values of σ. Second, the average length of a pattern (ϕ_l) is always larger in group $\bar{\mathcal{A}}$, while the average number of alternations

---

[3] http://kdd.ics.uci.edu/databases/eeg/eeg.html

Table 4. Group specific means for different features computed from patterns found in an EEG data set with different values of σ and γ. We observe that the group of alcoholic subjects differs from the control group in every case.

group	σ	γ	$E[\phi_n]$	$E[\phi_s]$	$E[\phi_l]$	$E[\phi_a]$
$\mathcal{A}$	6	6	0.20	0.02	-0.15	0.06
$\bar{\mathcal{A}}$	6	6	-0.34	-0.04	0.27	-0.10
$\mathcal{A}$	12	8	-0.17	0.01	-0.15	0.10
$\bar{\mathcal{A}}$	12	8	0.30	-0.01	0.26	-0.18
$\mathcal{A}$	24	6	-0.21	-0.02	-0.14	0.06
$\bar{\mathcal{A}}$	24	6	0.37	0.03	0.24	-0.11

between a $+$ and a $-$ (ϕ_a) is always slightly larger in group $\mathcal{A}$. This result indicates that the sets of patterns found in measurements of alcoholic subjects differ from the sets of patterns found in measurements of subjects belonging to the control group.

As a next step we build a simple nearest mean classifier using these features. Let $\mathcal{A}_{\text{train}}$ and $\bar{\mathcal{A}}_{\text{train}}$ be the sets of alcoholic and control subjects used for training. The training data consists of the feature vectors f_{ij} for all j and for all $i \in \{\mathcal{A}_{\text{train}} \cup \bar{\mathcal{A}}_{\text{train}}\}$. First we normalize the features in the training data to zero mean and unit variance. Then we compute the group specific means $E[\mathcal{A}_{\text{train}}]$ and $E[\bar{\mathcal{A}}_{\text{train}}]$. A new subject k is classified by considering the feature vectors f_{kj} separately for each j. The classifier assigns subject k to group $\mathcal{A}$ if the number of vectors f_{kj} that are closer to $E[\mathcal{A}_{\text{train}}]$ is larger than the number of vectors f_{kj} that are closer to $E[\bar{\mathcal{A}}_{\text{train}}]$, and to the class $\bar{\mathcal{A}}$ otherwise.

We test the approach using 1-fold cross-validation, i.e., one subject is classified at a time with a training set that contains all other subjects but not the one we are classifying. This experiment is repeated using patterns computed with the same values of σ and γ used above in Table 4. In every case we get the following confusion matrix:

$$
\begin{array}{c|cc}
 & \mathcal{A} & \bar{\mathcal{A}} \\
\hline
\mathcal{A} & 77 & 0 \\
\bar{\mathcal{A}} & 1 & 44
\end{array}
$$

All alcoholic subjects are classified correctly, and one control subject is classified incorrectly to group $\mathcal{A}$. Moreover, the parameters used when mining the patterns seem to have a negligible effect on the classifier. This result is not surprising considering that in [5] it is reported that alcoholic and control subjects differ significantly in a number of traditional features used in EEG analysis. However, our experiments show that simple and general, *non-application specific features* based on local correlation patterns can also be very effective for classification.

5   Related Work

Mining patterns in streams and time series is a well studied topic. To the best of our knowledge, the local correlation patterns introduced here have not appeared previously in literature. However, methods for mining several other kinds of patterns from time series and data streams have been proposed.

In [2] an algorithm is presented for finding rules in streams. These are frequently occurring patterns of the form "if A occurs then B occurs within a certain time". A related approach is that of finding "motifs" [8], where one seeks parts of the time series that occur repeatedly. In some applications it is not enough to have a pattern repeat itself, but it must do so in a certain period. Such patterns are mined for example in [4,14]. Yet another pattern class is defined by so called surprising patterns [7] that are parts of a stream that are unexpected given some previously observed history. Another related line of research is about mining frequent patterns in streams. Examples include frequent itemsets [9,3], sequential patterns [11], and trees [1].

A common characteristic of these examples is that they are concerned with finding repeated occurrences of a pattern within a window of the stream. In this work we want to find *simultaneous* occurrences of the pattern in multiple parallel streams.

6   Conclusion

We have introduced local correlation patterns as a method for analyzing multidimensional time series. We model such time series as sets of sequences over some finite alphabet. A local correlation pattern is defined as a point t in time together with a string of symbols from the alphabet. The pattern tells that starting from time t a subset of the sequences all simultaneously output the specified string. In practice we want to find patterns where the sequence is at least of length γ and the subset of sequences is of size at least σ. Moreover, we are only interested in maximal local correlation patterns, i.e., patterns where the sequence is not the prefix nor suffix of any other local correlation pattern.

We proposed an algorithm that mines maximal local correlation patterns from a set of sequences in an online setting. The algorithm works by maintaining a trie of candidate patterns that is updated at every step when new symbols are read from the sequences. The complexity of our algorithm is $O(|\Sigma|n)$ for each timestep, where n is the number of sequences and $|\Sigma|$ the size of the alphabet. New patterns are output by the algorithm as soon as they are found.

We conducted experiments that show the algorithm is fast. A simple implementation running on a regular PC can process up to thousands of steps per second. Even for small support thresholds and a large number of sequences (say, $n = 1000$), the algorithm is capable of processing over a hundred steps per second. We also show that the local correlation patterns can be used for classifying EEG time series. In our experiment a simple nearest mean classifier based on the patterns had nearly 100 percent accuracy.

A problem with some practical applications is that we cannot assume the sequences to be perfectly aligned. In such situations we must allow small variations in the starting time of the pattern in a sequence. That is, a sequence s would support the pattern (i, p) if p occurs as a substring in s at the position $j = i \pm \delta$.

Another question is how to incorporate wildcards into the pattern string, or how to allow (a small number of) mismatches in the supporting sequences.

References

1. Asai, T., Arimura, H., Abe, K., Kawasoe, S., Arikawa, S.: Online algorithms for mining semi-structured data stream. In: Proceedings of the 2002 IEEE International Conference on Data Mining, p. 27 (2002)
2. Das, G., Lin, K.-I., Mannila, H., Renganathan, G., Smyth, P.: Rule discovery from time series. In: Proceedings of the 4th International Conference on Knowledge Discovery and Data Mining, pp. 16–22 (1998)
3. Giannella, C., Han, J., Pei, J., Yan, X., Yu, P.S.: Mining Frequent Patterns in Data Streams at Multiple Granularities. In: Data Mining: Next Generation Challenges and Future Directions. MIT Press, Cambridge (2004)
4. Han, J., Dong, G., Yin, Y.: Efficient mining of partial periodic patterns in time series database. In: Proceedings of the 15th International Conference on Data Engineering (ICDE 1999), pp. 106–115 (1999)
5. Kannathal, N., Acharya, U., Lim, C., Sadasivan, P.: Characterization of eeg – a comparative study. Computer Methods and Programs in Biomedicine 80(1), 17–23 (2005)
6. Kärkkäinen, J., Sanders, P., Burkhardt, S.: Linear work suffix array construction. Journal of the ACM 53(6), 918–936 (2006)
7. Keogh, E., Leonardi, S., Chiu, B.: Finding surprising patterns in a time series database in linear time and space. In: Proceedings of the eighth ACM SIGKDD international conference on Knowledge discovery and data mining, pp. 550–556 (2002)
8. Lin, J., Keogh, E., Lonardi, S., Patel, P.: Finding motifs in time series. In: Proceedings of the Second Workshop on Temporal Data Mining (2002)
9. Manku, G.S., Motwani, R.: Approximate frequency counts over data streams. In: Proceedings of the 28th international conference on Very Large Data Bases, pp. 346–357 (2002)
10. McCreight, E.M.: A space-economical suffix tree construction algorithm. Journal of Algorithms 23(2), 262–272 (1976)
11. Raïssi, C., Poncelet, P., Teisseire, M.: Speed: Mining maximal sequential patterns over data streams. In: Proceedings of the 3rd International IEEE Conference on Intelligent Systems, pp. 546–552 (2006)
12. Ukkonen, E.: On-line construction of suffix trees. Algorithmica 14(3), 249–260 (1995)
13. Weiner, P.: Linear pattern matching algorithms. In: Proceedings of the 14th IEEE Annual Symposium on Switching and Automata Theory, pp. 1–11 (1973)
14. Yang, J., Wang, W., Yu, P.S.: Mining asynchronous periodic patterns in time series data. IEEE Transactions on Knowledge Engineering 15(3), 613–628 (2003)
15. Zhang, X.L., Begleiter, H., Porjesz, B., Wang, W., Litke, A.: Event related potentials during object recognition tasks. Brain Research Bulletin 38(6), 531–538 (1995)

Subspace Discovery for Promotion: A Cell Clustering Approach[*]

Tianyi Wu and Jiawei Han

University of Illinois at Urbana-Champaign, USA
{twu5,hanj}@illinois.edu

Abstract. The promotion analysis problem has been proposed in [16], where ranking-based promotion query processing techniques are studied to effectively and efficiently promote a given object, such as a product, by exploring ranked answers. To be more specific, in a multidimensional data set, our goal is to discover interesting subspaces in which the object is ranked high. In this paper, we extend the previously proposed promotion cube techniques and develop a cell clustering approach that is able to further achieve better tradeoff between offline materialization and on-line query processing. We formally formulate our problem and present a solution to it. Our empirical evaluation on both synthetic and real data sets show that the proposed technique can greatly speedup query processing with respect to baseline implementations.

1 Introduction

The *promotion analysis problem* [16] aims to search for interesting subspaces for some user-specified target object such as a person or a product item so that it can be promoted in the subspaces discovered. Such a function is called *ranking-based promotion query*. Given a user-specified object, a promotion query should return the most interesting subspaces in the multidimensional lattice, where "interesting" intuitively means that the object is among the top ranked answers in a particular subspace. It has been shown that many OLAP and decision support applications can potentially benefit from such a function, as data analysts may find it useful to explore the search space to promote a target object. For example, suppose a car model is given as the target object, then interesting locations may be subsequently discovered such that the car model is highly ranked in terms of sales or customer review in those locations, which brings up opportunities for promotion.

To process promotion queries efficiently, however, is a challenging problem. In this paper, we focus on the offline side of the problem. We extend the previously studied promotion cube technique [16] and propose a general *cell clustering approach* to further achieve better query execution time vs. materialization

---

[*] The work was supported in part by the U.S. National Science Foundation grants IIS-08-42769 and BDI- 05-15813, and the Air Force Office of Scientific Research MURI award FA9550-08-1-0265.

J. Gama et al. (Eds.): DS 2009, LNAI 5808, pp. 362–376, 2009.

A_1	A_2	O	M
a_1^1	a_2^2	o_1	0.3
a_1^1	a_2^2	o_3	0.2
a_1^1	a_3^2	o_1	0.5
a_1^1	a_3^2	o_3	0.6
a_1^1	a_3^2	o_1	0.8
a_2^1	a_3^2	o_2	0.5
a_2^1	a_2^2	o_3	0.6
a_2^1	a_2^2	o_2	0.3
a_2^1	a_2^2	o_1	0.3

Fig. 1. Multidimensional table

tradeoff. Our observation is that promotion cells can be clustered together in a coherent representation, while at the online processing step the uninteresting subspaces can still be effectively pruned out.

Example 1. *Before introducing our proposed techniques, let us first examine a concrete example of promotion analysis. Figure 1 shows an example multidimensional table with two dimensions A_1 and A_2, each having 2 distinct values. In addition, the object dimension is represented by O, and the measure dimension is a numerical dimension represented by M. Each row in this table is called a base tuple and by enforcing selection conditions over A_1 and A_2 we can obtain different object subspaces. Figure 2 shows all nine subspaces obtained from the example multidimensional table. Among these subspaces, the first $\{*, *\}$ is a special one called full space. For each of these subspaces, we show its ranked list of aggregates based on the AVG aggregation. For example, for the subspace $\{A_1 = a_1^1, A_2 = *\}$, o_1's AVG can be computed as $(0.3 + 0.5 + 0.8)/3 = 0.53$.*

A straightforward offline method for processing the promotion query is to materialize all aggregates, e.g., store Table 2 completely (notice that the objects are shown in the table for reference, but they do not need to be stored). Now, given a target object, say o_3, one can easily locate the most interesting subspaces $\{a_2^1, *\}$, $\{*, a_2^2\}$, and $\{a_2^1, a_2^2\}$, because it is ranked the very first in those subspaces. Similarly, given the aggregate measure, any other target object can be processed in the same fashion using these completely precomputed results.

Since there could be a very large number of subspaces as well as objects even for a moderately large data set, the precompute-all strategy can be very costly at the offline stage. To this end, a general promotion cube framework has been proposed as a partial precomputation strategy that avoids such high cost. In a promotion cube, each promotion cell consists of a set of summary aggregates to help bound the target object's rank. However, in most cases it is even unnecessary to store all promotion cells since object aggregates tend to be similar across different subspaces. This leads to our proposed clustering approach to further remove redundancy. After clustering, the number of materialized aggregates can be largely reduced, while the efficiency of online query execution can still be guaranteed. Another advantage is that such a clustered cube structure does not

A_1	A_2	Ranked list of aggregates
$*$	$*$	(o_1) 0.48, (o_3) 0.47, (o_2) 0.4
a_1^1	$*$	(o_1) 0.53, (o_3) 0.4
a_2^1	$*$	(o_3) 0.6, (o_2) 0.4, (o_1) 0.3
$*$	a_2^2	(o_3) 0.4, (o_1) 0.3, (o_2) 0.3,
$*$	a_3^2	(o_1) 0.65, (o_3) 0.6, (o_2) 0.5
a_1^1	a_2^2	(o_1) 0.3, (o_3) 0.2
a_1^1	a_3^2	(o_1) 0.65, (o_3) 0.6
a_2^1	a_2^2	(o_3) 0.6, (o_1) 0.3, (o_2) 0.3
a_2^1	a_3^2	(o_2) 0.5

Fig. 2. Fully precomputed results using AVG aggregation

restrict the capability of promotion analysis in that any promotion query can be supported for a given type of aggregation. In this paper, we

- extend the promotion cube framework to a cell clustering approach to further achieve better balance between offline materialization cost and online query execution cost;
- formally define the promotion cube structure with clustered cells;
- discuss the hardness of the problem and develop a greedy algorithm to perform clustering so that the quality of result is guaranteed; and
- conduct empirical evaluation on synthetic as well as real-world data sets to verify the performance of the cell clustering approach, and show that the approach is superior to baseline approaches.

The remainder of the paper is organized as follows. Section 2 discusses related work. Section 3 formulates the problem of the paper. In Section 4 we propose the cell clustering approach based on the promotion cube framework. Section 5 reports experimental results, and finally, Section 6 concludes this work.

2   Related Work

The promotion analysis problem is originally studied in [16], which proposes the promotion query model and its query processing techniques; also, a statistical method is discussed to prevent spurious promotion results. Our paper can be regarded as a follow-up study of it in that the underlying multidimensional data model as well as the promotion query model remain the same. However, the main focus of this study is different from the previous work; that is, the major goal here is to propose a new offline strategy to perform clustering over promotion cells, while being able to answer online promotion queries on multidimensional data in an efficient way.

Besides, this work is related to several previous studies on database ranking techniques. The ranking cube method is first studied in [17] for answering top-k queries which allows users to enforce conditions on database attributes.

Subsequently, [15] discusses the problem of processing top-k cells from multidimensional group-by's for different types of measures. Both of the above methods leverage offline precomputation to achieve better online performance. Moreover, the threshold algorithms [6] represent yet another family of ranking techniques, where the objective is to efficiently produce the top answers by aggregating scores from multiple ranked lists. Other variants of ranking algorithms have also been studied and applied in a variety of applications including Web-accessible databases [12], supporting expensive predicates [2], keyword search [10], and visualization [14]. Unfortunately, none of these techniques can handle our problem.

Various clustering problems are also related to our study. [13] presents the *RankClus* method for integrating ranking and clustering such that the results of both can be mutually enhanced. Earlier papers like [5,18] have discussed methods for compressing and/or clustering data points to facilitate further data mining. However, these methods are different from the one in this paper since our clustered cells are for pruning. In addition, our work shares similar objectives as some previous clustering algorithms. For example, [3,1,4] propose approximation algorithms to clustering data points so as to minimize a particular objective (*e.g.*, the sum of cluster diameters). Our cell clustering problem can fit into their problem setting but their results are mainly of theoretical interests and may not scale to large data. There are also other related studies in that multidimensional analysis is conducted for data mining [11,7], but none of them can be applied toward our context.

3   Problem Formulation

To ensure that the discussion be self-contained, we present the formulation of the promotion query problem in this section. Consider a multidimensional table T consisting of a collection of base tuples. There are three categories of columns: d categorical dimensions, $A_1, \ldots, A_d$, which are called *subspace dimensions*, a column O storing objects called *object dimension*, and a numerical *score dimension*, M. We use O to represent also the set of objects. Based on the multidimensional schema, we call $S = \{a_1, a_2, \ldots, a_d\}$ a *subspace*, where a_i is either a dimension value of A_i or it is "any" value denoted by "*". moreover, the set of all subspaces is denoted by $\mathbf{U}$.

Given an aggregate function $\mathcal{M}$ (*e.g.*, AVG), one can derive for each subspace the ranked list of objects and their aggregate values.

The promotion analysis problem can be formulated as follows. Given a *target object* τ for promotion ($\tau \in O$), we let $\mathbf{V} = \{S | \tau$ occurs in subspace $S \wedge S \in \mathbf{U}\}$, *i.e.*, the set of subspaces where τ occurs. Our goal is to *discover the top-k subspaces in* $\mathbf{V}$ *such that* τ *is the most highly ranked in these k subspaces*, where k is a user parameter.

Example 2. *In the running example shown in Table 1, we can see that A_1 and A_2 are the subspace dimensions, O is the object dimension, and M is the score dimension. Table 2 shows* $\mathbf{U}$*, i.e., the set of all 9 subspaces. Now suppose the target object is* $\tau = o_3$ *and the aggregate function is* $\mathcal{M} = AVG$*. Then* $|\mathbf{V}| = 8$

because o_3 has 8 subspaces containing τ ($\{a_2^1, a_3^2\}$ is not in $\mathbf{V}$ because o_3 does not appear in it). If $k = 3$, the top-k subspaces for o_3 should be $\{a_2^1, *\}$, $\{*, a_2^2\}$, and $\{a_2^1, a_2^2\}$ because o_3 is ranked first in all of them but it has lower rank in any other subspace.

4 The Cell Clustering Approach

To handle the subspace discovery problem efficiently, we propose to study a cell clustering approach within the promotion cube framework. We present the cube structure in Section 4.1, followed by a brief description of the complementary online algorithm in Section 4.2. Section 4.3 discusses an optimization technique for clustering.

4.1 Promotion Cube with Clustered Cells

To support data analysis for large-scale applications, computing the top-k subspaces from scratch may not be a good solution. On the other extreme, materializing all cells in a promotion cube may suffer from excessive space requirement due to a blow-up of the number of subspaces (which can be up to $O(\prod_{i=1}^d |A_i|)$, where A_i denotes the cardinality of dimensions A_i). To tackle this problem, we observe that aggregates in different subspaces tend to be similar, so it would be wasteful to materialize all cells separately. Thus, we propose a clustering approach to reduce the space overhead of the promotion cube. This structure presents a summary of aggregated data and meanwhile is able to facilitate online exploration.

Given multidimensional table T and aggregate function $\mathcal{M}$, The definition of the cube structure with clustered cells is as follows.

Definition 1. *Given any subspace $S \in \mathbf{U}$, a clustered cell, $CluCell(S)$, is defined as a collection of upper and lower bound aggregates; it specifically consists of $2n$ values, $(\bar{c}_1, \underline{c}_1; \bar{c}_2, \underline{c}_2; \ldots; \bar{c}_n, \underline{c}_n)$, where $\bar{c}_1$ and $\underline{c}_1$ ($1 \le i \le n$) are the upper and lower bounds for the $((i-1) \cdot w + 1)$-th largest object aggregate value in subspace S. Here n and w are positive integers.*

The promotion cube with clustered cells, *denoted by $CluPromoCube$, is defined as a collection of clustered cells in the format $(S, CluCell, Sup)$ for all subspaces qualifying some* minsup *threshold; that is, $CluPromoCube = \{S : CluCell(S), Sup(S) \mid Sup(S) > minsup\}$.*

Here $Sup(S)$ refers to how many objects subspace S has. In the above definitions n and w are user-specified parameters that determine the size of each materialized cell, whereas $minsup$ is another parameter that dictates which cells to materialize. These parameters are application-dependent. When $minsup = 0$, $n = |\mathcal{O}|$, $w = 1$, and $\bar{c}_i = \underline{c}_i$, the promotion cube would degenerate to fully precomputing all results since all subspaces and all object aggregate scores are precomputed. In fact we often have $minsup > 0$ and $n \ll |\mathcal{O}|$ (i.e., only a small fraction of aggregate values will be materialized), and $\bar{c}_i > \underline{c}_i$ (i.e., several promotion cells will be clustered together).

$Subspace$	$CluCell(S)$	$Sup(S)$
S_1	(8.9,8.8; 7.0,6.8; 3.4,3.0; 1.1,0.9; 0.3,0.3)	200
S_2	(5.1,5.0; 3.6,3.2; 1.4,1.2; 0.8,0.5; 0.05,0.01)	150
S_3	(9.1,8.5; 7.2,6.1; 3.2,2.9; 1.1,1.0; 0.2,0.02)	260
S_4	(9.0,8.8; 6.9,6.0; 3.3,3.0; 0.9,0.5; 0.2,0.1)	220
$\ldots$	$\ldots$	$\ldots$

Fig. 3. An example $CluPromoCube$ with $n = 5$, $w = 10$, and $minsup$=100

Example 3. *Figure 3 illustrates an example of clustered cells (not following the previous examples), where for each subspace passing $minsup = 100$, a clustered cell ($n = 5$, $w = 10$) is materialized. For instance, subspace S_3 contains $2 \times n = 10$ aggregate values, where $\bar{c}_2 = 7.2$, $\underline{c}_2 = 6.1$ indicate that the $(i - 1) \cdot w + 1 = (2 - 1) \cdot 10 + 1 = 11$th largest aggregate is no more than 7.2 and no less than 6.1.*

4.2   Online Algorithm

Given the offline data structure $CluPromoCube$, we briefly describe the complementary online algorithm $PromoRank$ to produce top-k subspaces for any input τ [16]. Notice that other online algorithms may also work with $CluPromoCube$ in a similar way but they are beyond the scope of this study. $PromoRank$ first computes $\mathbf{V}$, the subspaces containing τ along with τ's aggregate values. Second, a *candidate* set of subspaces is generated based on $CluPromoCube$. Third, the candidate set is aggregated to produce the correct and complete set of top-k results. This algorithm is depicted in Table 1.

At the beginning, $\mathbf{V}$ is obtained and for each $S \in \mathbf{V}$, τ's aggregate value is computed. For these $|\mathbf{V}|$ subspaces, we denote τ's aggregates as $M_1, M_2, \ldots, M_{|\mathbf{V}|}$. To compute the aggregates, a depth-first enumeration of all τ's tuples would be enough. Then, a candidate subspace set is generated whereas non-candidate subspaces are pruned. The generation of the candidate set requires us to first compute $HRank_s$ (i.e., the highest possible rank) and $LRank_s$ (ie, the lower possible rank) for each subspace S_s ($1 \leq s \leq |\mathbf{V}|$ and $S_s \in \mathbf{V}$) based on $CluPromoCube$.

When some S_s does not meet the $minsup$ threshold, we cannot derive τ's rank in it since it becomes unbounded (if the final top-k subspaces are also required to satisfy the $minsup$ condition, S_s can be pruned immediately). Otherwise, given S_s's clustered promotion cell $CluCell(S_s) = (\bar{c}_1, \underline{c}_1; \bar{c}_2, \underline{c}_2; \ldots; \bar{c}_n, \underline{c}_n)$, $HRank$ and $LRank$ can be computed as in the following cases:

- Let i be the smallest value in $\{1, 2, \ldots, n\}$ such that $M_s > \bar{c}_i$, then $LRank_s = (i - 1) \cdot w$. If such i does not exist, we have $LRank_s = Sup(S_s)$;
- Let j be the largest value in $\{1, 2, \ldots, n\}$ such that $\underline{c}_j > M_s$, then $HRank_s = (j - 1) \cdot w + 2$. If such j does not exist, we have $HRank_s = 1$.

Note that the above computation assumes that there is no duplicate aggregate value within any subspace. In the presence of duplicate aggregates, the computation can be easily extended. Let R_k be the k-th largest $LRank_s$ for $1 \leq s \leq |\mathbf{V}|$.

Table 1. The online processing algorithm

```
Algorithm 1
    (Aggregate V)
 1  Compute M₁, M₂, ..., M|V|, the aggregate values of τ in each subspace
    in V;
    (Prune out non-candidate subspaces)
 2  for s = 1 → V do compute HRankₛ and LRankₛ using
    CluPromoCube;
 3  Rₖ ← the k-th largest value among {LRankₛ|1 ≤ s ≤ |V|};
 4  Prune out all subspaces having HRankₛ > Rₖ;
    (Generate the complete set of results)
 5  Call the recursive procedure below on ({*}, T, 0) to compute τ's rank
    in each unpruned subspace;
 6  Return the top-k subspaces where τ has the highest ranks;
    (Below is a recursive procedure on (S, T, d₀) )
 7  if S is not pruned then compute τ's exact rank in S;
 8  for d' ← d₀ + 1 to d do
 9      Sort T's tuples based on d'-th subspace dimension;
10      for each distinct value v ∈ A_d' do
11          S' ← S ∪ {d' : v};          /* next subspace */
12          T' ← T's tuples having v on A_d'
13          Recursively call (S', T', d') if S' contains τ;
14      end
15 end
```

Any subspace S_s with $HRank_s$ lower than R_k (i.e., $HRank_s > R_k$) must not be a top-k result and thus can be pruned. As a result, the unpruned subspaces form the candidate set which must be a superset of the final top-k subspaces. Notice that if the exact order of the top-k results is not required, one may directly output the subspaces whose $LRank$ is greater than the k-th highest $HRank$ without adding them into the candidate set.

Example 4. *Given the CluPromoCube in Example 3, Figure 4 illustrates the computation of HRank and LRank for some τ. For instance, suppose τ's aggregate value for subspace S_2 has been computed as $M_2 = 3.5$. Based on the corresponding clustered cell for S_2 in Figure 3, we have that $M_2 > \bar{c}_3$, meaning that $i = 3$ and $LRank_2 = (i - 1) \cdot w = 20$; also we have $HRank_2 = (j - 1) \cdot w + 2 = 2$ because*

Subspace	M_s	$HRank$	$LRank$
S_1	0.2	42	200
S_2	3.5	2	20
S_4	5.0	12	20
...	...	...	

Fig. 4. Example $HRank$ and $LRank$ computation based on $CluPromoCube$

$\underline{c}_j > M_2$ holds when $j = 1$. Similarly for S_1 and S_4 we obtain their $HRank$ and $LRank$ respectively as shown in the figure. Thus, S_1 can be pruned when $k = 2$ because R_k would be no more than 20 while for S_1, its $HRank$ is 42.

Finally, the exact ranks of τ are evaluated for the unpruned subspaces using a recursive procedure displayed in Table 1 (Lines 7–15). It starts with the full space and recursively sorts data to generate children subspaces in a depth-first manner. The multidimensional table T is iteratively sorted according to the d'-th dimension (Line 9), such that each dimension value extends to a new child subspace S' (Line 11). For the current subspace S, if it is in the candidate set, τ's exact rank would be derived by aggregating objects in the input table T (Line 7). In this way, all subspaces in $\mathbf{V}$ can be enumerated. Although the bottleneck of the online algorithm lies in the recursive procedure, where the worst-case time complexity could be $O(|\mathbf{V}| \cdot T)$, $CluPromoCube$ is able to help prune many uninteresting subspaces, thereby bringing down the total online cost. Note that the size of $CluPromoCube$ would only affect candidate generation and thus the performance.

4.3   Clustered Cell Generation

We now turn to the generation of clustered cell at the offline stage. Given the multidimensional table, two parameters n and w, we can generate promotion cells for all subspaces by aggregating this table and then select n aggregates at every w-th position. Doing multidimensional aggregation and generating such selected aggregates for each subspace have been well-studied [7] and thus we do not provide details here. Our focus is on how to generate clustered cells.

For each subspace, let us call its promotion cell, i.e., the n selected aggregates, an n-dimensional *point* in the Euclidean space. Therefore, our problem is to cluster $|\mathbf{U}|$ points into a few clusters. For example, given two 5-dimensional points $(5.0, 4.0, 3.0, 2.0, 1.0)$ and $(5.5, 3.6, 3.1, 1.8, 0.05)$, they can be clustered into a $CluCell$, which is $(5.5, 5.0; 4.0, 3.6; 3.1, 3.0; 2.0, 1.8; 1.0, 0.05)$. In principle, a $CluCell$ can be generated from multiple promotion cells by taking the maximum and minimum aggregate value at each of the n dimensions. Note that such a clustering approach does not affect the correctness and completeness of results by the definition of $CluCell$. Since clustering two distant points may adversely affect the online algorithm, we use the following distance function:

Definition 2. *Given two n-dimensional points in the Euclidean space, $P_1 = (p_1^1, p_2^1, \ldots, p_n^1)$ and $P_2 = (p_1^2, p_2^2, \ldots, p_n^2)$, let their distance be the L_1-distance: $dist(P_1, P_2) = \|P_1, P_2\|_1 = \sum_{i=1}^{n} |p_i^1 - p_i^2|$. (Euclidean, or L_2, distance may be used alternatively to prevent large variance along different dimensions.)*

A typical clustering approach like k-means may fail to yield desired tradeoff because it would be difficult to specify the number of clusters, and also the radii of clusters may be affected by outliers. In order to guarantee the quality of clusters, we transform our clustering problem to the following optimization problem.

Definition 3. *Given a collection of* $|\mathbf{U}|$ *points in an n-dimensional Euclidean space (promotion cells),* $\Phi = \{P_1, P_2, \ldots, P_{|\mathbf{U}|}\}$, *and a user-specified radius* r (≥ 0). *A* P_i-*cluster is defined as* $\{P_j | 1 \leq j \leq |\mathbf{U}|, dist(P_i, P_j) \leq r\}$, *i.e., all points with* L_1-*distance to* P_i *no more than* r. *The problem asks for the minimum number of clusters to cover all points in* Φ.

Based on the definition, we can see that when $r = 0$, each cluster contains identical points while the pruning power would remain exactly the same as the promotion cube; when $r = \infty$, all cells are packed into a single cluster, but it is unlikely to help prune any subspace. We hence recommend to set r to about a small percentage, such as 10%, of the average gap between the aggregates of promotion cells. Unfortunately, to solve the cluster selection problem optimally turns out to be difficult given the following hardness result.

Lemma 1. *The cluster selection problem is NP-hard (see Appendix for proof sketch).*

Thus, we resort to a greedy algorithm as follows. First, given the set of points, compute for each point P_i the P_i-cluster. Second, iteratively select clusters until all points are covered. At each iteration, greedily select the cluster containing the maximum number of uncovered points. Finally, output the clusters selected and generate the corresponding *CluPromoCube*. This greedy algorithm has a worst-case time complexity of $O(n \cdot |\mathbf{U}|^2)$ that is affordable for a large number of subspaces. After clusters have been selected, *CluPromoCube* will materialize their corresponding *CluCells* by merging the points. Thus, multiple subspaces can share one clustered cell, instead of storing a promotion cell for each.

5   Experiment

In this section we report our experimental results. First we will introduce our implementation methodology in Section 4.1. Then, comprehensive experiments will be conducted on both synthetic data set as well as real-world data set. Our performance evaluation shows that the clustering approach can achieve an order of magnitude speedup while using much smaller storage space than a baseline method does.

5.1   Implementation Methodology

We compare three methods in terms of query execution time and/or materialization cost. They are: (i) the clustering-based promotion cube strategy denoted by *CluPromoCube*; (ii) a naive strategy that fully materializes all results in every subspace passing a given *minsup*, which we denote using *PrecomputeAll*; and (iii) the *OnlineExec* method where queries are answered from scratch using the method described in Section 4.2. Among these three methods, *PrecomputeAll* can be regarded as the naive strategy for the materialization cost measure. On the other hand, *OnlineExec* can be considered a bottom line implementation for

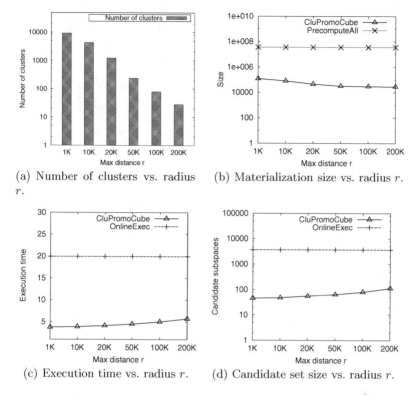

(a) Number of clusters vs. radius r.

(b) Materialization size vs. radius r.

(c) Execution time vs. radius r.

(d) Candidate set size vs. radius r.

Fig. 5. Performance results on the synthetic data set

the query execution time measure as no precomputation is needed. The query execution time is measured in terms of seconds whereas the materialization cost is measured in terms of the total number of values materialized.

Our experiments were carried out on PC with a Pentium 3GHz processor, 2GB of memory, and 160G hard disk drive. The programs for the implementation were all written in Microsoft Visual C# 2008 in the Windows XP operating system. All the programs ran in the main memory and we did not count any time for loading the precomputed files before query processing since these files can be usually placed in the memory to answer multiple queries.

5.2   Evaluation on Synthetic Data Set

We first produced a synthetic data set using a random data generator. The data set generated has $1M$ rows and 8 subspaces dimensions, whose cardinalities fall in range $(1, 35)$, and the average cardinality is 11. There are 10000 distinct objects and the score dimension contains real numbers. By default we fix k to 10 and $\mathcal{M}$ to SUM. We also fix the minimum support threshold $minsup$ to 2000 to filter out a large number of less interesting subspaces. For $CluPromoCube$ we let $n = 10$, $w = 50$, and $r = 20K$ by default. All performance results are reported by averaging over 5 random target objects.

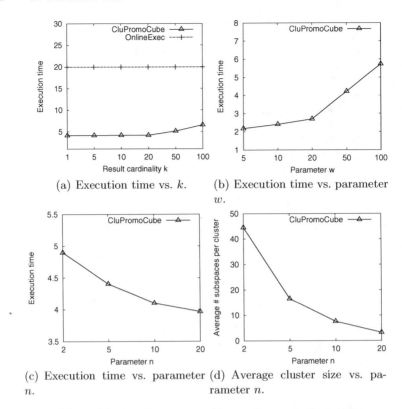

(a) Execution time vs. k.

(b) Execution time vs. parameter w.

(c) Execution time vs. parameter n.

(d) Average cluster size vs. parameter n.

Fig. 6. More performance results on the synthetic data set

We first report the clustering results of *CluPromoCube*. Figure 5(a) displays the number of clusters generated when varying r, the radius parameter of the clustering algorithm. There are nearly $10K$ subspaces passing *minsup* in total; after clustering, however, the number of clustered subspaces dramatically reduce by orders of magnitude. For example, when $r = 50K$, only 239 clusters are needed to cover all points, which is about $\frac{1}{40}$ of the total number of subspaces; when $r = 100K$, the total number of clusters is only 79. Figure 5(b) illustrates the actual materialization cost of *PrecomputeAll* vs. *CluPromoCube* at different r values. *PrecomputeAll* need to materialize more than $35M$ values, while *CluPromoCube* requires less than $130K$. Observe that, as opposed to Figure 5(a), this curve tends to be flat when r is large. For example, the size at $r = 100K$ is only larger than the size at $r = 200K$ by 3.3% of the latter. This is because each subspace has an overhead for maintaining a pointer to the clustered cell that accounts for a fixed amount of materialization cost. The relation between r and execution time is depicted in Figure 5(c). *CluPromoCube* is at least 3.4 times faster than *OnlineExec*. When decreasing r, *CluPromoCube* becomes faster as expected; when $r = 1K$, *CluPromoCube* is 5 times faster than *OnlineExec*. To justify the execution time gain, Figure 5(d) shows the unpruned number of subspaces vs. r. It is not surprising to see this curve is correlated with the one in Figure 5(c). However, we can see that reducing space requirement only incurs a small amount of online cost, which verifies the effectiveness of the proposed method.

Now we vary the other parameters. Figure 6(a) shows the execution time vs. k. For any $k \leq 100$, $CluPromoCube$ outperforms $OnlineExec$ by at least 3 times. At $k = 1$, $CluPromoCube$ is 4.8 times more efficient. Notice that here $CluPromoCube$ has $r = 20K$, meaning that its total size is no more than $50K$. When k increases, the execution time also increases because of a weaker threshold R_k. In Figure 6(b) we show the performance by varying w. Note that varying w does not change the space requirement of $CluPromoCube$. We can see that a larger w leads to more execution time. In particular, the execution time for $w = 5$ is more than 2 times faster than the case for $w = 100$. This is because when w is larger, the materialized aggregates are widely spread and thus the bound becomes less tight. Figure 6(c) further shows the performance vs. n. As n increases, the execution time becomes smaller and smaller. On the other hand, as shown in Figure 6(d), the clustering becomes "harder" in the sense that more clusters are needed when the points (promotion cells) are higher-dimensional, i.e., the average number of subspaces per cluster decreases.

5.3   Evaluation on the DBLP Data Set

In this subsection we evaluate the $CluPromoCube$ approach on the DBLP data set[1]. We constructed a fact table containing $1.7M$ tuples from it. The fact table has 6 subspace dimensions including $Venue$, $Year$ and 4 other dimensions corresponding to 4 research areas. The cardinalities of the 6 dimensions vary from 50 to more than $100K$. There are totally $450K$ distinct authors considered as objects, and we set $\mathcal{M}$ to $COUNT$. All results are reported by averaging 5 random authors.

To evaluate the implemented methods, we constructed a $CluPromoCube$ using $minsup = 1000$, $w = 5$, $n = 20$, and $r = 0$. Surprisingly, the overall number of 6847 subspaces can be packed into 4817 clusters even at $r = 0$ (summarized in Table 2), meaning that each cluster represents an average of 1.42 completely same points (or promotion cells). The offline cost for this $CluPromoCube$ amounts to 207,601, only 0.54% of the corresponding $PrecomputeAll$, which has a materialization cost of over $38M$. Next, we compare $CluPromoCube$ with $OnlineExec$

Table 2. Materialization cost of $CluPromoCube$ with different radii r.

	CluPromoCube					PrecomputeAll
Radius	$r = 0$	$r = 0.1$	$r = 5$	$r = 10$	$r = 100$	-
Num. of subspaces	6847	6847	6847	6847	6847	-
Num. of clusters	4817	4817	1785	905	505	-
Avg. cluster size	1.42	1.42	3.84	7.57	13.56	-
Materialization cost	207,061	207,061	107,341	59,271	33,226	38,010,069

[1] http://www.informatik.uni-trier.de/~ley/db/

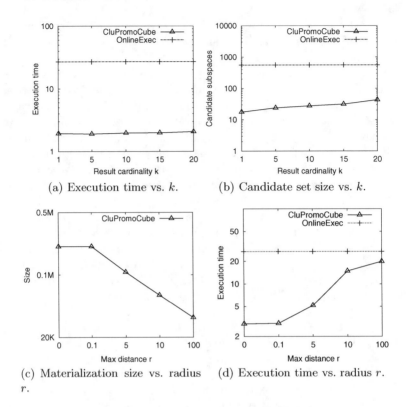

(a) Execution time vs. k.

(b) Candidate set size vs. k.

(c) Materialization size vs. radius r.

(d) Execution time vs. radius r.

Fig. 7. Performance results on the DBLP data

in terms of efficiency. Figure 7(a) displays the results when k is varied from 1 to 20. We can see that *OnlineExec* uses constant time since it is not aware of the query parameter k, whereas *CluPromoCube* is faster by over an order of magnitude when k is not large (e.g., 13 times faster at $k = 1$). Furthermore, to justify the efficiency improvement of *CluPromoCube*, we plot the number of unpruned candidate subspaces with respect to k in Figure 7(b). We can see that *CluPromoCube* need to compute much less candidate subspaces. These results therefore verify that the *CluPromoCube* approach not only performs well on real-world as well as synthetic data set.

To see how the clustering radius would affect *CluPromoCube*'s online and offline costs, we vary the parameter r in order to construct different cubes. Table 2 summarizes the clustering results. We set r to 5 values respectively, namely 0, 0.1, 5, 10, and 100, while keeping other parameters fixed. Recall that a larger value of r would lead to less clusters because each cluster is able to pack more cells. When r is set to 100, the number of clustered generated in *CluPromoCube* is 505, indicating that on average each cluster represents 13.56 subspaces. The resulting materialization cost is also significantly reduced to only 33,226; in other words, in this case *CluPromoCube* incurs only $\frac{0.87}{1000}$ of the materialization cost of *PrecomputeAll*, or $\frac{1}{6.2}$ of that of the *CluPromoCube* at $r = 0$. These results show that there indeed exists many subspaces sharing very similar aggregates that can be clustered effectively. Note that if each aggregate value is stored in a double type using 8 bytes, the $CluPromoCube$ at $r = 0$ would consume less than $1.6MB$ space. We thus conclude that even for large data sets it would be feasible to maintain $CluPromoCube$ completely in memory. It is also worth mentioning that the greedy clustering algorithm is in fact scalable: the offline construction time of $CluPromoCube$ for any r on the DBLP data set is within 10 minutes, so we do not further analyze its complexity empirically.

In Figure 7(c), we graphically present the linear relation between r and the materialization cost shown in Table 2. We also show in Figure 7(d) the runtime of $CluPromoCube$ and $OnlineExec$ when varying r. As expected, the result shows that decreasing the size of $CluPromoCube$ by generating less clusters would lead to more execution time, due to less effective bounding as explained earlier. Overall, although there is no universally accepted way to determine the r parameter, $CluPromoCube$ is often able to achieve satisfactory performance for a wide range of parameters.

6   Conclusions

In this paper we have studied the promotion analysis problem and extended the promotion cube framework through a cell clustering approach in order to achieve better balance between the offline and online processing of promotion queries. We formally defined the promotion cube augmented with the clustered cell structure, as well as its complementary online algorithm. We transformed the cell clustering problem to an optimization problem and formally discussed its hardness result. A greedy algorithm is developed for generating clustered cells efficiently. By evaluating our approach against baseline strategies, we verified that the performance of our cell clustering approach is significantly better on both synthetic and real data sets. Extensions of this approach to handle other application domains such as social network analysis will be interesting directions for future study.

References

1. Arkin, E.M., Barequet, G., Mitchell, J.S.B.: Algorithms for two-box covering. In: Symposium on Computational Geometry, pp. 459–467 (2006)
2. Chang, K.C.-C., Hwang, S.-w.: Minimal probing: supporting expensive predicates for top-k queries. In: SIGMOD Conference, pp. 346–357 (2002)
3. Charikar, M., Panigrahy, R.: Clustering to minimize the sum of cluster diameters. In: STOC, pp. 1–10 (2001)
4. Doddi, S.R., Marathe, M.V., Ravi, S.S., Taylor, D.S., Widmayer, P.: Approximation algorithms for clustering to minimize the sum of diameters. In: Halldórsson, M.M. (ed.) SWAT 2000. LNCS, vol. 1851, pp. 237–250. Springer, Heidelberg (2000)
5. DuMouchel, W., Volinsky, C., Johnson, T., Cortes, C., Pregibon, D.: Squashing flat files flatter. In: KDD, pp. 6–15 (1999)

6. Fagin, R., Lotem, A., Naor, M.: Optimal aggregation algorithms for middleware. J. Comput. Syst. Sci. 66(4), 614–656 (2003)
7. Han, J., Kamber, M.: Data mining: concepts and techniques, 2nd edn. Morgan Kaufmann, San Francisco (2006)
8. Hochbaum, D.S. (ed.): Approximation algorithms for NP-hard problems. PWS Publishing Co., Boston (1997)
9. Hochbaum, D.S., Maass, W.: Approximation schemes for covering and packing problems in image processing and vlsi. J. ACM 32(1), 130–136 (1985)
10. Hristidis, V., Gravano, L., Papakonstantinou, Y.: Efficient ir-style keyword search over relational databases. In: VLDB, pp. 850–861 (2003)
11. Li, C., Ooi, B.C., Tung, A.K.H., Wang, S.: Dada: a data cube for dominant relationship analysis. In: SIGMOD, pp. 659–670 (2006)
12. Marian, A., Bruno, N., Gravano, L.: Evaluating top- queries over web-accessible databases. ACM Trans. Database Syst. 29(2), 319–362 (2004)
13. Sun, Y., Han, J., Zhao, P., Yin, Z., Cheng, H., Wu, T.: Rankclus: integrating clustering with ranking for heterogeneous information network analysis. In: EDBT, pp. 565–576 (2009)
14. Wu, T., Li, X., Xin, D., Han, J., Lee, J., Redder, R.: Datascope: Viewing database contents in google maps' way. In: VLDB, pp. 1314–1317 (2007)
15. Wu, T., Xin, D., Han, J.: Arcube: supporting ranking aggregate queries in partially materialized data cubes. In: SIGMOD Conference, pp. 79–92 (2008)
16. Wu, T., Xin, D., Mei, Q., Han, J.: Promotion analysis in multi-dimensional space. In: PVLDB (2009)
17. Xin, D., Han, J., Cheng, H., Li, X.: Answering top-k queries with multi-dimensional selections: The ranking cube approach. In: VLDB, pp. 463–475 (2006)
18. Zhang, T., Ramakrishnan, R., Livny, M.: Birch: A new data clustering algorithm and its applications. Data Min. Knowl. Discov. 1(2), 141–182 (1997)

Appendix

Proof Sketch for Lemma 1 in Section 4.3. We can reduce the NP-hard MINIMUM GEOMETRIC DISK COVER problem [9,8] to our problem. In an instance of the former problem, we are given a set of points on a 2d Euclidean plane and a radius r, and the goal is to compute a subset of points with minimum cardinality such that every point in the full set can be covered by a disk which centers at some point in the subset and has radius r. We transform each 2d point (x, y) to a promotion cell as $[x + \alpha, y]$ in polynomial time, where α is a large constant s.t. $x + \alpha > y$ always holds. The radius r remains unchanged. We can prove that the optimal solutions are equal for both problem instances. $\square$

Contrasting Sequence Groups by Emerging Sequences

Kang Deng and Osmar R. Zaïane

Department of Computing Science, University of Alberta
Edmonton, Alberta, T6G 2E8
{kdeng2,zaiane}@cs.ualberta.ca

Abstract. Group comparison per se is a fundamental task in many scientific endeavours but is also the basis of any classifier. Contrast sets and emerging patterns contrast between groups of categorical data. Comparing groups of sequence data is a relevant task in many applications. We define Emerging Sequences (ESs) as subsequences that are frequent in sequences of one group and less frequent in the sequences of another, and thus distinguishing or contrasting sequences of different classes. There are two challenges to distinguish sequence classes: the extraction of ESs is not trivially efficient and only exact matches of sequences are considered. In our work we address those problems by a suffix tree-based framework and a similar matching mechanism. We propose a classifier based on Emerging Sequences. Evaluating against two learning algorithms based on frequent subsequences and exact matching subsequences, the experiments on two datasets show that our model outperforms the baseline approaches by up to 20% in prediction accuracy.

Keywords: Emerging Sequences, Classification, Sequence Similarity.

1   Introduction

Any science inevitably calls for comparison. Group comparison has always been a scientific endeavour in Statistics [14] and since the early days of Data Mining such as discriminant rule discovery [7]. Contrast sets [2] and emerging patterns [4] contrast between groups of categorical data. Comparing groups of sequence data is a relevant task in many applications, such as comparing amino acid sequences of two protein families, distinguishing good customers from churning ones in e-business, or contrasting successful and unsuccessful learners of e-learning environments, are typical examples where contrasting sequence groups is crucial.

To contrast sequences, a straightforward approach is to extract discriminative patterns and contrast groups by those patterns. We opt to use subsequences as discriminative patterns, because they are helpful in classification. Borrowing an example from [9], the subsequences "having horns", "faces worship" and "ornaments price" appear several times in the Book of Revelation, but never in the Book of Genesis. Biblical scholars might be interested in those subsequences.

However, there are two main challenges to contrast sequence groups using subsequences. First, the mining of discriminative subsequences is hard. Wang

J. Gama et al. (Eds.): DS 2009, LNAI 5808, pp. 377–384, 2009.

et al. proved that the complexity of finding emerging patterns is MAX SNP-hard [18]. As a more complex pattern, the mining of subsequences cannot be done in polynomial time. Another problem is during the classification stage: as subsequences become long, an approximative match is desired instead of an exact match when subsequences are compared against discriminative patterns.

In this paper, we first define Emerging Sequences (ESs) as subsequences that are frequent in sequences of one group and less frequent in the sequences of another, and thus distinguishing or contrasting sequences of different classes. Then, a similar ES-based classification framework is proposed. In this framework, ESs are mined more efficiently by a suffix tree-based approach; and similar subsequences are also considered in the classification stage. Our proposed similar ES-based learning model can be divided into four stages:

1. Preprocess the sequence datasets and extract Emerging Sequence candidates.
2. Select the most discriminative Emerging Sequences.
3. Transform the sequences into tokenized transactional datasets.
4. Train the classifier by Emerging Sequences.

To validate our learning model, we perform experiments on two datasets, one from software engineering and another from bioinformatics. We compare our approach to two other techniques to illustrate the discriminative power of ESs and the performance of the similar matching mechanism. The experiments show that our similar ES-based classification model outperforms the other two approaches by up to 20% in prediction accuracy. When our algorithm is trained by using *jumping emerging sequences* (i.e. subsequences present in a group and totally absent or negligible in others), the best performance can be achieved.

In the next section, we introduce some terminologies. In Section 3, we describe the sequence mining algorithm and the feature selection strategy. Section 4 presents the classification based on ESs. We present the prediction performance of our proposed approach in Section 5. Finally, Section 6 presents our conclusions.

2   Preliminaries

Let $I = \{i_1, i_2, \ldots, i_k\}$ be a set of all items, or the alphabet, a sequence is an ordered list of items from I. Given a sequence $S = \langle s_1, s_2, \ldots, s_n \rangle$ and a sequence $T = \langle t_1, t_2, \ldots, t_m \rangle$, we say that S is a subsequence of T or T contains S, denoted as $S \sqsubseteq T$, if there exist integers $1 \leq j_1 < j_2 < \ldots < j_n \leq m$ such that $s_1 = t_{j_1}$, $s_2 = t_{j_2}, \ldots, s_n = t_{j_n}$.

Definition 1 (Gap Constraint). *It is specified by a positive integer g. Given a sequence S and a subsequence S' of S, an occurrence of S' is a sequence of indices $o_s = \{i_1, i_2, \ldots, i_m\}$, whose items represent the positions of elements in S. $\forall k \in [1, m-1]$, if $i_{k+1} - i_k \leq g + 1$, we say o_s fulfills the g-gap constraint.*

For instance, if sequence $S = \langle B, C, A, B, C \rangle$, and its subsequence $S' = \langle B, C \rangle$. There are 3 occurrences of S': $\{1, 2\}$, $\{1, 5\}$, and $\{4, 5\}$. The occurrences of S' $\{1, 2\}$ and $\{4, 5\}$ fulfill the 1-gap constraint (also 0-gap) but $\{1, 5\}$ does not.

Definition 2 (Count and Support). *Given a sequence dataset $\mathcal{D}$, $\mathcal{D}$ consists of a set of sequences. The count of a sequence α, denoted as $count(\alpha, \mathcal{D})$, is the number of sequences in $\mathcal{D}$ containing α; while the support $support(\alpha, \mathcal{D})$ is the ratio between its count and the number of sequences in $\mathcal{D}$.*

Definition 3 (Emerging Sequences). *Given two contrasting sequence classes, Emerging Sequences (ESs) are subsequences that are frequent in sequences of one group and less frequent in the sequences of another, and thus distinguishing or contrasting sequences of different classes.*

Definition 4 (Edit Distance). *Edit Distance between two sequences is the minimum number of operations needed to transform one sequence into the other, where an operation is an insertion, deletion, or substitution of a single item.*

3   Sequence Mining and Feature Selection

In this section, we explain how we first preprocess the datasets and extract the ESs candidates; then implement a dynamic feature selection to mine the most discriminative subsequences.

3.1   ES Candidates Mining

To find the Emerging Sequence candidates, the following domain-and-classifier-independent heuristics are useful for selecting sequences to serve as features [12]:

- Features should be frequent.
- Features should be distinctive of at least one class.

Let $\mathcal{D}_{pos}$ and $\mathcal{D}_{neg}$ to be two classes of sequences; the supports of a ES candidate α in both classes, denoted as $support(\alpha, \mathcal{D}_{pos})$ and $support(\alpha, \mathcal{D}_{neg})$, need to meet the following conditions:

$$support(\alpha, \mathcal{D}_{pos}) > \theta, support(\alpha, \mathcal{D}_{neg}) \leq \theta$$

where θ is the minimum support threshold.

As sequence mining is well developed, many existing algorithms, such as SPADE [19] and PrefixSpan [16] can extract frequent subsequences easily. However, there are two problems by extracting ESs with those algorithms. One challenge is the low efficiency: the support thresholds in mining distinguishing patterns need to be lower than those used for mining frequent patterns [4], which means the minimum support offers very weak pruning power on the large search spaces [10]. Another problem of previous algorithms is that, items do not have to be appearing closely with each other in the original sequence, while the gaps between items are significant in comparing sequences. Hence, we implement a Generalized Suffix Tree (GST) [6] based algorithm to extract ES candidates.

The advantage of this framework is that ES candidates mining can be done in linear time. However, only subsequences fulfilling the 0-gap constraint are mined, i.e. items have to be appearing immediately next to each other in the original sequence. To handle the low gap constraint subsequences, we propose a similar matching mechanism; more information is provided in Section 4.2.

3.2   Feature Selection

After preprocessing, numerous ES candidates are extracted. In this section, we refine the result and select the most discriminative subsequences as ESs. To evaluate the discriminative power of subsequences, a similar mechanism with Contrast Sets [2] is applied. Given two sequence groups $\mathcal{D}_{pos}$ and $\mathcal{D}_{neg}$, the ES candidates are ranked by the supports difference:

$$sup\_diff = support(\alpha, \mathcal{D}_{pos}) - support(\alpha, \mathcal{D}_{neg})$$

The selected features should be representative enough so that every original sequence can be covered. To avoid numerous ESs, a dynamic feature selection strategy is adopted [8] (Algorithm 1). For any sequence, only the top-m subsequences, based on sup_diff, are kept. It guarantees that each sequence can be represented by at least m ESs (the high-ranked ones) and the database does not become too large due to the possible sheer number of candidate subsequences.

Input: the sequence dataset $\mathcal{D}$, the sorted set of Emerging Sequence
          candidates ES_c, the minimum subsequence number m
Output: The set of Emerging Sequences ES
1 **foreach** *sequence* $\in \mathcal{D}$ **do**
2 | *count* $\leftarrow 0$;
3 | **foreach** *candidate* $\in ES_c$ **do**
4 | | **if** *candidate* $\sqsubseteq$ *sequence* **then**
5 | | | *count* $\leftarrow$ *count* $+ 1$;
6 | | | mark the *candidate* ;
7 | | **end**
8 | | **if** *count* $= m$ **then**
9 | | | break;
10 | | **end**
11 | **end**
12 **end**
13 $ES \leftarrow$ all marked subsequences in ES_c;

Algorithm 1. Dynamic Feature Selection.

Figure 1 presents the 4 stages of our proposed learning model. The minimum support θ in Stage 1 is set to 50% as an example. The numbers in the brackets after ESs are their supports in the positive and the negative class respectively.

4   Transformation and Classification

In this section, the sequence datasets are transformed to transactional datasets in order to be in a suitable form for learning algorithms. Then, a classification algorithm trained by ESs is proposed. The transactions are simple sets of tokens representing ESs. Each ES is represented by a token (i.e. a simple ID) used within transactions (See Fig 1).

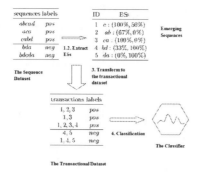

Fig. 1. Four stages of classification. In Stage 1 and 2, ESs fulfilling 0-gap constraint are extracted. We transform the sequence dataset to a transactional dataset in Stage 3. The classification is performed in Stage 4.

4.1 Transformation to Transactional Datasets

To transform a sequence using the Emerging Sequence set representation, we implement a similar matching mechanism. Given a sequence S and an emerging sequence es of length l_e, if S contains a subsequence, which is similar with es, the corresponding transaction should contain the token representing es. To compare the emerging sequence and the extracted subsequences, we introduce a maximum difference $\gamma \in [0, 1]$. If the edit distance between sequences is equal to or lower than $\gamma \times l_e$, we say they are similar.

This strategy can be implemented by dynamic programming algorithms (e.g. local aligement). We adopt a sliding window algorithm, because the lengths of ESs are short, and more complicated algorithms are not necessary. Due to the space limitation, we move this algorithm to a technical report [3].

4.2 Classification

On the classification stage, we implement a Naïve Bayes (NB) classifier based on ESs. Trained by representative features, NB outperforms other state-of-art learning algorithms [13].

A NB classifier [11] assumes that all features are independent. Given a sequence S and a set of independent subsequences, the sequence S can be represented by a set of subsequence-value pairs: $S = \{seq_1 = v_1, seq_2 = v_2, ..., seq_n = v_n\}$, where v_i is either *true* or *false*. When C is the class set, according to the Bayes rules, the probability that sequence S is in the class c is: $p(c|S) = \frac{p(S|c)p(c)}{p(S)}$, where $p(S|c)$ is the conditional probability of sequence S when class label c is known, and $c \in C$. Due to the independence of subsequences, $p(S|c)$ can be rewritten as: $p(S|c) = \prod_i p(seq_i = v_i|c)$. The class label predicted by NB is:

$$predict(S) = arg\ max_{c \in C} p(c) \times \prod_i p(seq_i = v_i|c) \qquad (1)$$

When features are independent, the NB may not have the best prediction accuracy [17]. Therefore, in the Emerging Sequences Naïve Bayes (es-NB), we do not assume the independence of features. To convert the original NB to es-NB, Equation 1 is still used to predict labels, while the feature set is built by ESs.

5   Experimental Results

5.1   Evaluation Methodology

Our proposed similar Emerging Sequence-based algorithm (Similar ES) can be divided into four stages (See Figure 1):

1. Subsequences fulfilling the support conditions are chosen as candidates.
2. ESs are selected, so each sequence can be covered by the top-m ESs.
3. Transform sequences to transactions by the similar matching mechanism.
4. Train an es-NB classifier.

For comparison, we design two other models, one based on frequencies, where frequent subsequences in the positive class are considered discriminant (it also does exact matches), and one identical to our approach but doing exact matches (Exact ES). The motivation for the frequency-based algorithm is that if we rank subsequences according to frequency, those discriminative ones usually have high ranks [15]. We can evaluate the effect of ESs according to the comparison between this approach and the Exact ESs-based Algorithm. Exact ESs-based Algorithm is used to test the performance of the similar matching mechanism.

We apply the F-measure to evaluate the prediction performance. The F-measure is a harmonic average between precision and recall. Finally, we perform 6-fold cross validation, and the average F-measure of the 6 folds is reported.

5.2   Comparisons on Two Types of Datasets

The first type of datasets is the UNIX commands dataset [1], which contains 9 sets of user data. In each experiment, 2 users' commands are chosen, and the F-measures and standard deviations are presented in Row 1-3 of Table 1. The second dataset is the epitope data, which are short linear peptides generated by cleavage of proteins [5]. To contrast the binding and non-binding peptides, we perform the test on several groups of data, and the results are presented in Row 4-6 of Table 1. For more characterization about the datasets, please refer to [3].

We observe that, our similar ESs-based algorithm achieves satisfactory accuracies, comparing with the other two simpler approaches. By comparing the first two approaches (frequent subsequences versus ESs), we find that ESs play a significant role in classification: the F-measures are improved by up to 15%. The similar matching mechanism improve the F-measures as well (by up to 5%). However, its improvement also depends on the datasets. An extreme example is the result of user 2 and 3 (Row 3), where the last two algorithms have similar F-measures. The reason for that is that users 2 and 3 have one length-1 ES

Table 1. Classification performances of three algorithms. Row 1-3: $\theta = 0.01$, $\gamma = 0.1$, Row 4-6: $\theta = 0.05$, $\gamma = 0.2$

Datasets	Frequency-based	Exact ESs-based	Similar ESs-based
user 0 and 3	0.891992 ± 0.0280118	0.953464 ± 0.0121005	0.962192 ± 0.00864717
user 0 and 5	0.869967 ± 0.0250203	0.939128 ± 0.0107074	0.940028 ± 0.0118862
user 2 and 3	0.965973 ± 0.0225253	0.984494 ± 0.00798139	0.984516 ± 0.00600006
I-Ek	0.760296 ± 0.0299356	0.859395 ± 0.0237948	0.862556 ± 0.023571
HLA-DQ2	0.661909 ± 0.0620348	0.811726 ± 0.0936836	0.864592 ± 0.0300073
HLA-DQ4	0.712941 ± 0.0581559	0.789487 ± 0.0922259	0.821227 ± 0.055203

Table 2. Classification performances of different minimum supports

θ	user 0 and 3	user 7 and 8	user 2 and 7
0.01	0.962192 ± 0.00864717	0.853639 ± 0.0233047	0.969818 ± 0.00861439
0.09	0.920798 ± 0.0184558	0.799181 ± 0.0159464	0.923499 ± 0.0162301
0.17	0.865009 ± 0.0182603	0.718475 ± 0.0192607	0.851387 ± 0.0126434
0.25	0.84761 ± 0.0222	0.629819 ± 0.0100243	0.777707 ± 0.031677
0.33	0.785189 ± 0.0258754	0.573587 ± 0.0102116	0.703857 ± 0.0172083

respectively. When γ is set to 0.1, our framework always seeks exact matching subsequences, in other words, both approaches become literally identical.

5.3   Performances of Varying Minimum Support

In this sub-section, we test the performance on UNIX command dataset by varying the minimum support θ. Table 2 presents the results on 3 datasets. With the increase of θ, the classification accuracy degrades. When θ is set to 0.01, the Similar ESs-based model achieves the highest accuracy. Given the target group and the contrasting group, Stage 1 ensures that the ESs candidates hardly appear in the contrasting group, while Stage 2 selects the high-frequent candidates in the target group. In other words, the most ESs are frequent in the target group, while they (almost) cannot be found in the contrasting group. We name this type of subsequences *jumping emerging sequences* (JESs). In conclusion, our proposed algorithm achieves the best performance when the classifier is trained by JESs.

6   Conclusion

In this paper, we define Emerging Sequences (ESs) as subsequences that are frequent in sequences of one group and less frequent in the sequences of another, and thus distinguishing or contrasting sequences of different classes. There are two challenges to distinguish sequence classes: the extraction of ESs is not trivially efficient and only exact matches of sequences are considered. In our work we address those problems by a suffix tree-based framework and a similar matching mechanism. We propose a classifier for sequence data based on ESs.

Evaluating against two learning algorithms based on frequent subsequences and exact matching subsequences, the experiments on two datasets show that our similar ESs-based classification model outperforms the baseline approaches by up to 20% in prediction accuracy. When our algorithm is trained using *jumping emerging sequences*, the best performance can be achieved.

References

1. Asuncion, A., Newman, D.: UCI machine learning repository (2007)
2. Bay, S.D., Pazzani, M.J.: Detecting change in categorical data: Mining contrast sets. In: KDD, pp. 302–306 (1999)
3. Deng, K., Zaïane, O.R.: Technical report, Department of Computing Science, University of Alberta (2009),
   http://www.cs.ualberta.ca/~kdeng2/postscript/deng09.pdf
4. Dong, G., Li, J.: Efficient mining of emerging patterns: discovering trends and differences. In: KDD, pp. 43–52. ACM Press, New York (1999)
5. EL-Manzalawy, Y., Dobbs, D., Honavar, V.: On evaluating mhc-ii binding peptide prediction methods. PLoS ONE 3(9), e3268 (2008)
6. Gusfield, D.: Algorithms on Strings, Trees, and Sequences: Computer Science and Computational Biology. Cambridge University Press, Cambridge (1997)
7. Han, J., Kamber, M.: Data Mining, Concepts and Techniques. Morgan Kaufmann, San Francisco (2001)
8. Jazayeri, S.V., Zaïane, O.R.: Plant protein localization using discriminative and frequent partition-based subsequences. In: ICDM Workshops, pp. 228–237 (2008)
9. Ji, X., Bailey, J., Dong, G.: Mining minimal distinguishing subsequence patterns with gap constraints. Knowl. Inf. Syst. 11(3), 259–286 (2007)
10. Ramamohanarao, J.B.K., Dong, G.: tutorial Contrast Data Mining: Methods and Applications. In: ICDM (2007)
11. Langley, P., Iba, W., Thompson, K.: An analysis of bayesian classifiers. In: National Conference on Artificial Intelligence, pp. 223–228 (1992)
12. Lesh, N., Zaki, M.J., Ogihara, M.: Mining features for sequence classification. In: KDD, pp. 342–346 (1999)
13. Li, J., Yang, Q.: Strong compound-risk factors: Efficient discovery through emerging patterns and contrast sets. IEEE Transactions on Information Technology in Biomedicine 11(5), 544–552 (2007)
14. Liao, T.F.: Statoistical Group Comparison. Wiley's Series in probability and Statistics (2002)
15. Lo, D., Cheng, H., Han, J., Khoo, S.-C.: Classification of software behaviors for failure detection: A discriminative pattern mining approach. In: KDD (2009)
16. Pei, J., Han, J., Mortazavi-Asl, B., Pinto, H., Chen, Q., Dayal, U., Hsu, M.: Prefixspan: Mining sequential patterns by prefix-projected growth. In: ICDE, pp. 215–224 (2001)
17. Rish, I.: An empirical study of the naive bayes classifier. In: IJCAI workshop (2001)
18. Wang, L., Zhao, H., Dong, G., Li, J.: On the complexity of finding emerging patterns. Theor. Comput. Sci. 335(1), 15–27 (2005)
19. Zaki, M.J.: Efficient enumeration of frequent sequences. In: CIKM, pp. 68–75 (1998)

A Sliding Window Algorithm for Relational Frequent Patterns Mining from Data Streams

Fabio Fumarola, Anna Ciampi, Annalisa Appice, and Donato Malerba

Dipartimento di Informatica, Università degli Studi di Bari
via Orabona, 4 - 70126 Bari - Italy
{ffumarola,aciampi,appice,malerba}@di.uniba.it

Abstract. Some challenges in frequent pattern mining from data streams are the drift of data distribution and the computational efficiency. In this work an additional challenge is considered: data streams describe complex objects modeled by multiple database relations. A multi-relational data mining algorithm is proposed to efficiently discover approximate relational frequent patterns over a sliding time window of a complex data stream. The effectiveness of the method is proved on application to the Internet packet stream.

1 Introduction

A data stream is a sequence of time-stamped transactions which arrive on-line, at consecutive time points. The large volume of data continuously generated in short time and the change over time of statistical properties of data, make traditional data mining techniques unsuitable for data streams. The main challenges are avoiding multiple scans of the entire data sets, optimizing memory usage, and mining only the most recent patterns. In this work, we consider a further issue: the stream is a sequence of complex data elements, composed of several objects of various data types are someway related. For instance, network traffic in a LAN can be seen as a stream of connections, which have an inherent structure (e.g., the sequence of packets in the connection). The structure of complex data elements can be naturally modeled by means of multiple database relations and foreign key constraints (*(multi-)relational representation*). Therefore, we face a problem of *relational data stream mining*.

The task considered in this paper is frequent pattern mining. The proposed approach is based on the *sliding window model*, which completely discard stale data, thus saving memory storage and facilitating the detection of the distribution drift. This model is common to several algorithms for frequent pattern mining in data streams [9,11,6,13]. However, all these algorithms work on a single database relation (*propositional representation*) and are not able to deal directly with complex data stored in multiple database relations.

Although it is possible to "propositionalize" relational data, i.e., transform them into a propositional form by building features which capture relational properties of data, this transformation can cause information loss.

J. Gama et al. (Eds.): DS 2009, LNAI 5808, pp. 385–392, 2009.
© Springer-Verlag Berlin Heidelberg 2009

Multi-relational data mining (MRDM) algorithms [4], which can navigate the relational structure in its original format, generate potentially new forms of evidence (*relational patterns*), which are not readily available in a propositional representation [5]. Several MRDM systems allow frequent patterns mining. Two representative examples of the state-of-the-art are WARMR [3] and SPADA [7], which both represent relational data and domain (or background) knowledge *à la* Datalog [2]. However, these systems are not designed to efficiently process data streams and to capture the possible drift of data distribution.

In this work, we propose a novel MRDM algorithm, called SWARM (**S**liding **W**indow **A**lgorithm for **R**elational Pattern **M**ining), which discovers approximate frequent relational patterns over a sliding time window of a relational data stream. SWARM is a false positive oriented algorithm, i.e., it does not discover any false negative frequent pattern. The contributions of SWARM are threefold. First, the multi-relational approach to complex data stream mining. Second, the use of the SE-tree to efficiently store and retrieve relational patterns. Third, the efficient and accurate approximation of the support of the frequent patterns over the sliding time window.

The paper is organized as follows. Section 2 introduces some preliminary concepts. The algorithm is described in Section 3, while experiments on an Internet packet stream are reported in Section 4. Finally, conclusions are drawn.

2   Preliminary Concepts and Definitions

In this work, objects stored in distinct relations of a database D play different roles. We distinguish between the set S of *reference* (or target) objects, which are the main subject of analysis, and the sets R_k, $1 \leq k \leq M$, of *task-relevant* (non-target) objects, which are related to the former and can contribute to define the units of analysis. It has been proved that this "individual centered" representation has several computational advantages, both theoretical (e.g., PAC-learnability) and practical (efficient exploration of the search space) [1].

Henceforth, we adopt a logic framework for the representation of units of analysis, and we categorize predicates into three classes. The unary *key predicate* identifies the reference objects in S (e.g., connection in Example 1). Binary *structural predicates* either relate task-relevant objects (e.g., next) or relate reference objects with task-relevant objects (e.g., packet) in the same unit of analysis. *Property predicates* define the value taken by a property. They can be either binary, when the attribute represents a property of a single object (e.g., nation source), or ternary, when the attribute represents a property of a relationship between two objects (e.g., distance between consecutive packets).

Example 1. A unit of analysis formed by a connection c (reference object) and a sequence of packets $p1$, $p2$, ... (task-relevant objects) is reported below:
  connection(c), time(c,12:05), sourceNation(c, japan), ...,packet(c,p1),
    time(p1,12:05), number(p1,1), packet(c,p2),   time(p2,12:06), number(p2,2),
    next(p1,p2), distance(p1,p2,1), packet(c,p2), ...

A relational pattern is a set of atoms (atomset). An atom is a predicate applied to a tuple of terms (variables or constants). Variables denote objects in S or some R_k, while constants denote values of property predicates.

Definition 1 (Relational pattern). *A relational pattern P is a set of atoms*
$$p_0(t_0), \{p_i(t_{i_1}, t_{i_2})\}_{i=0,...,n}, \{p_j(t_{j_1}, t_{j_2}, t_{j_3})\}_{j=0,...,m}$$
where p_0 is the key predicate, p_i $(i = 0, \ldots, n)$ are either structural predicates or binary property predicates, p_j $(j = 0, \ldots, m)$ are ternary property predicates.

Example 2. A relational pattern is reported below:
"*connection(C), packet(C,P), number(P,4), next(P,Q), distance(P,Q,3ms), number(Q,2), next (Q,R)*".

The *support* of a relational pattern P, denoted as $sup(P|_D)$, is the percentage of units of analysis in D "covered" (i.e., logically entailed) by P. P is frequent if $sup(P)$ is greater than a user-defined threshold σ.

Following the sliding window model, the units of analysis in D depend on a time-sensitive sliding window.

Definition 2 (Time-sensitive sliding-window). *Given a time point p, the set of units of analysis arriving in the period $[t - p + 1, t]$ forms a slide B. Let B_i be the i-th slide, the time-sensitive sliding-window W_i associated with B_i is the set of w consecutive slides from B_{i-w+1} to B_i.*

The window moves forward by a certain amount of unit of analysis by adding the new slide (B_{i+1}) and dropping the expired one (B_{i-w+1}). The number of units of analysis that are added to (and removed from) each window is $|B_i|$. We assume that a unit of analysis is associated with a timestamp and data elements forming a single unit of analysis flow in the stream at the same time.

3   The Algorithm

A buffer continuously consumes the stream units of analysis and pours them slide-by-slide into SWARM system. After a slide goes through SWARM, it is discarded. SWARM operations consist of discovering relational patterns over a slide, maintaining relational patterns over a window and approximating frequent relational patterns over a window. Input parameters are: the minimum support threshold σ, the maximum support error ϵ ($\epsilon < \sigma$), the period p of a slide, the number w of slides in a window, and the maximum depth $MaxDepth$ of patterns.

3.1   Relational Pattern Discovery over a Slide

Once a slide flows in the buffer, relational patterns are locally discovered by exploring the lattice of relational patterns ordered according to a generality order ($\geq$). This generality order is based on θ-subsumption [10] and is monotonic with respect to support. The search proceeds in a Set Enumerated tree (SE-tree) search framework [12], starting from the most general pattern (the one with only the key predicate), and iteratively alternating the candidate generation and candidate evaluation as in the level-wise method [8]. The SE-tree search framework has several advantages. First, the SE-tree enumerates all possible patterns by allowing a complete search. Second, it prevents the generation and evaluation of candidates which are equivalent under θ-subsumption. Third, it effectively exploits the monotonicity property of $\geq$ to prune the search space.

A node of the SE-tree is associated with a progressive natural index and it is represented by the *head* and the *tail*. The head of the root is the pattern that contains only the key predicate. The tail is the ordered set of atoms which may be appended to the head by the downward refinement operator ρ.

Definition 3 (Downward refinement operator). *Let P be a relational pattern. Then $\rho(P) = \{P \cup \{p(\ldots)\} | p$ is either a structural predicate or a property predicate that shares at least one argument with one of the atoms in $P\}$.*

Let $n[head, tail]$ be a node of the SE-tree and $q(\ldots)$ be an atom in $tail(n)$. Then n has a child $n_q[head, tail]$ whose head is defined as follows:

$$head(n_q) = head(n) \cup q(\ldots). \tag{1}$$

If q is based on a property predicate, its tail is defined as follows:

$$tail(n_q) = \Pi_{>q} tail(n) \tag{2}$$

where $\Pi_{>q} tail(n)$ is the order set of atoms stored after q in $tail(n)$. Differently, if q is based on a structural predicate, its tail is defined as follows:

$$tail(n_q) = \Pi_{>q} tail(n) \cup \{r(\ldots)\} \tag{3}$$

where $\{r(\ldots)\}$ is a set of atoms $r(\ldots)$. Each $r(\ldots)$ is an atom that belongs to one of the refinement $\rho(head(n_q))$ under the conditions that $r(\ldots)$ shares variables with $q(\ldots)$ and $r(\ldots)$ is not included in $tail(n)$. When $r(\ldots)$ is based on a structural predicate, one of its arguments must be a new variable.

The monotonicity property of $\geq$ with respect to support makes the expansion of infrequent nodes (i.e., nodes whose local support is less than ϵ) useless. In addition, we prevent the expansion of nodes at a depth greater than $MaxDepth$.

3.2   Relational Pattern Maintenance over a Window

Distinct sets of relational patterns are discovered for each slide. The naive solution is to keep in memory a distinct SE-tree for each slide of the window. This would lead to enumerate several times relational patterns which are discovered in distinct slides. To reduce memory usage, a single SE-tree is maintained on the window. At this aim, each node n of the SE-tree maintains a w sized sliding vector $sv(n)$, which stores one support for each slide in the window. By default, the local support values which are stored in $sv(n)$ are set to unknown. According to the sliding model when a new slide flows in the buffer, the support vector is shifted on the left in order to remove the expired support. In this way, only the last w support values are maintained in the nodes of the SE-tree.

The maintenance of the SE-tree proceeds as follows. When a relational pattern P_n is discovered over a slide B, we distinguish between two cases, namely, P_n is enumerated in the SE-tree or not. In the former case, the SE-tree is expanded with the new node n which enumerates P_n, while in the latter case the node n already exists in the SE-tree and $sv(n)$ is shifted on the left. In both cases, the value of support $sup(P_n|_B)$ is computed over B and is then stored in the last position of $sv(n)$. Finally, nodes are pruned when they enumerate relational patterns which are unknown on each slide of the window.

3.3 Relational Frequent Pattern Approximation over a Window

A relational pattern P_n is identified as approximately frequent over W iff the approximate support $sup_A(P_n|_W)$ estimated over W is greater than σ. The approximate support of P_n is computed on the basis of the local support values which are stored in $sv(n)$.

$$sup_A(P_n|_W) = \sum_{i=1}^{w} (sv(n)[i] \times |B_i|) / \sum_{i=1}^{w} |B_i| \qquad (4)$$

When the local support $sv(n)[i]$ is unknown over a slide B_i, it is estimated by using the known support of an ancestor of P_n. In particular, the pattern Q_m is found such that Q_m is the most specific ancestor of P_n in the SE-tree with a known support value over B_i. Theoretically, the complete set of at worst $2^k - 1$ ancestors should be explored, where k denotes the pattern length. This solution may be impractical for high value of k. To improve efficiency, only the ancestors along the path from n to the root are truly explored. This way, the time complexity of this search is $O(k)$.

Since the SE-tree enumerates patterns discovered by using the maximum support error ϵ as support threshold, Q_m can either be *infrequent* ($sup(Q_m|_{B_i}) < \epsilon$), or *sub-frequent* ($\epsilon \le sup(Q_m|_{B_i}) < \sigma$) or *frequent* ($sup(Q_m|_{B_i}) \ge \sigma$). In the first case, the support of Q_m is used to estimate the support of P_n. In the other two cases the support of P_n is correctly determined as zero. Indeed, the fact that a pattern is refined until it is not *infrequent*, except when ρ refinements of a pattern have zero valued support over the slide, and the monotonicity property of $\ge$ with respect to support, ensure that $sup(P_n|_{B_i}) = 0$.

4 Experiments

We evaluate SWARM on a real Internet packet stream that was logged by the firewall of our Department, from June 1st till June 28th, 2004. This stream consists of 380,733 ingoing connections for a total of 651,037 packets. A connection is described by means of six properties (e.g. service, protocol, ...). A packet is described by means of the order of arrival of the packet within the connection. This order of arrival allows us to represent a relationship of sequentiality between 270,304 pairs of consecutive packets. The time distance between two packets is a property of each pair of consecutive packets. The stream is segmented in time slides with a period p and approximate relational frequent patterns are discovered on sliding windows covering w consecutive slides. Experiments are run by varying p ($p = 30, 60$ minutes), w ($w = 6h/p, 12h/p, 18h/p$) and ϵ ($\epsilon = 0.5, 0.7$). σ is set to 0.7 and $MaxDepth$ is set to 8.

Relational patterns discovered by SWARM are compared with relational patterns discovered by a multi-relational implementation that we have done of the algorithm SW [6]. Initially, we analyze the total number of false positive patterns which are discovered over the sliding windows of the entire stream. No false negative pattern is discovered by both SWARM and SW due to the overestimation of the support. The number of false positive patterns is reported in Table 1. False positive are those approximate patterns which are not included in the set of true frequent patterns we have directly discovered over the entire windows. These results confirm that SWARM discovers a lower number of false positive than SW by providing a more significant approximation of support when local support values are unknown. Additionally, the number of false positive patterns is significantly lower when sub-frequent candidates ($\epsilon = 0.5 < \sigma = 0.7$) are locally generated at slide level. As expected, the number of false positive patterns increases by enlarging the window size and/or reducing the slide period.

Further considerations are suggested by the analysis of the absolute error of the approximated support, averaged over the true positive patterns. Only the sliding windows where the error is greater than zero are plotted in Figure 1. Due to space limitations, the plot concerns only the parameter setting $p = 30, 60$ minutes and $w = 12h/p$, but the considerations we report below can be extended to other settings we tried. We observe that SWARM always exhibits a lower error rate than SW. Additionally, the discovery of sub-frequent local patterns ($\epsilon < \sigma$) makes more accurate the approximation of the support.

A different perspective of the results is offered by the comparison of the relational patterns discovered by both SWARM and SW over the sliding windows that cover the same portion of the data stream, but are generated with different slide period. Although the same number of false positive patterns is discovered independently from the slide period, some differences are observed in the error rate

Table 1. The total number of false positive patterns discovered on the entire stream: comparison between SWARM and SW. $\sigma = 0.7$

Experimental Setting	SWARM $\sigma = 0.7$ $\epsilon = 0.5$	SWARM $\sigma = \epsilon = 0.7$	SW
$p = 30$ min $w = 12$	1	42	145
$p = 30$ min $w = 24$	6	68	203
$p = 30$ min $w = 36$	8	42	240
$p = 60$ min $w = 6$	0	27	68
$p = 60$ min $w = 12$	3	28	104
$p = 60$ min $w = 18$	3	23	135

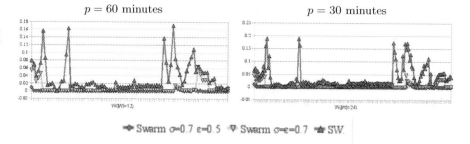

Fig. 1. Average absolute error rate: SWARM with $\epsilon = 0.5$ and $\epsilon = 0.7$ vs. SW

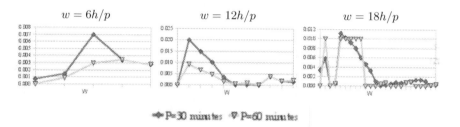

Fig. 2. Average absolute error rate of SWARM ($\theta = 0.7$ and $\epsilon = 0.5$): $p = 60$ minutes vs. $p = 30$ minutes

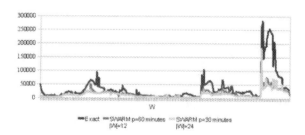

Fig. 3. Elapsed time: discovering approximate frequent patterns on a slide-by-slide basis vs. discovering exact frequent patterns on the entire window

plotted in Figure 2. The general trend is that the error decreases by enlarging the period of a slide. Few exceptions are observed with greater values of w.

Statistics on the elapsed time are shown in Figure 3. The discovery of approximate frequent patterns on a slide-by-slide basis is more efficient than the discovery of exact frequent patterns on the entire window. As expected, elapsed time decreases by reducing the slide period.

5   Conclusions

We present a novel multi-relational data mining algorithm for approximate frequent relational pattern discovery over sliding time windows of a data stream.

The algorithm is evaluated in a real Internet packet stream. Experiments prove that our algorithm is both accurate and efficient. In a future work, we intend to investigate the quality of the approximation the unknown local support of a pattern when it is based on *all* ancestors of the pattern and not only the most specific ancestor along the path to the top of SE-tree.

Acknowledgments

This work is supported by both the Project "Scoperta di conoscenza in domini relazionali" funded by the University of Bari and the Strategic Project PS121 "Telecommunication Facilities and Wireless Sensor Networks in Emergency Management" funded by Apulia Region.

References

1. Blockeel, H., Sebag, M.: Scalability and efficiency in multi-relational data mining. SIGKDD Explorations 5(1), 17–30 (2003)
2. Ceri, S., Gottlob, G., Tanca, L.: Logic Programming and Databases. Springer, New York (1990)
3. Dehaspe, L., De Raedt, L.: Mining association rules in multiple relations. In: Džeroski, S., Lavrač, N. (eds.) ILP 1997. LNCS, vol. 1297, pp. 125–132. Springer, Heidelberg (1997)
4. Džeroski, S., Lavrač, N.: Relational Data Mining. Springer, Heidelberg (2001)
5. Kramer, S.: Relational Learning vs. Propositionalization: Investigations in Inductive Logic Programming and Propositional Machine Learning. PhD thesis (1999)
6. Lin, C., Chiu, D., Wu, Y.: Mining frequent itemsets from data streams with a time-sensitive sliding window. In: Proc. of the SIAM Int. Data Mining Conf. (2005)
7. Lisi, F.A., Malerba, D.: Inducing multi-level association rules from multiple relations. Machine Learning 55(2), 175–210 (2004)
8. Mannila, H., Toivonen, H.: Levelwise search and borders of theories in knowledge discovery. Data Mining and Knowledge Discovery 1(3), 241–258 (1997)
9. Mozafari, B., Thakkar, H., Zaniolo, C.: Verifying and mining frequent patterns from large windows over data streams. In: Proc. Int. Conf. on Data Engineering, pp. 179–188. IEEE Computer Society Press, Los Alamitos (2008)
10. Plotkin, G.D.: A note on inductive generalization. Machine Intelligence 5, 153–163 (1970)
11. Ren, J., Li, K.: Find recent frequent items with sliding windows in data streams. In: Proc. 3rd Int. Conf. on Information Hiding and Multimedia Signal Processing, pp. 625–628. IEEE Computer Society Press, Los Alamitos (2007)
12. Rymon, R.: An SE-tree based characterization of the induction problem. In: Proc. Int. Conf on Machine Learning, pp. 268–275. Morgan Kaufmann, San Francisco (1993)
13. Silvestri, C., Orlando, S.: Approximate mining of frequent patterns on streams. Intelligent Data Analysis 11(1), 49–73 (2007)

A Hybrid Collaborative Filtering System for Contextual Recommendations in Social Networks

Jorge Gonzalo-Alonso[1], Paloma de Juan[1], Elena García-Hortelano[1],
and Carlos Á. Iglesias[2],*

[1] Departamento de Ingeniería de Sistemas Telemáticos
Universidad Politécnica de Madrid
{jgonzalo,paloko,elenagh}@dit.upm.es
[2] Germinus XXI, Grupo Gesfor
cif@germinus.com

Abstract. Recommender systems are based mainly on collaborative filtering algorithms, which only use the ratings given by the users to the products. When context is taken into account, there might be difficulties when it comes to making recommendations to users who are placed in a context other than the usual one, since their preferences will not correlate with the preferences of those in the new context. In this paper, a hybrid collaborative filtering model is proposed, which provides recommendations based on the context of the travelling users. A combination of a user-based collaborative filtering method and a semantic-based one has been used. Contextual recommendation may be applied in multiple social networks that are spreading world-wide. The resulting system has been tested over 11870.com, a good example of a social network where context is a primary concern.

1 Introduction

This article addresses contextual recommendation, which is a new research area in the field of recommender systems [1]. Our definition of context is based on the representational view proposed by Dourish [2]. According to this definition, the context is presented as a series of attributes representing the features of a user's situation. In our case, these attributes were modelled using an ontology.

In particular, this work is devoted to the geographical contextualization of recommendations, although our system has been built so it can be easily adapted to any other definition of context by just adding attributes to the ontology. For example, supposing that a user has only rated restaurants in her city and wants to find a restaurant in a city she is visiting, the purpose of our work is to suggest

---

* This research project is partly funded by the Spanish Government under the $R\&D$ projects CONTENIDOS A LA CARTA (Plan AVANZA I+D TSI-020501-2008-114) and AdmiTI2 (Plan AVANZA I+D TSI-020100-2009-527).

J. Gama et al. (Eds.): DS 2009, LNAI 5808, pp. 393–400, 2009.

an item (the restaurant) based on its compatibility to her profile, but restricting the results to the new context (the city that she is going to visit).

A recommender system based on *user-based or item-based collaborative filtering* [3, 4] only uses the ratings given by the users to the products, making recommendations from the evaluation of the similarity between the profiles of different users, i.e. people that tend to rate the same items will have similar tastes. Therefore, when contextual components are added to the items, it turns very difficult to find similar user profiles in different contexts. Using the previous example of the user who only rates restaurants in her city, if that user wanted to travel abroad, her profile in a user-based collaborative filtering system would be uncorrelated with the profiles of users in the city she intends to visit.

Recommendation based on geographical context may be applied to multiple social networks, such as the one chosen to test the results: 11870.com, a supervised social network where users store and review services they like and share them with the rest of the network community. Contextualization is vital to this community, since these networks provide users with recommendations based on geography, meaning that items rated by users are actual companies or services in their respective cities.

The rest of the article is organized as follows: In *Sect. 2*, all the profiles and concepts used throughout the paper are introduced. In *Sect. 3*, the core of the proposed solution is explained. In *Sect. 4*, an analysis of the solution is performed providing the results over a controlled and a real-world scenario. Finally related work and conclusions will be presented in *Sect. 5* and *6*.

2   Context-Aware Recommendation Framework

In order to develop the proposed recommender system, the User Profile and the Decontextualized User Profile are introduced.

User Profile

- Let $U = \{u_0, u_1, u_2 \ldots u_{|U|}\}$ be the set of users within the application. Where $|U|$ is the total number of users within the application.
- Let $P = \{p_0, p_1, p_2 \ldots p_{|P|}\}$ be the set of products. Where $|P|$ is the total number of products of the application.
- Let R be the set of ratings.
- Let UP be the User Profile, a function: $U \times P \longrightarrow R$

A matrix will be built in the plane UP where the position (i, j) will be the rating $r \in R$ given by the user $u_i \in U$ to the product $p_j \in P$.

Decontextualized User Profile. Another profile is created using a formal conceptualization of the domain in which the products are framed. Every product P in the system will be classified so that a new profile can be built laying a semantic layer over the preferences of every user, e.g. instead of knowing that one user likes a specific comic book shop the aim, rather, is to know that she likes the category of Comic Book Shops. In this study, a taxonomy is used to classify the products and the profile is built over its categories.

- Let $C = \{c_0, c_1, c_2 \ldots c_{|C|}\}$ be the set of all the categories defined in the taxonomy
  . Where $|C|$ is the total number of the categories in the taxonomy.
- Let S be the set of scores given to one category. This score will be obtained for
  one user using the ratings given to P.
- Let DUP be the Decontextualized User Profile, a function: $U \times C \longrightarrow S$

The **decontextualization** needed in the system is related to the semantic layer
that will be laid upon the set of items over which the recommendation is made.
The UP will be dependant on the context whereas the DUP will be context free
as long as the taxonomy used is correctly designed. A final **contextualization**
is then needed to adapt the final recommendations to the current context of the
user.

3 Context-Aware Recommendation Process

In this section the core of the solution proposed is explained. The
recommendation process is depicted in Fig. 1. The system comprises the following
blocks:

Collaborative Filtering Recommender System (block 1). This system
carries out the computation of a recommendation list based on the UP.
Collaborative filtering (CF) is the method chosen for recommendations in most
web applications. The broadness and diversity of the products treated by any
application make it very difficult to use typical content-based methods [5].
Recommendations are provided by studying the correlation between users'
ratings. Therefore, the results are content agnostic and independent of the
domain [3, 4]. A basic user-based CF algorithm with Pearson's correlation [6]
is used to compute the recommendations in this branch of the system.

Semantic Recommender System (block 2). This system carries out the
computation of a second recommendation list using the DUP. The semantic
structure used will be a taxonomy which categorizes all the products in the
system. The user's taste vector is now the rating for each category, based on the
ratings of the items included in each category, i.e. the DUP. It will consist of

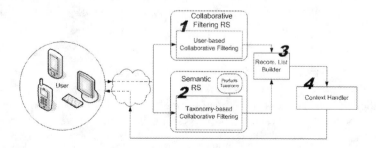

Fig. 1. Overview of the proposed solution

scores given to categories in the taxonomy C using Ziegler's method to distribute the score [7]. Once the DUP is generated, similarities are computed using the cosine of the angle between the vectors [6]. Then, the users whose profiles are most closely correlated to the given user will be taken into consideration to compute recommendations using the relevance formula in [7].

Recommendation List Builder (block 3). This module merges the recommendation lists provided by both recommender systems. This block is tuned depending on the accuracy given by each of the branches of the hybrid system. The accuracy will not only depend on the actual system but also on the quality of the structures used, i.e. the taxonomy. Ziegler [7] tested the taxonomy-based CF algorithm using Amazon's taxonomy and proved that it performed better than the user-based CF one, hence, in this case, the recommendations given by the taxonomy-based branch should be prioritized. But there are other studies which show the opposite behavior [8] and that is the case of the dataset that we will be using, 11870.com. The tests run proved that the recall [9] is between 2 and 3 times better with the user-based CF and its precision [9] triples the case of the taxonomy-based CF algorithm. This is the case that makes the hybrid system really useful because it makes it possible for us to take advantage of the diversity introduced by the user-based CF system and also rely on the taxonomy-based CF algorithm to overcome the sparsity problems [10] [8] or the uncorrelation between profiles of users in different contexts.

Context Handler (block 4). This module gathers all the contextual information concerning both the user and each of the recommended items and processes it according to a context ontology, using a reasoner that also processes the data contained in a knowledge base according to a set of rules that allow the generation of context-related entailments. Every time a user asks for a recommendation, instances representing both her profile and her context are generated, according to the corresponding descriptions available in the ontology. These instances are put into the knowledge base. Once a decontextualized recommendation has been generated, instances representing the context of each recommended item are also put into the knowledge base. The reasoner can then build a new list with all the recommended items matching the user's situation and attending the imposed rules.

Note that in the aforementioned example of the traveler, most of the products recommended by block 1 will be filtered out when contextualization is performed. In this case, block 2 will be the one providing the output to the user.

4 Impact of the Contextual Hybrid System

Two different experiments have been carried out to evaluate the performance of the proposed hybrid recommender system. The first experiment was done using a restricted set of users, which is shown in Fig. 2. This experiment was performed to prove that the hybrid system solves the contextual problems explained throughout the paper. After that, a second experiment was developed using a real social network to test how the hybrid system would perform in a real-world scenario. In both experiments, only the geographical features of the context have been considered.

Instead of using precision and recall as metrics for our experiments, we will count the number of possible recommendations that each branch can provide. Precision and recall are measured over a test set extracted out of the products rated by the user [9] and that set would form the perfect recommendation list. But in the example that we are using, based on geographical contextualization, the aim is to prove that we are able to provide recommendations anywhere, specially in a geographical context where the user has never rated a product. Therefore the recommendations we are looking for will never appear in any test set chosen.

4.1   Results over a Restricted Dataset

The dataset used for this experiment consists of 10 users in 3 different countries, rating 17 services that will be classified according to a small taxonomy (left box of Fig. 2). As we can see in this example, 3 users in Beijing and 4 users in Madrid are uncorrelated because they do not have any products in common.

This experiment aims to prove that with a hybrid system, we can take advantage of the basic CF but in the case of non-correlation between users in context-based recommendations, we are still able to provide an accurate recommendation, which will satisfy the user's criteria.

We used a similarity threshold in order to limit the users who could be in the Top-M of most similar users, given that the DUP could make all users similar because they all have at least one category (the root one) in common. In this experiment, the similarity threshold was set very high, to 0.5. Even at this threshold setting, the main issue was proved.

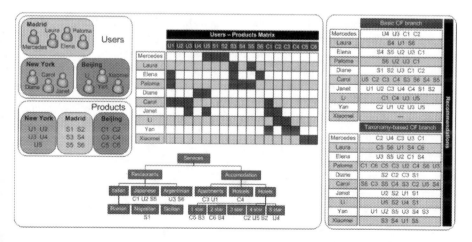

Fig. 2. Left box: Restricted dataset, UP matrix, taxonomy and classification of the products. Right box: Recommendations obtained for both systems

The result of the recommendations obtained for every user in the data set is shown in the right box of Fig. 2. The user Xiaomei, is not only uncorrelated using the UP with the rest of users in a different country but also with the users in the same geographical context. In this extreme situation, the basic CF algorithm is not able to provide recommendations whereas the taxonomy branch is. Also we can observe that the basic CF algorithm provides diversity to the first recommended items (see the user Li). On the other hand, the taxonomy-based algorithm will be more focused on the semantically related products, especially at the top of the recommendation list.

4.2   Results over a Social Web: 11870.com

The social network 11870.com has three versions, depending on the language setting, but is situated mainly in Spain. Therefore, it is important to consider that 16.12% of the site's services are set outside of Spain and 3.9% of users can be considered to originate from other countries because they do not have any reviews in Spain. These people outside of Spain review an average of 2.61 products or services. This analysis of the data in this social network allows us to see that this scenario is not as international as it should be to check the proper use of the hybrid system, but makes a good scenario for testing the performance with authentic data which is location dependant.

Two different experiments will compare the basic CF algorithm and the taxonomy-based one. The first will study users within Spain travelling to other cities in Spain where they do not have any service stored. The second experiment will do the same but uses data from cities outside of Spain. For these experiments, the similarity threshold used in the taxonomy branch is $0, 5$.

The charts in Fig. 3 show the number of possible recommendations given by each branch against the number of users taken into account when obtaining the most similar users to the one getting the recommendation.

The results show that the number of possible recommendations grows as the Top-M grows, being the taxonomy-based branch the one providing more recommendations (Fig. 3). This is specially remarkable in the case of cities within Spain. The reason is that users in the Top-M of the user-based CF system may have travelled to the city we are considering, but there may also be many similar

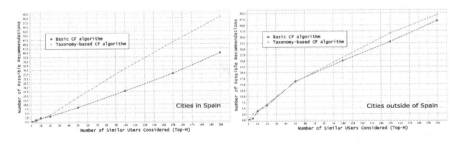

Fig. 3. Number of recommendations shown (user-based vs. taxonomy-based CF)

users in the Top-M of the taxonomy-based CF system who were born in that city. When a user wants a recommendation when travelling abroad, it might be more interesting to provide recommendations using the taxonomy branch because they will fit better with her usual interests.

Both experiments were performed using averages among most populous cities, first in Spain, then worldwide. In a broader dataset these results will be amplified, especially if the cities chosen are less important, which would make it more difficult to correlate in the UP plane.

5   Related Work

Hybrid approaches using content-based and CF recommendations have been used to solve the sparsity and cold-start problems. Examples of these hybrid systems are those presented in [10], [8] and [11], but none of these have been used to contextualize the recommended products.

In contextualizing recommendations, interesting work is being done assuming that the preferences of one user may vary over time [1] or that the context is not a fixed set of attributes [12], as according to Dourish's interactional view [2].

In [1] and [13], a reduction-based approach is presented. Both works propose that the contextualization should be carried out on the dataset before any recommendation is generated, in order to achieve a dimensional reduction of the original space of products.

Finally, [14] and [15] propose a different solution to contextualization, based on the use of user feedback to produce content-based recommendations from her interaction with the system.

6   Conclusions

Recommender systems have proven to be a key element of new web applications. But the internationalization of any social network brings with it new problems for these systems. A very serious problem appears when the items treated are context dependant.

In this paper, and as a result of the work done over the social network 11870.com, a novel approach to contextual recommendations is proposed. In this case, a hybrid two-branch system is introduced. The first branch utilizes a user-based CF algorithm; the second uses this same algorithm but does so over a semantic structure, e.g. a taxonomy.

Finally, a semantic context handler has been designed and implemented, in order to perform context-aware recommendations. The concepts of descontextualization and contextualization have been discussed in this paper.

The techniques explained above have been proposed as a novel solution for the social network 11870.com to tackle the problem of non-correlation among the tastes of users from disparate geographical areas.

References

1. Adomavicius, G., Sankaranarayanan, R., Sen, S., Tuzhilin, A.: Incorporating contextual information in recommender systems using a multidimensional approach. ACM Trans. Inf. Syst. 23(1), 103–145 (2005)
2. Dourish, P.: What we talk about when we talk about context. Personal Ubiquitous Comput 8(1), 19–30 (2004)
3. Resnick, P., Iacovou, N., Suchak, M., Bergstorm, P., Riedl, J.: GroupLens: An Open Architecture for Collaborative Filtering of Netnews. In: Proc. of ACM 1994 Conference on Computer Supported Cooperative Work, pp. 175-186. Chapel Hill, North Carolina (1994)
4. Breese, J.S., Heckerman, D., Kadie, C.: Empirical analysis of predictive algorithms for collaborative filtering. In: Proceedings of the Fourteenth Conference on Uncertainty in Artificial Intelligence, pp. 43–52 (1998)
5. Adomavicius, G., Tuzhilin, A.: Towards the next generation of recommender systems: A survey of the state-of-the-art and possible extensions. IEEE Transactions on Knowledge and Data Engineering 17(6), 734–749 (2005)
6. Herlocker, J.L., Konstan, J.A., Borchers, A., Riedl, J.: An algorithmic framework for performing collaborative filtering. In: SIGIR 1999: Proc. of the 22nd annual international ACM SIGIR conference on Research and development in information retrieval, pp. 230–237. ACM Press, New York (1999)
7. Ziegler, C.-N., Lausen, G., Schmidt-Thieme, L.: Taxonomy-driven computation of product recommendations. In: Proc. of the 2004 ACM CIKM Conference on Information and Knowledge Management, pp. 406–415. ACM Press, Washington (2004)
8. Weng, L.-T., Xu, Y., Li, Y., Nayak, R.: Exploiting item taxonomy for solving cold-start problem in recommendation making. In: ICTAI 2008: Proceedings of the 2008 20th IEEE International Conference on Tools with Artificial Intelligence, pp. 113–120. IEEE Computer Society, Washington (2008)
9. Herlocker, J.L., Konstan, J.A., Terveen, L.G., Riedl, J.T.: Evaluating collaborative filtering recommender systems. ACM Trans. Inf. Syst. 22(1), 5–53 (2004)
10. Weng, L.-T., Xu, Y., Li, Y., Nayak, R.: Improving recommendation novelty based on topic taxonomy. In: WI-IATW 2007: Proc. of the 2007 IEEE/WIC/ACM International Conferences on Web Intelligence and Intelligent Agent Technology - Workshops, pp. 115–118. IEEE Computer Society, Washington (2007)
11. Cho, Y.H., Kim, J.K.: Application of web usage mining and product taxonomy to collaborative recommendations in e-commerce. Expert Systems with Applications 26(2), 233–246 (2004)
12. Anand, S., Mobasher, B.: Contextual recommendation, pp. 142–160 (2007)
13. Woerndl, W., Schlichter, J.: Introducing context into recommender systems. In: Proc. AAAI 2007 Workshop on Recommender Systems in e-Commerce, Vancouver, Canada (2007)
14. Yap, G.-E., Tan, A.-H., Pang, H.-H.: Dynamically-optimized context in recommender systems. In: MDM 2005: Proceedings of the 6th international conference on Mobile data management, pp. 265–272. ACM, New York (2005)
15. Kim, S., Kwon, J.: Efective context-aware recommendation on the semantic web. International Journal of Computer Science and Network Security 7(8), 154–159 (2007)

Linear Programming Boosting
by Column and Row Generation

Kohei Hatano⋆ and Eiji Takimoto

Department of Informatics, Kyushu University
{hatano,eiji}@i.kyushu-u.ac.jp

Abstract. We propose a new boosting algorithm based on a linear programming formulation. Our algorithm can take advantage of the sparsity
of the solution of the underlying optimization problem. In preliminary
experiments, our algorithm outperforms a state-of-the-art LP solver and
LPBoost especially when the solution is given by a small set of relevant
hypotheses and support vectors.

1 Introduction

Learning of sparse classifiers has been popular these days in Machine Learning
and related fields. For example, in text classification applications, the number
of features is typically more than millions, but only a small fraction of features
are likely to be relevant. In such cases, sparse classifiers are useful not only for
classification but for feature selection.

A major approach for learning sparse classifiers is to formulate problems as ℓ_1
soft margin optimization, which learns a linear classifier enlarging the margin by
regularizing ℓ_1 norm of the weight vector [12]. Large margin theory guarantees that
this approach is robust in classification (see, e.g., [12]). Recently, the ℓ_1 soft margin optimization is also applied in learning with "similarity functions" [6,8,1,14].
Since the ℓ_1 soft margin optimization is a linear program, standard optimization
methods such as simplex methods or interior point methods can solve the problem. However, solving the problem directly might need much computation time
even when the number of features or examples goes beyond ten thousands.

LPBoost, proposed by Demiriz et al., is a popular boosting algorithm designed
to solve the soft margin optimization problem [5]. Although its iteration bound
is not known and a worst case lowerbound is exponentially worse than other
boosting algorithms, it is very fast in in most practical cases (earlier results of
LPBoost for hard margin optimization are appeared in [7]). Given m labeled
instances and n hypotheses, consider the $m \times n$ matrix in which each component
is $u_{ij} = y_i h_j(\boldsymbol{x}_i)$ for $i = 1, \ldots, m$ and $j = 1, \ldots, n$. Note that each row or column
corresponds to an example or a hypothesis, respectively. Instead of solving the
soft margin LP problem directly, LPBoost works repeatedly as follows: For each
iteration t, it finds a "good" hypothesis h_t w.r.t. the current distribution $\boldsymbol{d}_t$

⋆ Supported by MEXT Grand-in-Aid for Young Scientists (B) 21700171.

J. Gama et al. (Eds.): DS 2009, LNAI 5808, pp. 401–408, 2009.

over instances and construct the next distribution d_{t+1} by solving the reduced soft margin LP problem restricted to the hypotheses set $\{h_1, \ldots, h_t\}$. The final hypothesis is given by a linear combination of past chosen hypotheses, whose coefficients are Lagrange multipliers of the reduced problem. In the view point of the matrix, it generates columns and solve LPs repeatedly. In fact, LPBoost can be viewed as a LP solver using the column generation approach (e.g., [11]), which is a classical technique in Optimization literature.

LPBoost, however, does not seem to fully exploit the sparsity of the underlying problem. In fact, the ℓ_1 soft margin optimization problems have two kinds of sparsity. First sparsity arises in hypotheses. As explained above, only relevant hypotheses have nonzero coefficients in the optimal solution. The other sparsity appears in examples. More precisely, only some relevant examples (often called "support vectors") affect the optimal solution and the solution does not change even if other examples are removed.

In this paper, we propose a new boosting algorithm which take advantage of the sparsity of both hypotheses and examples. Our algorithm, Sparse LPBoost, takes a "column and row" generation approach. Sparse LPBoost generates seemingly relevant columns (hypotheses) and rows (examples) and solves the linear programs repeatedly. We prove that, given precision parameter $\varepsilon > 0$, Sparse LPBoost outputs the final combined hypothesis with soft margin larger than $\gamma^* - \varepsilon$, where γ^* is the optimal soft margin. Further, we propose some heuristics for choosing hypotheses and examples to make the algorithm faster. In our preliminary experiments, Sparse LPBoost solves ℓ_1 soft margin problems faster than the standard LP solver and LPBoost both in artificial and real data. Especially, for large datasets with ten thousands hypotheses and examples, Sparse LPBoost runs more than seven times faster than other algorithms.

There are some related researches. Warmuth et al. proposed Entropy Regularized LPBoost [16], a variant of LPBoost that approximately solves the soft margin optimization problem. Entropy Regularized LPBoost provably runs in $O(\log(m/\nu)/\varepsilon^2)$ iterations, while a lowerbound of iterations of LPBoost is $\Omega(m)$ [15].

The algorithms proposed by Mangasarian [10] and Sra [13] add a quadratic term into the linear objective in the original LP problem and solve the modified quadratic program by Newton methods and Bregman's method (see, e.g., [3]), respectively. Their methods, unlike ours, does not take advantage of the sparsity of the underlying problem.

Bradley and Mangasarian [2] also proposed an algorithm that decomposes the underlying linear program into smaller ones, which seems similar to our idea. However, this algorithm only generates columns (hypotheses) as done in LPBoost.

2   Preliminaries

Let $\mathcal{X}$ be the domain of interest. Let $S = ((\boldsymbol{x}_1, y_1), \ldots, (\boldsymbol{x}_m, y_m))$ be the given set of m examples, where each $\boldsymbol{x}_i$ is in $\mathcal{X}$ and each y_i is -1 or $+1$ $(i = 1, \ldots, m)$. Let $\mathcal{H}$ be the set of n hypotheses, where each hypothesis is a function from $\mathcal{X}$

to $[-1, +1]$. For any integer k, let $\mathcal{P}^k$ be the set of probability simplex, that is, $\mathcal{P}^k = \{\boldsymbol{p} \in [0,1]^k : \sum_{i=1}^{k} p_i = 1\}$. The margin of an example $(\boldsymbol{x}, y)$ w.r.t. a (normalized) hypothesis weighting $\boldsymbol{\alpha} \in \mathcal{P}^n$ is defined as $y_i \sum_{j=1}^{n} \alpha_j h_j(\boldsymbol{x}_i)$. Also, the margin of the set S of examples w.r.t. $\boldsymbol{w} \in \mathcal{P}^n$ is defined as the minimum margin of examples in S. The edge of a hypothesis $h \in \mathcal{H}$ w.r.t. a distribution $\boldsymbol{d} \in \mathcal{P}^m$ is defined as $\sum_{i=1}^{m} y_i d_i h(\boldsymbol{x}_i)$. For convenience, we denote $\gamma_{\boldsymbol{d}}(h)$ as the edge of a hypothesis h w.r.t. a distribution $\boldsymbol{d}$.

2.1   Linear Programming

A soft margin optimization problem with ℓ_1 regularization is formulated as follows (see, e.g., [5,16]):

$$\max_{\rho, \boldsymbol{\alpha}, \boldsymbol{\xi}} \rho - \frac{1}{\nu} \sum_{i=1}^{m} \xi_i \quad (1)$$

$$\text{sub.to}$$

$$y_i \sum_j \alpha_j h_j(\boldsymbol{x}_i) \geq \rho - \xi_i \ (i = 1, \ldots, m),$$

$$\boldsymbol{\alpha} \in \mathcal{P}^n,$$

$$\min_{\gamma, \boldsymbol{d}} \gamma \quad (2)$$

$$\text{sub.to}$$

$$\sum_i d_i y_i h_j(\boldsymbol{x}_i) \leq \gamma \ (j = 1, \ldots, n),$$

$$\boldsymbol{d} \leq \frac{1}{\nu} \mathbf{1}, \boldsymbol{d} \in \mathcal{P}^m,$$

where the primal problem is given as (1) and the dual problem is given as (2), respectively. Let $(\rho^*, \boldsymbol{\alpha}^*, \boldsymbol{\xi})$ be an optimizer of the primal problem (1) and let $(\gamma^*, \boldsymbol{d}^*)$ be an optimizer of the dual problem (2). Then, by the duality of the linear program, $\rho^* - \frac{1}{\nu} \sum_{i=1}^{m} \xi_i^* = \gamma^*$. A notable property of the solution is its sparsity. By KKT conditions, an optimal solution satisfies the following property.

$$d_i^* \left(y_i \sum_j \alpha_j^* h_j(\boldsymbol{x}_i) - \rho^* + \xi_i^* \right) = 0 \ (i = 1, \ldots, m).$$

$$d_i^* \geq 0, \ y_i \sum_j \alpha_j^* h_j(\boldsymbol{x}_i) - \rho^* + \xi_i^* \geq 0 \ (i = 1, \ldots, m).$$

$$\xi_i^* (1/\nu - d_i^*) = 0, \ \xi_i^* \geq 0, \ d_i^* \leq 1/\nu \ (i = 1, \ldots, m).$$

This property implies that (i) If $y_i \sum_j \alpha_j^* h_j(\boldsymbol{x}_i) > \rho^*$, then $d_i^* = 0$. (ii) If $0 < d_i^* < 1/\nu$, then $y_i \sum_j \alpha_j^* h_j(\boldsymbol{x}_i) = \rho^*$. (iii) If $\xi_i^* > 0$, then $d_i^* = 1/\nu$. Especially, an example $(\boldsymbol{x}_i, y_i)$ s.t. $d_i^* \neq 0$ is called a "support vector". Note that the number of inseparable examples (for which $\xi_i^* > 0$) is at most ν, since, otherwise, $\sum_i d_i^* > 1$. Further, the primal solution has sparsity as well: (iv) If $\boldsymbol{d}^* \cdot \boldsymbol{u}_j < \gamma^*$, then $\alpha_j^* = 0$. We call hypothesis h_j relevant if $\alpha_j^* > 0$. So, we can reconstruct an optimal solution by using only support vectors and relevant hypotheses.

3   Algorithms

In this section, we describe algorithms for solving the problem (2) in details.

Algorithm 1. LPBoost(S,ε)

---

1. Let d_1 be the uniform distribution over S.
2. For $t = 1, \ldots,$
    (a) Choose a hypothesis h_t whose edge w.r.t. d_t is more than $\gamma_t + \varepsilon$.
    (b) If such a hypothesis doe not exist in $\mathcal{H}$, let $T = t - 1$ and break.
    (c) Solve the soft margin optimization problem (2) w.r.t. the restricted hypothesis set $\{h_1, \ldots, h_t\}$. Let (γ_{t+1}, d_{t+1}) be a solution.
3. Output $f(x) = \sum_{t=1}^{T} \alpha_t h_t(x)$, where each α_t ($t = 1, \ldots, T$) is a Lagrange dual of the soft margin optimization problem (2).

3.1   LPBoost

First, we review LPBoost [5]. Given the initial distribution d_1 which is uniform over examples, LPBoost works in iterations. At each iteration t, LPBoost choose a hypothesis h_t with edge larger than $\gamma_t + \varepsilon$ w.r.t. d_t, and add a new constraint $d \cdot u_t$, where $u_{t,i} = y_i h_t(x_i)$ for $i = 1, \ldots, m$ to the current optimization problem and solve the linear program and get d_{t+1} and γ_{t+1}. We summarize the description of LPBoost in Figure 1.

For completeness, we show a proof that LPBoost can approximately solve the optimization problem (2).

Theorem 1 *LPBoost outputs a hypothesis whose soft margin is at least $\gamma^* - \varepsilon$.*

Proof. By definition of the algorithm, when LPBoost outputs the final hypothesis, it holds that $\gamma_T \geq \max_{h \in \mathcal{H}} \gamma_{d_T}(h) - \varepsilon$. Further, since d_t is a feasible solution of the dual problem (2), we have $\max_{h \in \mathcal{H}} \gamma_{d_T}(h) \geq \gamma^*$. Combining these facts, we obtain $\gamma_T \geq \gamma^* - \varepsilon$.

3.2   Our Algorithm

Now we describe our algorithm Sparse LPBoost. Sparse LPBoost is a modification of LPBoost. There are two main differences. Fist difference is that the support of the distribution does not cover the entire set of examples, but covers the examples which have low margin with respect to the current hypothesis weighting. The second difference is that Sparse LPBoostcan choose more than two hypotheses at each iteration. The details of Sparse LPBoost is shown in Figure 2.

Then we prove the correctness of Sparse LPBoost.

Theorem 2 *Sparse LPBoost outputs a hypothesis whose soft margin is at least $\gamma^* - \varepsilon$.*

Proof. Let $C = S - S_T$. Since there is no hypothesis whose edge is more than $\gamma_T + \varepsilon$, $\gamma_T + \varepsilon \geq \max_{h \in \mathcal{H}} \gamma_{d_T}(h)$. Further, since $d_T \in \mathcal{P}^{|S_T}$ is a feasible solution of the problem (2) w.r.t. S_T, $\max_{h \in \mathcal{H}} \gamma_{d_T}(h) \geq \gamma^*$, which implies $\gamma_T \geq \gamma^* - \varepsilon$. Now,

---

Algorithm 2. Sparse LPBoost(S,ε)

---

1. (initialization) Pick up ν examples arbitrarily and put them into S_1. Let $f_1(\boldsymbol{x}) = 0$ and let $\gamma_1 = 1$.
2. For $t = 1, \ldots,$
    - (a) Choose a set S'_t of examples with margin w.r.t. f_t less than ρ_t.
    - (b) If there exists no such S'_t, then let $T = t - 1$ and break.
    - (c) Let $S_{t+1} = S_t \cup S'_t$.
    - (d) For $t' = 1, \ldots,$
        - i. Choose a set $H'_{t'}$ of hypotheses whose edge is larger than $\gamma_t + \varepsilon$. Let $H_t = H_t \cup H'_{t'}$.
        - ii. If there exists no such $H'_{t'}$, then let $f_t = f_{t'}$, $\rho_t = \rho_{t'}$ and break.
        - iii. Solve soft margin LP problem (2) with respect to S_t and H_t. Let $f_{t'+1}(\boldsymbol{x}) = \sum_{h \in H_t} \alpha_h h(\boldsymbol{x})$, where each α_h is a Lagrange dual of the problem (2), and $\rho_{t'+1}$ be a solution of the primal problem (1).
3. Output $f_T(\boldsymbol{x}) = \sum_{h \in H_T} \alpha_h h(\boldsymbol{x})$.

---

consider the distribution $\boldsymbol{d}'_T = (\boldsymbol{d}_T, 0, \ldots, 0) \in \mathcal{P}^{|S|}$, which puts zero weights on examples in C. Then, it is clear that $(\gamma_T, \boldsymbol{d}'_T)$ satisfies the KKT conditions w.r.t. S.

3.3 Heuristics for Choosing Hypotheses and Examples

So far, we have not specified the way of choosing hypotheses or examples. In this subsection, we consider some heuristics.

Threshold: Choose a hypothesis with edge larger than $\gamma'_t + \varepsilon$ and an example with margin less than ρ_t.

Max/min-one: Choose a hypothesis with maximum edge and an example with minimum margin.

Max/min-exponential: Let $\widehat{\mathcal{H}}_{t'}$ be the set of hypotheses whose edges with respect to $\boldsymbol{d}_{t'}$ are more than $\gamma'_t + \varepsilon$, and let $\widehat{S}_t$ be the set of examples whose margin is less than ρ_t. Then, choose the top K hypotheses with highest edges among $\widehat{\mathcal{H}}_{t'}$ and the top L examples with lowest edges among $\widehat{S}_t$, where K is $\min\{|\widehat{\mathcal{H}}_{t'}|, 2^{t'}\}$ and L is $\min\{|\widehat{S}_t|, 2^t\}$.

Let us consider which strategies we should employ. Suppose that $\nu = 0.2m$ and we use a linear programming solver which takes time is m^k, where k is a constant. Note that the value of ν is a reasonable choice since we allow at most 20% of examples to be misclassified.

If we take Threshold or Max/min-one approach, the computation time of Sparse LPBoost needs at least $\sum_{t=1}^{\nu} t^k > \int_{t=1}^{\nu} t^k dt = \frac{\nu^{k+1} - 1}{k+1} = \Omega(m^{k+1})$.

On the other hand, suppose we choose Max/min-exponential approach and the algorithm terminates when the number of chosen examples is cm ($0 < c < 1$). Then, the computation time is at most

$$\sum_{t=1}^{\lceil \log(cm) \rceil} (\nu + 2^t)^k = \sum_{s=0}^{k} \binom{k}{s} \nu^s \sum_{t=1}^{\lceil \log(cm) \rceil} 2^{t(k-s)} \leq \sum_{s=0}^{k} \binom{k}{s} \nu^s (c'm)^{k-s} = O(m^k).$$

Similar arguments hold for choosing hypotheses as well.

Therefore, we conclude that, among these approaches, Max/min-exponential approach is a more robust choice. So, from later on, we assume that Sparse LPBoost uses Max/min-exponential approach. Note that, the advantage of Sparse LPBoost is its small constant factor.But, even if the improvement is only by a constant factor, it might still influence the performance significantly.

4   Experiments

We compare LP, LPBoost and Sparse LPBoost for artificial and real datasets. Our experiments are performed on a workstation with a 8Gb RAM and Xeon 3.8GHz processors. We implemented our experiments with Matlab and CPLEX 11.0, a state-of-the art LP solver.

Our artificial datasets contain from $m = 10^3$ to 10^6 instances in $\{-1, +1\}^n$. We fix a linear threshold function $f(\boldsymbol{x}) = x_1 + x_2 + \cdots + x_k + b$, where $x_1, \ldots, x_k$ are the first k dimensions of $\boldsymbol{x}$ and $b \in \mathbb{R}$ is a bias constant. The function f assigns a label(-1 or $+1$) of each instance $\boldsymbol{x}$ with the sign of $f(\boldsymbol{x}) \in -1, +1$. We set $n = 100$, $k = 10$ and $b = 5$. For each data set, we generate instances randomly so that positive and negative instances are equally likely. Then we add 0% or 5 % random noise on labels.

For each dataset, we prepare $n + 1$ weak hypotheses. First n hypotheses correspond to the n th dimensions, that is $h_j(\boldsymbol{x}) = x_j$ for $j = 1, \ldots, n$. The last hypothesis corresponds to the constant hypothesis which always answers $+1$. We set $\nu = 1$ and $\nu = 0.2m$ for noise-free datasets and noisy datasets, respectively. For LPBoost and Sparse LPBoost, we set $\varepsilon = 0.01$.

We summarize the results for noise-free data and noisy data in Table 1. Sparse LPBoost tend to run faster than others while approximating the solutions well. Note that, compared to other algorithms, the result of Sparse LPBoost is more robust with respect to the choice of ν. In addition, one can observe that, for both noise-free or noisy datasets, Sparse LPBoost picks up fewer examples than the total size m. As a result, the number of variables in the underlying problem is reduced, which makes computation faster. Also, as can be seen, Sparse LPBoost has fewer non-zero d_is when setting $\nu = 1$. On the other hand, Sparse LPBoost's computation time tends to increase when $\nu = 0.2m$. This is because the optimal distribution needs at least ν non-zero components.

Then we show experimental results for some real datasets. As real datasets, we use Reuters-21578[1] and RCV1 [9]. For Reuters-21578, we use the modified Apte("ModApte") split which contains 10170 news documents labeled with topics. We create a binary classification problem by choosing a major topic "acq" as positive and regarding other topics as negative. As hypotheses, we prepare about

---

[1] http://www.daviddlewis.com/resources/testcollections/reuters21578

Table 1. Summary of results for artificial data with $n = 100$, $\nu = 1$ and $0.2m$. For LPBoost (LPB) and Sparse LPBoost (SLPB), the numbers of chosen hypotheses or examples are shown in parentheses. LP, LPBoost and Sparse LPBoost obtained the same objective values γ.

m	Alg.	$\nu = 0$ (noise-free)			$\nu = 0.2m$ (noisy)		
		time(sec.)	$\#(d_i > 0)$	$\#(w_j > 0)$	time(sec.)	$\#(d_i > 0)$	$\#(w_j > 0)$
10^3	LP	**0.98**	96	96	**0.46**	217	46
	LPB	5.46	83	83(84)	6.95	237	65(66)
	SLPB	2.38	26(520)	26(98)	4.80	243(655)	79(82)
10^4	LP	29.66	101	101	**7.01**	2035	70
	LPB	267.85	67	67(67)	21.51	2012	29(29)
	SLPB	**6.78**	25(5250)	25(98)	65.76	2031(6551)	58(58)
10^5	LP	132.99	101	101	321.54	20046	92
	LPB	1843.1	97	97(99)	71.65	200007	11(11)
	SLPB	**62.89**	22(50515)	21(94)	**60.51**	20006(64810)	11(11)
10^6	LP	2139.3	101	101	39923	200031	60
	LPB	17435	97	97(97)	**1179**	2000004	11(11)
	SLPB	**632.29**	22(439991)	22(100)	1281.1	200004(648771)	11(11)

Table 2. Summary of results for real datasets. For LPBoost (LPB) and Sparse LP-Boost (SLPB), the numbers of chosen hypotheses or examples are shown in parentheses.

Reuters-21578 (m=10,170,n=30,839)	time(sec.)	ρ ($\times 10^{-3}$)	γ ($\times 10^{-3}$)	$\#(d_i > 0)$	$\#(w_j > 0)$
LP	381.18	4.8	0.633	2261	463
LPB	804.39	4.8	0.633	2158	452(528)
SLPB	**52.16**	4.8	0.633	2262(6578)	458(613)
RCV1 (m=20,242,n=47,237)	time(sec.)	ρ ($\times 10^{-3}$)	γ ($\times 10^{-3}$)	$\#(d_i > 0)$	$\#(w_j > 0)$
LP	2298.1	1.9	0.267	8389	639
LPB	2688.1	1.9	0.261	8333	454(465)
SLPB	**235.63**	1.9	0.262	8335(16445)	480(518)

$30,839$ decision stumps corresponding to words. That is, each decision stumps answers $+1$ if a given text contains the associated word and answers 0, otherwise.

For RCV1 data, we use the data provided by LIBSVM tools [4]. In the data, we consider binary classification problem by regarding the labels CCAT and ECAT as positive and labels GCAT and MCAT as negative. Each hypothesis is associated with a word and there are 47236 hypotheses in total. The output of hypothesis is given by the tf-idf weighting.

For both datasets, we add the constant hypothesis -1. We set $\varepsilon = 10^{-4}$ as the precision parameter of LPBoost and Sparse LPBoost. We specify $\nu = 0.2m$ and $\nu = 0.4m$ for Reuters-21578 and RCV1, respectively.

The results are summarized in Table 2. Sparse LPBoost runs several times faster than other algorithms. Like previous results for artificial datasets, Sparse LPBoost uses fewer examples (as many as about $0.6m$ to $0.8m$). Further, Sparse

LPBoost seems to take advantage of the sparsity of relevant hypotheses as well. In both datasets, Sparse LPBoost chooses only about 600 hypotheses among more than 30,000 hypotheses.

5   Conclusion

In this paper, we proposed a decomposition algorithm that approximately solves ℓ_1 soft margin optimization problems. Our algorithm performs faster than the standard LP solver using CPLEX and LPBoost by exploiting the sparsity of the underlying solution with respect to hypotheses and examples.

One of our future work is to modify Sparse LPBoost so as to have a theoretical guarantee of iteration bounds. Also, as a practical viewpoint, better heuristics for choosing hypotheses and examples should be investigated.

References

1. Balcan, N., Blum, A., Srebro, N.: A theory of learning with similarity functions. Machine Learning 72(1-2), 89–112 (2008)
2. Bradley, P.S., Mangasarian, O.L.: Massive data discrimination via linear support vector machines. Optimization Methods and Software 13(1), 1–10 (2000)
3. Censor, Y., Zenios, S.A.: Parallel Optimization: Theory, Algorithms, and Applications. Oxford University Press, Oxford (1998)
4. Chang, C.C., Lin, C.J.: Libsvm: a library for support vector machines (2001), http://www.csie.ntu.edu.tw/~cjlin/libsvm
5. Demiriz, A., Bennett, K.P., Shawe-Taylor, J.: Linear programming boosting via column generation. Machine Learning 46(1-3), 225–254 (2002)
6. Graepel, T., Herbrich, R., Schölkopf, B., Smola, A., Bartlett, P., Müller, K., Obermayer, K., Williamson, R.: Classification on proximity data with LP-machines. In: International Conference on Artificial Neural Networks, pp. 304–309 (1999)
7. Grove, A.J., Schuurmans, D.: Boosting in the limit: Maximizing the margin of learned ensembles. In: AAAI 1998, pp. 692–698 (1998)
8. Hein, M., Bousquet, O., Schölkopf, B.: Maximal margin classification for metric spaces. JCSS 71, 333–359 (2005)
9. Lewis, D.D., Yang, Y., Rose, T.G., Li, F.: Rcv1: A new benchmark collection for text categorization research. JMLR 5, 361–397 (2004)
10. Mangasarian, O.: Exact 1-norm support vector machines via unconstrained convex differentiable minimization. JMLR 7, 1517–1530 (2006)
11. Nash, S., Sofer, A.: Linear and Nonlinear Programming. McGraw-Hill, New York (1996)
12. Schapire, R.E., Freund, Y., Bartlett, P., Lee, W.S.: Boosting the margin: a new explanation for the effectiveness of voting methods. The Annals of Statistics 26(5), 1651–1686 (1998)
13. Sra, S.: Efficient large scale linear programming support vector machines. In: Fürnkranz, J., Scheffer, T., Spiliopoulou, M. (eds.) ECML 2006. LNCS, vol. 4212, pp. 767–774. Springer, Heidelberg (2006)
14. Wang, L., Sugiyama, M., Yang, C., Hatano, K., Fung, J.: Theory and algorithms for learning with dissimilarity functions. Neural Computation 21(5), 1459–1484 (2009)
15. Warmuth, M., Glocer, K., Rätsch, G.: Boosting algorithms for maximizing the soft margin. In: NIPS 20, pp. 1585–1592 (2008)
16. Warmuth, M.K., Glocer, K.A., Vishwanathan, S.V.N.: Entropy regularized lPBoost. In: Freund, Y., Györfi, L., Turán, G., Zeugmann, T. (eds.) ALT 2008. LNCS, vol. 5254, pp. 256–271. Springer, Heidelberg (2008)

Discovering Abstract Concepts to Aid Cross-Map Transfer for a Learning Agent

Cédric Herpson[*] and Vincent Corruble

Laboratoire d'Informatique de Paris 6, Université Pierre et Marie Curie,
104 avenue du président Kennedy 75016 Paris, France
{cedric.herpson,vincent.corruble}@lip6.fr

Abstract. The capacity to apply knowledge in a context different than the one in which it was learned has become crucial within the area of autonomous agents. This paper specifically addresses the issue of transfer of knowledge acquired through online learning in partially observable environments. We investigate the discovery of relevant abstract concepts which help the transfer of knowledge in the context of an environment characterized by its 2D geographical configuration. The architecture proposed is tested in a simple grid-world environment where two agents duel each other. Results show that an agent's performances are improved through learning, including when it is tested on a map it has not yet seen.

Keywords: Concept Discovery, Online Learning, Transfer Learning.

1 Motivation and Related Work

Learning and transfer of knowledge is a cross-discipline issue for those interested in understanding or simulating intelligent behavior. Knowledge transfer opens up the possibility of improving both learning speed and global performance of a system on a problem close to a known one. It has been addressed in particular by research in cognitive psychology and neuroscience [7]. In Artificial Intelligence, and with the notable exception of Case Base Reasoning (CBR), until recently little effort had been put into the transfer of learned knowledge. Indeed, the need for an architecture which integrates learning and transfer capacities has become crucial in recent years with the development of new application domains using, or open to the use of, autonomous systems, such as video games, military simulations or general public robotics.

Work in the area of strategy games is one concrete example; it has led to techniques which let agents learn strategies through playing [1, 2]. Yet, learned strategies are only relevant to the game context in which they have been learned, that is to say a scenario, and more pointedly a specific "game map". Hence, each new scenario requires a new phase of learning since previous experience is not put to use.

To go beyond these limitations, an important part of recent work tackles the learning and transfer of knowledge problem by means of the Reinforcement Learning (RL) framework. To be able to generalize and to use RL in complex worlds, researchers

---

[*] Corresponding author.

J. Gama et al. (Eds.): DS 2009, LNAI 5808, pp. 409–416, 2009.

propose reducing the considered state space. Thus, [3] assumes that states of the modeled Markovian Decision Process (MDP) are of few different types (determined *a priori*). With a similar idea, i.e. to generalize across states, Dzeroski & Driessen [4] have proposed the Relational Reinforcement Learning approach. This attractive combination of RL and Inductive Logic Programming is based upon the relational structure of the problem to abstract from the current goal. In the area of Real-Time Strategy games (RTS), [5] represents the structure of the problem as a relational MDP to tackle the planning problem on different maps. However, the required full description of the state-action space does not allow the use of these approaches in open or complex environments.

More recently, [6] addresses the transfer problem in RTS proposing an interesting architecture where an agent controls from a central point of view all the agents from its side using a combination of CBR and RL. Results obtained with this approach indicate its ability to reuse learned knowledge when initial positions and/or number of units vary. However, the fact that the game state description on which decisions are made is completely unrelated to the context (including its topology) constitutes a major obstacle to more ambitious transfer. Thus, a map of higher complexity, or a complete change of environment will not impact the state description and lead therefore to only one high-level description for two distinct situations. As a result, only one action will be chosen where two different actions have to be selected.

The goal of this paper is to present one aspect of our agent architecture that allows concept learning, aiming to improve agent performances in the transfer learning task in open and/or complex environments. The discovered abstract concepts help the transfer of knowledge learned on a given topology to a different one, yet unseen. Relevance of a learned concept is evaluated from the standpoint of the agent's performances.

In the following, we describe the proposed architecture and the preliminary evaluation of it in a simple game environment. Finally, we discuss experimental results and current limitations of our approach before concluding with some perspectives.

2   An Architecture for a Cross-Map Transfer

We consider a representation using the notion of a situated agent. It lets one change from a central, globalized point of view, often used in work for strategy games to an agent-centered perspective.

Definition 1 (Situation). *A situation is the world view as perceived by the agent from its sensors at a given moment.*

A situation is the basic level of information; data obtained from sensors are expressed as a set of attribute-value couples. Our first hypothesis is that the elimination of some perceived information (out of sensors' field of view) offers the possibility for the agent's reasoning to be independent of map-specific (x,y) coordinates and of the environment's complexity. However, we lose some possibly relevant state information and enter the realm of partially observable environments. Thus, if the environment contains multiple agents, they become unobserved and unpredictable; which means that the environment is not stationary anymore. These properties are usually working hypotheses in most learning agent frameworks, as they are necessary to guarantee convergence with typical learning algorithms.

2.1 The Duel Example

Before presenting in detail our architecture, we introduce an example that will be used both to illustrate our following explanations as well as a test bed in the experimental section. Consider a grid-world type of environment where each location on the grid is characterized by its altitude as well as its (x, y) coordinates. The effective field of view of an agent evolving on this map takes into account the obstacles present. As a consequence, the field of view of an agent located on a high position is better than the one of an agent located downhill.

The chosen scenario consists of a duel between two agents evolving within the above grid-world. Each agent's goal is to become the last survivor. They have available a range weapon and have to hit their competitor twice to win. The probability of hitting when shooting depends on the shooter-target distance, as well as the angle of incidence of the shooting. Thus, when two agents are located at different altitudes, the one positioned higher will have a hit probability much higher than the lower one.

Final goal : $alone(x) \land alive(x)$
State : $ammo(x) \in \{true, false\}$ and Status $\in \{hurt, dead, ok\}$
Rules : $ok(x) \lor hurt(x) \rightarrow alive(x)$; $dead(x) \rightarrow alone(y)$ [with $x \neq y$] :
$hurt(x) \land hit(x) \rightarrow dead(x)$; $ok(x) \land hit(x) \rightarrow hurt(x)$
Actions : $ammo(x) \land see(x,y) \rightarrow Shoot(x,y)$;
move and rotation towards the 4 cardinal directions.

Fig. 1. A simplified version of available rules for a duel

2.2 Architecture

Our architecture is based on a perception-action loop involving several components. As shown in Figure 2, an agent has a memory to save facts, learned concepts and rules about the environment (*Concepts* and *Environment rules' Databases*).

Definition 2 (Concept) *Let* **pre** $\longrightarrow$ **Action** *be a rule.*
A concept is a couple **<pre, descr>** *where* **descr** *is the agent's representation of the precondition* **pre,** *learned from its world view (the situations).*

Thus, the learning mechanism (*Concept Learning*), using both supervised and unsupervised methods, continuously extracts patterns from perceived situations and associates them to the premise of the agent's actions/decision rules (red process in Fig 2). An inference engine (*Action Selection*) uses the relations between these different elements in order to select the actions that can lead the agent toward its current goal. The automatic recognition in the current situation of a prior learned pattern (*Identification*) helps reaching better decisions. Thus, learned concepts are used as an interface between the physical sphere and the representation sphere.

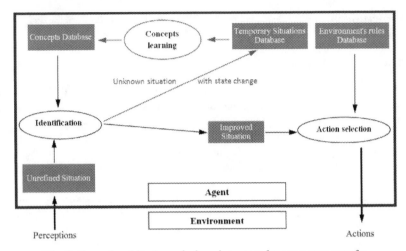

Fig. 2. General architecture of a learning agent for cross-map transfer

2.2.1 Learning Phase

As shown in the duel's rules example, Figure 1, the agent's internal state is defined by one or more variables. An agent's change of internal state corresponds to a change in one of its attributes. A change occurs as a consequence of the agent's last action or of another agent's one. Without prior knowledge, the learning or discovery of a new concept proceeds in three stages:

1. Save in the *Temporary Situation Base* (TSB) all situations newly encountered when a change in the agent's internal state occurs. Each saved situation is associated to the precondition part of the last action rule used.
2. Apply a learning algorithm to infer a pattern when the number of situations associated with the same precondition is above a threshold N.
3. Add the resulting couple < precondition, pattern,> into the *Concept Base* (CB).

In our duel example, the *Shoot* action can only be used when an agent perceives its enemy. After a successful shot has killed its enemy, a shooter-agent's final goal is reached. This internal state change triggers the learning algorithm. Thus, the current situation is associated with the pre-condition $see(x,y)$. After N duels, the agent discovers that being located, for example, at a high point, improves the chances of seeing the enemy and therefore of winning.

Environments in which agents evolve are, just as the real world, stochastic. There is therefore a significant amount of noise in the data from which we wish to extract a relevant situation's pattern. To tackle this, our algorithm (below) firstly removes some noise from data using a majority vote process (line 1 to 9). After this step, it then infers a pattern among remaining situations using a probabilistic classification tree[1] based on the C4.5 algorithm.

---

[1] The classification tree was selected after having tested a number of learning algorithms including Clustering, Boosting on Multi-layer Perceptron, KNN, Kmoy and Support Vector Machines.

Let VS: VictorySituations,DS: DefeatSituations and pre: precondition

```
 1 :  VS ← {vs ∈ TSB | vs ∈ <pre,descr>}
 2 :  DS ← {ds ∈ TSB | ds ∈ <pre,descr>}
 3 :  ∀ x ∈ VS ⋃ DS
 4 :      ∀ y ∈ VS \{x}
 5 :          If (dist(x, y) < η) then x.nbPos ← x.nbPos+1
 6 :      ∀ y ∈ DS \{x}
 7 :          If (dist(x, y) < θ) then x.nbNeg ← x.nbNeg+1
 8 :      If ((x.nbPos/|VS| < x.nbNeg/|DS|) ∧ x ∈ DS) then nDS←nDS ⋃ {x}
 9 :      If ((x.nbPos/|VS| > x.nbNeg/|DS|) ∧ x ∈ VS) then nVS←nVS ⋃ {x}
10 :  pattern ← C4.5(nVS,nDS)
11 :  Concept Base ← Concept Base ⋃ <pre,pattern>
12 :  TSB ← TSB \(VS ⋃ DS)
```

Algorithm 1. Learning a concept related to the final goal

Once these two stages have been realized, the new association between a pattern and a precondition is added to the *Concept Base* (CB). Finally, the situations associated to the precondition of the action having triggered the learning phase are removed from the *Temporary Situation Base* (TSB).

2.2.2 Decision Phase

The inference engine, prolog, is interfaced with the agent's learning and action mechanisms (in Java). At each time step, the situation perceived by the agent updates the set of facts available to the inference engine. Prolog then uses its model of the environment and known facts so as to 'prove' the desired goal and to select the best action.

During the identification stage, at line 10, the agent considers virtually all the positions it can perceive in its field of view. The classification tree computes for each

```
 1 :  While (!goalReached)
 2 :      currentSit ← agent.updateCurrentSit();
 3 :      agent.updateFactToProlog(currentSit);
 4 :      action ← agent.prolog.prove();
 5 :      If (action!=null) then
 6 :          agent.doAction(action);
 7 :      else
 8 :          descr ← agent.CB.search(agent.prolog.getMissingPre());
 9 :          If (descr!=null) then
10 :              location ← agent.searchInFov(descr);
11 :              x ← randomInt();
12 :              If (x > ε ∧ location!=null) then
13 :                  agent.doMove(location);
14 :          else
15 :              agent.randomMove();
16 : EndWhile
```

Algorithm 2. Decision-Action loop

situation its probability of relevance to the concept the agent is looking for (where 1 means certainty and 0 the opposite). To limit the risk of misclassification, only a virtual situation with a confidence value above a threshold β can be selected.

3 Experimental Evaluation

In this section, we purpose an empirical evaluation of the learning and transfer capacity of our architecture. Towards this aim, we set up three maps with different topologies, sizes and maximum altitudes. Changes of internal states leading to the storing of situations and to the learning of a concept are only related to end-game situations. We therefore consider that ammunitions are unlimited. Thus, the impact on the performances of learning a single concept can be measured. N, the threshold that triggers the learning algorithm is set to 20. The β parameter is set to 0,8. η, θ and ε, the different parameters in algorithms 1 and 2 are set to 0,1. Performances are evaluated based on 2 criteria: The percentage of victories obtained by each agent and the average number of time steps needed to end an episode (i.e. when an agent wins)

The experiment evaluates the ability of an agent to transfer knowledge to a new environment. To obtain a baseline, 1000 duels between two random agents are run for the three considered environments. Next, one learning agent runs a series of episodes against a random one on the learning environment until it learns one concept. Then, the learning step is deactivated and we run a series of 1000 episodes opposing a random agent and the trained agent on the 3 environments. In the presentation of all results, we refer to the two following types of agents:

-   *Random* : baseline agent, with no knowledge nor learning mechanism.
-   *Intel_1* : agent having discovered/learned a concept related to the end goal.

3.1 Results

The tree learned in the first experiment (fig. 3 below) shows that the agent will favor situations with a good field of view, or, if the field of view is considered average, situations where average altitude around the agent's location is lower than its own. Thus the learned concept drives the agent to search and follow the map's ridge paths.

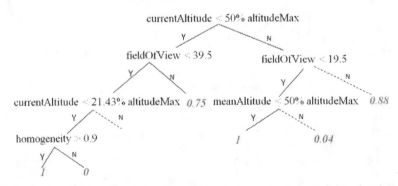

Fig. 3. Tree automatically generated and associated to the *see(x,y)* precondition. A value above 0.75 indicates a favorable situation. Dotted lines indicate sub-trees that were manually pruned from the figure for better clarity.

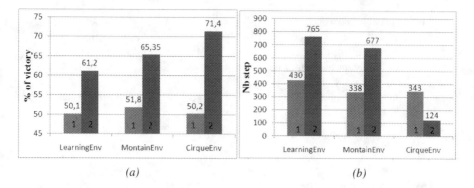

(a)   (b)

Fig. 4. graphic *(a)* represents the percentage of wins for agents *random* (1), and *Intel_1* (2) respectively, all against a random agent - graphic *(b)* shows the average duration of a duel, in step number, for the two types of agents facing a random agent. In each case, results are given after 1000 duels.

The results obtained on the three environments are presented in figure 4. They show that the concept learned is sufficiently relevant to improve performance on the training environment. Moreover, the results obtained on the two test environments are even better than those from the training environment.

4 Discussion

Though our results provide evidence regarding the validity of the principles behind our learning architecture for knowledge transfer in partially observable environments, significant work remains ahead of us. First of all, the topology plays a large part in determining the ability of the agents. Thus, an in-depth study in more various and more complex environments has to be done to prove our architecture's robustness when scaling up. Next, on this simplified environment, the threshold N necessary to be reached by the agent to learn a concept was manually set to 20 episodes. If, compared to the length of the training phase in RL approach, our architecture may offer an interesting alternative, significant improvement can be made. A learning grounded on an incremental approach rather than on a discontinuous one could allow an agent to adapt more quickly to new environments and new associated concepts.

5 Conclusion and Perspectives

We have proposed a new learning agent architecture based on the discovery of relevant abstract concepts for cross-map transfer without requiring a full description of the state-action space. Initial results are encouraging. They show that our architecture does lead to a relevant knowledge transfer in stochastic environments regarding topological information, which proves efficient when the environment changes. Besides some necessary improvements discussed in section 4, one short-term perspective is to evaluate the efficiency of our learning architecture facing the Reinforcement

Learning approach. This experiment will allow us to compare both adaptability and efficiency of our approach to the RL-based ones in the transfer context. With a long-term view, an important perspective for this work is to extend the architecture so that the agent no longer needs the environment's rules and is able to discover actions that it is possible to achieve. Thus, our architecture will allow an agent to be totally autonomous in a new environment, discovering both available actions and related concepts.

References

1. Madeira, C., Corruble, V., Ramalho, G.: Designing a Reinforcement Learning-based Adaptive AI for Large-Scale Strategy Games. In: AIIDE (2006)
2. Ponsen, M., Munoz-Avila, H., Spronck, P., Aha, D.: Automatically Generating Game Tactics through Evolutionary Learning. AI Magazine (2006)
3. Leffler, B.R., Littman, M., Edmunds, T.: Efficient Reinforcement Learning With Relocatable Action Models. In: Proc. of AAAI (2007)
4. Driessens, K., Givan, R., Tadepalli, P.: RRL: An overview. In: Proc. of ICML (2004)
5. Guestrin, C., Koller, D., Gearhart, C., Kanodia, N.: Generalizing Plans to New Environments in Relational MDPs. In: IJCAI (2003)
6. Sharma, M., Holmes, M., Santamaria, J.C., Irani, A., Lee, C., Ram, I.J.A.: Transfer Learning in Real-Time Strategy Games using hybrid cbr/rl. In: IJCAI (2007)
7. Thrun, S., Pratt, L.: Learning to learn. Norwell edn. Kluwer Academic Publishers, Dordrecht (1998)

A Dialectic Approach to Problem-Solving

Eric Martin[1] and Jean Sallantin[2]

[1] School of Computer Science and Eng., UNSW Sydney NSW 2052, Australia
emartin@cse.unsw.edu.au
[2] LIRMM, CNRS UM2, 161, rue Ada, 34 392 Montpellier Cedex 5, France
js@lirmm.fr

Abstract. We analyze the dynamics of problem-solving in a framework which captures two key features of that activity. The first feature is that problem-solving is a social game where a number of problem-solvers interact, rely on other agents to tackle parts of a problem, and regularly communicate the outcomes of their investigations. The second feature is that problem-solving requires a careful control over the set of hypotheses that might be needed at various stages of the investigation for the problem to be solved; more particularly, that any incorrect hypothesis be eventually refuted in the face of some evidence: all agents can expect such evidence to be brought to their knowledge whenever it holds. Our presentation uses a very general form of logic programs, viewed as sets of rules that can be activated and fire, depending on what a problem-solver is willing to explore, what a problem-solver is willing to hypothesize, and what a problem-solver knows about the problem to be solved in the form of data or background knowledge.

Our framework supports two fundamental aspects of problem-solving. The first aspect is that no matter how the work is being distributed amongst agents, exactly the same knowledge is guaranteed to be discovered eventually. The second aspect is that any group of agents (with at one end, one agent being in charge of all rules and at another end, one agent being in charge of one and only one rule) might need to sometimes put forward some hypotheses to allow for the discovery of a particular piece of knowledge in finite time.

1 Introduction

The last century has seen the advent of a number of frameworks that place rationality at the heart of the process of scientific discovery; still none of those frameworks has endowed epistemology with a definitive mathematical foundation. The seminal research of Herbert Simon on the logical theorist and McCarthy's research on commonsense reasoning are two prominent examples of general attempts at interfacing the thinking of rational agents and the dynamics of scientific discovery, before more specific approaches, tackling more specific problems, have appeared. Departing more or less from Boole's framework, a plethora of logics, mainly developed in the AI community, have grounded various rational approaches to problem solving; some of them have been validated

J. Gama et al. (Eds.): DS 2009, LNAI 5808, pp. 417–424, 2009.

by the implementation of tools, successfully applied to the resolution of a broad class of problems. Formal Learning Theory, PAC learning and Query learning, developed around the prominent work of Gold [4], Valiant [8] and Angluin [2], offer theoretical concepts, rooted in recursion theory, statistics and complexity theory, to describe the process of data generalization. Induction has been studied from different angles, in particular by Pierce and Suppes, before the Inductive Logic Programming community suggested a more practical approach. Numerous investigations on automatic or semi-automatic scientific discovery have taken place [5,7,3,1]. Finally, let us mention the more recent work on the relationships between scientific discovery and game theory; but these pointers do not by far exhaust the whole body of work on the relationship between rationality and scientific discovery.

Our approach is based on an extension of Parametric logic [6], a new framework that unifies logic and formal learning theory, developed along three dimensions, two of which are illustrated in this paper.

- The first dimension is formal. It equates logical discovery with theorem proving, in a logical setting where the work of scientists boils down to inferring a set of theorems. We consider two binary categories of agents. The first category opposes *independent agents*, who work alone, to *social agents*, who share the work. The second category opposes *theoretical agents*, who have no time nor space restrictions on the inferences they can perform, including the ability to perform transfinite inferences, to *empirical agents*, whose inferences must be performed in finite (but unbounded) time with finite (but unbounded) memory.
- The second dimension is cognitive, and applies to the way theorems can be derived, based on two key notions: *postulates* and *hypotheses*. Postulates are what agents use when they organize their work; they represent statements whose validity will be assessed "later." Postulates allow for particular scheduling or "outsourcing" of the work. Hypotheses are what agents assume in order to seed or activate a proof. Hypotheses can turn out to be confirmed or refuted, they can end up being plausible or paradoxical.

  The categorization of agents is based on how they deal with postulates and hypotheses. Theoretical agents do not need hypotheses whereas empirical agents might. Independent agents might not need postulates whereas social agents do.

2   The Logical Framework

2.1   An Illustrative Example

Imagine the following game. Countably many copies of every card in a deck of 52 cards are available to a *game master*. The game master chooses a particular ω-sequence of cards. For instance, she might choose the sequence consisting of nothing but the ace of spades. Or she might choose the sequence where the queen of hearts alternates with the four of spades, starting with the latter. A number of *players*, who do not know which sequence has been chosen by the game master, can make requests and ask her to reveal the nth card in the sequence, for some natural number n. The players aim at eventually discovering which sequence of cards has been chosen by the game master, or to discover some of its properties.

The game illustrates the process of *scientific discovery*, with the game master playing the role of Nature, and the players the role of the scientists. A feature of the game is that unless the game master has explicitly ruled out a large number of possible sequences, the players usually cannot, at any point in time, know whether their guesses are correct: they might at best be able to *converge in the limit* to correct guesses. We have not precisely defined what a "guess" is. There has to be a language where some properties of a sequence of cards can be described, and the expressive power of the language is crucial in circumscribing what the players can or cannot achieve. Let us refer to such a description as a *theory*, in analogy to the work of scientists whose aim is to discover theories that correctly describe or predict some aspects of the field of study. In this paper, we will let *logic programs* play the role of theories.

2.2 Logical Background

$\mathbb{N}$ denotes the set of natural numbers and Ord the class of ordinals. We consider a finite *vocabulary* $\mathcal{V}$ consisting of a constant $\overline{0}$, a unary function symbol s, the *observational* predicate symbols, namely, the unary predicate symbols

hearts spades diamonds clubs ace two ... ten jack queen king

and a number of other predicate symbols. For all nonzero $n \in \mathbb{N}$, we denote by $\overline{n}$ the term obtained from $\overline{0}$ by n successive applications of s; $\overline{n}$ will refer to the nth card. We denote by $\mathrm{Prd}(\mathcal{V})$ the set of predicate symbols in $\mathcal{V}$. Given $n \in \mathbb{N}$, we denote by $\mathrm{Prd}(\mathcal{V}, n)$ the set of members of $\mathrm{Prd}(\mathcal{V})$ of arity n. We fix a countably infinite set of (first-order) variables and a repetition-free enumeration $(v_i)_{i \in \mathbb{N}}$ of this set. We need a notation for the set of all possible sequences of cards.

Definition 1. *We call* possible game *any set T of closed atoms such that:*

- *for all $n \in \mathbb{N}$, T contains one and only one member of $hearts(\overline{n})$, $spades(\overline{n})$, $diamonds(\overline{n})$, $clubs(\overline{n})$;*
- *for all $n \in \mathbb{N}$, T contains one and only one member of $ace(\overline{n})$, $two(\overline{n})$, ..., $ten(\overline{n})$, $jack(\overline{n})$, $queen(\overline{n})$, $king(\overline{n})$;*
- *T contains no other atom.*

We consider a notion of logical consequence that is best expressed on the basis of a forcing relation $\Vdash$, based on both principles that follow.

- The intended interpretations are *Herbrand structures*: every individual has a unique name (a numeral); this is because intended interpretations are ω-sequences of cards—$\overline{n}$ being the name of the nth card in the sequence.

- Disjunction and existential quantification are constructive: an agent will derive a disjunction iff she has previously derived one of the disjuncts, and she will derive an existential sentence iff she has previously derived one of the closed instances of the sentence's matrix.

We denote by $\mathcal{L}_{\omega\omega}(\mathcal{V})$ the set set *sentences*, that is, closed first-order formulas over $\mathcal{V}$. Given two sets S and T of sentences, we write $S \Vdash T$ iff S forces all members of T.

2.3   Logic Programs and Occurrence Markers

A formal logic program provides, for every $n \in \mathbb{N}$ and $\wp \in \mathrm{Prd}(\mathcal{V}, n)$, two rules: one whose head is $\wp(v_1, \ldots, v_n)$, and one whose head is $\neg\wp(v_1, \ldots, v_n)$. This is at no loss of generality since the left hand side of the rules can contain equality and the intended interpretations are Herbrand. So to define a formal logic program, we only need the left hand side of both rules associated with a predicate symbol and its negation. It is convenient, and fully general as well, to assume that all variables that occur free on the left hand side of a rule also occur on the right hand side of the rule.

Definition 2. *A logic program (over $\mathcal{V}$) is defined as a family of pairs of formulas over $\mathcal{V}$ indexed by $\mathrm{Prd}(\mathcal{V})$, say $((\varphi_\wp^+, \varphi_\wp^-))_{\wp \in \mathrm{Prd}(\mathcal{V})}$, such that for all $n \in \mathbb{N}$ and $\wp \in \mathrm{Prd}(\mathcal{V}, n)$, $\mathrm{fv}(\varphi_\wp^+) \cup \mathrm{fv}(\varphi_\wp^-)$ is included in $\{ v_1, \ldots, v_n \}$.*

An important particular kind of logic program is the folllowing.

Definition 3. *Let a logic program $\mathcal{P} = ((\varphi_\wp^+, \varphi_\wp^-))_{\wp \in \mathrm{Prd}(\mathcal{V})}$ be given. We say that $\mathcal{P}$ is symmetric iff for all $\wp \in \mathrm{Prd}(\mathcal{V})$, $\varphi_\wp^- = {\sim}\varphi_\wp^+$.*

To distinguish between agents, we need the key notion of *occurrence marker*, which intuitively is a function that selects some occurrences of literals in some formulas. Let ψ be a nullary predicate symbol or the negation of a nullary predicate symbol. An agent could select an occurrence o of ψ in a formula φ because she wants to (provisionally) assume that ψ is either true or false, at least in the particular context of ψ occurring in φ at occurrence o. We will see that social agents will make use of the opportunity of assuming that ψ is false, whereas empirical agents will make use of the opportunity of assuming that ψ is true. Actually, ψ does not have to be nullary for these ideas to be developed (we will need more generality anyway), so the definitions that follow deal with arbitrary literals, not only literals built from a nullary predicate symbol. The underlying idea is the same, though it was more easily explained under the assumption that ψ is nullary.

We want to be able to select occurrences of literals in the left hand sides of the rules of a logic program. This justifies the definition that follows.

Definition 4. *Let a logic program $\mathcal{P} = ((\varphi_\wp^+, \varphi_\wp^-))_{\wp \in \mathrm{Prd}(\mathcal{V})}$ be given. An occurrence marker for $\mathcal{P}$ is a sequence of the form $((O_\wp^+, O_\wp^-))_{\wp \in \mathrm{Prd}(\mathcal{V})}$ where for all members $\wp$ of $\mathrm{Prd}(\mathcal{V})$, $O_\wp^+$ and $O_\wp^-$ are sets of occurrences of literals in $\varphi_\wp^+$ and $\varphi_\wp^-$, respectively.*

What we need is to be able to replace some occurrences of literals in some formulas by some other formulas. Given a formula φ and a partial function ρ from the set of occurrences of literals in φ to $\mathcal{L}_{\omega\omega}(\mathcal{V})$, we denote by $\varphi[\rho]$ the result of applying ρ to φ. For instance, if ρ is the function that maps the first occurrence of p in $\varphi = p \wedge (q \vee p)$ to $r \wedge s$, then $\varphi[\rho] = (r \wedge s) \wedge (q \vee p)$.

3   Independent and Social Agents

Let a logic program $\mathcal{P}$ and an occurrence marker Ω for $\mathcal{P}$ be given. Suppose that $\mathcal{V}$ contains n predicate symbols for some nonzero $n \in \mathbb{N}$, so there are $2n$ rules in $\mathcal{P}$, n positive rules and n negative rules, say $R_0, \ldots R_{2n-1}$. Imagine that for all $m < 2n$, R_m is 'under the responsability' of some agent A_m (a single agent might be responsible for many rules in $\mathcal{P}$, possibly all of them). Let $m < 2n$ be given. Some occurrences of literals in R_m might be marked by Ω. Intuitively, these are the occurrences of literals that A_m 'does not bother to' or 'is not able to' directly deal with: a marked occurrence of literal in R_m is assumed by A_m to be false *unless A_m is told otherwise* (expectedly by another agent, but possibly by himself...), for instance because those literals are not under A_m's responsibility—they are instances of rules whose right hand side are under the responsibility of other agents. The definitions that follow formalize these ideas.

Definition 5. *Let a formula φ, a set O of occurrences of literals in φ, and a set E of literals be given. Let ρ be the function from O into $\mathcal{L}_{\omega\omega}(\mathcal{V})$ such that for all $o \in O$, $n \in \mathbb{N}$, $\wp \in \mathrm{Prd}(\mathcal{V}, n)$ and terms $t_1, \ldots, t_n,$[1]*

$$\rho(o) = \begin{cases} \bigvee\{ \bigwedge_{1 \le i \le n} t_i = t'_i \mid \wp(t'_1, \ldots, t'_n) \in E \} & \text{if } \wp(t_1, \ldots, t_n) \in o, \\ \bigvee\{ \bigwedge_{1 \le i \le n} t_i = t'_i \mid \neg\wp(t'_1, \ldots, t'_n) \in E \} & \text{if } \neg\wp(t_1, \ldots, t_n) \in o. \end{cases}$$

We let $\odot_E^O \varphi$ denote $\varphi[\rho]$.

Definition 6. *Let a logic program $\mathcal{P} = ((\varphi_\wp^+, \varphi_\wp^-))_{\wp \in \mathrm{Prd}(\mathcal{V})}$, a possible game T, and an occurrence marker $\Omega = ((O_\wp^+, O_\wp^-))_{\wp \in \mathrm{Prd}(\mathcal{V})}$ for $\mathcal{P}$ be given. We inductively define a family $([\mathcal{P}, T, \Omega]_\alpha)_{\alpha \in \mathrm{Ord}}$ of sets of closed literals as follows. For all ordinals α, $[\mathcal{P}, T, \Omega]_\alpha$ is the $\subseteq$-minimal set of literals that contains T and such that for all $n \in \mathbb{N}$, $\wp \in \mathrm{Prd}(\mathcal{V}, n)$ and closed terms $t_1, \ldots, t_n,$*

$- \wp(t_1, \ldots, t_n) \in [\mathcal{P}, T, \Omega]_\alpha$ *iff* $[\mathcal{P}, T, \Omega]_\alpha \Vdash \odot_{\bigcup_{\beta < \alpha}[\mathcal{P}, T, \Omega]_\beta}^{O_\wp^+} \varphi_\wp^+[t_1/v_1, \ldots, t_n/v_n];$

$- \neg\wp(t_1, \ldots, t_n) \in [\mathcal{P}, T, \Omega]_\alpha$ *iff* $[\mathcal{P}, T, \Omega]_\alpha \Vdash \odot_{\bigcup_{\beta < \alpha}[\mathcal{P}, T, \Omega]_\beta}^{O_\wp^-} \varphi_\wp^-[t_1/v_1, \ldots, t_n/v_n].$

We set $[\mathcal{P}, T, \Omega] = \bigcup_{\alpha \in \mathrm{Ord}}[\mathcal{P}, T, \Omega]_\alpha$.

The independent agent does everything by herself; she does not rely on anyone. If we assume that she works 'nonstop' then her behavior is captured by the empty occurrence marker.

---

[1] In case $n = 0$, the replacing expression is $\bigvee\{ \bigwedge \varnothing \}$ if $\wp \in E$, and $\bigvee \varnothing$ if $\wp \notin E$. Note that $\bigvee\{ \bigwedge \varnothing \}$ is logically equivalent to $\bigwedge \varnothing$.

Definition 7. *Let a logic program* $\mathcal{P} = ((\varphi_\wp^+, \varphi_\wp^-))_{\wp \in \mathrm{Prd}(\mathcal{V})}$ *and a possible game* T *be given. Let* $\Omega = ((O_\wp^+, O_\wp^-))_{\wp \in \mathrm{Prd}(\mathcal{V})}$ *be the occurrence marker for* $\mathcal{P}$ *such that for all* $\wp \in \mathrm{Prd}(\mathcal{V})$, $O_\wp^+$ *and* $O_\wp^-$ *are empty. We write* $[\mathcal{P}, T]$ *for* $[\mathcal{P}, T, \Omega]$.

The next result shows that social agents, irrespective of how their responsibility has been defined, discover the same information, no less, not more, as the independent agent.

Proposition 1. *For all logic programs* $\mathcal{P}$, *possible games* T *and occurrence markers* Ω *for* $\mathcal{P}$, $[\mathcal{P}, T, \Omega] = [\mathcal{P}, T]$.

4   Theoretical and Empirical Agents

In the previous section, we have allowed agents to interact transfinitely many times: in $[\mathcal{P}, T, \Omega]_\alpha$, we allow α to be an infinite ordinal. In this section, we tackle the following issue: is it possible to derive all derivable information in finite time, irrespective of how social agents share their work, or of how single agents organize their work? Obviously, this requires a way of 'working' different to what the concepts that have been defined so far accept. In this section, we will allow agents to make *hypotheses*. If an agent can assume that some literals in the bodies of some rules are true, she might be able to speed up the derivations she can perform. Such hypotheses should abide stringent conditions. We suggest that a hypothesis should eventually either be *confirmed*, that is, proved correct, or *refuted*, that is, proved wrong. Let us first precisely define what 'making a hypothesis' means. A pleasant feature of this notion is that it is again based on the notion of occurrence marker. This time, we use occurrence markers to select some occurrences of literals on the left hand side of some rules to make them the targets of some hypotheses.

Definition 8. *Let a formula* φ, *a set* O *of occurrences of literals in* φ, *and a set* E *of literals be given. Let* ρ *be the function from* O *into the set of formulas such that for all* $o \in O$, $n \in \mathbb{N}$, $\wp \in \mathrm{Prd}(\mathcal{V}, n)$ *and terms* $t_1, \ldots, t_n$, $\rho(o)$ *is equal to* $\bigvee\{\wp(t_1, \ldots, t_n), \bigwedge_{i=1}^n t_i = t_i' \mid \wp(t_1', \ldots, t_n') \in E\}$ *if* $\wp(t_1, \ldots, t_n) \in o$, *and to* $\bigvee\{\neg\wp(t_1, \ldots, t_n), \bigwedge_{i=1}^n t_i = t_i' \mid \neg\wp(t_1', \ldots, t_n') \in E\}$ *if* $\neg\wp(t_1, \ldots, t_n) \in o$. *We let* $\odot_E^O \varphi$ *denote* $\varphi[\rho]$.

An agent willing to assume that the literals in E are true provided that they occur on the left hand side of the rules of a logic program $\mathcal{P}$, as selected by the occurrence marker Ω for $\mathcal{P}$, essentially decides to work on the basis of the logic program $\mathcal{P} +_\Omega E$ introduced in the definition that follows.

Definition 9. *Let a logic program* $\mathcal{P}$ *and an occurrence marker* Ω *for* $\mathcal{P}$ *be given. Write* $\mathcal{P} = ((\varphi_\wp^+, \varphi_\wp^-))_{\wp \in \mathrm{Prd}(\mathcal{V})}$ *and* $\Omega = ((O_\wp^+, O_\wp^-))_{\wp \in \mathrm{Prd}(\mathcal{V})}$. *Given a set* E *of literals, the sequence* $((\odot_E^{O_\wp^+} \varphi_\wp^+, \odot_E^{O_\wp^-} \varphi_\wp^-))_{\wp \in \mathrm{Prd}(\mathcal{V})}$ *is denoted* $\mathcal{P} +_\Omega E$.

Our aim is to show that making hypotheses can pay off.

Definition 10. *A logic program* $\mathcal{P} = ((\varphi_\wp^+, \varphi_\wp^-))_{\wp \in \mathrm{Prd}(\mathcal{V})}$ *is* acceptable *iff the following holds. Let* $\mathcal{V}^*$ *be* $\mathcal{V}$ *without the observational predicate symbols.*

- *For all possible games* T, $[\mathcal{P}, T]$ *is a complete set of literals.*
- *The restriction of* $\mathcal{P}$ *to* $\mathcal{V}^*$ *is symmetric.*
- *For all* $\wp \in \mathrm{Prd}(\mathcal{V})$,
  - *if* $\wp$ *is observational then both* $\varphi_\wp^+$ *and* $\varphi_\wp^-$ *are equal to* $\bigvee \varnothing$,
  - *either* $\wp$ *is nullary or no quantifier occurs in* $\varphi_\wp^+$, *and*
  - *all quantified formulas that occur in* $\varphi_\wp^+$ *have one quantifier only.*

Here is an example of part of an acceptable logic program.

$$\forall v_1 ((\mathrm{hearts}(v_1) \vee \mathrm{diamonds}(v_1)) \to \mathrm{red}(v_1))$$
$$\forall v_1 ((\mathrm{spades}(v_1) \vee \mathrm{clubs}(v_1)) \to \mathrm{black}(v_1))$$
$$\forall v_0 (\mathrm{red}(v_0) \leftrightarrow \mathrm{black}(s(v_0))) \to \mathrm{alternatedColors}$$
$$(\forall v_0 \, \mathrm{red}(v_0) \vee \exists v_0 (\mathrm{queen}(v_0) \wedge \mathrm{clubs}(v_0))) \to \mathrm{allRedsOrAQofC}$$

The proposition that follows shows that it is possible to enrich $\mathcal{V}$ into a vocabulary $\mathcal{V}'$, transform $\mathcal{P}$ into a logic program $\mathcal{P}'$ over $\mathcal{V}'$, and make some assumptions such that all possible games T, all members of $[\mathcal{P}, T]$ can be derived after a finite number of steps. Moreover, $\mathcal{P}'$ is such that it is safe to make any set of assumptions; indeed, any set of assumptions that is inconsistent with $\mathcal{P}'$ and a possible game will proved inconsistent after finitely many inferences.

Proposition 2. *Let* $\mathcal{V}^*$ *be* $\mathcal{V}$ *without the observational predicate symbols. For all acceptable logic programs* $\mathcal{P}$, *there exists a finite set* E *of nullary predicate symbols that do not belong to* $\mathcal{V}$ *and there exists a logic program* $\mathcal{P}'$ *over* $\mathcal{V} \cup E$ *whose restriction to* $\mathcal{V}^* \cup E$ *is symmetric such that for all possible games* T, *there exists an occurrence marker* Ω *for* $\mathcal{P}'$ *with the following properties.*

- $[\mathcal{P}, T]$ *and the restrictions of* $[\mathcal{P}', T]$ *and* $[\mathcal{P}' +_\Omega E, T]$ *to* $\mathcal{V}$ *are equal;*
- *for all occurrence markers* Ω' *for* $\mathcal{P}'$, $[\mathcal{P}' +_\Omega E, T] = \bigcup_{n \in \mathbb{N}} [\mathcal{P}' +_\Omega E, T, \Omega']_n$;
- *for all possible games* T *and for all occurrence markers* Ω' *and* Ω'' *for* $\mathcal{P}'$, *if* $[\mathcal{P}' +_{\Omega''} E, T] \neq [\mathcal{P}', T]$ *then* $\bigcup_{n \in \mathbb{N}} [\mathcal{P}' +_{\Omega''} E, T, \Omega']_n$ *is inconsistent.*

The transformation of $\mathcal{P}$ to $\mathcal{P}'$ amounts to replacing some complex formulas in the bodies of some rules of $\mathcal{P}$ by some new nullary predicate symbols, themselves defined thanks to a new pair of rules—a form of predicate invention—that can play the role of hypotheses and enjoy a refutation property. With the previous example of acceptable logic program, E could consist of two nullary predicate symbols, say p and q, and $\mathcal{P}'$ could be defined as

$$\forall v_1 ((\mathrm{hearts}(v_1) \vee \mathrm{diamonds}(v_1)) \to \mathrm{red}(v_1))$$
$$\forall v_1 ((\mathrm{spades}(v_1) \vee \mathrm{clubs}(v_1)) \to \mathrm{black}(v_1))$$
$$\forall v_0 (\mathrm{red}(v_0) \leftrightarrow \mathrm{black}(s(v_0))) \to p$$
$$p \to \mathrm{alternatedColors}$$
$$\forall v_0 \, \mathrm{red}(v_0) \to q$$
$$(q \vee \exists v_0 (\mathrm{queen}(v_0) \wedge \mathrm{clubs}(v_0))) \to \mathrm{allRedsOrAQofC}$$

An agent would then have four options, depending on whether she would assume p or q in the bodies of the 4th and 6th rules, respectively. For any possible game T, one of these options would be appropriate and allow the agent to discover whether T is a sequence of cards where black and red alternate, or whether T is a sequence consisting of nothing but red cards, unless it contains a queen of clubs. Any wrong set of hypotheses would be guaranteed to be eventually refuted in the limit on the basis of a finite subset of T.

5   Conclusion

We have presented a framework where fundamental questions about the nature of scientific discovery can be formulated and studied. The basic working hypothesis is that a purely logical approach to scientific discovery and problem solving is possible, in a way that can shed light on the nature of those activities. We believe that our approach can address a whole range of questions related to the nature of scientific discovery or problem solving, always within the boundaries of a pure logical setting. For instance, Angluin proposes a binary categorization of agents, with learners and teachers, and she proves robustness results about their interaction; how does this categorization translate into our setting? Starting from a fixed language, we have to a certain extent accounted for predicate invention in the last proposition, allowing agents to make a rational use of hypotheses expressed in an extension of the original language, but how does predicate invention relate to postulates? Surely, logic is not an iron collar, but it can potentially strive far beyond the territories where it has been confined to.

References

1. Afshar, M., Dartnell, C., Luzeaux, D., Sallantin, J.: Aristotle's square revisited to frame discovery science. Journal of Computers 2(5), 54–66 (2007)
2. Angluin, D., Krikis, M.: Learning from different teachers. Machine Learning 51(2), 137–163 (2003)
3. Chavalarias, D., Cointet, J.-P.: Bottom-up scientific field detection for dynamical and hierarchical science mapping—methodology and case study. Scientometrics 75(1) (2008)
4. Gold, M.E.: Language identification in the limit. Information and Control 10(5), 447–474 (1967)
5. Langley, P.W., Bradshaw, G.L., Simon, H.A.: Rediscovering chemistry with the BACON system. In: Michalski, R.S., Carbonell, J.G., Mitchell, T.M. (eds.) Machine Learning: An Artificial Intelligence Approach, Springer, Heidelberg (1984)
6. Martin, E., Sharma, A., Stephan, F.: Deduction, induction and beyond in parametric logic. In: Friend, M., Goethe, N.B., Harizanov, V.S. (eds.) Induction, Algorithmic Learning Theory, and Philosophy. Logic, Epistemology and the Unity of Science, vol. 9. Springer, Heidelberg (2007)
7. Soldatova, L.N., Clare, A., Sparkes, A., King, R.D.: An ontology for a robot scientist. Bioinformatics 22(14), e464–e471 (2006)
8. Valiant, L.L.: A theory of the learnable. Communications of the ACM 27(11), 1134–1142 (1984)

Gene Functional Annotation with Dynamic Hierarchical Classification Guided by Orthologs

Kazuhiro Seki, Yoshihiro Kino, and Kuniaki Uehara

Kobe University
1-1 Rokkodai, Nada, Kobe 657-8501, Japan
seki@cs.kobe-u.ac.jp

Abstract. This paper proposes an approach to automating Gene Ontology (GO) annotation in the framework of hierarchical classification that uses known, already annotated functions of the orthologs of a given gene. The proposed approach exploits such known functions as constraints and dynamically builds classifiers based on the training data available under the constraints. In addition, two unsupervised approaches are applied to complement the classification framework. The validity and effectiveness of the proposed approach are empirically demonstrated.

Keywords: Gene ontology, String matching, Information retrieval.

1 Introduction

Since the completion of the Human Genome Project, a large number of studies have been conducted to identify the roles of individual genes, which would help us understand critical mechanisms of human bodies, such as aging and disorders. The active research in the domain has been producing numerous publications. Although they are rich intellectual resources, it is extremely labor-intensive to collect all the information relevant to a given user information need, such as "a list of functions of gene X" or "a list of genes having function Y", since such information can be only accessed by extensive reading. To remedy the problem, numbers of organizations have been working to annotate each gene of model organisms with controlled vocabularies, called Gene Ontology (GO) terms, based on the contents of published scientific articles. GO is defined as a directed acyclic graph (DAG), and organized under three top level nodes: molecular function (MF), cellular component (CC), and biological process (BP). Currently, there are nearly 30,000 GO terms in total.

The effort of GO annotation has enabled uniform access to different model organism databases, including FlyBase, Mouse Genome Database (MGD), and Saccharomyces Genome Database, by the common vocabularies. However, the annotation requires trained human experts with extensive domain knowledge. With limited human resources and the ever-growing literature, it was reported that it would never be completed at the current rate of production [1].

Motivated by the background, this study proposes an approach to automatic GO annotation, which exploits the structure of GO and applies hierarchical

J. Gama et al. (Eds.): DS 2009, LNAI 5808, pp. 425–432, 2009.

classification. In addition, we take advantage of orthologous genes and use their known gene functions as constraints to enable efficient learning. Moreover, we apply string matching-based and information retrieval model-based approaches to deal with the case where sufficient training data are not available.

2   Related Work

Due to the large number of genes, gene functions, and scientific articles, manual GO annotation is inevitably labor-intensive. In addition, because of the highly specialized contents, it requires skilled professionals with expertise in the domain. To alleviate the burden, TREC 2004 Genomics Track [2] and BioCreative [3] targeted automatic GO domain/term annotation.

The Genomics Track attempted to automate the process of assigning the first level of GO (i.e., MF, CC, BP), called "GO domains". The participants of the workshop were given a mouse gene and an article in which the gene appears and were expected to annotate zero to three GO domains with the gene based on the contents of the article. For this task, Seki and Mostafa [4] developed an approach featuring flexible gene mention extraction techniques based on a synonym dictionary and approximate name match. They used gene-centered representation by extracting fragments of an article mentioning the target gene and applied k nearest neighbor (kNN) classifiers with supervised term weighting.

In contrast to the Genomics Track only targeting GO domains, BioCreative aimed at assigning specific GO terms to human genes. Among others, Ray and Craven [5] looked at the occurrences of GO terms and their related terms to assign GO terms. Stoica and Hearst [6] took advantage of orthologs of a given gene and considered the GO terms already associated with them as candidates. Orthologs are genes in different species rooted from the same gene of their common ancestor and often have the same functions. Stoica and Hearst associated a given human gene with its mouse ortholog, and if the majority of terms consisting of each GO term assigned to the ortholog appeared in a given article, they assigned the GO term to the human gene. In addition, they used GO term co-annotation to prevent false positives. Their idea was based on the observation that there were cases where some GO terms were not usually co-annotated together to the same gene because annotating them together was illogical. For instance, "transcription (GO:0006350)" and "extracellular (GO:0005576)" are not likely to be co-annotated as transcription cannot happen outside of a cell.

Comparing the approaches taken at the Genomics Track and BioCreative, the participants for the former reported the effectiveness of supervised classification techniques. On the other hand, those for the latter mainly adopted string matching-based approaches. Such different strategies attributed to the fact that the former considers only three categories (i.e., GO domains), whereas the latter takes account of nearly 30,000 GO terms; dealing with less and general classes is more suitable for text categorization in terms of available training data and overfitting.

This study takes a classification approach to GO annotation by leveraging a limited amount of training data, where the GO structure and orthologous genes are used for guiding efficient classification. In addition, we complementarily use other unsupervised approaches when there is only insufficient training data so as to boost the coverage of GO annotation.

3   Proposed Approach

3.1   Overview

Our approach assigns appropriate GO terms for a given pair of gene g and an article d based on a set of text fragments mentioning g extracted from d. If there are multiple functions of g reported in d, we assign multiple GO terms corresponding to them. Roughly, our approach consists of the following steps: 1) Assign GO domains, 2) Obtain GO terms already assigned to the ortholog of the given gene g, 3) Assign GO terms by hierarchical classification, 4) Assign GO terms based on unsupervised approaches. Each step is described below.

3.2   Assigning GO Domains

For GO domain annotation, we follow the approach proposed by Seki and Mostafa [4] who have reported the best performance in the literature. Simply put, for a given pair of gene g and article d, they first extract paragraphs mentioning g. Then, from the set of extracted paragraphs, a term vector is constructed to represent the input pair $\langle d, g \rangle$. Based on the representation, they assign GO domains by a variant of kNN.

3.3   Obtaining GO Terms Annotated with Orthologs

After assigning GO domains, we identify promising GO term candidates in order to enable both effective and efficient GO term annotation. This study adapts the approach by Stoica and Hearst [6] using orthologs; That is, we consider only GO terms already assigned to the ortholog g' of a given gene g as GO term candidates. By this constraint, we can drastically reduce the number of GO terms to be considered from around 30,000 to only dozens at most. For instance, a mouse gene Sox21 has an ortholog in human genome, called SOX21, and the human gene has been already annotated with GO terms, including "RNA polymerase II transcription factor activity (GO:0003702)" and "establishment or maintenance of chromatin architecture (GO:0006325)", where the numbers in the parentheses are corresponding GO codes. Because these two genes are orthologous and are likely to have the same functions, we can expect higher precision by focusing only on these GO terms. Of course, it is also possible that true GO terms are not found in these GO term candidates. We will empirically investigate how often such cases occur in Section 4.2.

For the sources of the information regarding orthologs and their known gene functions (GO terms), this study uses two existing databases, MGD and Gene Ontology Annotation (GOA).

3.4   GO Term Annotation by Dynamic Hierarchical Classification

Using the GO term candidates obtained through the ortholog of the given gene, we then assign specific GO terms by taking advantage of the structure of GO. For the above-mentioned example of Sox21, we consider only the GO terms already annotated with its ortholog as possible classes and train classifiers for them. However, as the number of the training instances with the classes (i.e., the GO terms) is often limited as discussed in Section 2, we enhance the training data set based on the GO structure. That is, for the candidate GO terms, we first identify their least common ancestor (LCS) and then train classifiers for the GO terms immediately under the LCS, where we consider only GO terms which have any candidates as descendants. For training data, we use not only the instances having the exact GO terms immediately under the LCS but also those having more specific GO terms under them. This way, one can use more training data and diminish the influence of the overfitting problem. Although this approach is similar to the hierarchical classification approach by McCallum et al. [7], a difference is that this study does not take into account all the classes in a given structure but only the limited number of the GO terms associated with a given gene through its ortholog. Also, training instances are dynamically harvested at each step of classification based on the GO term candidates, so as to learn classifiers on the fly.

A more precise algorithm of our dynamic hierarchical classification for GO term annotation is presented in Fig. 1, where the input is a test instance b, a set of training instances T, a set of GO term candidates C, and a set of GO domains assigned as described in Section 3.2; and the output is a set of GO terms F with which b is annotated. For each GO domain s, we identify GO term candidates C_s in the GO domain. If the number of the candidates $|C_s|$ equals 1, we unconditionally add the sole GO term candidate to the output F considering the fact that the GO domain s is already assigned and the GO term candidate is the only possible one to assign in the domain s. If $|C_s|$ is greater than 1, the following steps are carried out. First, we identify a set of GO terms C'_s immediately under the LCS and then, for each GO term in C'_s, we collect all the instances having GO terms under it. If the number of training instances for every GO term in C'_s is greater than a predefined threshold τ, we train classifier $\mathcal{F}$ and set per-class thresholds $\Theta = \theta_1, \ldots, \theta_{|C'_s|}$ to maximize F_1 score for each class $c'_i \in C'_s$ using the training instances. If classifier's output p_i for c'_i exceeds the threshold θ_i, c'_i is added to F in the case where c'_i is one of the GO term candidates, or we recursively apply the same procedure using c'_i as if it were a GO domain. If the number of training instances is below the threshold τ for any c'_i, we resort to the unsupervised approaches to avoid the overfitting problem as described next.

3.5   Unsupervised Approaches to GO Term Annotation

In order to deal with the classes with insufficient training data (less than threshold τ), we make use of a string matching-based approach and an approach using

1 **Input:** test instance b, set of training instances T, set of GO term candidates C, set of predicted GO domains D;
2 **Output:** set of predicted GO terms F for b;
3 **Variables:** set of GO terms/domains S, prediction $p_i \in \mathbb{R}$ for a GO term c_i', threshold τ for training data size;

4 $S = D$
5 **while** S is not empty **do**
6   Take any GO term/domain out from S and set it to s
7     $C_s = \{c \,|\, c \in C$ under $s\}$
8   **if** $|C_s| = 1$ **then add** C_s **to** F
9   **else if** $|C_s| > 1$ **then**
10         $C_s' = \{c' \,|\, $GO terms immediately below $s\}$
11   **for each** $c' \in C_s'$ **do**
12             $T_{c'} = \{t \,|\, t \in T$ assigned any GO term under $c'\}$
13   **if** $\forall c', |T_c'| > \tau$ **then**
14   Build a classifier $\mathcal{F} \mapsto (p_1, \ldots, p_{|C_s'|})$
15   Determine per-class thresholds $\Theta = \theta_1, \ldots, \theta_{|C_s'|}$
16   **for each** $c_i' \in C_s'$ **do**
17   **if** p_i (predicted by $\mathcal{F}$ for b) $> \theta_i$ **then**
18   **if** $c_i' \in C_s$ **then add** c_i' **to** F
19   **else add** c_i' **to** S

Fig. 1. Dynamic hierarchical GO term annotation algorithm

an information retrieval model. These approaches were adapted from the related work in BioCreative and others.

String matching-based approach. Since GO terms are concise descriptions of gene functions in natural language, if a text contains a certain GO term, the text may be describing the corresponding gene function. This is not necessarily the case for general GO terms located at the higher level of the GO tree, such as "behaviour (GO:0007610)", but is likely to apply to more specific ones, such as "regulation of transcription from RNA polymerase II promoter (GO:0006357)". In this study, we use the edit distance to deal with some writing variations and differences. The edit distance basically counts the number of edit operations (i.e., insert, delete, and substitution) to convert a string (i.e., GO term) to another string (i.e., actual expression found in text). Also, to consider the different importance of words, we define different penalty costs for different words based on document frequencies (DF). We define the DF of a word w as the logarithm of the total number of GO terms containing w.

Information retrieval model-based approach. Another unsupervised approach has been proposed by Ruch [8]. We take a similar approach as him and assign GO terms based on a vector space model. Simply put, this approach measures the cosine similarity between a GO term and text and assigns the GO term if the similarity between them exceeds a predefined threshold. Essentially, this approach is similar to the string matching-based approach above except that this approach is less restrictive, not considering word orders.

4 Evaluation

4.1 Experimental Settings

For evaluation, we use the data set provided for the TREC 2004 Genomics Track supplemented by GO term information. The data set consists of 849 training instances and 604 test instances, where each instance is a triplet of an article d represented by PubMed ID, a gene g mentioned in d, and a GO term f which is reported in d as a function of g. This data set is a subset of MGD, and thus, only dealing with mouse genes.

As an evaluation metric, we use F_1 score for direct comparison with the previous work, i.e., Genomics Track and BioCreative which used the same metric. F_1 is defined as a harmonic mean of recall (R) and precision (P). P is defined as the number of correct GO terms assigned divided by the number of GO terms assigned, and R is the number of correct GO terms assigned divided by the number of GO terms in the test data.

The proposed GO term annotation framework is general and by design does not depend on a particular classifier. Although the following experiments used kNN as it has been shown effective in the related work [4], it can be easily replaced with other classifiers.

4.2 Validity of the Use of Orthologs for GO Annotation

As orthologs, we experimentally chose human and rat genes to annotate mouse genes. Our first experiment examined the validity of the use of those orthologs for GO term annotation. To be precise, we simply annotated input mouse genes with all the GO term candidates obtained from their orthologs *without* classification. This experiment reveals the coverage of the GO term candidates obtained through different species.

When comparing two species, human and rat, the latter works better for all of recall (0.800), precision (0.045), and F_1 (0.086). This is expected, as rat is genetically closer to mouse than human. Using rat genes, the recall was found 0.800, which means that 80.0% of true GO terms annotated to the test data are found in the GO terms already assigned to the rat orthologs. Differently put, this is the upper bound of recall for our framework to look only at GO term candidates obtained from orthologs. In this study, we focus on the 80.0% and recovering the remaining 20.0% are left for the future work.

4.3 GO Term Annotation by Hierarchical Classification

Table 1 shows the results for GO term annotation when our proposed approach (denoted as "Hierarchical") based on hierarchical classification was applied, where we used only rat genes as orthologs based on the observation in Section 4.2. In addition, the table shows for reference the results reported by Stoica and Hearst [6] and Chiang and Yu [9] on the BioCreative data set. Also, the results for standard flat classification without considering GO structure is included

Table 1. Comparison of the performance of GO term annotation

Approaches	Precision	Recall	F_1
Stoica & Hearst [6]	0.168	0.121	0.140
Chiang & Yu [9]	0.332	0.051	0.089
Hierarchical (proposed approach)	0.248	0.210	0.227
Flat	0.041	0.551	0.075

Table 2. Results of GO annotation when hierarchical classification and unsupervised approaches are combined

Approaches	Precision	Recall	F_1
Hierarchical	0.248	0.210	0.227
Hierarchical + Edit	0.238	0.245	0.242
Hierarchical + IR	**0.256**	0.275	**0.265**
Hierarchical + Edit + IR	0.236	**0.282**	0.257

(denoted as "Flat"). Note that "Flat" also looked at only GO term candidates obtained from orthologs and thus can be used to evaluate the effect of the use of the GO structure.

Comparing with the results by Stoica and Hearst [6] and Chiang and Yu [9], our proposed approach obtained the best performance in F_1. This result indicates that, if we can restrict the number of GO terms to be considered, supervised classification approaches can be effective even for GO term annotation for which a large number of classes otherwise exist. In addition, we can observe that the flat classification produced poor performance, which means that it is not sufficient only to restrict the number of possible classes.

4.4   GO Term Annotation by Unsupervised Approaches

In general, classification performance in precision improves up to some point as the training data size increases. However, because the GO terms with a large number of instances are limited, recall inevitably decreases with higher τ, the threshold for the number of instances. To improve recall, we apply two unsupervised approaches described in Section 3.5 when there are insufficient training data (less than τ). The results are shown in Table 2, where "Hierarchical" is the result by hierarchical classification taken from Table 1, "Edit" and "IR" denote approaches based on string matching and the IR model, respectively.

When Hierarchical is combined with one of Edit and IR, recall improved to 0.245 (+16.7%) and 0.275 (+31.0%), respectively. This result confirms the effectiveness of the unsupervised approaches and indicates that they work complementarily with our hierarchical classification approach. Especially, the IR approach resulted in a significant boost in recall and even improved precision as compared with Hierarchical alone. In addition, when both Edit and IR are combined with Hierarchical, recall further improved to 0.282 (+34.3%), which, however, decreased precision. (It is attributed to the different value of τ used, which was chosen to maximize F_1 for each configuration.) Focusing on F_1, Hierarchical+IR was the best combination achieving an F_1 of 0.265. For comparison, when only IR without classification was applied for GO term annotation, F_1 was found to be around 0.150 (not shown in the table). This result also confirms that combining classification and unsupervised approaches is effective for GO term annotation.

5   Conclusions

This study proposed an approach to GO term annotation using orthologs to effectively guide hierarchical classification. In addition, two unsupervised approaches were applied when sufficient training data were not available. From the experiments on the Genomics Track data, we observed that 1) by using rat genes as orthologs, up to 80% of correct GO terms can be annotated; 2) using the GO term candidates obtained from orthologs, our hierarchical classifiers were able to annotate mouse genes at an F_1 of 0.227; and 3) by combining the hierarchical classification and the IR model-based approach, the performance improved up to 0.265. For future work, we aim to recover the remaining 20% of true GO terms not covered by the ortholog-based framework. This could be partly done by exploiting other homologs, e.g., paralogous and xenologous genes.

Acknowledgments. This work is partially supported by KAKENHI #21700169.

References

1. Baumgartner, Jr.,W.A., Cohen, K.B., Fox, L.M., Acquaah-Mensah, G., Hunter, L.: Manual curation is not sufficient for annotation of genomic databases. Bioinformatics 23(13), 41–48 (2007)
2. Hersh, W., Bhuptiraju, R.T., Ross, L., Cohen, A.M., Kraemer, D.F.: TREC 2004 genomics track overview. In: Proc. of the 13th Text REtrieval Conference (2004)
3. Blaschke, C., Leon, E., Krallinger, M., Valencia, A.: Evaluation of BioCreAtIvE assessment of task 2. BMC Bioinformatics 16(1), S16 (2005)
4. Seki, K., Mostafa, J.: Gene ontology annotation as text categorization: An empirical study. Information Processing & Management 44(5), 1754–1770 (2008)
5. Ray, S., Craven, M.: Learning statistical models for annotating proteins with function information using biomedical text. BMC Bioinformatics 6(1), S18 (2005)
6. Stoica, E., Hearst, M.: newblock Predicting gene functions from text using a cross-species approach. In: Proc. of the Pacific Symposium on Biocomputing, pp. 88–99 (2006)
7. McCallum, A., Rosenfeld, R., Mitchell, T.M., Ng, A.Y.: Improving text classification by shrinkage in a hierarchy of classes. In: Proc. of the 15th International Conference on Machine Learning, pp. 359–367 (1998)
8. Ruch, P.: Automatic assignment of biomedical categories: toward a generic approach. Bioinformatics 22(6), 658–664 (2006)
9. Chiang, J., Yu, H.: Extracting functional annotations of proteins based on hybrid text mining approaches. In: Proc. of the BioCreAtIvE (2004)

Stream Clustering of Growing Objects

Zaigham Faraz Siddiqui and Myra Spiliopoulou

Otto-von-Guericke-University of Magdeburg,
Magdeburg 39106, Germany
{siddiqui,myra}@iti.cs.uni-magdeburg.de

Abstract. We study incremental clustering of objects that grow and accumulate over time. The objects come from a multi-table stream e.g. streams of `Customer` and `Transaction`. As the Transactions stream accumulates, the Customers' profiles *grow*. First, we use an incremental propositionalisation to convert the multi-table stream into a single-table stream upon which we apply clustering. For this purpose, we develop an online version of K-Means algorithm that can handle these swelling objects and any new objects that arrive. The algorithm also *monitors* the quality of the model and performs re-clustering when it deteriorates. We evaluate our method on the PKDD Challenge 1999 dataset.

1 Introduction

The rapid developments in hardware technology have enabled generation of massive information in the financial institutions, scientific laboratories, communication networks and everyday life. Data is produced continuously and is referred to as *data streams*. Such data may contain objects that are also dynamic i.e. grow over time. Such objects are the result of combining multi-relational data across several streams, e.g. streams of Customers and Transactions. As transactions arrive, more information about customers' preferences and purchases becomes available. We use the term *growing objects* for objects that acquire more and more information and change their definitions over time. By *change of definition* we mean that a customer whose bank balance was 10€is now 55€, or he did 10 transactions, now has 55 transactions that arrived in the meanwhile. To perform stream clustering, we first convert the multi-table stream into a single stream and then use a stream clustering algorithm to find groups of similar objects.

Most of the clustering algorithms over data streams assume that data stream is an ordered sequence of data points $x_1, \ldots, x_i, \ldots x_n$, read in increasing number of indices i [1]. Each data point is *unique* and with a *unique identifier*. In other words data streams are assumed to be generated by a dynamic process but consist of objects that are *static* themselves.

As explained above the objects may themselves be dynamic. Hence, we propose a new stream mining framework which assumes a data stream to be generated by a dynamic process and consist of objects that are dynamic and independent of each other. More formally, a data stream contains an infinite sequence of objects $\mathcal{X} = \langle x_1, \ldots x_j \ldots \rangle$ arriving at timepoints $\langle t_1, \ldots t_i, \ldots \rangle$. These objects

J. Gama et al. (Eds.): DS 2009, LNAI 5808, pp. 433–440, 2009.

arrive as $\mathcal{S} = \langle x_{(1,t_1)}, x_{(2,t_1)}, \ldots x_{(2,t_i)}, x_{(3,t_i)}, \ldots \rangle$ and are read in the increasing order of indices t_i. For any two occurrences $x_{(j,t_i)}, x_{(j,t_k)}$ of x_j where $t_i \neq t_k$, $x_{(j,t_i)} \neq x_{(j,t_k)}$. For any x_j that appears more than once, the one with the higher t index replaces the old occurrence, i.e. if $t_i > t_k$, $x_{(j,t_i)}$ replaces $x_{(j,t_k)}$.

In this study we concentrate on discovering patterns over objects that change their definitions with time. We propose a variant of K-Means algorithm, extended to deal with growing objects and able to incorporate any newer objects as they arrive. The rest of the paper paper is organized as follows. In Section 2 we discuss related work. In Section 3 we present the clustering algorithm. Our experimental findings are reported in Section 4. We conclude with directions for future work.

2   Related Work

Most stream clustering algorithms process and then discard tuples. Bradley et al identify and store data of interest inside a buffer [2] in a compressed way. Farnstrom et al points out that compression is an expensive strategy and does not necessarily improves clustering [3]. Callaghan et al use a buffer to store points into batches of m points [4]. After clustering, K centres are retained (with statistics) and buffer is refilled with new points. Guha et al maintain at each moment the m most recent tuples and K medians that stand for $K \times m$ tuples seen in the past [1].

Aggarwal et al warn that one pass algorithm over can be dominated out dated data and say that stream exploration over different time windows can provide a deeper understanding [5]. Their micro-clustering approach keeps clusters in multiple snapshots of pyramidal time frame. In all the works, the problem is to cluster objects that are static and are from one stream. This is different from the problem we address.

Incremental clustering of the time-series is closest to incremental clustering of growing objects. Beringer et al present an elaborate method that aims at clustering multiple streams [6]: for each individual stream a window w with m blocks of v points is defined that stands for the $m \times v$ most recent values. Distance is computed incrementally using Discrete Fourier Transformation before k-means clustering. It also utilises a fuzzy approach to dynamically update the optimal number of clusters. The Clustering on Demand (COD) framework clusters multiple data streams using a single online scan for statistics collection and an offline step to define clustering structure with adaptive window sizes [7]. In [8] methods are provided to add and delete static data points. Neither work supports the growth of dynamic objects, though.

3   Framework

Our method consists of two parts. The first part is an incremental propositionalisation algorithm. We briefly describe it here; the complete description is in [9]. The second part is an incremental clustering algorithm that uses Online K-Means [6] as basis but can deal with growing and changing objects.

3.1   Combining Multiple Streams

As discussed earlier (c.f. Section 1), growing objects come from a multiple inter-related streams that need to be transformed into a single stream. We associate each stream of data that *may be forgotten* with a *window*; data that *may not be forgotten* are kept in a *cache*.

The incremental propositionalisation begins by specifying *target stream* T_0. For each stream T_j that is in 1-to-m or m-to-n relationship to T_0, each object $x \in T_0$ is associated with the set of matching objects $matches(x) \subset T_j$. The objects in this set are summarized in a single *sub-object*. For the summarisation of the values of each numerical attribute A among the $matches(x)$, four attributes are generated to accommodate the *min, max, count* and *average* of the A values seen in $matches(x)$. For summarisation of each nominal attribute A, as many columns(r_A) for A are generated as there are distinct values in $\bigcup_x matches(x)$ at t_0.

The domain of this nominal attribute A may change when a new value v arrives at some later timepoint. The value v is either assigned a column that is lying vacant due to the disappearance of some value v'. If no column is vacant, similar nominal values are clustered into k groups (where $k = r_A$ i.e. number of reserved columns), with values in a cluster sharing one column.

The propositionalisation is done only on the contents of windows and caches. At timepoint t_i, a sliding window contains objects seen since t_{i-L}, where L is the length of the window and is defined separately for each table. At each time-point t_i, the Cache Update Algorithm [9] calculates statistics for the objects and retains the ones that are frequently referenced. At timepoint t_i, the current contents of T_0 in the cache or window of T_0 are "propositionalised" to accommodate the records in the cache or window of each connected stream T_j.

3.2   Stream K-Means for Growing Objects

In the previous section we presented the method for the update and maintenance of dynamic object streams. In this section we present an incremental version of K-Means. Our method is inspired by the Online K-Means proposed in [6]. However, there are fundamental differences between the two. We discuss them at the end of this section, but they can be summarized into saying that the Online K-Means solves a different problem.

The pseudo-code of the incremental clustering algorithm is shown in (c.f. Algorithm 1). It starts by initialising k cluster centres (c.f. Line1). Parameter w denotes the window size; the algorithm considers only objects that are inside the sliding window. The parameter $\mathcal{X}^B$ contains the schema, the specification of the target stream T_0 and the size of the cache and windows. It is important to point here that the schema $\mathcal{X}^B$ may change as new nominal values show up and old ones are forgotten. As a result, the schema for the output stream $\mathcal{S}$ also changes.

At timepoint t_i data arrives in the streams, is first transformed from multi-table stream into a single-table stream S_i (c.f. Line3). It is important to stress here that Cache Update Algorithm in [9] prefers objects referenced frequently over those that are not: the objects having more references are likely to carry more information and be more *mature* or *grown*. This makes more information available and is likely to result in better clustering. Objects with fewer are treated as noise and are kept away until they mature.

Algorithm 1. Incremental K-Means
Input : $\mathcal{X}^B, k, w$
1 init k centres for model ζ_0
2 **for** $i = 1$ **to** *STREAM_END* **do**
3 $\quad$ $\mathcal{S}_i \leftarrow$ IncProp($\mathcal{X}^B$)
4 $\quad$ $\mathcal{W}_i \leftarrow$ UpdWin($\mathcal{W}_{i-1}, \mathcal{S}_i, w$)
5 $\quad$ $\zeta_i \leftarrow$ UpdateClu($\mathcal{W}_i, \zeta_{i-1}$)
6 $\quad$ $\rho \leftarrow$ JaccardCoeff(ζ_i, ζ_{i-1})
7 $\quad$ **if** $\rho < \tau$ **then**
8 $\qquad$ **for** $j = 1 \rightarrow MAX\_MDL$ **do**
9 $\qquad\quad$ init k centres for ζ_i
10 $\qquad\quad$ $\zeta_T \leftarrow$ UpdateClu($\mathcal{W}_i, \zeta_i$)
11 $\qquad\quad$ $\zeta_B \leftarrow \zeta_T$
12 $\qquad\quad$ **if** Q(ζ_T)>Q(ζ_B) **then**
13 $\qquad\qquad$ $\zeta_B \leftarrow \zeta_T$
14 $\qquad$ $\zeta_i \leftarrow \zeta_B$

Function. UpdateClustering
Input : W, ζ
Output: ζ
1 $j = 0$
2 **while** $j <MAX\_STEPS \wedge \zeta$ *not stable* **do**
3 $\quad$ assign data pts to the clusters
4 $\quad$ adjust cluster centres
5 $\quad$ Increment(j)
6 **return** ζ

After transformation of the multi-table stream, the algorithm updates window $\mathcal{W}_i$ by replacing outdated objects with new ones (Line4). After the data pre-processing step, the clustering structure ζ_{i-1}(i.e. membership and vectors of the centroids) from t_{i-1} is taken as the initialisation for incremental clustering at t_i. After each iteration, the clustering structure ζ_i at t_i is compared with that of ζ_{i-1} at t_{i-1} (c.f. Line6). The comparison is done using the Jaccard Co-efficient [10]. If the calculated value is less than a user defined threshold τ, the current clustering ζ_i is discarded and re-clustering is performed over the $\mathcal{W}_i$.

When model a is to be created from scratch i.e. during re-clustering, we create a set of models $C = \{\zeta^1, \zeta^2 \ldots \zeta^q\}$ with different initial centroids. The value of $q =$ MAX_MDL. The quality of each model is evaluated and the best one is chosen. The *sum of squared errors* (SSE) or *silhouette coefficient* if the k is small, can be used as a measure for quality [10].

As we have already mentioned, our method is similar to Online K-Means [6]. We would like to point out the differences here. Online K-mean works with multiple time-series. Where as focus of our method is on clustering of growing objects that come from a multi-table stream and only the latest version of the object is considered. Online K-Means only considers flats objects, while our method allows adding and deleting objects referenced by them. It can also handle the addition and deletion of target objects and does so smoothly by preferring objects that are likely to hold more information over those that do not (c.f. Section 3.1). Unlike Online K-Means our method does not allow for dynamic updating of the number of cluster over time. We rely on quality comparison of the clustering ζ_i at t_i with that of ζ_{i-1} at t_{i-1}.

4   Experiments

In [9], we used two datasets for our evaluation. Here our objective is to study the performance of our method over dynamic objects, so we use only the Financial dataset[1]. Since Financial dataset contains labelled data, we have tested our strategies against the ground truth. We designed a variety of experiments that deals with the effect of window size on quality.

			Card
w Transaction 191,556			170
c District 77	C Account 682 (606 & 76)		
w Order 1513			Client 827

Fig. 1. Evaluation strategies

Table 1. Evaluation strategies

Acronym	REF	FIN1	FIN2	FIN3
Accounts	∞	100	200	300
District	∞	20	40	50
r	∞	3	3	3

In Fig. 1 we give the statistics of the dataset; the target table is highlighted. We mark with C each stream associated with a cache, while W stands for window. Already during the competition, the classes A&C and B&D were merged into *loan-trusted* and *loan-risk* respectively. We do the same.

This dataset puts forwards a difficult learning problem. The class distributions are not only very skewed to begin with, they also reflect the state of accounts when they have matured. Class labels become applicable later than when the objects were introduced. Because of this it is infeasible to propagate labels to the beginning of time. To exploit them efficiently, we use the last 30 months from the stream. This chunk of stream is repeated three times.

4.1   Experimental Settings

Our hypothesis is that the amount of information remembered as the multi-table stream progresses has an impact upon the quality of the clustering results. The specification of cache size and reserved columns is a "cache strategy". The strategies we used are given in Table 1.

Our reference strategy has unlimited storage and knows the future. Also, the number of reserved columns is large enough to accommodate all nominal values that will come in the future. For clustering we used the cosine similarity with $K = 9$. With the above strategies we report experiments with two different settings. In the first one we consider the whole of transaction stream and use a window size of $w = 30$. For the second one, we consider transactions from

---

[1] http://lisp.vse.cz/challenge/

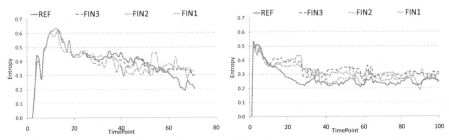

Fig. 2. Entropy for $K = 9$ and $\tau = 0.7$ **(left)** $w = 30$ **(right)** $w = 24$

last 30 months only. The window size and value of τ is the varied for this set of experiments. We used window $w = 24, 18$ and $\tau = 0.7, 0.8$. As the data is labelled, we use *entropy* to evaluate a clustering against explicit class labels [10].

4.2   Experimental Results

Each account arrives with zero transactions and *evolves* into either "loan-trusted" or "loan-risk" class. To avoid learning our models on data that arrive early but are not relevant (and would thus blur our results in an undisciplined way), we have trained a classifier (J4.8 [11]) to identify a subset of predictive attributes to reduce the noise.

In the left of Figure 2 we show the entropy for each strategy. At the beginning, as there are less accounts and all belong to the same class, all strategies have an entropy value of zero. As accounts from the other class arrive, the entropy rises sharply.

It must be stressed here that accounts are dynamic objects. Initially, as there is little or no transaction information associated, they are clustered on the basis of their static properties. As the class labels reflect their *final* state i.e. after many transactions have been done, strategies perform poorly.

Around timepoint t_{10}, the entropy of FIN1 (with the smallest cache) starts dropping. Because of fixed cache size accounts that have more transactions are preferred. For FIN1 this means that accounts with fewer transactions are not in the cache. The other strategies have larger caches and also store the accounts which cannot be easily classified. At timepoints t_{23} and t_{30}, FIN2 and FIN3 reach their cache limits, respectively. As they keep only mature accounts inside the cache, their performance improves. From timepoint t_{32} until t_{55}, the reference strategy with its infinite cache shows the worst performance. The lesson learned is that in stream mining it is not always desirable to remember all the data. For the Financial dataset, oblivion is best: FIN1 that has the smallest cache size outperforms all other strategies.

The last account arrives at t_{60}. After that existing accounts keep evolving as new transactions arrive. Since there are no noisy accounts and no information loss due to memory limitations, the reference strategy outperforms all others.

In the right side of Figure 2, drawn for $K = 9$, $w = 24$ and $\tau = 0.70$, we show the entropy for each cache strategy. Because of the large window size (i.e. from

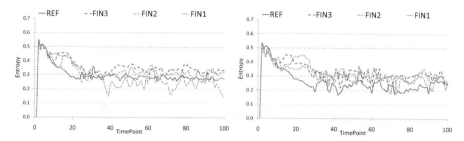

Fig. 3. Entropy for $K = 9$ **(left)** $w = 18, \tau = 0.70$ **(right)** $w = 24, \tau = 0.85$

last 30 timepoints), almost all of the available information gets enveloped. As the transactions accumulate, FIN1 and FIN2 are the first to show improvement. However, after timepoint t_{12} they are over taken by the reference. From t_1 to t_{12} about 400 accounts are active. The main advantage that the smaller strategies draw comes from their ability to prefer objects that have grown substantially and are likely to carry more information. Due to the richness of information in the subset of transactions (i.e. last 30 timepoints), almost all of these accounts are growing simultaneously by the timepoint t_{15}. Therefore the reference strategy that can cache all objects shows best performance.

By timepoint t_{30} all accounts mature. As the transactions repeat, the reference strategy shows a strong periodic behaviour with lowest entropy. FIN2 is the second most competitive strategy during timepoints $t_{30} \ldots t_{60}$. FIN1 also has a periodic behaviour and shows improved performance as the transaction data are repeated.

In the left side of Figure 3, we show the entropy of cache strategies drawn for $K = 9$, $w = 18$ and $\tau = 0.70$. Till timepoint t_{18} performance is similar to that for $w = 24$. The reference strategy shows best. As the window size is reached at t_{18}, the reference stabilises and does not show any significant improvement after that. By timepoint t_{30} all strategies have somewhat comparable performance. As the transactions are repeated, the performance of FIN1 starts improving. As we have pointed out earlier, the cache strategies draw advantage by focussing on objects that are more mature than others. Because of the smaller window size, accounts with less transactions and contain less information are dropped. FIN1 improves its performance by focusing on the accounts that have done more transactions and shows shows the best performance as the stream gets repeated over and over.

In the right side of Figure 3, we show the entropy of cache strategies drawn for $K = 9$, $w = 24$ and $\tau = 0.85$. The graph is comparable to Figure 3(right), however is a bit more fragile because it is re-clusters more due to higher value of $\tau = 0.85$. The strategies, specially the reference, show slightly better performance at various timepoint due to stricter threshold.

5   Conclusion

Our method first transforms a multi-table stream, which contains dynamic objects, into a single stream. This generated stream is passed to incremental version of k-means clustering algorithm that can handle objects that grow over time. The clustering quality is monitored at each timepoint and if it drops below a certain threshold, re-clustering is performed. To study the performance of our approach we have designed a reference strategy that knows the future and has unlimited resources. We have shown that our approach approximates the reference well and even outperforms it in those cases where oblivion is preferable. Oblivion, expressed through small cache sizes, means here remembering grown objects and forgetting those that contain little information. This kind of oblivion competitive even in the case of periodicity.

As a next step, we want to study the potential of data sampling, investigate more elaborate caching strategies and minimize the schema size for the incremental propositionalisation algorithm. We further intend to extend the incremental clustering algorithm to dynamically update the number of cluster towards values that ensure a better model.

References

1. Guha, S., Meyerson, A., Mishra, N., Motwani, R., O'Callaghan, L.: Clustering data streams: Theory and practice. IEEE TKDE 15(3), 515–528 (2003)
2. Bradley, P.S., Fayyad, U.M., Reina, C.: Scaling clustering algorithms to large databases. In: KDD, pp. 9–15 (1998)
3. Farnstrom, F., Lewis, J., Elkan, C.: Scalability for clustering algorithms revisited. SIGKDD Explorations 2, 51–57 (2000)
4. Streaming-Data Algorithms for High-Quality Clustering. In: IEEE ICDE (2001)
5. Aggarwal, C., Han, J., Wang, J., Yu, P.: A framework for clustering evolving data streams. In: Proc. of Int. Conf. on Very Large Data Bases, VLDB 2003 (2003)
6. Beringer, J., Huellermeier, E.: Online clustering of parallel data streams. Data & Knowledge Engineering 58(2), 180–204 (2006)
7. Dai, B.R., Huang, J.W., Yeh, M.Y., Chen, M.S.: Adaptive clustering for multiple evolving streams. IEEE TKDE 18(9), 1166–1180 (2006)
8. Elghazel, H., Kheddouci, H., Deslandres, V., Dussauchoy, A.: A partially dynamic clustering algorithm for data insertion and removal. In: Corruble, V., Takeda, M., Suzuki, E. (eds.) DS 2007. LNCS (LNAI), vol. 4755, pp. 78–90. Springer, Heidelberg (2007)
9. Siddiqui, Z.F., Spiliopoulou, M.: Combining multiple interrelated streams for incremental clustering. In: Proceedings of SSDBM 2009 (2009)
10. Tan, P.N., Steinbach, M., Kumar, V.: Introduction to Data Mining. Wiley, Chichester (2004)
11. Witten, I.H., Frank, E.: Data Mining: Practical machine learning tools and techniques, 2nd edn. Morgan Kaufmann, San Francisco (2005)

Finding the *k*-Most Abnormal Subgraphs from a Single Graph

JianBin Wang, Bin-Hui Chou, and Einoshin Suzuki⋆

Department of Informatics, ISEE, Kyushu University, Fukuoka 819-0395, Japan
{jianbin.wang,chou}@i.kyushu-u.ac.jp, suzuki@inf.kyushu-u.ac.jp
http://www.i.kyushu-u.ac.jp/{~suzuki/choue.html, ~suzuki}

Abstract. In this paper, we propose a discord discovery method which finds the *k*-most dissimilar subgraphs of size *n* among the subgraphs of the same size of an input graph, where the values of *k* and *n* are given by the user. Our algorithm SD3 (Subgraph Discord Detector based on Dissimilarity) exploits a dynamic index structure and its effectiveness is demonstrated through experiments using graph data in chemical-informatics and bioinformatics.

1 Introduction

[1,2] have introduced a new problem of finding the *k*-most abnormal subsequences of size *n* to other subsequences of the same size from a time series sequence and proposed efficient solutions. The abnormal subsequences are called discords and are defined as the subsequences that have the *k*-largest distances to the corresponding nearest subsequences among all subsequences of length *n* in a given time-series sequence. The discord discovery requires only two parameters, i.e., *k* and *n*, in its definition and models anomalies in various domains.

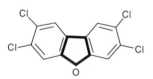

Fig. 1. Example of a discord in a chemical graph

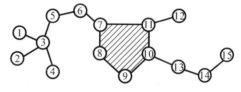

Fig. 2. Examples of self-match

In this paper, we tackle a problem of finding the top *k*-discords which are defined as the subgraphs that are *k*-most dissimilar among all subgraphs of size *n* from a given graph and propose an efficient solution. Figure 1 shows an example of a graph and its discord. Since the chemical structure is left-right

---

⋆ A part of this research was supported by the Strategic International Cooperative Program funded by Japan Science and Technology Agency (JST) and the grant-in-aid for scientific research on fundamental research (B) 21300053 from the Japanese Ministry of Education, Culture, Sports, Science and Technology.

J. Gama et al. (Eds.): DS 2009, LNAI 5808, pp. 441–448, 2009.

symmetric, the most distinct subgraph of the graph is intuitively considered to be the pentagon in bold, which matches our definition. The discord can be found by a brute force search, which is essentially an "all-to-all" comparison. As the required time-complexity is quadratic to the number of the substructures, this approach is prohibitive when the number of substructures is huge.

To circumvent the corresponding problem, [1] have proposed two heuristics and a data structure and have achieved three to four orders of magnitude speedup. Obviously the sliding window, which is successfully used in [1] for time series data, is not directly applicable to our problem. Moreover, subgraph isomorphism is an NP problem, so a cautious implementation is necessary for an efficient solution. For these problems, we propose an efficient algorithm using a dynamic index structure.

Few works [3,4,5,6,7] have tackled anomaly detection from graph data. [5] used the MDL principle and the idea that subgraphs containing many common substructures are generally less anomalous than subgraphs with few common substructures. Three algorithms for graph-based anomaly detection were proposed in [3]. Two of them are based on the idea of "cost of transformation" which is similar to the edit distance, and they use the MDL principle as [5]. The third algorithm is a probabilistic algorithm. Our method shares the fundamental motivation but adopts a more intuitive notion of similarity between subgraphs.

Sun et al. [4] focus on anomalous nodes by computing the normality of each node to the other nodes. [7] discovers unusual links, paths, loops, and significant nodes. Our method is different from [4,7] as we discover substructures. By using a matrix which represents the relation between edges and substructures, an approach called Grafil [8] was proposed for graph similarity filtering. Unlike us, Grafil is based on the number of selected features contained in the query graph.

2   Discord Discovery Problem from a Single Graph

A graph G is represented by a tuple $G = < V, E >$, where V is a non-empty finite set of vertices and its size is defined as the cardinality $|V|$ of V. E is a set of edges where an edge is a binary relation of an unordered pair of distinct elements of V. In this paper, a subgraph of a graph is its induced subgraph.

A graph g of size n may be transformed into a string of length $(_nC_2 + n)$ called a canonical form, where $_nC_2$ represents the binary coefficient. The first $_nC_2$ symbols represent the string obtained by concatenating the upper triangular elements of the adjacency matrix of the graph when the matrix has been symmetrically permuted such that this string becomes the lexicographically smallest string among the strings obtained from all such permutations [9]. The latter n elements of the string consist of the attributes of the nodes.

For the definition of the abnormality, we first consider introducing the degree of similarity between graphs. In [1], the Euclidean distance is used for measuring the degree of dissimilarity between subsequences of time series data but it cannot be used for graph data in a straightforward manner. The Hamming distance between a pair of canonical forms is inadequate due to the permutation of the order of the nodes in obtaining a canonical form, which results in counter-intuitive results. We may transform graphs into vectors in an n-dimensional space and use the graph edit distance as in [10], but the issue of appropriately selecting prototypes must be resolved. In [11], the metric graph space is equipped with the Euclidean distance of graphs induced by the Shur-Hadamard Inner product, which is not intuitive and depends on the weights of attributes.

We define the similarity degree $s(g, h)$ of two graphs g and h as $s(g, h) \equiv |c(g, h)|$, where $c(g, h)$ represents the maximum common subgraph of g and h. This similarity degree is considered to be natural as its high value implies that g and h share a large portion of themselves. To formally define our problem, we need a degree $\Delta(g, G, GS)$ of similarity of a subgraph g of size n in the set GS of all subgraphs of size n of G. Simply using $|c(g, h)|$ for $\forall h \in GS$, $h \neq g$ may give counterintuitive results as we show in Figure 2 with $k = 1, n = 6$. The pentagon connected with a node may appear abnormal but as the graph contains 3 subgraphs each of which consists of the pentagon and a node, and thus it is not judged as a discord. To resolve this problem we borrow the concept of self-match from [1]. Given two subgraphs g_i, $g_j \in G$, if and only if g_i and g_j share at least one vertex, we say that g_i is a self-match to g_j. We define $\Delta(g, G, GS) \equiv |c(g, h)|$ where $h \in GS$ and h is not a self-match to g.

Given a graph G, we define the subgraphs $g_1, g_2, \ldots, g_k$ which have the k smallest $\Delta(g_i, G, GS)$ as the top k discords. Our discord discovery problem is defined as, given the values for n, k, and a graph data G, to output the top k discord. The lowest ranked tie-breaks may not be output to keep the number of discords no more than k.

3   Our Algorithm SD3

3.1   SD3 (Subgraph Discord Detector Based on Dissimilarity)

Given all subgraphs of size n of G, we begin by comparing their subgraphs of size $(n - 1)$ and eliminating the most similar subgraphs that share at least a subgraph of size $n - 1$. If the number of subgraphs which are not eliminated yet is larger than k, we iterate the same process for the size $(n-2), (n-3), \ldots$ until the number of the remaining candidates is no greater than k. Figure 3 shows a simple example of five subgraphs $g_1, g_2, \ldots, g_5$ of size seven with $k = 1$. The subgraphs in the middle column are their subgraphs of size six. g_1 and g_2 are most similar as they share a subgraph of size six and thus they are both marked as non-candidates, and so g_3 does. The graphs in the rightmost column are the subgraphs of $g_1, g_2, \ldots, g_5$ of size five. Similarly g_4 is marked as a non-candidate and finally g_5 is output as the top 1 discord.

The pseudo code of SD3 is shown below. SD3 can be decomposed into two phases: the first phase lists out the subgraphs of G as their canonical forms, which are the candidates of the discords. The second phase compares the subgraphs of the candidates and finds out the discords by eliminating similar candidates.

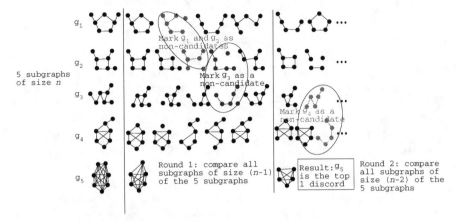

Fig. 3. An example of finding the most abnormal graph in five subgraphs $g_1, g_2, \ldots, g_5$. Note that g_5 is a clique and thus has only one kind of subgraph for each size

```
algorithm SD3
```
INPUT. G: the input graph; n: the size of a discord; k: the maximum number of discords for output;

OUTPUT. T: the set of top d discords $(d \le k)$

   0. Canonical-form-trie $C \leftarrow$ empty

   1. FOR EACH subgraph g_i of G, $|g_i| = n$

      A. Transform g_i into its canonical-form c_i

      C. IF c_i does not exist in C // index by C

         Add c_i into C, then add g_i as a tag in the tail of c_i

         Add g_i and c_i into Table T and c_i.status $\leftarrow$ candidate;

      ELSE

         Add g_i in the list in the tail of c_i which contains its identical subgraph c_j.status $\leftarrow$ non-candidate

   2. FOR $x=n$-1 to 3 //because all subgraphs are identical when $x = 1$ or 2

      A. FOR EACH $c_i \in T$, c_i.status = candidate

         a. FOR EACH $c_j \in T$, $i \ne j$

            (i). IF subgraph c_j is not a self-match to c_i AND $\exists c_{ip}$ matches with $\exists c_{jq}$, where $|c_{ip}| = |c_{jq}| = x$, c_{ip} is a subgraph of c_i, and c_{jq} is a subgraph of c_j

               c_i.status $\leftarrow$ non-candidate; c_j.status $\leftarrow$ non-candidate;

      B. IF the number of candidates in $T \le k$

         BREAK;

   3. $T' \leftarrow \emptyset$

      FOR EACH $c_i \in T$

         IF c_i.status = candidate

            $T' \leftarrow T' \cup \{c_i\}$

   4. Return T'

In the first phase we list out all the possible subgraphs which are of size n, then we transform them into canonical forms in step 1. This approach is similar to [1], in which time series data are transformed into strings. After the transformation, we mark the canonical forms which have no duplicate to be candidates, and mark other canonical forms as non-candidates to make an initial candidate list stored in a table T. Then in the second phase we keep marking the candidates once we found that thay have bigger $\Delta(g, G, T)$ as non-candidates, until the number of candidates in T is no greater than k.

A major difficulty of graph mining comes from the enormous number of subgraphs even for a moderate size of graph (e.g., in our experiments, 261073 subgraphs of size 9 were found from a graph of size 409). For this problem, we adopt two strategies, the first one exploits data structures to speed-up the search process. The node identifiers recorded in a trie and an adjacency list of the graph G accelerate the listing in the first phase, and linked lists in ascending order are used to speed up the non-self-match checking and the duplicate checking. The second strategy exploits a dynamic index based on the trie data structure. We mark a subgraph as a candidate only when it has no duplicate.

We use this trie, which is called a canonical form trie, for efficiently judging whether we already have a subgraph which is transformed into the same canonical form. This trie is also used as an index in the phase 2 for checking whether the currently processed candidate is a self-match of any member in a specific non-candidate subgraph. Figure 4 shows an example of the canonical form trie. The trie contains, from its root to its leaves, the information of edges, the attributes of the nodes, and the subgraphs. The former two of a subgraph is its canonical form, where 1 and 0 in the edge information represent the existence and the absence of an edge, respectively. The table T stores the initial candidate list for further processing in the phase 2. In the table, NC and C represent a non-candidate and a candidate, respectively. We dynamically update the result in the table T where we store the current candidates of discord.

Phase 2 starts from step 2. From step 2, we start to sieve the candidates in T by checking pairs of subgraphs which are not self-match. For eliminating candidates g' from the candidates list in T, we begin by checking the subgraphs of size $x = (n - 1)$ of each subgraph of size n, then continue this checking in descending order. Once we found a candidate g_i has a subgraph of size x matching to a subgraph of g_j which is a member in T, which means the similarity of this candidate is $\Delta(g_i, G, T) = x$, then we mark g_i and g_j non-candidate. In step 2, the candidate list for each value of x is also output. This intermediate result is expected to help the user to settle a more appropriate value of k if necessary. For instance, if no discord is discovered, the user can use the intermediate result to increase the value of k.

3.2   Examples of Using the Dynamic Index Structure

In the phase 2, the brute force algorithm will perform "all-to-all" comparison for all subgraphs of G. If the number of the subgraphs is s, the complexity is $O(s^2{}_nC_2)$, which is unacceptable because s is typically huge. We use the trie

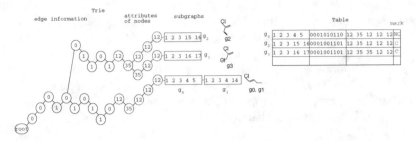

Fig. 4. Example of the trie and the table of Figure 1 in the phase 1

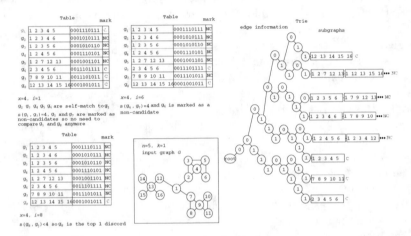

Fig. 5. Example of the updates of the trie and the table in the phase 2

data structure in each phase and the trie in the phase 1 improves the efficiency of checking whether the current subgraph is already listed out. In the step 1.C in the pseudo code, if the canonical form is new in the trie, we add it into the trie and table T and we mark the canonical form as a candidate. Otherwise we add it into the trie and mark the identical one in T as a non-candidate. Figure 4 shows an example of the trie and the table, where subgraphs g_0, g_1, g_2, g_3 are examined, in the phase 1. In this example, g_1 is identical to g_0. Consequently, g_0 is marked as a non-candidate in the table and g_1 is inserted right after g_0 in the trie.

The trie in the phase 2 is used in cooperation with a table which stores the candidates list of discords to update the content. In Figure 5, an example is shown for demonstrating the trie in the step 2, where $x = 4$. In the first table of Figure 5, four candidates g_1, g_6, g_7, g_8 and four non-candidates g_2, g_3, g_4, g_5 are stored in table T and $i = 1$ (cf. phase 2 in the pseudo code). Here each of g_2, g_3, g_4, g_5, g_6 is a self-match to g_1 and thus are not considered and it turned out $s(g_1, g_7) = 4$. Therefore, both of g_1 and g_7 are marked as non-candidates and we do not need to compare g_8 with g_1 for round $i = 1$, since g_1 is no longer a candidate. The second table is processed similarly and it turned out $s(g_6, g_7) = 4$

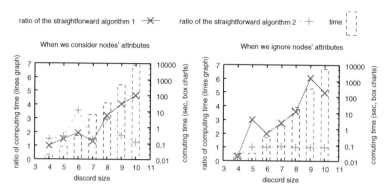

Fig. 6. Results of comparison to straightforward algorithms

and thus we mark g_6 as a non-candidate. In the third table, $s(g_8, g_j) < 4$ and consequently g_8 is output as the top 1 discord.

4 Experiments

We perform our experiments on a PC with Vine Linux 4.2, Intel(R) Core(TM)2 Quad CPU Q6600 2.40GHz, 3 GB memory. DNA fragment data was used as the input graph G. We omitted one particular chemical bond $e(164, 165)$ to mimic a damage in DNA. The size of G is only 166 but the number of subgraphs is 12036, which makes this problem relatively challenging. As the DNA data is composed of 6 kinds of substructures, i.e., Adenine, Thymine, Guanine, Cytosine, 3'end, and 5's end, and each of these substructures have at least one duplicate in G, the subgraph which contains the omitted bond should be found as anomaly. Actually, this subgraph was detected as the discord with $n = 10, k = 1$ as expected[1].

Using this data, we compare SD3 with two straightforward algorithms. The first one is identical to SD3 except for a pruning of the canonical form transformation. The other one just performs the quadratic comparison among all subgraphs. As shown in Figure 6, we tested the cases when the attributes of the nodes are ignored and considered. A ratio of a straightforward algorithm represents its computation time divided by that of SD3. A higher ratio represents a superior performance of SD3 to the corresponding straightforward algorithm. A dotted box represents the computation time of SD3 in logarithmic scale.

It is obvious that the number of candidates of discords in the former case is larger than that in the latter case. As shown in the left plot, the two speed-up methods of SD3 are almost equally effective in the former case while in the latter case, the acceleration of phase 1 is much more important, which explains the higher ratios of the straightforward algorithm 1. We can see that SD3 is about 3 times faster than the two straightforward algorithms when $n = 8$ and the

---

[1] We applied SD3 to a graph G' obtained by omitting two more bonds, which are $e(73, 74)$ and $e(106, 107)$, and G' is a disconnected graph. As expected, SD3 is still able to effectively find out the discord.

attributes of the nodes are considered. It is 4 to 6 times faster when the attributes of the nodes are ignored, $n = 9$ or $n = 10$. In another series of experiments presented below, SD3 is 4.2 times faster than the first straightforward algorithm when $n = 9$.

We also applied SD3 to a fragment of a chemical structure data of a virus protein. The input graph G is a fragment of a chemical structure data of a dengue virus protein, where $|G| = 409$. In G, every 4 to 14 chemical element consists of biological substructure defined by experts such as LYS, GLN, PRO. SD3 detects a portion of TRP when $n = 9$, $k = 6$ and the attributes of the nodes are ignored. As there exists only one TRP is G, this result makes sense. It should be noted that the node number is 2.46 times of the first experiment while the running time is 37.7 times of the corresponding case. We believe that this is an excellent result as the number of subgraphs is huge with this G.

In SD3, the canonical transformation, i.e., the phase 1, dominates the computation time when n is large. It should be noted that each subgraph requires $n!$ processing for determining its canonical form. In the experiments, we found that usually over 99% of running time is spent for the transformation of canonical forms for large n when the attributes of the nodes are ignored. The computation time is more than one day if the size of discord is over 11 with the DNA fragment data. Using other definitions of subgraphs and matching is left for future work.

References

1. Keogh, E., Lin, J., Lee, S.H., Herle, H.V.: Finding the Most Unusual Time Series Subsequence: Algorithms and Applications. KAIS 11(1), 1–27 (2006)
2. Bu, Y., Leung, O.T.W., Fu, A.W.C., Keogh, E.J., Pei, J., Meshkin, S.: WAT: Finding Top-K Discords in Time Series Database. In: Proc. Seventh SDM (2007)
3. Eberle, W., Holder, L.B.: Discovering Structural Anomalies in Graph-based Data. In: Proc. 7th IEEE Int. Conf. on Data Mining Workshops, pp. 393–398 (2007)
4. Sun, J., Qu, H., Chakrabarti, D., Faloutsos, C.: Relevance Search and Anomaly Detection in Bipartite Graphs. ACM SIGKDD Explorations 7(2), 48–55 (2005)
5. Noble, C.C., Cook, D.J.: Graph-based Anomaly Detection. In: Proc. 9th ACM SIGKDD Int. Conf. on Knowledge discovery and data (KDD), pp. 631–636 (2003)
6. Chakrabarti, D.: AutoPart: Parameter-free graph partitioning and outlier detection. In: Boulicaut, J.-F., Esposito, F., Giannotti, F., Pedreschi, D. (eds.) PKDD 2004. LNCS (LNAI), vol. 3202, pp. 112–124. Springer, Heidelberg (2004)
7. Lin, S., Chalupsky, H.: Unsupervised Link Discovery in Multi-relational Data via Rarity Analysis. In: Proc. 3rd ICDM, pp. 171–178 (2003)
8. Yan, X., Yu, P.S., Han, J.: Substructure Similarity Search in Graph Database. In: Proc. ACM Int. Conf. on Management of Data (SIGMOD), pp. 766–777 (2005)
9. Cook, D.J., Holder, L.B.: Mining Graph Data. John Wiley & Sons, Chichester (2007)
10. Riesen, K., Neuhaus, M., Bunke, H.: Graph Embedding in Vector Spaces by Means of Prototype Selection. In: Escolano, F., Vento, M. (eds.) GbRPR 2007. LNCS, vol. 4538, pp. 383–393. Springer, Heidelberg (2007)
11. Jain, B.J., Wysotzki, F.: Central Clustering of Attributed Graphs. Machine Learning 56(1-3), 169–207 (2004)

Latent Topic Extraction from Relational Table for Record Matching

Atsuhiro Takasu[1], Daiji Fukagawa[1], and Tatsuya Akutsu[2]

[1] National Institute of Informatics, 2-1-2 Hitotsubashi, Chiyoda, Tokyo, Japan
[2] Kyoto University, Gokasho, Uji, Kyoto 611-0011, Japan
{takasu,daiji}@nii.ac.jp, takutsu@kuicr.kyoto-u.ac.jp

Abstract. We propose a latent feature extraction method for record linkage. We first introduce a probabilistic model that generates records with their latent topics. The proposed generative model is designed to utilize the co-occurrence among the attributes of the record. Then, we derive a topic estimation algorithm using the Gibbs sampling technique. The estimated topics are used to identify records. The proposed algorithm works in an unsupervised way; i.e., we do not need to prepare labor-intensive training data. We evaluated the proposed model using bibliographic records and proved that the proposed method tended to perform better for records with more attributes by utilizing their co-occurrence.

1 Introduction

Record linkage is an important task in information integration and data mining. It has a wide range of applications such as citation matching, name disambiguation, and data cleaning. Although a lot of research has been conducted on this task, it is still a challenging problem due to a variety of discrepancies.

Record similarity plays a very important role in record linkage. It is usually measured by combining the similarities at the level of attributes of the record. Researchers have applied machine learning techniques to both the similarity measurement at the attribute level and the combination of similarities. For example, Bilenko et al. [2] proposed a linkage method that uses a learnable string similarity that is applied to the similarity measurement at the attribute level. Supervised learning techniques such as support vector machines are frequently used to combine the similarities at the attribute level.

There are two kinds of similarities at the attribute level: notational and semantic similarities. Attribute values are represented in various ways due to abbreviations, acronyms, etc. For example, the "International Conference on Data Mining" is usually abbreviated "ICDM". To handle this notational similarity, various kinds of string similarities are used [2,5].

Semantic similarity is also useful for record linkage. For example, "ICDM" and "KDD" are similar, but are different conferences. Latent topics are often used in information retrieval to handle this semantic similarity. For example, the latent semantic index (LSI) extracts latent topics from documents using singular value

J. Gama et al. (Eds.): DS 2009, LNAI 5808, pp. 449–456, 2009.

Table 1. Example of records for authors

Author Name	Journal	Affiliation	Abstract
M. Jordan	ACM TKDD	A Univ.	{linkage, generative model}
M. Jordan	IEEE TKDE	A Univ.	{LDA, EM algorithm}
M. Jordan	Physical Review	A Univ.	{relativity, quantum}
M. Jordan	ACM TKDD	B Univ.	{HCI, user model}

decomposition. The probabilistic latent semantic index (PLSI) has features that are similar to LSI using a statistical framework. Latent topics usually create low dimensional feature space, and semantically similar values tend to be mapped onto neighboring points in the feature space. In this way, latent topics are used to handle a kind of semantic similarity.

Researchers have mainly been handling the notational similarity in record linkage. However, some researchers have recently introduced latent topic models for the record linkage. For example, Bhattacharya and Getoor [1] proposed a probabilistic model that exploited the co-authorship of papers. Song et al. [9] introduced a latent topic model that generated both authors and a document written by them from a latent topic. Shu et al. [8] proposed a latent topic model that is similar to Song's model and applied it to three kinds of entity resolution problems, namely, the name sharing, name variant, and name mixing problems. These methods used generative models that are an extension of the latent Dirichlet allocation (LDA) [3] to exploit the co-occurrence of multiple entity names or the co-occurrence of entity names and related documents.

This paper proposes a new latent topic model for record linkage. It is an extension of the LDA, like many other latent topic models [1,9,8], however, it focuses on the data in the record structure consisting of multiple attributes and utilizes the co-occurrence of their attribute values.

2   A Record Model

2.1   Notations

We first define the concepts and notations using the example in Table 1. In a relational table, one attribute is selected as the target for identification. We refer to this attribute as a *target attribute*. In Table 1, "Author Name" is the target attribute. An attribute that has a single value is called a *scalar attribute*. The "Author Name", "Journal", and "Affiliation" in Table 1 are examples of scalar attributes. An attribute that has a text value is called a *text attribute*. The "Abstract" in Table 1 is an example of a text attribute. In the proposed record model, we handle a table consisting of a target attribute, a set of scalar attributes, and a set of text attributes.

To handle the semantic similarity such that "ACM TKDD" and "IEEE TKDE" are journals from a similar research field, we introduce a latent topic $\tilde{r}$ for each attribute value r. For example, $\tilde{r}$ is "Data Mining" for an attribute value $r =$ "ACM TKDD". For a table consisting of a target attribute, n scalar attributes, and $m - n$ text attributes, we represent a record using

$$\boldsymbol{r} = (r_0, r_1, \cdots, r_n, \boldsymbol{r}_{n+1}, \cdots, \boldsymbol{r}_m, \tilde{r}_0, \tilde{r}_1, \cdots, \tilde{r}_n, \tilde{r}_{n+1}, \cdots, \tilde{r}_m) \qquad (1)$$

where r_i and $\tilde{r}_i$ $(0 \le i \le n)$ stand for the observed attribute value and its corresponding latent topic of the ith scalar attribute, respectively. $\boldsymbol{r}_i = (r_{i1}, \cdots, r_{il_i})$ and $\tilde{r}_i$ $(n < i \le m)$, on the other hand, denote the bag of words and their corresponding latent topic of the ith text attributes, respectively. The symbol l_i represents the number of words in $\boldsymbol{r}_i$. Although l_i is a function of a record $\boldsymbol{r}$, we omit $\boldsymbol{r}$ to simplify the description.

For the ith attribute, we use D_i and $\tilde{D}_i$ to denote the set of attribute values and latent topics, respectively. $|D_i|$ and $|\tilde{D}_i|$ denote their cardinality.

2.2 Probabilities of the Record Model

The record model is a probabilistic model that generates a table with a certain probability. In this section, we introduce three kinds of probability distributions for defining the probability that a record $\boldsymbol{r}$ is produced by the model.

A *target probability*, denoted as $p(\tilde{v})$, is the probability that a topic of the target attribute is $\tilde{v}$ in $\tilde{D}_0$.

The second probability is the conditional probability of a topic of the ith attribute for a topic of the target attribute. For a topic $\tilde{v}$ in $\tilde{D}_0$ and a topic $\tilde{w}$ in $\tilde{D}_i$ of the ith attribute, it is denoted as $p(\tilde{w} \mid \tilde{v})$. We refer to this probability as a *topic correlation probability*. It is introduced to propagate the topic information among attributes.

The third probability is the conditional probability of an attribute value for a topic for each attribute. For a topic $\tilde{v}$ in $\tilde{D}_i$ and an attribute value v in D_i, it is denoted as $p(v \mid \tilde{v})$. This probability is referred to as a *topic-value probability*.

Note that topics are not specific values such as "data mining" but are like those in the principal components of PCA and the topics of LSI and PLSI. Actually, we do not need to define specific topics, but only to determine how many there are.

2.3 A Generative Model

We assume that the three kinds of probabilities introduced in the previous subsection are multinomial distributions generated by a Dirichlet distribution. Then, the generative model consists of two phases, the generation of multinomial distributions and a set of records.

In the first phase, the model generates the parameters of the target, the topic correlation, and topic-value probabilities by using the following steps:

1. Choose $\boldsymbol{\theta}_0 \sim \text{Dir}(\boldsymbol{\alpha}_0)$ // target probability
2. For each topic $\tilde{v}$ of the target attribute and each ith attribute $(1 \le i \le m)$:
   (a) Choose $\boldsymbol{\theta}_{i\tilde{v}} \sim \text{Dir}(\boldsymbol{\alpha}_i)$ // topic correlation probability

3. For each ith attribute $(0 \le i \le m)$: // topic-value probability
   (a) For each topic $\tilde{v}$ of the ith attribute:
       i. Choose $\phi_{i\tilde{v}} \sim \text{Dir}(\beta_i)$

where $\text{Dir}(\alpha)$ denotes the Dirichlet distribution of the parameter α. The first step generates the parameters θ_0 of a multinomial distribution of the target probability according to $\text{Dir}(\alpha_0)$. Similarly, the second and third steps generate the parameters of the topic correlation and topic-value probabilities for each attribute, respectively.

In the second phase, the model generates each of the specified number of records using the multinomial probability distributions generated in the first phase:

1. Choose a topic $\tilde{r}_0 \sim \text{Multi}(\theta_0)$
2. Choose a value $r_0 \sim \text{Multi}(\phi_{0\tilde{r}_0})$
3. For each ith scalar attribute $(1 \le i \le n)$:
   (a) Choose a topic $\tilde{r}_i \sim \text{Multi}(\theta_i)$
   (b) Choose a value $r_i \sim \text{Multi}(\phi_{i\tilde{r}_i})$
4. For each ith text attribute $(n < i \le m)$:
   (a) Choose a topic $\tilde{r}_i \sim \text{Multi}(\theta_i)$
   (b) For each word of the attribute:
       i. Choose a word $r_{ij} \sim \text{Multi}(\phi_{i\tilde{r}_i})$,

where $\text{Multi}(\theta)$ denotes the multinomial distribution of parameter θ.

We denote the parameters of the Dirichlet distributions as

$$\Lambda \equiv \{\alpha_0, \cdots, \alpha_m, \beta_0, \cdots, \beta_m\} ,$$

and the parameters of the multinomial distributions as

$$\Delta \equiv \{\theta_0, \Theta_1, \cdots, \Theta_m, \Phi_0, \cdots, \Phi_m\} ,$$

where

$$\Theta_i = \{\theta_{i\tilde{v}} \mid \tilde{v} \in \tilde{D}_0\} \quad \text{for } 1 \le i \le m$$
$$\Phi_i = \{\phi_{i\tilde{v}} \mid \tilde{v} \in \tilde{D}_i\} \quad \text{for } 0 \le i \le m .$$

For a topic $\tilde{v}$ (resp. value v), we denote the component for the topic (resp. value) of the parameters of the Dirichlet (resp. multinomial) distribution α (resp. θ) as $\alpha[\tilde{v}]$ (resp. $\theta[v]$).

The probability that the record model generates a record r according to Δ is

$$p(r \mid \Delta) = \theta_{0\tilde{r}_0} \left[\prod_{i=1}^{m} \theta_{i\tilde{r}_0\tilde{r}_i} \right] \left[\prod_{i=0}^{n} \phi_{i\tilde{r}_i r_i} \right] \left[\prod_{i=n+1}^{m} \prod_{j=1}^{l_i} \phi_{i\tilde{r}_i r_{ij}} \right] , \qquad (2)$$

where r is a record represented by Eq. (1).

3   Parameter Estimation

The proposed algorithm uses the Gibbs sampling technique which is widely used for the Bayesian parameter estimation of complex statistical models (e.g., [7], [10]). For each record r and topic $\tilde{r}_i{}^1$ for an ith attribute, let $p(\tilde{v} \mid \boldsymbol{R}_{-\tilde{r}_i}; \Lambda)$ denote the posterior probability, i.e., the probability that the topic of the ith attribute is $\tilde{v}$ when we observe all the records $\boldsymbol{R}$ except for the topic $\tilde{r}_i$. In Gibbs sampling, we repeatedly assign a topic of each attribute of each record based on the above-mentioned posterior probability distribution until convergence.

We need to estimate three kinds of posterior probability distributions in the proposed model, one each for the target, scalar, and text attributes. We introduce three kinds of count functions to derive these posterior probability distributions. The first function, denoted as $C(\boldsymbol{R}; T_i = \tilde{v})$, stands for the frequency with which a topic of the ith attribute is $\tilde{v}$ in $\boldsymbol{R}$. Let us consider the table $\boldsymbol{R}$ in Table 2. Then, $C(\boldsymbol{R}; T_0 = 1) = 1$, because only the first record satisfies the condition $T_0 = 1$.

The second function, denoted as $C(\boldsymbol{R}; T_0, T_i = \tilde{v}, \tilde{w})$, stands for the frequency with which a topic of the target attribute is $\tilde{v}$ and a topic of the ith attribute is $\tilde{w}$ in a record.

The third function, denoted as $C(\boldsymbol{R}; T_i, V_i = \tilde{v}, v)$, stands for the frequency with which a topic of the ith attribute is $\tilde{v}$ and the corresponding attribute value is v. For a text attribute, the same value appears more than once in a record.

Using these functions, the marginal probability of a record set $\boldsymbol{R}$ is given as

$$p(\boldsymbol{R}; \Lambda) \equiv \int p(\Delta; \Lambda) \prod_{r \in \boldsymbol{R}} p(r \mid \Delta)\, d\Delta$$

$$= \left[\mathcal{D}(\boldsymbol{\alpha}_0) \frac{\prod_{\tilde{v} \in \tilde{D}_0} \Gamma(C(\boldsymbol{R}; T_0 = \tilde{v}) + \alpha_{0\tilde{v}})}{\Gamma(|\boldsymbol{R}| + \sum \boldsymbol{\alpha}_0)} \right]$$

$$\left[\prod_{i=1}^{m} \prod_{\tilde{v} \in \tilde{D}_0} \mathcal{D}(\boldsymbol{\alpha}_i) \frac{\prod_{\tilde{w} \in \tilde{D}_i} \Gamma(C(\boldsymbol{R}; T_0, T_i = \tilde{v}, \tilde{w}) + \alpha_{i\tilde{w}})}{\Gamma(C(\boldsymbol{R}; T_0 = \tilde{v}) + \sum \boldsymbol{\alpha}_i)} \right]$$

$$\left[\prod_{i=0}^{n} \prod_{\tilde{v} \in \tilde{D}_i} \mathcal{D}(\boldsymbol{\beta}_i) \frac{\prod_{v \in D_i} \Gamma(C(\boldsymbol{R}; T_i, V_i = \tilde{v}, v) + \beta_{iv})}{\Gamma(C(\boldsymbol{R}; T_i = \tilde{v}) + \sum \boldsymbol{\beta}_i)} \right]$$

$$\left[\prod_{i=n+1}^{m} \prod_{\tilde{v} \in \tilde{D}_i} \mathcal{D}(\boldsymbol{\beta}_i) \frac{\prod_{v \in D_i} \Gamma(C(\boldsymbol{R}; T_i, V_i = \tilde{v}, v) + \beta_{iv})}{\Gamma(C(\boldsymbol{R}; T_i = \tilde{v}) + \sum \boldsymbol{\beta}_i)} \right] \qquad (3)$$

where $\mathcal{D}(\boldsymbol{\alpha})$ denotes the coefficient of a Dirichlet distribution, i.e., $\frac{\Gamma(\sum_v \alpha_v)}{\prod_v \Gamma(\alpha_v)}$ whereas $\sum \boldsymbol{\alpha}$ is an abbreviation of $\sum_v \alpha_v$.

For a record r in $\boldsymbol{R}$, let $\boldsymbol{R}_{-r}$ denote the set of records consisting of those except for r, i.e., $\boldsymbol{R} - \{r\}$. Furthermore, $\boldsymbol{R}_{-r_{ij<}}$ denotes the set of records where $r_{ij}, \cdots, r_{il_i}$ are removed from r. For example, for the first record in Table 2, $\boldsymbol{R}_{-r_{22<}}$

---

^{1} Latent topic $\tilde{r}_{ij}$ is scalar in case of a text attribute.

Table 2. A relation

T_0	V_0	T_1	V_1	T_2	V_2
1	1	2	1	1	{2,3}
2	2	1	1	3	{1,3}
2	2	2	1	3	{1,1}

is a table where the first record is replaced with $(1, 1, 2, 1, 1, \{2\})$. Then, for each record r in R and a topic $\tilde{v} \in \tilde{D}_0$, we can derive Eq. (4) from Eq. (3) denoting the posterior probability that the topic of target attribute is $\tilde{v}$:

$$
\begin{aligned}
&p(T_0 = \tilde{v} \mid R_{-\tilde{r}_0}; \Lambda) \\
&\propto \frac{C(R_{-r}; T_0 = \tilde{v}) + \alpha_{0\tilde{v}}}{|R| - 1 + \sum \alpha_0} \frac{C(R_{-r}; T_0, V_0 = \tilde{v}, r_0) + \beta_{0r_0}}{C(R_{-r}; T_0 = \tilde{v}) + \sum \beta_0} \\
&\quad \prod_{i=1}^{m} \frac{C(R_{-r}; T_0, T_i = \tilde{v}, \tilde{r}_i) + \alpha_{i\tilde{r}_i}}{C(R_{-r}; T_0 = \tilde{v}) + \sum \alpha_i} .
\end{aligned}
\tag{4}
$$

Similarly, for each record r in R and a topic $\tilde{v} \in \tilde{D}_i$ of the ith attribute, Eq. (5) is the posterior probability that the topic of the ith scalar attribute is $\tilde{v}$:

$$
\begin{aligned}
&p(T_i = \tilde{v} \mid R_{-\tilde{r}_i}; \Lambda) \\
&\propto \frac{C(R_{-r}; T_0, T_i = \tilde{r}_0, \tilde{v}) + \alpha_{i\tilde{v}}}{C(R_{-r}; T_0 = \tilde{r}_0) + \sum \alpha_i} \frac{C(R_{-r}; T_i, V_i = \tilde{v}, r_i) + \beta_{ir_i}}{C(R_{-r}; T_i = \tilde{v}) + \sum \beta_i} ,
\end{aligned}
\tag{5}
$$

whereas Eq. (6) is the posterior probability that the topic of the ith text attribute is $\tilde{v}$:

$$
\begin{aligned}
&p(T_i = \tilde{v} \mid R_{-\tilde{r}_i}; \Lambda) \\
&\propto \frac{C(R_{-r}; T_0, T_i = \tilde{r}_0, \tilde{v}) + \alpha_{i\tilde{v}}}{C(R_{-r}; T_0 = \tilde{r}_0) + \sum \alpha_i} \prod_{j=1}^{l_i} \frac{C(R_{-r_{ij<}}; T_i, V_i = \tilde{v}, r_{ij}) + \beta_{ir_{ij}}}{C(R_{-r_{ij<}}; T_i = \tilde{v}) + \sum \beta_i} .
\end{aligned}
\tag{6}
$$

Due to space limitations, we omit the derivation.

The estimation algorithm takes parameters Λ, the numbers of topics $|\tilde{D}_0|$, $\cdots, |\tilde{D}_m|$, and the attribute values of a table R, as inputs. Note that, at this point, only the attribute values are given and the corresponding topics are estimated in the algorithm. Therefore, the algorithm is unsupervised.

The algorithm first randomly assigns a topic to each attribute value in R and then modifies the topics according to Eqs. (4), (5), and (6). Reassignment of the topics is repeated until the posterior probabilities converge.

The output of the algorithm is the assignment of a latent topic to each attribute value. We can estimate the parameters Δ of the record model from these frequencies.

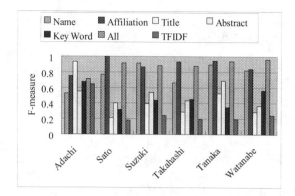

Fig. 1. Performance for name disambiguation problem

4   Experimental Results

We applied the record model to the author disambiguation problem [6,4,9] of the bibliographic database. The problem is identifying the authors appearing in a dataset. In this experiment, we used a bibliographic database from the digital library system NII-ELS[2]. This database contains bibliographic data on the academic papers published in journals in various research fields, such as natural sciences, engineering, medical science, social science, economics, and the arts. A bibliographic record consists of the author's name, affiliation, title, abstract, and key words. For a text, we chose important words based on the TFIDF weight as is usually done in information retrieval and text categorization. As for the title and abstract, we chose the top 5,000 words as important words. Then, we removed all the remaining words words from the titles and abstracts. Proper nouns are important for the affiliation. We chose the top 4,000 words based on the TFIDF weight for the affiliations. Almost all proper nouns were included in this word set.

We chose six popular author's names in the database and retrieved the records written by an author whose name was included in the six names. We manually identified authors in the retrieved records and applied the proposed method. We measured the f-measures using each of the five attributes alone as well as all the attributes. We also measured the performance of the TFIDF-based clustering method; i.e., we applied agglomerative single linkage clustering where the document and cluster similarity is measured by the cosine similarity weighted by TFIDF. Figure 1 shows the experimental results for each name. In the graph, the result for each attribute is labeled with its attribute name. Furthermore, "all" and "TFIDF" show the results obtained by using all the attributes and TFIDF based clustering, respectively.

First, the proposed method significantly outperformed the TFIDF based method. This result indicates that the latent topic is a good feature for

---

[2] http://reo.nii.ac.jp/journal/HtmlIndicate/html/index_en.html

identifying records. For this database, "affiliation" and "authors' name" are useful attributes for the name disambiguation, whereas "title", "abstract", and "key words" are not useful enough. Although for some names the proposed method using all the attributes degrades the performance compared with the best attribute, it performs very well with the information on the available attributes.

5   Conclusion

We have proposed a generative model for record linkage. The proposed method converts both scalar and text attributes into topic distributions and exploits the co-occurrence of multiple attributes. Experimental results showed that the proposed model outperformed the TFIDF model that directly uses the observed values. We evaluated the proposed method using only a small dataset. We plan to evaluate the model using a larger dataset as well as different types of records We also plan to research how to determine the proper hyper-parameters and number of topics. The combination of this with supervised techniques is useful to improve accuracy. This is another research direction we plan to pursue.

Acknowledgments

This paper is partly supported by MEXT Grant-in-Aid for Scientific Research (19300032).

References

1. Bhattacharya, L., Getoor, L.: A Latent Dirichlet Model for Unsupervised Entity Resolution. In: SIAM Intl. Conf. on Data Mining (2006)
2. Bilenko, M., Mooney, R.J.: Adaptive Duplicate Detection Using Learnable String Similarity Measures. In: International Conference on Knowledge Discovery and Data Mining, pp. 39–48 (2003)
3. Blei, D.M., Ng, A.Y., Jordan, M.I.: Latent dirichlet allocation. Journal of Machine Learning Research 3, 993–1022 (2003)
4. Bollegara, D., Matsuo, Y., Ishizuka, M.: Extracting key phrases to disambiguate personal name queries in web search. In: Coling-ACL Workshop on How Can Computational Liguistics Improve Information Retrieval (2006)
5. Cohen, W., Ravikumar, P., Fienberg, S.: A Comparison of String Distance Metrics for Name-matching Tasks. In: IIWeb 2003, pp. 73–78 (2003)
6. Fleischman, M., Hovy, E.: Multi-document Person Name Resolution. In: ACL Workshop on Reference Resolution (2004)
7. Griffiths, T.L., Steyvers, M.: Finding scientific topics. In: Proc. of the National Academy of Sciences, vol. 101 (suppl. 1), pp. 5228–5235 (2004)
8. Shu, L., Long, B., Meng, W.: A latent topic model for complete entity resolution. In: Intl. Conf. on Data Engineering, pp. 880–891 (2009)
9. Song, Y., Huang, J., Councill, I.G., Li, J., Giles, C.L.: Efficient topic-based unsupervised name disambiguation. In: Joint Conference on Digital Libraries, pp. 342–351 (2007)
10. Wang, X., Mohanty, N., McCallum, A.: Group and topic discovery from relations and text. In: Proc. of 3rd International Workshop on Link Discovery (LinkKDD), pp. 28–35 (2005)

Computing a Comprehensible Model for Spam Filtering

Amparo Ruiz-Sepúlveda, José L. Triviño-Rodriguez, and Rafael Morales-Bueno

Department of Computer Science and Artificial Intelligence
University of Málaga, Málaga, Spain
{amparo,trivino,morales}@lcc.uma.es
http://www.lcc.uma.es

Abstract. In this paper, we describe the application of the Desicion Tree Boosting (DTB) learning model to spam email filtering. This classification task implies the learning in a high dimensional feature space. So, it is an example of how the DTB algorithm performs in such feature space problems. In [1], it has been shown that hypotheses computed by the DTB model are more comprehensible that the ones computed by another ensemble methods. Hence, this paper tries to show that the DTB algorithm maintains the same comprehensibility of hypothesis in high dimensional feature space problems while achieving the performance of other ensemble methods. Four traditional evaluation measures (precision, recall, F1 and accuracy) have been considered for performance comparison between DTB and others models usually applied to spam email filtering. The size of the hypothesis computed by a DTB is smaller and more comprehensible than the hypothesis computed by Adaboost and Naïve Bayes.

1 Introduction

The increased availability of documents in digital form and the need to organize them has promoted the development of a large amount of techniques for automated text categorization.

Many algorithms for text categorization have been proposed and evaluated in last years. Some of the top performing techniques are based on decision trees [2], neural networks [3], nearest neighbor methods [4], Rocchio's method [5], support vector machines [6], naïve Bayes [7] and rule-based methods [8].

As a particular instance of the text categorization problem can be seen the Spam email filtering problem. in which a set of rules is created according to which messages are categorized as spam or legitimate mail.

During last years the amount of messages flowing throw the email servers has been increasing, some of them are useless and unwanted.

Automatic spam filtering systems can be done since spam messages can be identified because of their particular form, which can be found in the header or body of the messages, and its evaluation can be done by using common measures of automatic information retrieval (recall, precision, F1 and accuracy).

J. Gama et al. (Eds.): DS 2009, LNAI 5808, pp. 457–464, 2009.

Spamassasin [9] is one of the most popular automated filtering approach of Knowledge engineering to identify spam. It is an intelligent email filter which uses a diverse range of tests to identify unsolicited bulk email, spam.

The machine learning approach does not require specifying any rules explicitly. Instead, a set of pre-classified documents (training samples) is needed. A specific algorithm is then used to learn the classification rules from this data. Moreover, data mining is useful in discovering the classification rules, that can be used in software like Spamassasin.

Many data mining and machine learning researchers have worked on spam detection and filtering: naïve Bayes [7], support vector machine [6], memory based learning methods [10] and decision tree [11]. A comprehensive review of recent machine learning approaches to Spam filters was presented in [12].

In recent years, ensemble methods have gained popularity and boosting algorithms occupies special place in the classifier ensemble literature since have attractive theoretical properties, it has been applied to different domains and has also been shown to perform well experimentally on standard machine learning tasks [13], [14] and one of the best performers in spam email filtering [5], [11].

Focusing on comprehensibility without losing too much accuracy has been the main goal of several developments in machine learning in the last years.

Following this approach, Freund et. al. [15] suggest a new combination of decision trees with boosting called alternating decision trees (ADTrees). However, unlike decision trees, ADTrees are a scoring classification model that does not compute disjoint rules to classify a sample. The main difficulty in the application of ADTrees to text categorization is the high dimensionality of the input feature space typical for textual data.

Feature selection as a preprocessing step is frequently used in machine learning, in order to reduce dimensionality according to a certain evaluation criterion [16], [17]. In recent years data has become increasing in number of instances and number of features. So, spam filtering is likely to remain an active area of research in the next years.

In [1], a learning model that explore the area of trade-offs between accuracy and comprehensibility in ensemble machine learning methods is described using DTB.

Moreover, this models includes implicitly a feature selection scheme. So, it can be applied directly to high dimensionality problems like spam email filtering.

In this work, the DTB learning model applied to spam email filtering (Spanish Spam dataset) is evaluated.

This paper is organized as follows. Section 2 presents the DTB learning model. Section 3 describes the data collection and experimental results. Conclusions are given in Section 4.

2   Boosting with Decision Trees

In [18], Kearns et al. examine top-down induction decision tree (TDIDT) learning algorithms in the model of *weak learning* [19], that is, like a boosting algorithm. They proved that the standard TDIDT algorithms are in fact boosting algorithms.

From this point of view, the class of functions labeling the decision tree nodes acts as the set of weak hypothesis in the boosting algorithm. Thus, splitting a leaf in a top-down decision tree is equivalent to adding a new weak hypothesis in a boosting algorithm. A sample is classified by a decision tree by means of the vote of functions in the nodes of the tree. Every vote represents a step in a path from the root node of the tree to a leave of the tree. A sample is classified with the label of the leave at the end of the path defined by the votes of the functions in the internal nodes of the tree.

However, Dietterich et al. [20] have shown that top-down decision trees and Adaboost have qualitatively different behavior. Briefly, they find that although decision trees and Adaboost are both boosting algorithm, Adaboost creates successively *"harder"* filtered distributions, while decision trees create successively *"easier"* ones.

Nonetheless, both approaches could be combined in a learning method that has the accuracy and stability of boosting algorithms and the comprehensibility of decision trees hypothesis.

If we assume that the weak hypothesis learner is simple enough, for example a decision stump, then we could assume that all the weak hypotheses are comprehensible and that the difficulty in understanding the final hypothesis lies in the linear classifier over a space defined by weak learners.

The top-down decision tree can be a good substitute for the linear classifier of AdaBoost algorithm. It can be regarded as a boosting algorithm and it computes comprehensible hypotheses.

Thus, we could substitute the linear classifier of AdaBoost with a decision tree. This proposed approach is shown in algorithm 1. Although this approach could be seen as a kind of stacking, there are several features that make it closer to an AdaBoost approach than to the stacking approach described by Wolpert [21]: stacking approach splits the training dataset into several subsets each one used to train an independent weak learner, while each weak learner in AdaBoost is trained with the entire dataset. Moreover, Adaboost modifies the weight of samples of the learning dataset throughout the learning task. Finally, the accuracy of each weak learner is used to modify the weight of samples of the learning dataset in order to train the next weak learner; so each weak learner is influenced by the accuracy of the previous one, while stacking only considers the accuracy of the weak learners at the meta learning level.

Moreover, since decision stumps are classifiers that consider only a single variable of the input space, they are equivalent to the class of single variable projection functions usually used in the top-down decision tree learners [22]. If then the linear classifier of AdaBoost over decision stumps is substituted by a top-down decision tree learner, then the final hypothesis will be compounded by a tree with a query about a single variable of the input space in the internal nodes. Thus, this final hypothesis will be equivalent to a decision tree computed

---

Algorithm 1. The Decision Tree Boosting algorithm for binary classification problems returning $h_{final} : X \to Y$

Let:

- A training set E of m samples $E = \{(x_1, y_1), \ldots, (x_m, y_m)\}$ with $x_i \in X = \{$vectors of attribute values$\}$ and labels $y_i \in Y = \{-1, 1\}$
- A weak learning algorithm $WeakLearner$
- $T \in \mathbb{N}$ specifying number of iterations

1: Compute a set H of T weak hypothesis following steps 1 to 8 of the AdaBoost.M1 algorithm shown in [19]
2: Map every instance of E to the instance space $F = \{h_t(x)|x \in X\} \sqsubseteq Y^T$:

$$\bar{E} = \{(< h_1(x_i), \ldots, h_T(x_i) >, y_i)|(x_i, y_i) \in E\}$$

3: Let h_{final} be the decision tree learned from the training dataset $\bar{E}$

---

directly over the input space. However, as it is shown in [1], the mapping of the input space improves the accuracy and stability of the learned decision tree.

3 Empirical Evaluation

In order to evaluate the DTB algorithm for the anti-spam email filtering problem, a corpus with spam and legitimated emails messages have been collected. This corpus has been collected from emails accounts of the Department of Language and Computer Sciences of the University of Málaga. It consists of 762 messages: 665 of them are legitimate and the remaining 97 are spam. All the messages of this corpus are written in spanish.

After analyzing all the emails of the corpus, a dictionary with N words/features has been formed. During preprocessing, HTML tags of the emails and words with length less than four have been removed from the dictionary.

Every email has been represented as a feature vector including N elements, and the ith word of the vector is a number variable representing how many times this word is in this email. So. the feature set of the corpus is a bag-of-words model which consists in a set of 13252 features.

Several evaluation measures of the DTB, AdaBoost, Naïve Bayes and J48 over this corpus have been computed by means of a ten-fold cross validation. Let S and L be the number of spam and legitimate messages in the corpus, respectively; let S_+ denote the number of spam messages that are correctly classified by a system, and S_- the number of spam messages misclassified as legitimate. In the same way, let L_+ and L_- be the number of legitimate messages classified by a system as spam and legitimate, respectively. These four values from a contingency table which summarizes the behaviour of a system. The widely-used measures precision (p), recall (f), F1 and accuracy (Acc) are defined as follows:

$$p = \frac{S_+}{S_+ + L_+} \qquad F1 = \frac{2pr}{p+r} \qquad r = \frac{S_+}{S_+ + S_-} \qquad Acc = \frac{L_- + S_+}{L + S}$$

Boosting with Decision Trees method has been implemented and integrated into the Weka software package [23]. In the experiments then we will use the J48 implementation of the C4.5 algorithm as the Top-Down Decision Tree learning algorithm. Moreover, in order to compare our method with Boosting and Naïve Bayes, several tests have been carried out using the AdaBoostM1 implementation of Boosting in Weka and the Naïve Bayes classifier. Finally, the Decision-Stump learning algorithm of Weka has been taken as the weak learner algorithm.

The AdaBoostM1 learning models have a learning parameter that defines the number of weak learners computed throughout the learning task. Before comparing the results of these learning models over DTB, we optimized this parameter. So the results shown in this paper are the most accurate obtained from these learning algorithms after optimizing this learning parameter. However, this parameter has not been fully optimized for DTB and only two values for this parameter (10 and 100 weak learners) have been taken into account in order to avoid unfair advantage to DTB.

The results of this test and the mean hypothesis size throughout the ten iterations of this test are shown in table 1.

Naïve Bayes model has a precision and recall statistically significatively (using the t-test implemented in Weka) lower than the others models. However, the precision and recall of AdaBoostM1, J48 y DTB are note statistically significatively different. So, the best feature to compare these models is the accuracy. Table 1 shows that DTB has the same accuracy than Boosting. A t-test has proven that the difference between them is not statistically significative and it has shown that they perform better than J48 and Naïve Bayes.

Figure 1 shows the decision tree computed by the DTB learning model. Internal nodes in the tree query the number of times that a work appears in the email. This tree can be rewrite as a set of 13 disjoint rules. Results in table 1 show that DTB can achieve the same accuracy as Boosting with a size of

Table 1. Average accuracies (Acc.), IR Precision, IR Recall, F1 Measure and Hypothesis size (computed by Weka) of several learning models for the Spanish Spam dataset

LEARNING MODEL	ACCURACY	IR PRECISION	IR RECALL	F1 MEASURE	HYP. SIZE
NAÏVE BAYES	89,55	0,96	0,92	0,94	13252
J48	93,07	0,96	0,96	0,96	34
ADABOOSTM1 DS	95,00	0,96	0,99	0,98	200
ADABOOSTM1 J48	96,58	0,97	0,98	0,98	6792
DTB DS	95,08	0,96	0,98	0,97	25

```
''trouble'' = 0
|   ''clientes'' = 0
|   |   ''acción'' = 0
|   |   |   ''recibir'' = 0
|   |   |   |   ''gmail'' = 0
|   |   |   |   |   ''saludos'' > 0: Legitimate
|   |   |   |   |   ''saludos'' = 0
|   |   |   |   |   |   ''somos'' = 0
|   |   |   |   |   |   |   ''http'' <= 1: Legitimate
|   |   |   |   |   |   |   ''http'' > 1
|   |   |   |   |   |   |   |   ''list'' > 0: Legitimate
|   |   |   |   |   |   |   |   ''list'' = 0
|   |   |   |   |   |   |   |   |   ''wrote'' > 0: Legitimate
|   |   |   |   |   |   |   |   |   ''wrote'' = 0: Spam
|   |   |   |   |   |   ''somos'' > 0: Spam
|   |   |   |   ''gmail'' > 0
|   |   |   |   |   ''hola'' > 0: Legitimate
|   |   |   |   |   ''hola'' = 0: Spam
|   |   |   ''recibir'' > 0
|   |   |   |   ''hola'' > 0: Legitimate
|   |   |   |   ''hola'' = 0: Spam
|   |   ''acción'' > 0: Spam
|   ''clientes'' > 0: Spam
''trouble'' > 0: Spam
```

Fig. 1. Decision tree computed by the DTB learning model

hypothesis smaller than the size of the hypothesis of Boosting and J48 without a scoring classification scheme. Of course, the size of the hypothesis of a DTB is smaller and more comprehensible than the hypothesis of Boosting because the size of the latter is the number of the weak learners that is greater than 10 at best and usually greater than 100 weak hypotheses.

Rules in figure 1 show that there are words that clearly identify a email like spam. For example, "xlientes" and "acción" are not words used usually in mails written in the Department of Languages and computer Science and they denote spam. Instead, "saludos" is a word used usually to end a letter and it is not founded in spam messages.

4   Conclusion

Some of most important anti-spam filters uses a diverse range of tests to identify unsolicited bulk email. Data mining techniques are useful in discovering these classification rules, that can be used anti-spam filters.

Comprehensibility of the hypothesis model computed by these data mining techniques is a criterion that cannot be easily overlooked since knowledge engineers understand and accept it in order to embed it in the anti-spam filter.

In [1] is shown that the DTB learning model computes human readable and comprehensible hypothesis. but, it had not been tested in high dimensional feature space learning tasks like anti-spam filtering. Experimental evaluation carried out in this paper shows that the DTB algorithm maintains the same comprehensibility of hypothesis in high dimensional feature space problems while achieving the performance of other ensemble methods using for spam email filtering.

Experiments have shown that the hypothesis size computed by the DTB learning model is smaller than the size of the hypothesis of J48. Moreover, the accuracy of the DTB learning model improves statistically significatively the accuracy of the J48 and Naïve Bayes. DTB achieves statistically the same accuracy than learning model like AdaBoost with complex weak learners.

As a future research line, we would like to study the presented techniques in a larger corpus. Moreover, it could be take into account expressions or combinations of several words instead of simple words as feature space.

Another line for future research is the introduction of misclassification costs inside the DTB learning algorithm.

Acknowledgments

This reserarch has been partially funded by the SESAAME project, number TIN2008-06582-C03-03, of the MCyT, Spain.

References

1. Triviño-Rodriguez, J.L., Ruiz-Sepúlveda, A., Morales-Bueno, R.: How an ensemble method can compute a comprehensible model. In: Song, I.-Y., Eder, J., Nguyen, T.M. (eds.) 10th International Conference on Data Warehousing and Knowledge Discovery10th International Conference on Data Warehousing and Knowledge Discovery. LNCS, vol. 5182, pp. 268–378. Springer, Heidelberg (2008)
2. Cohen, W.W., Singer, Y.: Context-sensitive learning methods for text categorization. In: ACM Transactions on Information Systems, pp. 307–315. ACM Press, New York (1996)
3. Ruiz, M.E., Srinivasan, P.: Hierarchical neural networks for text categorization. In: Proceedings of the 22nd Annual International ACM SIGIR Conference on Research and Development in Information Retrieval, pp. 281–282 (1999)
4. Larkey, L.S., Croft, W.B.: Combining classifiers in text categorization. In: SIGIR 1996: Proceedings of the 19th annual international ACM SIGIR conference on Research and development in information retrieval, pp. 289–297. ACM, New York (1996)
5. Schapire, R.E., Singer, Y., Singhal, A.: Boosting and rocchio applied to text filtering. In: Proceedings of ACM SIGIR, pp. 215–223. ACM Press, New York (1998)
6. Drucker, H., Member, S., Wu, D., Member, S., Vapnik, V.N.: Support vector machines for spam categorization. IEEE Transactions on Neural Networks 10, 1048–1054 (1999)
7. Androutsopoulos, I., Paliouras, G., Karkaletsis, V., Sakkis, G., Spyropoulos, C.D., Stamatopoulos, P.: Learning to filter spam e-mail: A comparison of a naive bayesian and a memory-based approach. CoRR cs.CL/0009009 (2000)

8. Apte, C., Damerau, F., Weiss, S.M., Apte, C., Damerau, F., Weiss, S.M.: Automated learning of decision rules for text categorization. ACM Transactions on Information Systems 12, 233–251 (1994)
9. Spamassasin
10. Sakkis, G., Androutsopoulos, I., Paliouras, G., Karkaletsis, V., Spyropoulos, C.D., Stamatopoulos, P.: A memory-based approach to anti-spam filtering (2001)
11. Carreras, X., Marquez, L.S., Salgado, J.G.: Boosting trees for anti-spam email filtering. In: Proceedings of RANLP 2001, 4th International Conference on Recent Advances in Natural Language Processing, Tzigov Chark, BG, 58–64 (2001)
12. Guzella, T.S., Caminhas, W.M.: A review of machine learning approaches to spam filtering. Expert Systems with Application. Corrected Proof (in press, 2009)
13. Quinlan, J.: Bagging, boosting, and c4.5. In: Proc. of the 13th Nat. Conf. on A.I. and the 8th Innovate Applications of A.I. Conf., pp. 725–730. AAAI/MIT Press (1996)
14. Tretyakov, K.: Machine learning techniques in spam filtering. Technical report, Institute of Computer Science, University of Tartu (2004)
15. Freund, Y., Mason, L.: The alternating decision tree learning algorithm. In: Proc. 16th International Conf. on Machine Learning, pp. 124–133. Morgan Kaufmann, San Francisco (1999)
16. Méndez, J.R., Fdez-Riverola, F., Díaz, F., Iglesias, E.L., Corchado, J.M.: A comparative performance study of feature selection methods for the anti-spam filtering domain. In: Perner, P., Heidelberg, S.B. (eds.) ICDM 2006. LNCS (LNAI), vol. 4065, pp. 106–120. Springer, Heidelberg (2006)
17. Chen, C., Gong, Y., Bie, R., Gao, X.: Searching for interacting features for spam filtering. In: Sun, F., Zhang, J., Tan, Y., Cao, J., Yu, W. (eds.) ISNN 2008, Part I. LNCS, vol. 5263, pp. 491–500. Springer, Heidelberg (2008)
18. Kearns, M., Mansour, Y.: On the boosting ability of top-down decision tree learning algorithms. In: Twenty-eighth annual ACM symposium on Theory of computing, Philadelphia, Pennsylvania, United States, pp. 459–468 (1996)
19. Schapire, R., Freund, Y.: A decision-theoretic generalization of on-line learning and an application to boosting. In: Second European Conference on Computational Learning Theory, pp. 23–37. Springer, Heidelberg (1995)
20. Dietterich, T., Kearns, M., Mansour, Y.: Applying the weak learning framework to understand and improve C4.5. In: Proc. 13th International Conference on Machine Learning, pp. 96–104. Morgan Kaufmann, San Francisco (1996)
21. Wolpert, D.H.: Stacked generalization. Neural Networks 5(2), 241–259 (1992)
22. Quinlan, J.: C4.5: Programs for Machine Learning. Morgan Kaufmann, San Francisco (1993)
23. Witten, I., Frank, E.: Data Mining: Practical machine learning tools and techniques, 2nd edn. Morgan Kaufmann, San Francisco (2005)

Better Decomposition Heuristics for the Maximum-Weight Connected Graph Problem Using Betweenness Centrality

Takanori Yamamoto[1], Hideo Bannai[1,*], Masao Nagasaki[2,*], and Satoru Miyano[2]

[1] Department of Informatics, Kyushu University
744 Motooka, Nishi-ku, Fukuoka, 819–0395 Japan
{takanori.yamamoto@i,bannai@inf}.kyushu-u.ac.jp
[2] Human Genome Center, Institute of Medical Science, University of Tokyo
4-6-1 Shirokanedai, Minato-ku, Tokyo, 108–8639 Japan
{masao,miyano}@ims.u-tokyo.ac.jp

Abstract. We present new decomposition heuristics for finding the optimal solution for the maximum-weight connected graph problem, which is known to be NP-hard. Previous optimal algorithms for solving the problem decompose the input graph into subgraphs using heuristics based on node degree. We propose new heuristics based on betweenness centrality measures, and show through computational experiments that our new heuristics tend to reduce the number of subgraphs in the decomposition, and therefore could lead to the reduction in computational time for finding the optimal solution. The method is further applied to analysis of biological pathway data.

1 Introduction

The maximum-weight connected graph (MCG) problem is: given an undirected node-weighted graph $G = (V, E)$ and integer k, find a connected subgraph consisting of k nodes whose sum of weights is the largest of all such subgraphs. The constrained maximum-weight connected graph (CMCG) problem is the same as MCG but with a fixed vertex also given in the input, and the problem is to find the connected subgraph with largest sum of weights that includes this vertex. These problems were considered in [1,2], for potential applications in network design problems and facility expansion problems [3], and are known to be NP-hard [4].

In this paper, we propose a new application of MCG: core source component discovery in gene networks. One of the most important topics in Systems Biology is to find the difference between the normal cell and the mutant cell by analyzing the result of some high throughput experiments, e.g. microarray experiment. Once the differences (i.e. the set of genes with different expressions) are identified, the next challenge is to apply some pathway level analysis to identify the source

---

* Corresponding authors.

J. Gama et al. (Eds.): DS 2009, LNAI 5808, pp. 465–472, 2009.

components which are the true cause of the difference observed in the mutant. This is important since those identified components may be useful for clinical diagnoses by checking the expressions of the identified source components.

Recently, several pathway databases that includes signal transduction regulations, gene regulations and metabolic reactions are available, e.g. KEGG [5], Transpath [6], EcoCyc [7] and Reactome [8]. Additionally, many high-throughput gene expression data are publicly available, e.g. Gene Expression Omnibus [9] at NCBI. These pathway databases can be considered as directed or undirected graphs where each node represents a gene. Also, by using the gene expression data, we can give a weight to each node representing the difference in gene expression. It is thus natural to assume that a connected subgraph with large weights would be a good candidate for such source components.

An important aspect in this problem setting is that biologists are not satisfied with approximate solutions. Finding the optimal solution is very important to them, since the results will be used for various clinical applications, e.g. choosing the minimal set of marker genes. Therefore, our focus is on methods which solves the problem optimally. We follow the work by Lee and Dooly [2], which solves MCG optimally by decomposing the graph of the MCG problem to several subgraphs of smaller size. By solving CMCG optimally for the subgraphs, an optimal solution for MCG can be obtained. However, this decomposition is not unique, and Lee and Dooly present two heuristics to this end. The main contribution of this paper is to present new decomposition heuristics based on betweenness centrality measures of nodes and edges in a graph. Computational experiments show that the new heuristics reduce the number of CMCG problems to be solved for a given MCG problem, which could lead to a more efficient optimal solution to MCG.

2   Preliminaries

We consider undirected graphs with weighted nodes. Let $G = (V, E)$ represent an undirected graph where V is the set of nodes, and $E \subseteq \{\{x, y\} \mid x, y \in V, x \neq y\}$ is the set of edges. For any node $v \in V$, let $\deg(v) = |\{w \mid \{v, w\} \in E, \ v \neq w\}|$. For each node $v \in V$, its associated weight is denoted $weight(v)$.

Problem 1 (Maximum-Weighted Connected Graph (MCG)). Given a graph $G = (V, E)$ with weighted nodes and an integer k, find a subgraph of G with k nodes that is connected and maximizes the sum of weights of the k nodes.

Problem 2 (Constrained Maximum-Weighted Connected Graph (CMCG)). Given an undirected graph $G = (V, E)$ with weighted nodes, an integer k, and a node $u \in V$, find a subgraph of G with k nodes that includes u, is connected, and maximizes the sum of weights of the k nodes.

Next, we introduce the notion of betweenness centrality in graphs, which we will use in our algorithms. For a graph $G = (V, E)$ and any node $a, b \in V$, let σ_{ab} denote the number of distinct shortest paths between nodes a and b in G. For

---

Algorithm 1. Decomposing MCG into CMCG (Lee and Dooly [1,2])

---

   procedure MAIN$(G = (V, E), k)$ ▷ Solve MCG by Decomposition to CMCG
      $maxWeight := 0$
      for $i := k$ to $|V|$ **do**
        if $|\hat{V_i}| \geq k$ **then**   ▷ component is large enough
          Solve $V_{v_i}^* :=$ CMCG(G_i, k, v_i)
          if $maxWeight < \sum_{v \in V_{v_i}^*} weight(v)$ **then**   ▷ update best solution
             $maxWeight := \sum_{v \in V_{v_i}^*} weight(v)$
             $V_s := V_{v_i}^*$
          end if
        end if
      end for
      return $maxWeight, V_s$ ▷ return best score and subgraph
   end procedure

---

any $n \in V$, let $\sigma_{ab}(n)$ denote the number of distinct shortest paths between a and b that go through n. For edge $e \in E$, let $\sigma_{ab}(e)$ denote the number of distinct shortest paths between a and b that contain e.

Definition 1 (Betweenness). *The node betweenness* $\mathrm{NB}(G, v)$ *of a node* $v \in V$ *of a graph* $G = (V, E)$, *is defined as* $\mathrm{NB}(G, v) = \sum_{v_i \in V} \sum_{v_j \in V \setminus \{v_i\}} \frac{\sigma_{v_i v_j}(v)}{\sigma_{v_i v_j}}$. *The edge betweenness* $\mathrm{EB}(G, e)$ *of a node* $e \in E$ *of a graph* $G = (V, E)$, *is defined as* $\mathrm{EB}(G, e) = \sum_{v_i \in V} \sum_{v_j \in V \setminus \{v_i\}} \frac{\sigma_{v_i v_j}(e)}{\sigma_{v_i v_j}}$.

Intuitively, nodes or edges that have high betweenness are those that appear frequently in shortest paths between nodes. For example, nodes or edges that bridge two graphs will have higher betweenness values compared to those inside the two graphs. It is known that the betweeness of all nodes and edges can be computed in $O(|V| \cdot |E|)$ time and $O(|V| + |E|)$ space [10].

3   Algorithm

It was shown by Lee and Dooly [1,2] that any instance of the MCG problem can be decomposed into multiple CMCG problems, and by combining the optimal solutions to the decomposed CMCG problem gives an optimal solution to the MCG problem. Although both MCG and CMCG are known to be NP-hard, the decomposition helps in speeding up the search for the optimal solution to the MCG problem, since 1) the solution space of CMCG problems is restricted compared to MCG, and 2) the graph of the MCG problem is decomposed to smaller graphs.

    Algorithm 1 shows a general method to solve the MCG problem optimally by solving multiple CMCG problems. The decomposition of a graph $G = (V, E)$ depends on an ordering on the vertices. We will assume that the ordering on the vertices of $V = \{v_1, \ldots, v_n\}$ will be represented by their subscript. Further, let

$V_i = \{v_j \mid j \leq i\}$, and define $\hat{V}_i$ as the connected set of vertices of V_i that includes vertex v_i. Denote by G_i, the subgraph induced by $\hat{V}_i$, that is, $G_i = (\hat{V}_i, E_i)$ where $E_i = \{\{v_a, v_b\} \in E \mid v_a, v_b \in \hat{V}_i\}$. The algorithm decomposes the MCG problem to CMCG problems for the graphs $G_k = (\hat{V}_k, E_k)$ to $G_n = (\hat{V}_n, E_n)$, where the constrained vertex for G_i is vertex v_i ($k \leq i \leq n$), and if $|\hat{V}_i| \geq k$.

As noted earlier, the decomposition algorithm depends on an ordering of the vertices, which has a large effect on the number and size of the CMCG problems to be solved. Finding the best such ordering that reduces the number of CMCG problem to be solved is also known to be NP-hard [2].

3.1   ORIGINAL_DEGREE and UPDATED_DEGREE

For determining the ordering, Lee and Dooly presented two heuristics, ORIGI-NAL_DEGREE and UPDATED_DEGREE [2]. Both are essentially greedy algorithms based on local information. In ORIGINAL_DEGREE, the node with the smallest degree is chosen. More precisely, while CMCG calls are not caused (i.e. $|\hat{V}_i| < k$), nodes with smallest degree are chosen. If adding any node causes a CMCG call, then, the node that minimizes the size of the graph for which CMCG is called is chosen. When there are multiple such nodes, the node with smallest degree is again preferred. UPDATED_DEGREE is very similar to ORIG-INAL_DEGREE. The only difference is that in UPDATED_DEGREE, rather than using the degree of the node, the remaining degree of the node is used. Here, the remaining degree of a node v_i is $|\{\{v_i, v_j\} \in E \mid j > i\}|$, that is, the number of edges to other vertices that have not been added yet.

3.2   Betweenness Centrality

Next, we describe two new heuristics based on betweenness centrality measures. For the node betweenness ordering (NB), we determine the order greedily starting from v_n, down to v_1. For each $i = n, \ldots, 1$, v_i is chosen as the node that has the largest node betweenness in the graph induced by $V \setminus \{v_j \mid j > i\}$. The node betweenness of the graph is recalculated after each choice of v_i.

For the edge betweenness ordering (EB), we also determine the order greedily starting from v_n, down to v_1. For each $i = n, \ldots, 1$, v_i is chosen as follows: Let the edge $\{u, v\}$ be the edge with greatest edge betweenness in the graph induced by $V \setminus \{v_j \mid j > i\}$. Then, $v_i = \arg\max_{w \in \{u,v\}} \{\deg(w)\}$, that is, the node with the greater degree. When there are no edges left in the graph induced by $V \setminus \{v_j \mid j > i\}$, then v_i is the node with the largest remaining degree.

Algorithm 2 shows pseudocode for the node betweenness heuristics. The heuristics based on edge betweenness is similar and we omit the pseudocode.

An intuitive rationale behind the heuristics is as follows: As mentioned earlier, nodes and edges with high centrality are more likely to be those that "bridge" connected components in the graph. Therefore, if we choose the node ordering so that the nodes or edges with higher centrality come later, then we may hope to delay large connected components from being created in Algorithm 1.

Algorithm 2. Ordering based on Node Betweenness

---

procedure NODEBETWEENNESSORDERING($G = (V, E)$)
    $V' := V$
    for $i := |V|$ downto 1 **do**
        $G' = (V', E') :=$ the subgraph induced by V'
        $v_i := \arg\max_{v \in v'} \{ \mathrm{NB}(G', v) \}$
        remove vertex v_i from V'
    end for
    return $(v_1, \ldots, v_n)$.
end procedure

---

4 Computational Experiments

4.1 Random Data

We evaluate our heuristics as in [2]. Based on two parameters n, and x, we generate random graphs with n nodes and $x(n-1)$ edges for $n = 100, 300, 500$ and $x = 1, 3, 5$. The results for decomposing the MCG problem to the generated graphs with $k \doteq 10, 20, 40$ are shown in Table 1.

As in the previous paper [2], we compare the number of CMCG problems to be solved (count) and their average size (size) for 5 heuristics: NB (node betweenness), EB (edge betweenness), Original (ORIGINAL_DEGREE), Updated (UPDATED_DEGREE), and Random (random ordering). Each cell in Table 1 for each parameter n, x, k is an average of 10 random graphs.

We can see that the NB heuristics gives the smallest count for almost all parameter settings. Interestingly, the average CMCG size for the Random heuristics seems best. However, this is because Random decomposes the MCG problem into many CMCG problems where some of them can be fairly small, which reduces the average CMCG size.

4.2 Real Data

Here, we show results of applying our algorithm to real biological data. We selected the gene expression data of tumors of kidney cancer by Gumz M.L. et al. [11] in the Gene Expression Omnibus database (GEO) of NCBI [12]. The data compares the gene expressions between normal kidney tissue and stage I or II clear cell Renal Cell Carcinoma (cRCC) tumor tissue in human, using Affymetrix Human Genome U133A microarray platform (data GSE6344 in GEO[1]). Genomic profiling of cRCC patients indicated the loss of a negative regulator of the Wnt pathway, secreted frizzled-related protein 1 (sFRP1). As the target pathway model, we used Wnt pathway model that is generated by the method in [13]. The original source of the database is the TRANSPATH database [6], a manually curated high-quality pathway database.

---

[1] http://www.ncbi.nlm.nih.gov/geo/query/acc.cgi?acc=GSE7234

Table 1. Comparison of count and size of decomposed CMCG problems on random graphs with n nodes and $x(n-1)$ edges. The cell with bold figures is the best count among the five methods for the same n, x and k parameters. The cell with underline is the minimum size among the five methods for the same n, x and k parameters.

			NB		EB		Original		Updated		Random	
n	x	k	count	size	count	size	count	size	count	size	count	size
100	1	10	11.2	39.76	11.1	39.83	**11.0**	41.75	11.5	40.90	29.7	47.20
		20	**7.3**	58.03	7.9	55.97	8.0	54.04	8.1	55.74	24.7	58.39
		40	**5.2**	65.44	5.4	63.87	5.4	62.37	6.2	62.40	17.9	60.67
	3	10	**40.8**	68.03	44.1	64.11	42.7	67.35	45.0	65.30	71.8	60.98
		20	**36.6**	74.36	40.6	68.50	38.5	73.36	42.1	70.39	67.1	64.24
		40	**33.2**	79.09	35.7	74.27	40.0	71.77	41.5	73.50	57.3	70.79
	5	10	**55.6**	67.00	60.4	61.94	58.4	66.55	61.3	63.73	82.5	58.30
		20	**52.4**	69.69	56.3	65.20	55.4	68.14	57.8	67.20	77.2	61.43
		40	**47.3**	74.07	48.3	70.98	55.3	69.29	55.8	69.38	59.7	70.48
300	1	10	**34.8**	99.92	36.9	99.41	36.2	102.90	36.7	102.21	92.5	137.00
		20	**25.8**	138.25	26.4	132.42	27.8	132.35	28.4	133.02	80.4	150.31
		40	**20.3**	165.93	21.0	160.62	23.3	160.36	24.1	160.07	73.2	163.33
	3	10	**123.7**	199.71	130.1	189.76	132.4	190.06	132.9	192.67	221.1	177.95
		20	**114.1**	214.70	121.9	201.51	124.7	203.81	129.8	198.87	216.3	181.75
		40	**109.1**	225.77	119.2	207.19	120.7	210.91	129.4	204.74	207.6	187.11
	5	10	**166.8**	198.36	178.3	184.54	175.0	191.12	178.1	190.81	253.8	169.58
		20	**159.3**	206.77	172.5	191.73	170.1	198.38	177.6	193.61	248.1	173.33
		40	**153.6**	214.33	167.2	195.08	168.6	199.55	173.6	198.24	240.7	177.78
500	1	10	58.8	156.36	61.5	159.18	**58.7**	165.35	59.3	166.83	158.3	219.29
		20	**42.5**	214.92	44.2	210.63	46.0	208.54	48.0	211.34	138.6	246.15
		40	**34.9**	268.47	37.6	256.35	37.8	262.24	41.0	248.42	131.2	259.71
	3	10	**209.3**	334.77	216.0	321.19	216.9	322.51	228.4	312.82	375.0	292.57
		20	**191.9**	357.50	201.7	336.71	204.6	338.21	219.0	324.88	363.2	300.71
		40	**181.5**	376.00	195.7	348.19	196.6	350.91	213.8	333.32	356.4	305.06
	5	10	**279.2**	331.45	293.1	314.64	290.7	320.58	303.2	311.19	424.6	281.88
		20	**266.1**	347.80	283.5	323.71	284.2	326.75	299.1	316.61	418.0	285.94
		40	**258.5**	356.87	277.0	327.98	281.5	331.78	293.6	324.14	408.3	290.97

For the Wnt pathway network, we assign to each node (gene), absolute values of the difference in expression levels between normal tissue and tumor tissue, observed by the probe of the corresponding gene. In some case, several probes are mapped to the same gene in the microarray data. For this case, we selected the maximum value among them: i.e. for each gene, select the value $m(p) = \max |n(p) - c(p)|$ for $p \in P$, where P is the set of probes of the gene, $n(p)$ and $c(p)$ is the expression of normal cell and tumor expression of the probe p, respectively.

Nodes in the Wnt pathway network consists of mRNA, Protein and modified protein and complex proteins. We assigned the value $m(p)$ for mRNA and protein, the value $m(p)/2$ for modified protein, and zero for complex proteins.

Fig. 1. The MCG result for $k15$ applied to the Wnt pathway. The network contained all nodes of the MCG for $k = 5$ and 10. It can be observed that all nodes in the left-bottom and bottom subnetworks are not selected for the optimal subgraph.

The number of genes contained in this data was 56. The MCG obtained for $k = 15$ is shown in Fig. 1. The genes enclosed in squares indicate a solution to the MCG problem[2].

It can be observed that all nodes in the left-bottom and bottom subnetworks are not included in the optimal subgraph. Thus, we can infer that in cRCC tumor tissue, the Fz, Wnt and Dvl subnetwork and Frat1 subnetwork have less impact compared to other subnetworks in Wnt pathway, e.g. beta-catenin subnetwork.

5   Conclusion

In this paper, we introduced new heuristics for the decomposition of the maximum weighted connected graph problem (MCG) into the constrained maximum weighted connected graph problem (CMCG). The heuristics are based on betweenness centrality, which express the importance in terms of connectivity of

---

[2] Larger snapshots can be obtained from http://www.csml.org/download/ OptFinder/WntPathway_CMCG15_undirected.png

the nodes or edges in the graph. Experimental results show that our heuristics tend to reduce the number of CMCG problems to be solved. Therefore application of our heuristics could lead to finding the optimal solution more efficiently.

References

1. Lee, H.F., Dooly, D.R.: Algorithms for the constrained maximum-weight connected graph problem. Naval Research Logistics 43(7), 985–1008 (1996)
2. Lee, H.F., Dooly, D.R.: Decomposition algorithms for the maximum-weight connected graph problem. Naval Research Logistics 45(8), 817–837 (1998)
3. Lee, H.F., Dooly, D.R.: Heuristic algorithms for the fiber optic network expansion problem. Telecommunication Systems 7(4), 355–378 (1997)
4. Lee, H.F., Dooly, D.R.: The maximum-weight connected graph problem. Technical Report 93-4, Industrial Engineering, Southern Illinois University (1993) (revised November 1995)
5. Kanehisa, M., Araki, M., Goto, S., Hattori, M., Hirakawa, M., Itoh, M., Katayama, T., Kawashima, S., Okuda, S., Tokimatsu, T., Yamanishi, Y.: KEGG for linking genomes to life and the environment. Nucleic Acids Res. 36, D480–D484 (2008)
6. Choi, C., Crass, T., Kel, A., Kel-Margoulis, O., Krull, M., Pistor, S., Potapov, A., Voss, N., Wingender, E.: Consistent re-modeling of signaling pathways and its implementation in the TRANSPATH database. Genome Inform. 15(2), 244–254 (2004)
7. Keseler, I.M., Bonavides-Martínez, C., Collado-Vides, J., Gama-Castro, S., Gunsalus, R.P., Johnson, D.A., Krummenacker, M., Nolan, L.M., Paley, S., Paulsen, I.T., Peralta-Gil, M., Santos-Zavaleta, A., Shearer, A.G., Karp, P.D.: EcoCyc: A comprehensive view of Escherichia coli biology. Nucleic Acids Res. 37, D464–D470 (2009)
8. Vastrik, I., D'Eustachio, P., Schmidt, E., Joshi-Tope, G., Gopinath, G., Croft, D., de Bono, B., Gillespie, M., Jassal, B., Lewis, S., Matthews, L., Wu, G., Birney, E., Stein, L.: Reactome: a knowledge base of biologic pathways and processes. Genome Biology 8(R39) (2007)
9. Barrett, T., Troup, D.B., Wilhite, S.E., Ledoux, P., Rudnev, D., Evangelista, C., Kim, I.F., Soboleva, A., Tomashevsky, M., Edgar, R.: NCBI GEO: mining tens of millions of expression profiles – database and tools update. Nucleic Acids Res. 35, D760–D765 (2007)
10. Brandes, U.: A faster algorithm for betweenness centrality. Journal of Mathematical Sociology 25(2), 163–177 (2001)
11. Gumz, M., Zou, H., Kreinest, P., Childs, A., Belmonte, L., LeGrand, S., Wu, K., Luxon, B., Sinha, M., Parker, A., Sun, L., Ahlquist, D., Wood, C., Copland, J.: Secreted frizzled-related protein 1 loss contributes to tumor phenotype of clear cell renal cell carcinoma. Clin. Cancer Res. 13(16), 4740–4749 (2007)
12. Edgar, R., Domrachev, M., Lash, A.: Gene Expression Omnibus: NCBI gene expression and hybridization array data repository. Nucleic Acids Res. 30(1), 207–210 (2002)
13. Nagasaki, M., Saito, A., Li, C., Jeong, E., Miyano, S.: Systematic reconstruction of TRANSPATH data into cell system markup language. BMC Systems Biology 2(53), (2008)

Author Index